AQUAPORINS IN HEALTH AND DISEASE

New Molecular Targets
for Drug Discovery

AQUAPORINS IN HEALTH AND DISEASE

New Molecular Targets for Drug Discovery

edited by

Graça Soveral
Research Institute for Medicines, Faculty of
Pharmacy, University of Lisbon, Portugal

Søren Nielsen
Department of Health Science and Technology,
Aalborg University, Denmark

Angela Casini
School of Chemistry, Cardiff University, UK

with preface by Nobel Laureate **Peter Agre**

CRC Press
Taylor & Francis Group
Boca Raton London New York

CRC Press is an imprint of the
Taylor & Francis Group, an **informa** business

CRC Press
Taylor & Francis Group
6000 Broken Sound Parkway NW, Suite 300
Boca Raton, FL 33487-2742

First issued in paperback 2019

ISBN-13: 978-1-4987-0783-1 (hbk)
ISBN-13: 978-1-138-89401-3 (pbk)

Library of Congress Cataloging-in-Publication Data

Aquaporins in health and disease : new molecular targets for drug discovery / edited by Graça Soveral, Angela Casini, and Søren Nielsen.
    pages cm
  Includes bibliographical references and index.
  ISBN 978-1-4987-0783-1 -- ISBN 1-4987-0783-1 1. Aquaporins. 2. Drug targeting. 3. Drug development. 4. Drugs--Research. I. Soveral, Graça, editor. II. Casini, Angela, editor. III. Nielsen, Søren, 1962- editor.

  QP552.A65A685 2016
  572'.696--dc23

                                                                    2015024988

Visit the Taylor & Francis Web site at
http://www.taylorandfrancis.com

and the CRC Press Web site at
http://www.crcpress.com

Contents

Preface, vii

Contributors, ix

SECTION I **Mechanisms of Fluid Transport in Cells and Epithelia**

CHAPTER 1 ▪ Aquaporin Functional Detection: Basic Concepts on Water
and Solute Permeation   3

ANA MADEIRA, TERESA F. MOURA AND GRAÇA SOVERAL

CHAPTER 2 ▪ Aquaporin Discovery in the Genomic Era   19

RAFAEL ZARDOYA, IKER IRISARRI AND FEDERICO ABASCAL

CHAPTER 3 ▪ Aquaporin Structure and Selectivity   33

ARDESCHIR VAHEDI-FARIDI AND ANDREAS ENGEL

CHAPTER 4 ▪ Regulation of Eukaryotic Aquaporins   53

MARIA NYBLOM AND SUSANNA TÖRNROTH-HORSEFIELD

CHAPTER 5 ▪ Yeast Aquaporins and Aquaglyceroporins:
A Matter of Lifestyle   77

MIKAEL ANDERSSON AND STEFAN HOHMANN

SECTION II **Aquaporins in Health and Disease**

CHAPTER 6 ▪ Aquaporins in Health   103

AMAIA RODRÍGUEZ, LEIRE MÉNDEZ-GIMÉNEZ AND GEMA FRÜHBECK

CHAPTER 7 ▪ Renal Aquaporins: Role in Water Balance Disorders   125

TAE-HWAN KWON AND SØREN NIELSEN

CHAPTER 8 ▪ Vasopressin and the Regulation of Aquaporin-2 in Health and Disease   157

GIOVANNA VALENTI AND GRAZIA TAMMA

CHAPTER 9 ▪ Hepatobiliary, Salivary Glands and Pancreatic Aquaporins in Health and Disease   181

GIUSEPPE CALAMITA, CHRISTINE DELPORTE AND RAÚL A. MARINELLI

CHAPTER 10 ▪ Aquaporins within the Central Nervous System: Implications for Oedema Following Traumatic CNS Injury   205

ANNA V. LEONARD AND RENÉE J. TURNER

CHAPTER 11 ▪ Aquaporins in Carcinogenesis: Water and Glycerol Channels as New Potential Drug Targets   217

CHULSO MOON AND DAVID MOON

CHAPTER 12 ▪ Attacking Aquaporin Water and Solute Channels of Human-Pathogenic Parasites: New Routes for Treatment?   233

JULIA VON BÜLOW AND ERIC BEITZ

SECTION III   **Aquaporins as Drug Targets**

CHAPTER 13 ▪ Aquaporins: Chemical Inhibition by Small Molecules   249

VINCENT J. HUBER, SÖREN WACKER AND MICHAEL RÜTZLER

CHAPTER 14 ▪ Drug Discovery and Therapeutic Targets for Pharmacological Modulators of Aquaporin Channels   273

JINXIN V. PEI, JOSHUA L. AMELIORATE, MOHAMAD KOURGHI, MICHAEL L. DE IESO AND ANDREA J. YOOL

CHAPTER 15 ▪ Inorganic Compounds as Aquaporin Substrates or as Potent Inhibitors: A Coordination Chemistry Point of View   297

ANGELA CASINI AND ANDREIA DE ALMEIDA

EPILOGUE   319

ANGELA CASINI AND GRAÇA SOVERAL

INDEX, 321

Preface

WATER METABOLISM AND CELLULAR TRANSPORT of water is a prerequisite of life for humans, animals, plants and microbes. The existence of water-specific transport membrane proteins has been predicted for decades. Twenty-five years ago, the fundamental discovery and characterization of an abundant protein of the erythrocyte membrane termed Aquaporin-1 (AQP1, initially CHIP28 for channel-forming membrane protein of 28 kDa) represented a paradigm shift in the understanding of molecular, membrane and organism water transport. AQP1 has been shown through a series of studies undertaken worldwide on a member of a widespread and ancient family of water- and solute-transporting membrane proteins – the aquaporins. All eukaryotic organisms and most bacteria possess proteins of the aquaporin family (also referred to as the MIP [major intrinsic protein] family of proteins, the first protein identified more than 30 years ago). In addition to water-specific channels, the family comprises glycerol facilitators, which also transport other polyols, urea and related substances and proteins that transport both water and small solutes. The family is referred to as orthodox aquaporins (transporting water selectively) and aquaglyceroporins (conducting water and small solutes). Grouping has also included members of this family with less identified channel transport characteristics.

Since the discovery of AQP1 as a water channel, more than 2000 articles, reviews and chapters have been published in the scientific literature, mainly on mammalian aquaporins, and also aquaporins from plants and microorganisms. The wide expression of aquaporins among different organisms and functional and biological roles has documented their major and essential physiological importance. Dysregulation, missorting or mutations in aquaporins are also encountered in a number of human diseases, underscoring a fundamental role of aquaporins both in health and disease. This also concerns common and serious water balance disorders with reduced urinary and water balance conservation and states of water retention seen, for example, in cardiovascular and brain diseases. Thus, over the past years, structural, molecular biological, cell biological, physiological and pathophysiological studies, as well as studies on mutants lacking specific aquaporin channels, have revealed the importance of aquaporins in mammalian physiology and pathophysiology, as well as in plant and microbial biology. These studies have also revealed aquaporins as potential drug targets and as targets for the improvement of crop properties. This was highlighted in 2003 with the awarding of the Nobel Prize in Chemistry to Peter Agre for the discovery of aquaporins.

Over the past years, international meetings have been held focusing on aquaporins for presentation and discussions of the latest advances in research on aquaporins and other MIP channels. Contributors of these interactive and interdisciplinary conferences are among the authors of the chapters of this book. The book is divided into different key sections of focus. In Section I, Mechanisms of Fluid Transport in Cells and Epithelia, the general concepts of aquaporin channel function, genomic research, structure–function analysis of aquaporins and glycerol facilitators, as well as regulation by gating and trafficking, including yeast aquaporin regulation and function, are presented. In Section II, Aquaporins in Health and Disease, the physiological and pathophysiological roles of aquaporins in humans and microbes are discussed. Finally, in Section III, Aquaporins as Drug Targets, the concepts of the development of primarily inhibitors of aquaporin function are discussed. The book is concluded with a short account on future perspectives and directions mainly in the area of aquaporin-based diagnostics and therapeutics.

We hope that this book will stimulate future research on this important protein family and that the field of aquaporin research will develop further in the future into novel areas, leading to a paradigm shift in the understanding and roles of aquaporin membrane proteins in all biological settings. We also expect that this leads to novel approaches for the treatment of human diseases based on aquaporin function or dysfunction, including areas of developing, for example, drugs that block aquaporins and strategies to improve, for example, crop properties or microbial function.

Søren Nielsen

Peter Agre

Contributors

Federico Abascal
Structural Biology and Biocomputing
    Programme
Spanish National Cancer Research Centre
Madrid, Spain

Joshua L. Ameliorate
School of Medicine
and
Institute for Photonics and Advanced
    Sensing
and
Adelaide Centre for Neuroscience Research
The University of Adelaide
Adelaide, South Australia, Australia

Mikael Andersson
Department of Chemistry and Molecular
    Biology
University of Gothenburg
Gothenburg, Sweden

Eric Beitz
Pharmaceutical Institute
Christian-Albrechts-University of Kiel
Kiel, Germany

Giuseppe Calamita
Department of Biosciences,
    Biotechnologies and Biopharmaceutics
University of Bari "Aldo Moro"
Bari, Italy

Angela Casini
School of Chemistry
Cardiff University
Cardiff, United Kingdom
and
Department of Pharmacokinetics,
    Toxicology and Targeting
University of Groningen
Groningen, the Netherlands

Andreia de Almeida
Department of Pharmacokinetics,
    Toxicology and Targeting
University of Groningen
Groningen, the Netherlands

Michael L. De Ieso
School of Medicine
The University of Adelaide
Adelaide, South Australia, Australia

Christine Delporte
Laboratory of Pathophysiological and
    Nutritional Biochemistry
Université Libre de Bruxelles
Brussels, Belgium

Andreas Engel
Department of Bionanoscience
Delft University of Technology
Delft, the Netherlands

Gema Frühbeck
Metabolic Research Laboratory and
  Department of Endocrinology and
  Nutrition
University of Navarra Clinic
and
Centro de Investigación Biomédica en
  Red de Fisiopatología de la Obesidad y
  Nutrición
Instituto de Salud Carlos III
and
Obesity and Adipobiology Group
Instituto de Investigación Sanitaria de
  Navarra
Pamplona, Spain

Stefan Hohmann
Department of Chemistry and Molecular
  Biology
University of Gothenburg
Gothenburg, Sweden

Vincent J. Huber
Brain Research Institute
University of Niigata
Niigata, Japan

Iker Irisarri
Department of Biology
University of Konstanz
Konstanz, Germany

Mohamad Kourghi
School of Medicine
The University of Adelaide
Adelaide, South Australia, Australia

Tae-Hwan Kwon
Department of Biochemistry and
  Cell Biology
Kyungpook National University
Daegu, South Korea
and
Department of Health Science and
  Technology
Aalborg University
Aalborg, Denmark

Anna V. Leonard
Adelaide Centre for Neuroscience Research
and
School of Medicine
The University of Adelaide
Adelaide, South Australia, Australia

Ana Madeira
Research Institute for Medicines
Universidade de Lisboa
Lisboa, Portugal

Raúl A. Marinelli
Instituto de Fisiología Experimental
Universidad Nacional de Rosario
Santa Fe, Argentina

Leire Méndez-Giménez
Metabolic Research Laboratory
University of Navarra Clinic
and
Centro de Investigación Biomédica en
  Red de Fisiopatología de la Obesidad y
  Nutrición
Instituto de Salud Carlos III
and
Obesity and Adipobiology Group
Instituto de Investigación Sanitaria de
  Navarra
Pamplona, Spain

Chulso Moon
Department of Otolaryngology—Head and
    Neck Surgery
and
Department of Oncology
and
Sidney Kimmel Comprehensive Cancer
    Center
Johns Hopkins University
Baltimore, Maryland

David Moon
Department of Otolaryngology—Head and
    Neck Surgery
and
Department of Oncology
and
Sidney Kimmel Comprehensive Cancer
    Center
Johns Hopkins University
Baltimore, Maryland

Teresa F. Moura
Research Institute for Medicines
Universidade de Lisboa
Lisboa, Portugal

Søren Nielsen
Department of Health Science and
    Technology
Aalborg University
Aalborg, Denmark

Maria Nyblom
Department of Biochemistry and
    Structural Biology
Lund University
Lund, Sweden

Jinxin V. Pei
School of Medicine
and
Institute for Photonics and Advanced
    Sensing
The University of Adelaide
Adelaide, South Australia, Australia

Amaia Rodríguez
Metabolic Research Laboratory
University of Navarra Clinic
and
Centro de Investigación Biomédica en
    Red de Fisiopatología de la Obesidad y
    Nutrición
 Instituto de Salud Carlos III
and
Obesity and Adipobiology Group
Instituto de Investigación Sanitaria de
    Navarra
Pamplona, Spain

Michael Rützler
Department of Biochemistry and
    Structural Biology
Lund University
Lund, Sweden
and
Department of Health Science and
    Technology
Aalborg University
Aalborg, Denmark

Graça Soveral
Research Institute for Medicines
Universidade de Lisboa
Lisboa, Portugal

Grazia Tamma
Department of Biosciences,
    Biotechnologies and Biopharmaceutics
University of Bari "Aldo Moro"
Bari, Italy

Susanna Törnroth-Horsefield
Department of Biochemistry and
    Structural Biology
Lund University
Lund, Sweden

Renée J. Turner
Adelaide Centre for Neuroscience Research
and
School of Medicine
The University of Adelaide
Adelaide, South Australia, Australia

Ardeschir Vahedi-Faridi
Department of Physiology and Biophysics
Case Western Reserve University
Cleveland, Ohio

Giovanna Valenti
Department of Biosciences,
    Biotechnologies and Biopharmaceutics
University of Bari "Aldo Moro"
Bari, Italy

Julia von Bülow
Pharmaceutical Institute
Christian-Albrechts-University of Kiel
Kiel, Germany

Sören Wacker
Centre for Molecular Simulations
University of Calgary
Calgary, Alberta, Canada

Andrea J. Yool
School of Medicine
and
Institute for Photonics and Advanced
    Sensing
and
Adelaide Centre for Neuroscience
    Research
The University of Adelaide
Adelaide, South Australia, Australia

Rafael Zardoya
Department of Biodiversity and
    Evolutionary Biology
Museo Nacional de Ciencias Naturales
Madrid, Spain

I

Mechanisms of Fluid Transport in Cells and Epithelia

CHAPTER 1

Aquaporin Functional Detection

Basic Concepts on Water and Solute Permeation

Ana Madeira, Teresa F. Moura and Graça Soveral

CONTENTS

Abstract		3
1.1	Introduction	4
1.2	Methods to Assess AQP Function and Regulation	6
	1.2.1 Functional Analysis	6
	1.2.1.1 Models	6
	1.2.1.2 Technical Approaches	6
	1.2.2 Strategies for Assessing AQP Regulation	8
1.3	Permeability Analysis: Biophysics of Water and Solute Transport	9
	1.3.1 Channel, Cellular and Epithelial Permeabilities	9
	1.3.2 Osmotic and Solute Permeabilities	10
	1.3.3 Co-Existence of Osmotic and Hydrostatic Pressure Gradients	11
	1.3.4 Simplified Data Analysis for Permeability Evaluation	11
	1.3.5 Activation Energy to Indicate AQP Activity	12
1.4	Final Remarks	12
References		12

ABSTRACT

WATER IS THE MAJOR COMPONENT of cells and tissues in living organisms. Fluxes of water and solutes through cell membranes and epithelia (the wall of tubular organs) are considered essential for homeostasis. Aquaporins (AQPs) are crucial for cell function and regulation due to their involvement in the bidirectional transfer of water and small solutes across the cell membranes, yet their precise role in several pathophysiological aspects remains unclear. Suitable methods for membrane water and solute permeability analysis comprising optimal technical/cell system approaches and accurate permeability evaluations are crucial to validate AQP function, to assess regulation and to screen for activity modulators.

The present chapter describes established in vitro assays to assess AQP function in cells and tissues as well as the experimental strategies required to reveal functional regulation.

The basic principles involved in water and solute permeation through membranes are described along with the theoretical models to access parameters defining water and solute transport, such as osmotic water permeability (P_f), solute permeability (P_S) and Arrhenius activation energies (E_a).

1.1 INTRODUCTION

Water homeostasis is central to the physiology of all living cells. Exchanges of water and solutes between environment and intracellular compartments require their passage through a membrane barrier composed of a hydrophobic lipid bilayer with specific transmembrane proteins facilitating permeation of polar and charged species.

Channels that facilitate water permeation through cell membranes were first described on red blood cells in the late 1950s (Paganelli and Solomon 1957) and later on renal epithelia (Whittembury 1960). The first recognized water channel protein was identified in red blood cells by Agre and coworkers (Preston et al. 1992) and was named aquaporin 1 (AQP1). Now it is generally accepted that water crosses cell membranes by two parallel pathways, with distinct mechanisms for permeation: partition/diffusion of water molecules across the hydrophobic bilayer (with high activation energy for transport) and water molecule diffusion through aquaporins (AQPs) (with low activation energy) (Verkman 2000) (Figure 1.1).

AQPs belong to a highly conserved group of membrane proteins called the major intrinsic proteins that form a large family comprising more than 1700 integral membrane proteins found in virtually all living organisms (Abascal et al. 2014). AQPs can be divided into three subfamilies: (1) classical or orthodox AQPs, which are considered to be water selective; (2) aquaglyceroporins, which are permeable to glycerol and other small solutes in addition to water and (3) S-AQPs, the most recently identified members, also called subcellular AQPs (Benga 2012).

The number of AQP isoforms expressed varies significantly among organisms. For instance, *Escherichia coli* possesses one classical AQP (AqpZ) and one AQP-like sequence (glycerol facilitator GlpF) (Maurel et al. 1994; Calamita et al. 1995), whereas mammals possess 13 (Ishibashi et al. 2009) and plants up to 35 different isoforms (Maurel et al. 2008).

The most remarkable feature of AQP channels is their high selectivity and efficiency on water or glycerol permeation. AQPs allow water/glycerol to move freely and bidirectionally across the cell membrane in response to osmotic gradients, but exclude all ions including hydroxide, hydronium ions and protons (Murata et al. 2000), the latter being essential to preserve the electrochemical potential across the membrane. Apart from water and glycerol, a number of other permeants such as urea (Ma et al. 1997), hydrogen peroxide (Bienert et al. 2007), ammonia (Holm et al. 2005), nitrate (Ikeda et al. 2002), arsenite/antimonite (Sanders et al. 1997; Liu et al. 2002), nitric oxide (Herrera et al. 2006), silicon (Ma et al. 2006), carbon dioxide (Nakhoul et al. 1998) and even anions (Yasui et al. 1999) have been shown to permeate specific AQPs.

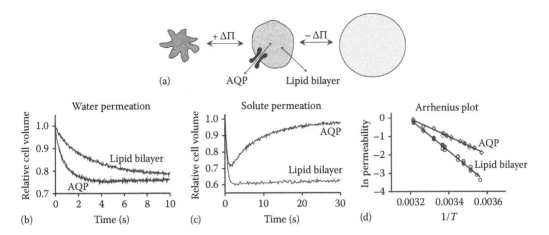

FIGURE 1.1 Water and solute transport across biological membranes. (a) Cell volume changes due to water fluxes after an imposed hyperosmotic (+ΔΠ) or hypoosmotic (−ΔΠ) gradient with an imperme- ant solute. Water crosses cell membranes simultaneously through the lipid bilayer and aquaporins until cells reach a new osmotic equilibrium. (b) Cell volume change after a hyperosmotic shock with an impermeant solute (stopped-flow experiment). A slower time course is depicted for cells without aquaporins (AQPs) where the lipid bilayer is the main pathway for water diffusion. The presence of functional AQPs increases the rate of cell shrinkage due to water outflow. (c) Cell volume change after a hyperosmotic shock with a permeant solute (e.g. glycerol). In the presence of functional aquaporins (AQPs), after the first fast cell shrinkage due to water outflow, glycerol influx in response to its chemi- cal gradient is followed by water influx with subsequent cell reswelling. In the absence of AQPs (lipid bilayer), glycerol influx is not detectable. (d) Representative Arrhenius plot for water transport for cells with or without aquaporins (AQPs) (lipid bilayer). The slope of the straight line is used to calculate E_a. A lower slope relative to a lipid bilayer (lower E_a value) indicates the presence of functional AQPs.

Eukaryotic AQPs are frequently regulated post-translationally either by gating, which controls the channels' rate of flux, or by trafficking, whereby AQPs are shuttled from intra- cellular storage sites to the plasma membrane (Tornroth-Horsefield et al. 2010). Gating of AQPs has been described for several cell systems. Factors like phosphorylation, pH, pres- sure, solute gradients, temperature and membrane tension, among others, were reported to affect the gating behaviour of yeast, plant and mammalian AQPs (Soveral et al. 1997a, 2008; Chaumont et al. 2005; Maurel 2007; Fischer et al. 2009; Tornroth-Horsefield et al. 2010; Leitao et al. 2012, 2014; Ozu et al. 2013). Additionally, in recent years, multiple com- pounds have been described as inhibitors of AQPs activity. AQP-based modulator drugs are predicted to be of broad utility in the treatment of several disorders, such as cerebral edema, cancer, obesity, wound healing, epilepsy, glaucoma and malaria (De Almeida et al. 2014; Verkman et al. 2014).

This chapter describes the methodologies to assess AQP function using permeability assays, detailing the biophysical basis and the experimental protocols for permeability evaluation in different cell systems. Detecting AQP function is crucial for assessing regu- lation and screening for new activity modulators that can be used for drug development.

1.2 METHODS TO ASSESS AQP FUNCTION AND REGULATION

1.2.1 Functional Analysis

1.2.1.1 Models

Due to the widespread distribution of AQPs in nature, water transport assays have been performed using isolated cells from different organisms, such as bacteria (Delamarche et al. 1999; Hubert et al. 2005; Mallo and Ashby 2006), yeast (Laize et al. 1999; Soveral et al. 2007; Madeira et al. 2010) and mammalian cells (Dobbs et al. 1998; Solenov et al. 2004). Intracellular vesicles (Coury et al. 1999; Meyrial et al. 2001) as well as plasma membrane vesicles obtained from animal tissues (mainly the kidney or intestinal epithelia) have been used to evaluate AQP activity either in intracellular organelles (Laize et al. 1995; Calamita et al. 2005) or through epithelia (Whittembury 1960; Mollajew et al. 2010). Other widely used approach consists of AQP heterologous expression in *Xenopus laevis* oocytes, which have very low intrinsic water permeability, to functionally characterize novel AQP isoforms (Zhang et al. 1990). In addition to oocytes, yeast cells lacking endogenous AQPs (Pettersson et al. 2006; Leitao et al. 2012), AQP-transfected cell lines (Ma et al. 1993; Madeira et al. 2013) and mRNA-injected zebra fish embryos (Ikeda et al. 2011) have also been used for heterologous expression. AQPs from different organisms have also been purified and reconstituted in liposomes, which allowed establishing their direct role in water/solute transport (van Hoek and Verkman 1992; Zeidel et al. 1992). Phenotypic analysis of transgenic mice lacking AQPs has also brought new insights into their mechanisms of permeation and revealed their involvement in multiple biological functions, including transepithelial fluid transport, cell migration, brain edema, neuroexcitation, cell proliferation, epidermal water retention and adipocyte metabolism (Verkman et al. 2014).

1.2.1.2 Technical Approaches

AQP function can be indirectly assessed by measuring water/solute permeability of biological membranes. Water permeability may be determined in native epithelial tissues (e.g. intestinal mucosa) or in epithelial cell monolayers cultured on permeable supports and mounted on Ussing chambers (Clarke 2009). In this experimental setting, apical and basolateral membranes of polarized cells face different half-chambers (Clarke 2009). By adding a membrane impermeant solute such as sucrose or mannitol to one half-chamber, the generated transepithelial water flux is measured by the height of the fluid in a capillary tube connected to the other half-chamber (Dorr et al. 1997).

Preparation of isolated vesicles from either basolateral or apical membranes is another widely used technique for AQP characterization in epithelia (Rigler et al. 1985; Donowitz et al. 1987). Assessing the time course of vesicle's volume change to rapidly imposed osmotic/solute gradients allows quantifying the permeability of membranes to water or to a specific solute (Verkman 2000). Equivalent strategies are also commonly used to assess the kinetics of transport in cells and proteoliposomes. Regardless of the biological model used, the characterization of AQP transport makes use of volume-dependent optical properties, such as light transmission (Zhang et al. 1990; Farinas and Verkman 1996; Farinas et al. 1997), absorbance (Levin et al. 2007) or scattering (Verkman et al. 1985; van Hoek and Verkman 1992) and fluorescence (Farinas et al. 1995; Galietta et al. 2001;

Hamann et al. 2002; Solenov et al. 2004). The possibility of combining different biological models (cells/vesicles/proteoliposomes) and optical detection systems offers many possibilities to conduct research on AQP function. The most commonly used approaches will be further discussed.

1.2.1.2.1 Osmotic Swelling Assay   AQP-mediated water transport can be evaluated by heterologous expression in *X. laevis* oocytes using an osmotic swelling assay (Zhang et al. 1990; Verkman 2000). Oocytes microinjected with AQP mRNA are subjected to hypoosmotic gradients, and the time course of cell swelling is followed by video microscopy. To test for solute permeability, an inwardly directed solute gradient is imposed, resulting in solute influx in response to the generated chemical gradient, followed by water influx and consequent oocyte swelling (Hansen et al. 2002; Beitz et al. 2009). This system is particularly advantageous for studying AQPs due to oocytes' low water permeability and hardly detectable permeability to glycerol and other solutes.

1.2.1.2.2 Microscopy Techniques   Over the years, many microscopy techniques have been employed to characterize AQP function, namely phase contrast (Preston et al. 1992), dark field/phase contrast (Farinas et al. 1997), interferometry (Farinas and Verkman 1996), confocal (Zelenina and Brismar 2000) and fluorescence microscopy (Chen et al. 1988; Verkman 2000; Solenov et al. 2004).

The most currently used microscopy approaches exploit the properties of volume-sensitive fluorescent dyes that undergo intracellular de-esterification and trapping (Chen et al. 1988; Verkman 2000). Cells are loaded with the membrane-permeant non-fluorescent precursor, which is cleaved intracellularly by non-specific esterases yielding the impermeable fluorescent form. Changes in fluorescence intensity resulting from osmotically induced volume changes can be monitored as the fluorescence of the fluorophore is quenched. Two hypotheses explain this phenomenon inside the cell: (1) the quenching is mediated by cytoplasmic proteins, whose concentrations change as cells shrink or swell (Solenov et al. 2004); (2) fluorophores undergo self-quenching, that is fluorescence intensity decreases with increasing fluorophore concentration (Hamann et al. 2002). In any case, water and/or solute permeability can be inferred from the linear relationship between cell volume and dye fluorescence intensity (Hamann et al. 2002).

Another method uses a genetically encoded optical sensor (yellow fluorescent protein YFP-H148Q-V163S), whose fluorescence is quenched by chloride (Galietta et al. 2001; Baumgart et al. 2012). Cell volume changes and consequent altered intracellular chloride concentration result in altered YFP emitted fluorescence.

More recently, a high-throughput system for automated water/solute permeability measurements using volume-sensitive fluorescent indicators has been optimized taking advantage of microplate readers (Fenton et al. 2010).

1.2.1.2.3 Stopped-Flow Fast Kinetics   The stopped-flow spectroscopy is a commonly used method to follow rapid volume changes in proteoliposomes (van Hoek and Verkman 1992; Zeidel et al. 1992), vesicles (Verkman et al. 1985; Soveral et al. 1997a,b) and cell suspensions

(Ma et al. 1993; Dobbs et al. 1998; Pettersson et al. 2006; Soveral et al. 2006, 2007). In the stopped-flow device, cell/vesicle suspensions are subjected to osmotic challenges by rapid mixing with an equal volume of hypo- or hyperosmotic solution (Figure 1.1a). Osmotic water fluxes produce changes in cell volume, with consequent alterations in scattered light intensity or fluorescence if the cell/vesicles are loaded with volume-sensitive fluorescent dyes. Due to a linear relation between the optical properties of the system and cell volume, water (Figure 1.1b) or solute (Figure 1.1c) movements can be followed until osmotic equilibrium is attained. Analysis of the relative volume changes due to osmotic or solute gradients allows quantification of osmotic permeability (P_f) or solute permeability (P_S) coefficients.

1.2.1.2.4 Computational Methods   The introduction of computational methods has shed new light on AQP's water and solute permeation mechanisms. These methods make use of the available high-resolution atomic structures for different AQP isoforms (Fu et al. 2000; Sui et al. 2001; Savage et al. 2003; Gonen et al. 2004; Harries et al. 2004; Lee et al. 2005; Tornroth-Horsefield et al. 2006; Horsefield et al. 2008; Newby et al. 2008; Fischer et al. 2009). Specifically, molecular dynamic simulations, which employ classical mechanics for the sampling of conformational changes in biomolecules, provided a unique dynamic insight onto AQPs structures (Hub et al. 2009). In the absence of a solved atomic structure for a specific AQP isoform, it is possible to assemble 3D models using experimentally determined structures of related family members as templates (Bordoli et al. 2009). The assembly of homology models, having GlpF channel as a template, allowed accessing the structural uniqueness of human AQP3 (Martins et al. 2012), AQP7 (Madeira et al. 2014) and AQP9 (Wacker et al. 2013).

1.2.2 Strategies for Assessing AQP Regulation

Within the gating mechanisms, it is well established that eukaryotic AQPs are often directly regulated by pH (Nemeth-Cahalan and Hall 2000; Leitao et al. 2012), phosphorylation (Maurel et al. 1995), divalent cations (Preston et al. 1993; Zelenina et al. 2003) and membrane stretching (membrane surface tension) (Soveral et al. 1997a; Soveral et al. 2008; Ozu et al. 2013; Leitao et al. 2014). These regulatory mechanisms directly affect the protein conformation, which in turn impacts its transport activity (Alleva et al. 2012). Control of AQP function by pH or specific inhibitors can be screened through simple measurements of permeability. In addition, site-directed mutagenesis is a valuable tool to identify specific residues involved in phosphorylation (Maurel et al. 1995; Johansson et al. 1998) and also in any other gating mechanism.

Levin et al. (2007) developed a simple screening method to identify inhibitors of AQP1 that involves measuring erythrocyte cell lysis using infrared light scattering. Erythrocytes expressing AQP1 and urea transporter (UT-B) are loaded with a urea analog, acetamide, which is transported by UT-B. Dilution of the cells into a hypoosmotic acetamide-free solution results in cell swelling and lysis. However, when AQP1 is inhibited, water influx is slowed and cell lysis prevented, as water influx is slower than the dissipation of the osmotic gradient by acetamide efflux (Verkman 2009).

To investigate inhibitors specifically binding to the intracellular side of AQPs, Ozu et al. (2005) developed an experimental setup where the cell membrane of an emptied-out *X. laevis* oocyte is used as a diaphragm between two independent chambers. In this configuration, water permeability can be continuously monitored while simultaneously controlling the media composition on both sides of the plasma membrane. Importantly, this approach also allows controlling the hydrostatic pressure on both sides of the cell membrane and investigating the gating by membrane surface tension of AQPs expressed in oocytes (Ozu et al. 2013).

Membrane tension is related to the force needed to deform a membrane, and according to Laplace's law for a sphere, the tension is directly related to the product of pressure and radius. This means that to reach the same membrane tension, larger cells need much lower levels of pressure than smaller cells (sometimes hardly discernible), making the disclosure of AQP modulation by tension difficult to assess. Systems that are able to sustain membrane tensions without rupture were devised and include membrane vesicles (Soveral et al. 1997a) and walled yeast cells (Soveral et al. 2008). Previous equilibration of vesicles or walled yeast cells in hypoosmotic media before applying the osmotic challenge induce vesicle/cell swelling with an increase in membrane surface tension prior to permeability assays. These approaches allowed the disclosure of AQP regulation by pressure on vesicles from kidney proximal tubule (Soveral et al. 1997a), yeasts (Soveral et al. 2008) and plants (Leitao et al. 2014).

In silico methods, in particular molecular docking and molecular dynamic simulations, have recently been used to screen large libraries of compounds as candidate AQP inhibitors (Seeliger et al. 2013). These methods are now considered essential for AQP drug discovery providing an educated guess for in vitro testing.

1.3 PERMEABILITY ANALYSIS: BIOPHYSICS OF WATER AND SOLUTE TRANSPORT

In the next section, the biophysical parameters used to define water and solute permeabilities that allow obtaining information on AQP function in several experimental conditions are described.

1.3.1 Channel, Cellular and Epithelial Permeabilities

In cells, the plasma membrane is the barrier that separates the intra- and the extracellular compartments, while in tubular organs, the epithelia is a more complex barrier separating the interior (plasma, interstitial fluid) of an organism from its environment (lumen), being a three compartment system. In the plasma membrane, the total permeability is the sum of the lipid and the protein (channel) paths, $P_{total} = P_{lipid} + P_{channel}$, and increases with the conductance and number of active channels. If total inhibition of channels is achieved, $P_{total} = P_{lipid}$. On the other hand, considering the epithelial wall as a monolayer of cells, the total epithelial permeability is the sum of its transcellular and junctional components, $P_{epithelial} = P_{cell} + P_{junction}$. The epithelial cells have a bipolar characteristic, with different transporters in the apical and basolateral membranes and so the total transcellular permeability results from the two membranes in series $P_{cell} = (P_{apical} \times P_{basolateral})/(P_{apical} + P_{basolateral})$, consequently $P_{cell} \ll P_{apical}$ or $P_{basolateral}$. An estimation of P_{cell} can be obtained by individual estimations of its components using apical and basolateral membrane vesicles as described previously.

1.3.2 Osmotic and Solute Permeabilities

When cells/vesicles equilibrated in an initial medium osmolarity $(osm_{out})_o$ are suddenly subjected to an osmotic challenge by the addition of a non-diffusible (ND) or diffusible (S) solute reaching a final osmolarity of $(osm_{out})_\infty$, the resulting changes in cell volume (V) depend on water and solute fluxes. The equations describing solvent flow through semi-permeable membranes proposed by Kedem–Katchalsky (Kedem and Katchalsky 1958) can be simplified if we consider that there are no convective fluxes (diffusible solute reflection coefficient $\sigma_S = 1$) and no hydrostatic pressure differences between the intra- and extra-cellular compartments ($\Delta P = 0$). Thus, the simplified flux equations for water (J_v) and diffusible solute (J_S) from the inner to the outer compartment and the resulting relative volume ($v_{rel} = V/V_o$) changes are

$$J_v = P_f \left(\frac{V_w A}{RT} \right)(-\Delta\pi_{Total}) \quad \text{cm}^3\text{s}^{-1} \tag{1.1}$$

$$J_S = P_S \Delta C_S A \quad \text{mol s}^{-1} \tag{1.2}$$

$$\frac{dv_{rel}}{dt} = -\frac{J_v}{V_o} \quad \text{s}^{-1} \tag{1.3}$$

In these equations, the water flow J_v is proportional to the total osmotic pressure differ-ence ($\Delta\pi_{Total} = RT(osm_{in} - (osm_{out})_\infty)$) due to the concentration gradients of diffusible (ΔC_S) and non-diffusible (ΔC_{ND}) solutes, $\Delta\pi_{Total} = RT(\Delta\pi_S + \Delta\pi_{ND})$, where $\Delta\pi_{S,ND} = RT[C_{S,ND\_in} - C_{S,ND\_out}]$. Solute flow J_S is proportional to its concentration gradient ΔC_S. In these equa-tions, P_f (osmotic water) and P_S (solute) are the permeability coefficients (cm/s), V_w is the partial molar volume of water, A is the membrane surface area, R is the gas constant and T is the absolute temperature.

The relative volume v_{rel} equals to the sum of the osmotic v_{osm} ($=V_{osm}/V_o$) and non-osmotic β ($=V_{Nosm}/V_o$) volume components. As β is constant, $dv_{rel}/dt = dv_{osm}/dt$. During the osmotic shock when v_{rel} is changing, the intracellular concentrations are also changing and can be calculated by their intracellular quantities Q_{S,ND_in} and V_{osm}, $C_{S,ND\_in} = Q_{S,ND\_in}/(V_{osm})$. While Q_{ND_in} is constant and always equal to $Q_{ND\_in} = (V_{osm})_o \cdot (osm_{out})_o$, Q_{S_in} changes and can be evaluated from J_S, $dQ_{S\_in}/dt = -J_S$, knowing its initial condition $(Q_{S,ND\_in})_o = 0$.

If the only gradient to be considered is from non-diffusible species $\Delta\pi_{ND}$, the param-eters to estimate from each experimental trace are reduced to two, P_f and β. If, on the other hand, there is also a diffusible solute gradient $\Delta\pi_S$ assuming $\sigma_S \approx 1$, the parame-ters to be estimated are the permeability coefficients P_f and P_S and β. However, if $\sigma_S < 1$, in Equation 1.1, the osmotic gradient due to solute S is reduced to $\sigma_S \Delta \pi_S$, and in Equation 1.2, the convective term cannot be ignored, increasing to four the number of parameters to be estimated (P_f, P_S, β and σ_S).

1.3.3 Co-Existence of Osmotic and Hydrostatic Pressure Gradients

In the presence of an initial hydrostatic pressure difference ($\Delta P = P_{in} - P_{out}$) when cells are in osmotic equilibrium, the osmotic and hydrostatic pressure differences are equal, $\Delta \Pi_o = \Delta P_o$. When applying an osmotic shock by abruptly changing the external media osmolarity (from $(osm_{out})_o$ to $(osm_{out})_\infty$), the initial osmotic pressure changes abruptly and the osmotic equilibrium is disturbed; β and Q_{ND_in} ($=osm_{in}\, V_{osm}$) remain constant, while v_{osm} (and v_{rel}) starts changing driving the changes of osm_{in} and ΔP till the new osmotic equilibrium is attained, $\Delta \Pi_\infty = \Delta P_\infty$. Under these conditions, the changes in v_{rel} (Equation 1.3) induced by the water flux in or out of cells (Equation 1.4), across the membrane area A, are proportional to P_f according to the following equation where $\Lambda = (osm_{out})_\infty/(osm_{out})_o$ and $p = \Delta P/[RT(osm_{out})_o]$:

$$\frac{dV_{rel}}{dt} = \frac{A}{V_o} P_f V_w (osm_{out})_o \left(\frac{1+p_o}{V_{osm}} - p - \Lambda \right) \tag{1.4}$$

In this equation, the only time-dependent variables are the volumes and p. When the final osmotic equilibrium is reached, all these variables reach their final equilibrium value, p_∞ and v_{osm_∞}, and the term in brackets vanishes, $\Lambda = [(1 + p_o)/v_{osm\_\infty}] - p_\infty$, and can be evaluated experimentally by plotting Λ as a function of final equilibrium osmotic volumes v_{osm_∞} as previously reported for vesicle systems (Soveral et al. 1997a,b) or yeast cells (Soveral et al. 2008).

1.3.4 Simplified Data Analysis for Permeability Evaluation

Simplifications of Equations 1.1 and 1.2 have been introduced for a simplified data analysis and are discussed by van Heeswijk and van Os (1986) for P_f and Verkman et al. (1985) for P_S. These include linear relations between permeability coefficients and rate constants of simple exponential fits to experimental traces. P_f is evaluated from the rate constant k of fitted traces obtained upon an osmotic challenge with an impermeable solute by $P_f = k(V_o/A)$ $(1/V_w(osm_{out})_\infty)$. Sometimes double exponential fits are needed that may reflect two different sample populations or may simply obscure the basic non-linearity that occurs in larger volume perturbations that may invalidate this approach; another drawback may result from the presence of hydrostatic pressure gradients disregarded in these calculations that will give overestimations of permeability values (Soveral et al. 1997a,b, 2008).

Solute permeabilities calculated from traces obtained when applying inward solute gradients have also been estimated from single exponential fits associated with cell/vesicle reswelling using the relation $P_S = k(V_o/A)$ (Verkman et al. 1985). This solution is only valid considering that P_S is much lower than P_f.

1.3.5 Activation Energy to Indicate AQP Activity

The activation energy (E_a, kcal/mol) constitutes another useful parameter to define water fluxes across membranes. Given that water permeates membranes both through the lipid and the channel pathways, E_a values provide a measure of the energy barrier to water movement indicating the predominant pathway for water permeation. E_a is defined by the relation $\ln P_f = E_a/RT + a$, where R is the gas constant, T is absolute temperature and a is an entropic term. E_a can be calculated from the slope of an Arrhenius plot ($\ln P_f$ as a function of $1/T$). High E_a values (>10 kcal/mol) suggest that the lipid pathway predominates, while lower values (<5 kcal/mol) indicate the presence of functional AQPs (Verkman 2000). An example of an Arrhenius plot used for E_a evaluation is depicted in Figure 1.1d. The stopped-flow technique is the most convenient method for the determination of E_a, due to the precise temperature control in different parts of the system, including sample suspension and test solution.

1.4 FINAL REMARKS

Since the discovery of AQP channels, new knowledge on their cell distribution, structure, regulation and pathophysiology is being progressively revealed. The characterization of selectivity and mechanisms of gating is essential to establish AQP contribution to homeostasis, which is critical to health, and further identify dysfunctions that may lead to phenotypes such as those found in disease. These aspects can only be supported if suitable methods for functional analysis are performed, including optimal technical/cell system approaches and accurate permeability evaluations. A well-designed experimental strategy for assessing AQP function and regulation allows discovering new AQP inhibitors and constitutes an important element in drug discovery.

REFERENCES

Abascal, F., I. Irisarri and R. Zardoya. 2014. Diversity and evolution of membrane intrinsic proteins. *Biochim Biophys Acta* 1840(5):1468–1481.

Alleva, K., O. Chara and G. Amodeo. 2012. Aquaporins: Another piece in the osmotic puzzle. *FEBS Lett* 586(19):2991–2999.

Baumgart, F., A. Rossi and A. S. Verkman. 2012. Light inactivation of water transport and protein–protein interactions of aquaporin-killer red chimeras. *J Gen Physiol* 139(1):83–91.

Beitz, E., D. Becker, J. von Bulow, C. Conrad, N. Fricke, A. Geadkaew, D. Krenc, J. Song, D. Wree and B. Wu. 2009. In vitro analysis and modification of aquaporin pore selectivity. *Handbook of Pharmacology*, Vol. 190:77–92.

Benga, G. 2012. On the definition, nomenclature and classification of water channel proteins (aquaporins and relatives). *Mol Aspects Med* 33(5–6):514–517.

Bienert, G. P., A. L. Moller, K. A. Kristiansen, A. Schulz, I. M. Moller, J. K. Schjoerring and T. P. Jahn. 2007. Specific aquaporins facilitate the diffusion of hydrogen peroxide across membranes. *J Biol Chem* 282(2):1183–1192.

Bordoli, L., F. Kiefer, K. Arnold, P. Benkert, J. Battey and T. Schwede. 2009. Protein structure homology modeling using SWISS-MODEL workspace. *Nat Protoc* 4(1):1–13.

Calamita, G., W. R. Bishai, G. M. Preston, W. B. Guggino and P. Agre. 1995. Molecular cloning and characterization of AqpZ, a water channel from *Escherichia coli*. *J Biol Chem* 270(49): 29063–29066.

Calamita, G., D. Ferri, P. Gena, G. E. Liquori, A. Cavalier, D. Thomas and M. Svelto. 2005. The inner mitochondrial membrane has aquaporin-8 water channels and is highly permeable to water. *J Biol Chem* 280(17):17149–17153.

Chaumont, F., M. Moshelion and M. J. Daniels. 2005. Regulation of plant aquaporin activity. *Biol Cell* 97(10):749–764.

Chen, P. Y., D. Pearce and A. S. Verkman. 1988. Membrane water and solute permeability determined quantitatively by self-quenching of an entrapped fluorophore. *Biochemistry* 27(15):5713–5718.

Clarke, L. L. 2009. A guide to using chamber studies of mouse intestine. *Am J Physiol Gastrointest Liver Physiol* 296(6):G1151–G1166.

Coury, L. A., M. Hiller, J. C. Mathai, E. W. Jones, M. L. Zeidel and J. L. Brodsky. 1999. Water transport across yeast vacuolar and plasma membrane-targeted secretory vesicles occurs by passive diffusion. *J Bacteriol* 181(14):4437–4440.

De Almeida, A., G. Soveral and A. Casini. 2014. Gold compounds as aquaporin inhibitors: New opportunities for therapy and imaging. *Med Chem Commun* 5:1444–1453.

Delamarche, C., D. Thomas, J. P. Rolland, A. Froger, J. Gouranton, M. Svelto, P. Agre and G. Calamita. 1999. Visualization of AqpZ-mediated water permeability in *Escherichia coli* by cryoelectron microscopy. *J Bacteriol* 181(14):4193–4197.

Dobbs, L. G., R. Gonzalez, M. A. Matthay, E. P. Carter, L. Allen and A. S. Verkman. 1998. Highly water-permeable type I alveolar epithelial cells confer high water permeability between the airspace and vasculature in rat lung. *Proc Natl Acad Sci USA* 95(6):2991–2996.

Donowitz, M., E. Emmer, J. McCullen, L. Reinlib, M. E. Cohen, R. P. Rood, J. Madara, G. W. Sharp, H. Murer and K. Malmstrom. 1987. Freeze–thaw and high-voltage discharge allow macromolecule uptake into ileal brush-border vesicles. *Am J Physiol* 252 (6 Pt 1):G723–G735.

Dorr, R. A., A. Kierbel, J. Vera and M. Parisi. 1997. A new data-acquisition system for the measurement of the net water flux across epithelia. *Comput Methods Programs Biomed* 53(1):9–14.

Farinas, J., M. Kneen, M. Moore and A. S. Verkman. 1997. Plasma membrane water permeability of cultured cells and epithelia measured by light microscopy with spatial filtering. *J Gen Physiol* 110(3):283–296.

Farinas, J., V. Simanek and A. S. Verkman. 1995. Cell volume measured by total internal reflection microfluorimetry: Application to water and solute transport in cells transfected with water channel homologs. *Biophys J* 68(4):1613–1620.

Farinas, J. and A. S. Verkman. 1996. Cell volume and plasma membrane osmotic water permeability in epithelial cell layers measured by interferometry. *Biophys J* 71(6):3511–3522.

Fenton, R. A., H. B. Moeller, S. Nielsen, B. L. de Groot and M. Rutzler. 2010. A plate reader-based method for cell water permeability measurement. *Am J Physiol Renal Physiol* 298(1):F224–F230.

Fischer, G., U. Kosinska-Eriksson, C. Aponte-Santamaria, M. Palmgren, C. Geijer, K. Hedfalk, S. Hohmann, B. L. de Groot, R. Neutze and K. Lindkvist-Petersson. 2009. Crystal structure of a yeast aquaporin at 1.15 angstrom reveals a novel gating mechanism. *PLoS Biol* 7(6):e1000130.

Fu, D., A. Libson, L. J. Miercke, C. Weitzman, P. Nollert, J. Krucinski and R. M. Stroud. 2000. Structure of a glycerol-conducting channel and the basis for its selectivity. *Science* 290(5491):481–486.

Galietta, L. J., P. M. Haggie and A. S. Verkman. 2001. Green fluorescent protein-based halide indicators with improved chloride and iodide affinities. *FEBS Lett* 499(3):220–224.

Gonen, T., P. Sliz, J. Kistler, Y. Cheng and T. Walz. 2004. Aquaporin-0 membrane junctions reveal the structure of a closed water pore. *Nature* 429(6988):193–197.

Hamann, S., J. F. Kiilgaard, T. Litman, F. J. Alvarez-Leefmans, B. R. Winther and T. Zeuthen. 2002. Measurement of cell volume changes by fluorescence self-quenching. *J Fluoresc* 12(2):139–145.

Hansen, M., J. F. Kun, J. E. Schultz and E. Beitz. 2002. A single, bi-functional aquaglyceroporin in blood-stage Plasmodium falciparum malaria parasites. *J Biol Chem* 277(7):4874–4882.

Harries, W. E., D. Akhavan, L. J. Miercke, S. Khademi and R. M. Stroud. 2004. The channel architecture of aquaporin 0 at a 2.2-A resolution. *Proc Natl Acad Sci USA* 101(39):14045–14050.

Herrera, M., N. J. Hong and J. L. Garvin. 2006. Aquaporin-1 transports NO across cell membranes. *Hypertension* 48(1):157–164.

Holm, L. M., T. P. Jahn, A. L. Moller, J. K. Schjoerring, D. Ferri, D. A. Klaerke and T. Zeuthen. 2005. NH_3 and NH_4^+ permeability in aquaporin-expressing *Xenopus* oocytes. *Pflugers Arch* 450(6):415–428.

Horsefield, R., K. Norden, M. Fellert, A. Backmark, S. Tornroth-Horsefield, A. C. T. van Scheltinga, J. Kvassman, P. Kjellbom, U. Johanson and R. Neutze. 2008. High-resolution x-ray structure of human aquaporin 5. *Proc Natl Acad Sci USA* 105(36):13327–13332.

Hub, J. S., H. Grubmuller and B. L. de Groot. 2009. Dynamics and energetics of permeation through aquaporins. What do we learn from molecular dynamics simulations? *Handbook of Pharmacology*, Vol. 190:57–76.

Hubert, J. F., L. Duchesne, C. Delamarche, A. Vaysse, H. Gueune and C. Raguenes-Nicol. 2005. Pore selectivity analysis of an aquaglyceroporin by stopped-flow spectrophotometry on bacterial cell suspensions. *Biol Cell* 97(9):675–686.

Ikeda, M., A. Andoo, M. Shimono, N. Takamatsu, A. Taki, K. Muta, W. Matsushita et al. 2011. The NPC motif of aquaporin-11, unlike the NPA motif of known aquaporins, is essential for full expression of molecular function. *J Biol Chem* 286(5):3342–3350.

Ikeda, M., E. Beitz, D. Kozono, W. B. Guggino, P. Agre and M. Yasui. 2002. Characterization of aquaporin-6 as a nitrate channel in mammalian cells. Requirement of pore-lining residue threonine 63. *J Biol Chem* 277(42):39873–39879.

Ishibashi, K., S. Hara and S. Kondo. 2009. Aquaporin water channels in mammals. *Clin Exp Nephrol* 13(2):107–117.

Johansson, I., M. Karlsson, V. K. Shukla, M. J. Chrispeels, C. Larsson and P. Kjellbom. 1998. Water transport activity of the plasma membrane aquaporin PM28A is regulated by phosphorylation. *Plant Cell* 10(3):451–459.

Kedem, O. and A. Katchalsky. 1958. Thermodynamic analysis of the permeability of biological membranes to non-electrolytes. *Biochim Biophys Acta* 27(2):229–246.

Laize, V., R. Gobin, G. Rousselet, C. Badier, S. Hohmann, P. Ripoche and F. Tacnet. 1999. Molecular and functional study of AQY1 from *Saccharomyces cerevisiae*: Role of the C-terminal domain. *Biochem Biophys Res Commun* 257(1):139–144.

Laize, V., G. Rousselet, J. M. Verbavatz, V. Berthonaud, R. Gobin, N. Roudier, L. Abrami, P. Ripoche and F. Tacnet. 1995. Functional expression of the human CHIP28 water channel in a yeast secretory mutant. *FEBS Lett* 373(3):269–274.

Lee, J. K., D. Kozono, J. Remis, Y. Kitagawa, P. Agre and R. M. Stroud. 2005. Structural basis for conductance by the archaeal aquaporin AqpM at 1.68 A. *Proc Natl Acad Sci USA* 102(52):18932–18937.

Leitao, L., C. Prista, M. C. Loureiro-Dias, T. F. Moura and G. Soveral. 2014. The grapevine tonoplast aquaporin TIP2;1 is a pressure gated water channel. *Biochem Biophys Res Commun* 450(1):289–294.

Leitao, L., C. Prista, T. F. Moura, M. C. Loureiro-Dias and G. Soveral. 2012. Grapevine aquaporins: Gating of a tonoplast intrinsic protein(TIP2;1) by cytosolic pH. *PLoS One* 7(3):e33219.

Levin, M. H., R. de la Fuente and A. S. Verkman. 2007. Urearetics: A small molecule screen yields nanomolar potency inhibitors of urea transporter UT-B. *FASEB J* 21(2):551–563.

Liu, Z., J. Shen, J. M. Carbrey, R. Mukhopadhyay, P. Agre and B. P. Rosen. 2002. Arsenite transport by mammalian aquaglyceroporins AQP7 and AQP9. *Proc Natl Acad Sci USA* 99(9): 6053–6058.

Ma, J. F., K. Tamai, N. Yamaji, N. Mitani, S. Konishi, M. Katsuhara, M. Ishiguro, Y. Murata and M. Ya. 2006. A silicon transporter in rice. *Nature* 440(7084):688–691.

Ma, T., A. Frigeri, S. T. Tsai, J. M. Verbavatz and A. S. Verkman. 1993. Localization and functional analysis of CHIP28k water channels in stably transfected Chinese hamster ovary cells. *J Biol Chem* 268(30):22756–22764.

Ma, T., B. Yang and A. S. Verkman. 1997. Cloning of a novel water and urea-permeable aquaporin from mouse expressed strongly in colon, placenta, liver, and heart. *Biochem Biophys Res Commun* 240(2):324–328.

Madeira, A., M. Camps, A. Zorzano, T. F. Moura and G. Soveral. 2013. Biophysical assessment of human aquaporin-7 as a water and glycerol channel in 3T3-L1 adipocytes. *PLoS One* 8(12):e83442.

Madeira, A., A. de Almeida, C. de Graaf, M. Camps, A. Zorzano, T. F. Moura, A. Casini and G. Soveral. 2014. A gold coordination compound as a chemical probe to unravel aquaporin-7 function. *Chembiochem* 15(10):1487–1494.

Madeira, A., L. Leitao, G. Soveral, P. Dias, C. Prista, T. Moura and M. C. Loureiro-Dias. 2010. Effect of ethanol on fluxes of water and protons across the plasma membrane of *Saccharomyces cerevisiae*. *FEMS Yeast Res* 10(3):252–258.

Mallo, R. C. and M. T. Ashby. 2006. AqpZ-mediated water permeability in *Escherichia coli* measured by stopped-flow spectroscopy. *J Bacteriol* 188(2):820–822.

Martins, A. P., A. Marrone, A. Ciancetta, A. Galan Cobo, M. Echevarria, T. F. Moura, N. Re, A. Casini and G. Soveral. 2012. Targeting aquaporin function: Potent inhibition of aquaglyceroporin-3 by a gold-based compound. *PLoS One* 7(5):e37435.

Maurel, C. 2007. Plant aquaporins: Novel functions and regulation properties. *FEBS Lett* 581(12):2227–2236.

Maurel, C., R. T. Kado, J. Guern and M. J. Chrispeels. 1995. Phosphorylation regulates the water channel activity of the seed-specific aquaporin alpha-TIP. *EMBO J* 14(13):3028–3035.

Maurel, C., J. Reizer, J. I. Schroeder, M. J. Chrispeels and M. H. Saier, Jr. 1994. Functional characterization of the *Escherichia coli* glycerol facilitator, GlpF, in *Xenopus* oocytes. *J Biol Chem* 269(16):11869–11872.

Maurel, C., L. Verdoucq, D. T. Luu and V. Santoni. 2008. Plant aquaporins: Membrane channels with multiple integrated functions. *Annu Rev Plant Biol* 59:595–624.

Meyrial, V., V. Laize, R. Gobin, P. Ripoche, S. Hohmann and F. Tacnet. 2001. Existence of a tightly regulated water channel in *Saccharomyces cerevisiae*. *Eur J Biochem* 268(2):334–343.

Mollajew, R., F. Zocher, A. Horner, B. Wiesner, E. Klussmann and P. Pohl. 2010. Routes of epithelial water flow: Aquaporins versus cotransporters. *Biophys J* 99(11):3647–3656.

Murata, K., K. Mitsuoka, T. Hirai, T. Walz, P. Agre, J. B. Heymann, A. Engel and Y. Fujiyoshi. 2000. Structural determinants of water permeation through aquaporin-1. *Nature* 407(6804):599–605.

Nakhoul, N. L., B. A. Davis, M. F. Romero and W. F. Boron. 1998. Effect of expressing the water channel aquaporin-1 on the CO_2 permeability of *Xenopus* oocytes. *Am J Physiol* 274(2 Pt 1):C543–C548.

Nemeth-Cahalan, K. L. and J. E. Hall. 2000. pH and calcium regulate the water permeability of aquaporin 0. *J Biol Chem* 275(10):6777–6782.

Newby, Z. E., J. O'Connell, 3rd, Y. Robles-Colmenares, S. Khademi, L. J. Miercke and R. M. Stroud. 2008. Crystal structure of the aquaglyceroporin PfAQP from the malarial parasite *Plasmodium falciparum*. *Nat Struct Mol Biol* 15(6):619–625.

Ozu, M., R. Dorr and M. Parisi. 2005. New method to measure water permeability in emptied-out *Xenopus* oocytes controlling conditions on both sides of the membrane. *J Biochem Biophys Methods* 63(3):187–200.

Ozu, M., R. A. Dorr, F. Gutierrez, M. T. Politi and R. Toria. 2013. Human AQP1 is a constitutively open channel that closes by a membrane-tension-mediated mechanism. *Biophys J* 104(1):85–95.

Paganelli, C. V. and A. K. Solomon. 1957. The rate of exchange of tritiated water across the human red cell membrane. *J Gen Physiol* 41(2):259–277.

Pettersson, N., J. Hagstrom, R. M. Bill and S. Hohmann. 2006. Expression of heterologous aquaporins for functional analysis in *Saccharomyces cerevisiae*. *Curr Genet* 50(4):247–255.

Preston, G. M., T. P. Carroll, W. B. Guggino and P. Agre. 1992. Appearance of water channels in *Xenopus* oocytes expressing red cell CHIP28 protein. *Science* 256(5055):385–387.

Preston, G. M., J. S. Jung, W. B. Guggino and P. Agre. 1993. The mercury-sensitive residue at cysteine 189 in the CHIP28 water channel. *J Biol Chem* 268(1):17–20.

Rigler, M. W., G. C. Ferreira and J. S. Patton. 1985. Intramembranous particles are clustered on microvillus membrane vesicles. *Biochim Biophys Acta* 816(1):131–141.

Sanders, O. I., C. Rensing, M. Kuroda, B. Mitra and B. P. Rosen. 1997. Antimonite is accumulated by the glycerol facilitator GlpF in *Escherichia coli*. *J Bacteriol* 179(10):3365–3367.

Savage, D. F., P. F. Egea, Y. Robles-Colmenares, J. D. O'Connell, 3rd and R. M. Stroud. 2003. Architecture and selectivity in aquaporins: 2.5 Å x-ray structure of aquaporin z. *PLoS Biol* 1(3):E72.

Seeliger, D., C. Zapater, D. Krenc, R. Haddoub, S. Flitsch, E. Beitz, J. Cerda and B. L. de Groot. 2013. Discovery of novel human aquaporin-1 blockers. *ACS Chem. Biol.* 8(1):249–256.

Solenov, E., H. Watanabe, G. T. Manley and A. S. Verkman. 2004. Sevenfold-reduced osmotic water permeability in primary astrocyte cultures from AQP-4-deficient mice, measured by a fluorescence quenching method. *Am J Physiol Cell Physiol* 286(2):C426–C432.

Soveral, G., R. I. Macey and T. F. Moura. 1997a. Membrane stress causes inhibition of water channels in brush border membrane vesicles from kidney proximal tubule. *Biol Cell* 89(5–6):275–282.

Soveral, G., R. I. Macey and T. F. Moura. 1997b. Water permeability of brush border membrane vesicles from kidney proximal tubule. *J Membr Biol* 158(3):219–228.

Soveral, G., A. Madeira, M. C. Loureiro-Dias and T. F. Moura. 2007. Water transport in intact yeast cells as assessed by fluorescence self-quenching. *Appl Environ Microbiol* 73(7):2341–2343.

Soveral, G., A. Madeira, M. C. Loureiro-Dias and T. F. Moura. 2008. Membrane tension regulates water transport in yeast. *Biochim Biophys Acta* 1778(11):2573–2579.

Soveral, G., A. Veiga, M. C. Loureiro-Dias, A. Tanghe, P. Van Dijck and T. F. Moura. 2006. Water channels are important for osmotic adjustments of yeast cells at low temperature. *Microbiology* 152(Pt 5):1515–1521.

Sui, H., B. G. Han, J. K. Lee, P. Walian and B. K. Jap. 2001. Structural basis of water-specific transport through the AQP1 water channel. *Nature* 414(6866):872–878.

Tornroth-Horsefield, S., K. Hedfalk, G. Fischer, K. Lindkvist-Petersson and R. Neutze. 2010. Structural insights into eukaryotic aquaporin regulation. *FEBS Lett* 584(12):2580–2588.

Tornroth-Horsefield, S., Y. Wang, K. Hedfalk, U. Johanson, M. Karlsson, E. Tajkhorshid, R. Neutze and P. Kjellbom. 2006. Structural mechanism of plant aquaporin gating. *Nature* 439(7077):688–694.

van Heeswijk, M. P. and C. H. van Os. 1986. Osmotic water permeabilities of brush border and basolateral membrane vesicles from rat renal cortex and small intestine. *J Membr Biol* 92(2):183–193.

van Hoek, A. N. and A. S. Verkman. 1992. Functional reconstitution of the isolated erythrocyte water channel CHIP28. *J Biol Chem* 18267–18269.

Verkman, A. S. 2000. Water permeability measurement in living cells and complex tissues. *J Membr Biol* 173(2):73–87.

Verkman, A. S. 2009. Aquaporins: Translating bench research to human disease. *J Exp Biol* 212(Pt 11):1707–1715.

Verkman, A. S., M. O. Anderson and M. C. Papadopoulos. 2014. Aquaporins: Important but elusive drug targets. *Nat Rev Drug Discov* 13(4):259–277.

Verkman, A. S., J. A. Dix and J. L. Seifter. 1985. Water and urea transport in renal microvillus membrane vesicles. *Am J Physiol* 248(5 Pt 2):F650–F655.

Wacker, S. J., C. Aponte-Santamaria, P. Kjellbom, S. Nielsen, B. L. de Groot and M. Rutzler. 2013. The identification of novel, high affinity AQP9 inhibitors in an intracellular binding site. *Mol Membr Biol* 30(3):246–260.

Whittembury, G. 1960. Ion and water transport in the proximal tubules of the kidney of *Necturus maculosus. J Gen Physiol* 43:43–56.

Yasui, M., A. Hazama, T. H. Kwon, S. Nielsen, W. B. Guggino and P. Agre. 1999. Rapid gating and anion permeability of an intracellular aquaporin. *Nature* 402(6758):184–187.

Zeidel, M. L., S. V. Ambudkar, B. L. Smith and P. Agre. 1992. Reconstitution of functional water channels in liposomes containing purified red cell CHIP28 protein. *Biochemistry* 31(33):7436–7440.

Zelenina, M. and H. Brismar. 2000. Osmotic water permeability measurements using confocal laser scanning microscopy. *Eur Biophys J* 29(3):165–171.

Zelenina, M., A. A. Bondar, S. Zelenin and A. Aperia. 2003. Nickel and extracellular acidification inhibit the water permeability of human aquaporin-3 in lung epithelial cells. *J Biol Chem* 278(32):30037–30043.

Zhang, R. B., K. A. Logee and A. S. Verkman. 1990. Expression of mRNA coding for kidney and red cell water channels in *Xenopus* oocytes. *J Biol Chem* 265(26):15375–15378.

CHAPTER 2

Aquaporin Discovery in the Genomic Era

Rafael Zardoya, Iker Irisarri and Federico Abascal

CONTENTS

Abstract		19
2.1	Introduction	19
2.2	Searching for Aquaporins in Sequence Databases	20
2.3	Aquaporin Discovery	22
2.4	Taxonomic Distribution of Aquaporins and Aquaglyceroporins	23
2.5	Phylogeny of Aquaporins	25
2.6	Data Availability	29
Acknowledgements		29
References		29

ABSTRACT

WATER IS ESSENTIAL FOR LIFE, and living organisms evolved a highly diverse family of membrane channel proteins termed aquaporins (AQPs) for controlling water intake. While pioneer studies mostly reported AQPs from vertebrate and angiosperm model systems, the discovery of AQPs in poorly known groups and non-model organisms has been greatly accelerated in recent years, thanks to the wealth of complete or ongoing genomic and transcriptomic projects. Yet the diversity of membrane intrinsic proteins is far from being completely characterized. The fast accumulation of AQP sequences from a myriad of organisms with very different evolutionary and life histories is providing a more complete portrait of how AQP diversity distributes among living organisms, which interpreted within a phylogenetic framework opens new possibilities for the study of the evolutionary processes underlying diversification of this fundamental protein family.

2.1 INTRODUCTION

Water is crucial to cell function, and cells contain in their membranes specific proteins for water transport known as AQPs or membrane intrinsic proteins (MIPs; hereafter both terms will be used indistinctly) (Connolly et al. 1998; Kruse et al. 2006; Maurel et al. 2008;

Finn and Cerda, 2015). These proteins are universal, abundant and highly diversified in both structure and function (they have been recently proposed to function also as transmembrane osmosensors; Hill and Shachar-Hill 2015). Therefore, besides their great physiological importance, MIPs constitute a sound model system for the study of protein family evolution through gene duplication and functional diversification (Zardoya 2005). AQPs are inserted in the cell membrane, forming channels composed of six bilayer-spanning domains that delimit a central pore through which water fluxes (Heymann and Engel 2000). Two highly conserved Asn-Pro-Ala (NPA) motifs that are located in opposite directions in the pore facilitate the passage of the water while they repel protons (Sui et al. 2001). In fact, these NPA motifs constitute the molecular signature of the MIP family (Connolly et al. 1998; Heymann and Engel 2000). In addition, certain amino acid residues differentially conserved in some MIP members are capable of increasing the size of the pore (Fu et al. 2000; Hub and de Groot 2008), allowing the transport of small neutral solutes such as glycerol, urea or ammonia besides water (Hove and Bhave 2011).

AQPs were first discovered in the membrane of human red cells and renal tubules, and their function was established after the injection of the complementary RNA into *Xenopus laevis* oocytes (Preston et al. 1992; Agre and Kozono 2003). Homologs found by sequence similarity searches and early phylogenetic analyses were then reported in other mammals, angiosperms, the fruit fly and bacteria (King and Agre 1996; Park and Saier 1996; Zardoya and Villalba 2001). With accumulating sequence data, it was soon clear that the different species usually had more than one copy of AQPs, showing a tissue-specific expression pattern in both angiosperms and mammals (Johanson et al. 2001; Agre and Kozono 2003; Carbrey and Agre 2009; Ishibashi et al. 2009). Another important early finding in the evolutionary characterization of this protein family was the existence of two distinct subfamilies: one including members that act mainly (but not exclusively) as water channels and another one including members that function mainly (but not only) as glycerol transporters (Park and Saier 1996; Heymann and Engel 1999; Agre and Kozono 2003; Zardoya 2005). In this chapter, we will use the abbreviations AQP (aquaporin *sensu stricto*) and GLP (aquaglyceroporin) to refer to the subfamilies of water and glycerol channels, respectively.

2.2 SEARCHING FOR AQUAPORINS IN SEQUENCE DATABASES

A simple *blastp* search at NCBI (National Center for Biotechnology Information; http://blast.ncbi.nlm.nih.gov/Blast.cgi) with the amino acid sequence of human AQP1 (GenBank accession No. P29972) as query is sufficient to find all human MIP paralogs except for AQP11 and AQP12 (the highest, least significant e-value is 1e−07 for AQP7, 28% sequence identity). Because the evolutionary relationships between human AQPs and GLPs trace back to the deepest node in the phylogenetic tree of MIPs, this simple BLAST exercise illustrates how easily most members of this family can be identified. In fact, human AQP1 is significantly similar to both *Escherichia coli* AqpZ (e-value of 9e−25) and GlpF (e-value of 1e−14). However, to identify highly divergent MIPs, like vertebrate AQP11 and AQP12 and plant small basic intrinsic proteins (SIPs), a more sensitive sequence profile-based search is necessary.

Highly divergent MIP sequences can be identified with tools like PSI-BLAST (NCBI; Bioinformatics Toolkit server at http://toolkit.tuebingen.mpg.de/psi_blast) or the HMMer package (JackHmmer server at http://hmmer.janelia.org/search/jackhmmer). In addition, a sequence or a set of sequences could be scanned against Pfam (which is internally based on the HMMer package, http://pfam.xfam.org/search) to detect significant hits. All human paralogs are easily detected with the Pfam+HMMer strategy, as the least significant e-value against the MIP profile is 1.5e−07, corresponding to AQP11.

An iterative profile hidden Markov model (HMM) search with JackHmmer against UniprotKB (http://www.uniprot.org/), using the human AQP1 as query, yielded in the first round a total of 24,516 significantly similar MIPs (e-value < 0.01). In the second round, 2,365 new additional MIPs were identified, including human AQP11 and AQP12 with e-values *ca.* 1e−11. In next rounds, very few new hits were identified (84, 24, 9, respectively), indicating that the search basically converged. The hits detected in the latest rounds include rare members of the MIP family, like an odd subfamily of arthropod AQPs, which presents CPY-NPV (Cys-Pro-Tyr Asn-Pro-Val) instead of the typical NPA–NPA boxes (Ishibashi 2006).

Searching a genome for MIPs is relatively straightforward. We devised two strategies to detect all the MIPs in a genome. One strategy consisted in preparing a reduced set of MIPs covering all or most of the MIP sequence diversity and then running a *tblastn* search (i.e. an amino acid query sequence is compared against a DNA sequence that is virtually translated into the six possible open reading frames) with each of the sequences against the reference genome database. A second strategy relied on profile-based searches. However, at present, profile-based searches between amino acid and virtually translated nucleotide sequences are not as easy as with protein–protein comparisons. Local searches using RPStblastn (from the BLAST package) might be a choice, although downloading and preparing the library of profiles requires some expertise. For DNA sequences of less than 200,000 bp, RPStblastn can be run through the NCBI's web server (http://www.ncbi.nlm.nih.gov/Structure/cdd/wrpsb.cgi).

Homology is defined as similarity due to common descent, and thus it is best determined through the reconstruction of phylogenetic relationships. The different sequences that are identified through BLAST similarity searches need to be subjected to phylogenetic analyses in order to cluster them into different subfamilies and groups of paralogy. Members of the family are then defined based on monophyletic groups (formed by all the sequences that share a common ancestor). Phylogenetic reconstruction based on sequences currently relies on probabilistic methods (maximum likelihood and Bayesian inference), which use models of nucleotide or amino acid evolution to infer the topology and branch lengths of a phylogenetic tree. Statistical robustness of reconstructed nodes is commonly measured using non-parametric bootstrapping and Bayesian posterior probabilities. Popular software for maximum likelihood and Bayesian inference are RaxML and MrBayes, respectively, and they can be run at the CIPRES web server (http://www.phylo.org/sub_sections/portal/). The alignment interpreted in a phylogenetic framework can be used to determine differentially conserved residues in the different paralogy groups and their evolution. For subfamily assignment, the alignments and phylogenies provided in Abascal et al. (2014) could be easily used as scaffold onto which newly discovered MIPs are added.

2.3 AQUAPORIN DISCOVERY

Given the important function of AQPs, during the 1990s much effort was devoted in finding and characterizing new members of the family, particularly within vertebrates and flowering plants. In these early years, studies focused on isolating AQPs from different tissues of model organisms and characterizing their function using the *Xenopus* oocyte system (Agre and Kozono 2003). A survey in UniProtKB (http://www.uniprot.org/; Figure 2.1a) shows that AQPs were described in 2–4 new species every year during 1990–1995 and that this rate increased to 12–19 new species per year in the second half of the 1990s. In the turn of the century, and coinciding with the annotation of the first genomes, the pace of newly discovered AQPs added to UniProtKB increased to 31–88 new species per year. It was after the implementation of next-generation sequencing (NGS) techniques in 2006 when the discovery of AQPs underwent a major boost and rapidly increased from hundreds to thousands of new species incorporated into UniProtKB each year (Figure 2.1a).

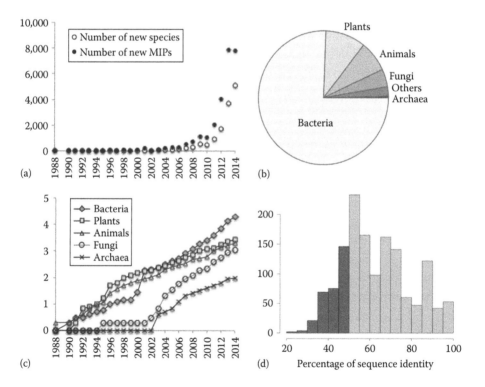

FIGURE 2.1 (a) Number of aquaporins (filled circles) and species (open circles) incorporated a new into UniProtKB database per year. (b) Aquaporins in the UniProtKB database as of 2014 sorted by taxonomic group. (c) First appearance and accumulation of aquaporins from new species per year in UniProtKB database sorted by taxonomic group. (d) Histogram showing differences in percentage of amino acid identity between *ca.* 1400 aquaporin sequences derived from metagenomic data and known aquaporins at the UniProt database estimated through BLAST searches.

As of 2014, the total number of AQPs found in this database using iterative profile HMM searches was 26,594 (see Section 2.2). Of these, *ca.* 75% were from Bacteria, plants being the second most abundant taxon (Figure 2.1b).

The majority of recently discovered AQPs derive directly from genomic and transcriptomic projects and are customarily identified through sequence similarity comparisons of the assemblies against annotated sequence databases (Johanson et al. 2001; Gupta and Sankararamakrishnan 2009; Anderberg et al. 2012; Verma et al. 2014). On the one hand, most AQPs derived from genomic projects are not biochemically and functionally characterized, which makes phylogenetic and evolutionary analyses particularly useful. On the other hand, these new procedures have the advantage of greatly accelerating the rate at which new AQPs are discovered, and more importantly, they provide an extraordinary increase in the taxonomic range of the newly determined proteins. In this regard, a clear shift is observed between the early 1990s, when most newly discovered AQPs were from model organisms (human, rat, cattle, *Caenorhabditis*, yeast, soybean, *Arabidopsis*, *Bacillus*) and after the appearance of NGS, when they were described in virtually any organism, as shown by the comparison of the first appearance of the different species in the UniProtKB database (Figure 2.1c). Archaea were first recorded in 2001 (Kozono et al. 2003), mosses and green algae in 2004 (Gustavsson et al. 2005), sponges and anemones in 2007, spikemosses in 2010 (Anderberg et al. 2012), mollusks in 2012 (Pieńkowska et al. 2014), myriapods and tardigrades in 2013 (Grohme et al. 2013) and several unicellular eukaryotes (Cryptophyta, Haptophyta, Rhizaria) between 2013 and 2014. Interestingly, the graph also indicates that none of the taxonomic groups has reached a plateau and thus that many more AQPs are yet to be discovered in the coming years (Figure 2.1c).

2.4 TAXONOMIC DISTRIBUTION OF AQUAPORINS AND AQUAGLYCEROPORINS

Many bacteria have a single copy of AQP and a single GLP, and together they fully cover the needs of the prokaryote cell for water and glycerol intake, respectively. However, several others live, remarkably, without any MIP. Bacteria have a very high surface-to-volume ratio and, by consequence, very rapid osmotic equilibration times (<100 ms). The simple diffusion of water in and out of the bacterial cell is high enough to explain why a certain number of bacterial genera do not possess MIP genes in their genome. In particular, it is remarkable that intracellular bacteria such as *Wolbachia*, *Rickettsia* and *Chlamydia*, or thermophilic bacteria such as *Thermus* and *Thermotoga*, lack both GLPs and AQPs (Abascal et al. 2014). Nitrogen-fixing bacteria (Rhizobiales) lack the GLP paralog, whereas other genera such as *Listeria*, *Clostridium*, *Yersinia* and *Borrelia* lack the AQP paralog. With regard to Archaea, thermophilic species have no MIPs; some Haloarchaea, which live in water saturated with salt, have only a single GLP copy, and methane-producing species have only one AQP paralog (Abascal et al. 2014).

In the last two years (2013–2014), the number of complete genomes of unicellular eukaryotes has increased considerably, improving our knowledge of the number and distribution of AQPs and GLPs among these organisms. The two subfamilies are heterogeneously distributed among the main groups of unicellular eukaryotes. Genera with a parasitic life-style such as *Giardia* (Fornicata) and *Trichomonas* (Parabasalia) lack MIPs (Abascal et al. 2014). The same status applies to *Tetrahymena* (Alveolata), in this case likely associated to its thermophilic lifestyle. In contrast, other genera such as *Phytophthora* (Oomycota), *Dictyostelium* (Amoebozoa), *Trypanosoma* and *Leishmania* (Euglenozoa), and *Paramecium* (Alveolata) have experienced bursts of duplications of AQP genes (Abascal et al. 2014). Interestingly, paralog expansions are not evenly distributed across taxa. For instance, GLPs are expanded in *Phytophthora*, whereas AQPs are the ones massively duplicated in *Dictyostelium*, *Leishmania* and *Paramecium* (Abascal et al. 2014). In Heterokontophyta, the expansion of AQPs included a putative new group of paralogy, which has been named large intrinsic proteins (Khabudaev et al. 2014).

Within Ophistokonta, the distribution patterns significantly differ between fungi, Filasterea (*Capsaspora*), Choanoflagellata (*Monosiga*) and Metazoa. In fungi, the different species nor-mally have one or several copies of both AQPs and GLPs (Abascal et al. 2014; Verma et al. 2014). In *Capsaspora*, there is a single copy of AQP and a single one of GLP (Abascal et al. 2014). In *Monosiga*, no GLPs are found and the three AQPs that are present are highly diver-gent (Abascal et al. 2014). In Metazoa, there has been a great expansion of both subfamilies (Abascal et al. 2014), which is particularly evident in vertebrates, thanks to the more extensive taxon coverage (Finn et al. 2014). Animal AQPs show high levels of diversity and sequence divergence (e.g. human has up to 13 distinct paralog groups, and AQP8, AQP11 and AQP12 are particularly divergent in their sequences). It is noteworthy that GLPs are absent in endop-terygote insects (AQP4 co-opted to transport glycerol in this insects; Finn et al. 2015) and that AQP8 is the only paralog found thus far in nematodes (Abascal et al. 2014).

Plants show the highest diversity of MIPs in terms of total number of major groups of paralogy (Danielson and Johanson 2010). Green algae have both GLPs and AQPs as well as several paralogs that are not found in any land plant (Anderberg et al. 2011). Within Embryophyta, diversity is highest in the early branching lineages. In particular, the moss *Physcomitrella* (Bryophyta) has a GLP that is lost in other land plants (Danielson and Johanson 2008). The spike moss *Selaginella* (Lycopodiophyta) has a paralog group (named hybrid intrinsic proteins [HIPs]) that is also found in mosses but not in seed plants (Anderberg et al. 2012). The exact number of MIP paralogs in ferns is yet unknown. The recent publication of the large genome of the loblolly pine (*Pinus taeda*) will help in deter-mining the exact MIP composition of gymnosperms. In angiosperms, paralogs show great levels of divergence and are tissue specific (Johanson et al. 2001).

In order to discern whether most of the diversity of MIPs is known or instead new groups of paralogy await discovery in poorly studied taxonomic groups, we conducted a search for MIP homologs in metagenomic sequence data (UniMES-UniProt). We found *ca.* 1,400 sequences of uncertain taxonomic adscription (archaea, bacteria or unicellular eukary-otes), of which 300 had less than 50% sequence identity with known MIPs (Figure 2.1d). Therefore, there is clear room for expanding the known diversity of the MIP family, par-ticularly in unicellular eukaryotes. Importantly, as shown in the following, a phylogenetic analysis revealed that many of these divergent MIPs form clearly defined monophyletic groups with high levels of intragroup diversity.

2.5 PHYLOGENY OF AQUAPORINS

Early phylogenies of AQPs were mostly restricted to bacteria, vertebrates and angiosperm model systems giving only a partial picture of the diversity of the family (Park and Saier 1996; Heymann and Engel 1999; Johanson et al. 2001; Zardoya and Villalba 2001). More recently, reconstructed phylogenies greatly extended taxon coverage providing a detailed account of the diversification of the different major groups of paralogy within many previously poorly studied taxonomic groups (Danielson and Johanson 2008; Abascal et al. 2014; Finn et al. 2014; Verma et al. 2014). In an attempt to establish deep homology relationships within the family, some studies opted for the reconstruction of simplified phylogenies based only on members with high amino acid identity from flowering plants and vertebrates (Soto et al. 2012; Perez Di Giorgio et al. 2014). These studies suggested that plant and animal AQPs derived from only four ancestral subfamilies, although the phylogenies were not reconstructed using probabilistic methods and they ignored the essential information provided by organisms other than flowering plants and vertebrates (Soto et al. 2012; Perez Di Giorgio et al. 2014). The universal distribution of AQPs should allow reconstructing a molecular phylogeny of these proteins that covers all living organisms (Figure 2.2).

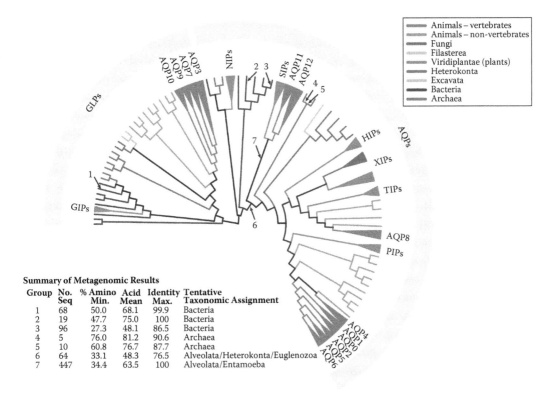

Summary of Metagenomic Results

Group No.	% Amino Acid Identity			Tentative
	Seq	Min.	Mean	Max.	Taxonomic Assignment
1	68	50.0	68.1	99.9	Bacteria
2	19	47.7	75.0	100	Bacteria
3	96	27.3	48.1	86.5	Bacteria
4	5	76.0	81.2	90.6	Archaea
5	10	60.8	76.7	87.7	Archaea
6	64	33.1	48.3	76.5	Alveolata/Heterokonta/Euglenozoa
7	447	34.4	63.5	100	Alveolata/Entamoeba

FIGURE 2.2   **(See colour insert.)** Phylogeny based on probabilistic analyses of a protein sequence data matrix of 162 membrane intrinsic proteins (MIPs). (From Abascal, F. et al., *Biochim. Biophys. Acta—Gen. Subjects*, 1840, 1468, 2014.) The putative phylogenetic position of MIP sequences mined from metagenomic data sets (groups 1–7) are indicated in the phylogeny, as well as the percentage of amino acid sequence identity to known MIPs and tentative taxonomic assignment.

However, few studies have attempted it (Zardoya 2005; Abascal et al. 2014) because the inference of robust evolutionary relationships among hundreds of AQP sequences from a wide range of organisms is an extremely challenging task, even more if we consider the limited phylogenetic information that can be contained in MIP sequences that are 250–300 amino acids long. The high sequence divergence among main paralog groups (and the associated saturation problems) combined with heterogeneity in rates of substitution among both paralogs and taxa complicates phylogenetic analyses even further, hindering the robust recovery of the short tree nodes corresponding to deep orthology relationships and rendering some instances of taxonomic incongruence (Abascal et al. 2014). Despite these shortcomings, the exhaustive taxon coverage of most recent phylogenies allowed finding interesting general patterns regarding the evolution of the protein family.

In agreement with the ancestral state of the family found in bacteria (one copy of AQP and one of GLP), a split between water and glycerol transporters is recovered at the base of the MIP phylogeny (evidenced by a long branch) (Figure 2.2) (Abascal et al. 2014; Finn et al. 2014). The GLP subfamily forms a compact group in which paralog diversification is relatively low, except in oomycetes (Heterokonta), nematodes (that possess at least four unnamed paralogs) and vertebrates in which up to five groups are recognized (AQP3, AQP7, AQP9, AQP10 and AQP13) (Finn et al. 2014). Instead, the AQP subfamily is highly diversified and contains many paralog groups. The number of unicellular eukaryote genomes sequenced is still low and taxonomically scattered to address a phylogeny-based classification of their AQPs and decipher the exact number of paralog groups and their evolutionary origin. According to the phylogeny, at least one instance of deep orthology is found with several proteins of the Dictyosteliidae (Amoebozoa) belonging to the X intrinsic protein (XIP) group of paralogy (Abascal et al. 2014). In Fungi, at least two different major paralogy groups are found, one corresponding to canonical AQPs and the other corresponding to XIPs, which are restricted to Ascomycota (Abascal et al. 2014; Verma et al. 2014). A third paralogy group has been described in Microsporidia and related to plant SIPs based on structural features but not on phylogenies (Verma et al. 2014). In animals, AQP diversification has led to up to three main groups of paralogy (Figure 2.2). One corresponds to highly divergent intracellular proteins (Ishibashi 2006). The lamprey has only one copy of these AQPs, whereas jawed vertebrates have two copies (AQP11 and AQP12) (Finn et al. 2014). A second group is AQP8, which is likely present in all metazoans (Abascal et al. 2014) and is most diverse in teleost fishes (Finn et al. 2014). AQP8 is found in the inner mitochondrial membrane of several animal tissues and has been involved in ammonia (Soria et al. 2010) and hydrogen peroxide (Bienert et al. 2007; Marchissio et al. 2012; Sies 2014; Chauvigne et al. 2015) transport. The third group includes canonical AQPs found in the plasmatic membrane of all animals (except nematodes) (Abascal et al. 2014). Non-vertebrate AQPs show few copies per taxon and remain unclassified in major paralogy groups due to the lack of phylogenetic resolution at this level and the taxonomically scattered data. Vertebrate AQPs likely result from two (or three, in teleost fishes) rounds of genome duplication that occurred at the early evolutionary history of this group, as well as from more recent tandem gene duplications, and are classified into eight groups (AQP0, AQP1, AQP2, AQP4, AQP5, AQP6, AQP14 and

AQP15; Abascal et al. 2014; Finn et al. 2014). The different vertebrate AQPs are expressed in a tissue-specific manner (Agre and Kozono 2003).

The reconstruction of the phylogeny of MIPs is most challenging in plants. A total of eight major groups of paralogy are found in plants: GlpF-like intrinsic proteins (GIPs), tonoplast intrinsic proteins (TIPs), plasma membrane intrinsic proteins (PIPs), NOD26-like intrinsic proteins (NIPs), SIPs, XIPs and HIPs (Figure 2.2). The different groups of plant paralogs are highly divergent in sequence and distributed throughout the general phylogeny of MIPs. GIPs are nested deep within the GLP subfamily and show a long branch. GIPs are found in diatoms (Khabudaev et al. 2014), green algae (Anderberg et al. 2011) and mosses (Danielson and Johanson 2008), but are lost in vascular plants (Abascal et al. 2014). They are recovered in the phylogenetic tree as sister group of bacterial and archaeal GLPs (Figure 2.2; Abascal et al. 2014). Within AQPs, PIPs are recovered as sister group of animal canonical AQPs, although without strong statistical support (Soto et al. 2012; Abascal et al. 2014). Two main groups (PIP1 and PIP2) are distinguished and are present from green algae to angiosperms (Johanson et al. 2001; Zardoya 2005; Danielson and Johanson 2008; Anderberg et al. 2011; Soto et al. 2012; Abascal et al. 2014). TIPs are clustered together with animal AQP8s, although without statistical support (Soto et al. 2012; Abascal et al. 2014). Mosses (Danielson and Johanson 2008) and spike mosses (Anderberg et al. 2012) have two TIP copies each, whereas up to five major TIP paralog groups could be recovered within seed plants (Johanson et al. 2001; Zardoya 2005; Soto et al. 2012; Abascal et al. 2014). HIPs are restricted to mosses (Danielson and Johanson 2008) and spike mosses (Anderberg et al. 2012) and placed as sister group of XIPs and TIPs plus AQP8s (Abascal et al. 2014). XIPs have orthologs not only in all plants (Danielson and Johanson 2008) but also in fungi and unicellular eukaryotes (Abascal et al. 2014). This wider taxon distribution of XIPs could indirectly support ancient homology of TIPs and AQP8s (Soto et al. 2012; Abascal et al. 2014). SIPs are present in the endoplasmic reticulum of all plants (Maeshima and Ishikawa 2008) and have experienced many substitutions due to its intracellular function and are characterized by extremely long branches (Johanson and Gustavsson 2002). The relative phylogenetic position of SIPs as sister group of AQP11 plus AQP12 (Soto et al. 2012; Abascal et al. 2014; Perez Di Giorgio et al. 2014) possibly reflects a long-branch attraction artifact of phylogenetic reconstruction (Abascal et al. 2014; but see below). Within SIPs, two distinct paralogy groups (SIP1 and SIP2) are recognized. Finally, NIPs (Johanson et al. 2001) have a peculiar relative position within the tree, at the base of all AQPs and as sister group of bacterial and archaeal AQPs (Zardoya 2005; Abascal et al. 2014). In fact, it has been hypothesized that these plant members could have a bacterial origin and would have been acquired through horizontal gene transfer (Zardoya et al. 2002). Within NIPs, up to four paralog groups (NIP1–NIP4) are recognized according to the phylogeny. Of these, three are present in all land plants and NIP1 is restricted to seed plants (Abascal et al. 2014). Interestingly, NIPs function as glycerol transporters in seed plants, which have lost the GIP ortholog (Zardoya et al. 2002).

In order to further characterize and classify unknown MIPs found in metagenomic data, we reconstructed a phylogeny that included the main known MIP lineages and the newly discovered MIPs (Figure 2.2). Interestingly, metagenomic MIPs are not distributed evenly across the phylogeny, but instead they are recovered clustered in few monophyletic groups (Figure 2.2). About half of the retrieved sequences likely correspond to unicellular eukaryotes (Alveolata/Entamoeba) and are recovered in the phylogenetic tree as closely related to SIPs, AQP11 and AQP12, although without statistical support (Figure 2.2). If confirmed in future studies, this phylogenetic relationship would be in support of a deep homology of intracellular MIPs (Soto et al. 2012). Most of the remaining sequences are placed close to bacterial AQPs and GLPs (Figure 2.2).

The alignments used to reconstruct the phylogeny of MIPs contain important information regarding different rates of evolution of the residues along the protein molecule. The easily recognizable common ancestry of MIPs mainly relies on the presence of several sites that are highly conserved. We recently identified up to 15 sites (including the two NPA boxes) that were conserved in more that 90% of the sequences from a diverse multiple alignment of 670 MIPs (Abascal et al. 2014). Such degree of conservation, atypical in membrane proteins, reflects strong functional and structural constraints acting on MIPs (Froger et al. 1998; Heymann and Engel 2000; Abascal et al. 2014). Highly conserved positions across taxa and groups of paralogy are shown in Figure 2.3. These residues have been shown to be crucial for the channel function using functional experiments and by deciphering their position and role in the 3D structure (Fu et al. 2000; Murata et al. 2000; Sui et al. 2001; Gonen and Walz 2006). Some of these motifs were thought to be invariable

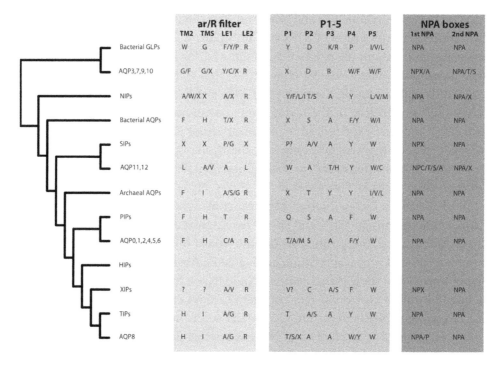

| | ar/R filter | | | | P1-5 | | | | | NPA boxes | |
	TM2	TMS	LE1	LE2	P1	P2	P3	P4	P5	1st NPA	2nd NPA
Bacterial GLPs	W	G	F/Y/P	R	Y	D	K/R	P	I/V/L	NPA	NPA
AQP3,7,9,10	G/F	G/X	Y/C/X	R	X	D	R	W/F	W/F	NPX/A	NPA/T/S
NIPs	A/W/X	X	A/X	R	Y/F/L/I	T/S	A	Y	L/V/M	NPA	NPA/X
Bacterial AQPs	F	H	T/X	R	X	S	A	F/Y	W/I	NPA	NPA
SIPs	X	X	P/G	X	P?	A/V	A	Y	W	NPX	NPA
AQP11,12	L	A/V	A	L	W	A	T/H	Y	W/C	NPC/T/S/A	NPA/X
Archaeal AQPs	F	I	A/S/G	R	X	T	Y	Y	I/V/L	NPA	NPA
PIPs	F	H	T	R	Q	S	A	F	W	NPA	NPA
AQP0,1,2,4,5,6	F	H	C/A	R	T/A/M	S	A	F/Y	W	NPA	NPA
HIPs											
XIPs	?	?	A/V	R	V?	C	A/S	F	W	NPX	NPA
TIPs	H	I	A/G	R	T	A/S	A	Y	W	NPA	NPA
AQP8	H	I	A/G	R	T/S/X	A	A	W/Y	W	NPA/P	NPA

FIGURE 2.3 Conserved residues (>90% of sequences in an alignment of 670 proteins) in the membrane intrinsic protein family. (From Abascal, F. et al., *Biochim. Biophys. Acta—Gen. Subjects*, 1840, 1468, 2014.) The main groups of paralogy within the family are shown. Amino acids use standard abbreviations. X means any amino acid and question marks indicates uncertainty.

such as the two NPA boxes, but as more MIPs are sequenced, it is clear that this molecular signature of the MIP proteins is not totally universal (Ishibashi et al. 2011; Abascal et al. 2014; Khabudaev et al. 2014).

2.6 DATA AVAILABILITY

The following data are available as supplementary material at http://pc16141.mncn.csic. es/AQPs_chapter/: all the MIPs identified in UniprotKB, all metagenomic MIPs identified in UniMES, maximum percentage of sequence identity for each of the metagenomic MIPs compared to known MIPs, multiple sequence alignment of 730 metagenomic MIPs and 1,489 MIPs from Abascal et al. (2014) and maximum likelihood phylogeny of the 730 metagenomic and 1,489 reference MIPs.

ACKNOWLEDGEMENTS

This study was partly funded by projects CGL2010-18216 and CGL2013-45211-C2-2-P of the Ministerio de Economía y Competitividad of Spain to RZ. It was supported by post-doctoral fellowships from the Alexander von Humboldt Foundation (application 1150725) and European Molecular Biology Organization (EMBO ALTF 440-2013).

REFERENCES

Abascal, F., I. Irisarri and R. Zardoya. 2014. Diversity and evolution of membrane intrinsic proteins. *Biochim. Biophys. Acta—Gen. Subjects* 1840:1468–1481.

Agre, P. and D. Kozono. 2003. Aquaporin water channels: Molecular mechanisms for human diseases1. *FEBS Lett.* 555:72–78.

Anderberg, H., J. Danielson and U. Johanson. 2011. Algal MIPs, high diversity and conserved motifs. *BMC Evol. Biol.* 11:110.

Anderberg, H.I., P. Kjellbom and U. Johanson. 2012. Annotation of *Selaginella moellendorffii* major intrinsic proteins and the evolution of the protein family in terrestrial plants. *Front. Plant Sci.* 3:33.

Bienert, G.P., A.L.B. Møller, K.A. Kristiansen, A. Schulz, I.M. Møller, J.K. Schjoerring and T.P. Jahn. 2007. Specific aquaporins facilitate the diffusion of hydrogen peroxide across membranes. *J. Biol. Chem.* 282:1183–1192.

Carbrey, J.M. and P. Agre. 2009. Discovery of the aquaporins and development of the field. In *Aquaporins*, ed. E. Beitz, pp. 3–28. Berlin, Germany: Springer.

Chauvigne, F., M. Boj, R.N. Finn and J. Cerda. 2015. Mitochondrial aquaporin-8-mediated hydrogen peroxide transport is essential for teleost spermatozoon motility. *Sci. Rep.* 5:7789.

Connolly, D.L., C.M. Shanahan and P.L. Weissberg. 1998. The aquaporins. A family of water channel proteins. *Int. J. Biochem. Cell Biol.* 30:169–172.

Danielson, J.A.H. and U. Johanson. 2008. Unexpected complexity of the Aquaporin gene family in the moss *Physcomitrella patens*. *BMC Plant Biol.* 8:1–15.

Danielson, J.A.H. and U. Johanson. 2010. Phylogeny of major intrinsic proteins. In *MIPs and Their Role in the Exchange of Metalloids*, eds. T.P. Jahn and G.P. Bienert, pp. 19–31. New York: Springer.

Finn, R.N. and J. Cerda. 2015. Evolution and functional diversity of aquaporins. *Biol. Bull.* 229:6–23.

Finn, R.N., F. Chauvigné, J.B. Hlidberg, C.P. Cutler and J. Cerdà. 2014. The lineage-specific evolution of aquaporin gene clusters facilitated tetrapod terrestrial adaptation. *PLoS ONE* 9:e113686.

Finn, R.N., F. Chauvigné, J.A. Stavang, X. Belles and J. Cerda. 2015. Insect glycerol transporters evolved by functional co-option and gene replacement. *Nat. Comm.* 6:7814.

Froger, A., D. Thomas, C. Delamarche and B. Tallur. 1998. Prediction of functional residues in water channels and related proteins. *Protein Sci.* 7:1458–1468.

Fu, D., A. Libson, L.J.W. Miercke, C. Weitzman, P. Nollert, J. Krucinski and R.M. Stroud. 2000. Structure of a glycerol-conducting channel and the basis for its selectivity. *Science* 290:481–486.

Gonen, T. and T. Walz. 2006. The structure of aquaporins. *Q. Rev. Biophys.* 39:361–396.

Grohme, M.A., B. Mali, W. Wełnicz, S. Michel, R.O. Schill and M. Frohme. 2013. The aquaporin channel repertoire of the tardigrade *Milnesium tardigradum*. *Bioinform. Biol. Insights* 7:153–165.

Gupta, A. and R. Sankararamakrishnan. 2009. Genome-wide analysis of major intrinsic proteins in the tree plant *Populus trichocarpa*: Characterization of XIP subfamily of aquaporins from evolutionary perspective. *BMC Plant Biol.* 9:134.

Gustavsson, S., A.-S. Lebrun, K. Nordén, F. Chaumont and U. Johanson. 2005. A novel plant major intrinsic protein in *Physcomitrella patens* most similar to bacterial glycerol channels. *Plant Physiol.* 139:287–295.

Heymann, J.B. and A. Engel. 1999. Aquaporins: Phylogeny, structure, and physiology of water channels. *Physiology* 14:187–193.

Heymann, J.B. and A. Engel. 2000. Structural clues in the sequences of the aquaporins. *J. Mol. Biol.* 295:1039–1053.

Hill, A.E. and Y. Shachar-Hill. 2015. Are aquaporins the missing transmembrane osmosensors? *J. Membr. Biol.* 248:753–765.

Hove, R. and M. Bhave. 2011. Plant aquaporins with non-aqua functions: Deciphering the signature sequences. *Plant Mol. Biol.* 75:413–430.

Hub, J.S. and B.L. de Groot. 2008. Mechanism of selectivity in aquaporins and aquaglyceroporins. *Proc. Natl. Acad. Sci. U.S.A.* 105:1198–1203.

Ishibashi, K. 2006. Aquaporin subfamily with unusual NPA boxes. *Biochim. Biophys. Acta— Biomembr.* 1758:989–993.

Ishibashi, K., S. Hara and S. Kondo. 2009. Aquaporin water channels in mammals. *Clin. Exp. Nephrol.* 13:107–117.

Ishibashi, K., S. Kondo, S. Hara and Y. Morishita. 2011. The evolutionary aspects of aquaporin family. *Am. J. Physiol. Reg. Integr. Comp. Physiol.* 300:R566–R576.

Johanson, U. and S. Gustavsson. 2002. A new subfamily of major intrinsic proteins in plants. *Mol. Biol. Evol.* 19:456–461.

Johanson, U., M. Karlsson, I. Johansson, S. Gustavsson, S. Sjövall, L. Fraysse, A.R. Weig and P. Kjellbom. 2001. The complete set of genes encoding major intrinsic proteins in *Arabidopsis* provides a framework for a new nomenclature for major intrinsic proteins in plants. *Plant Physiol.* 126:1358–1369.

Khabudaev, K., D. Petrova, M. Grachev and Y. Likhoshway. 2014. A new subfamily LIP of the major intrinsic proteins. *BMC Genomics* 15:173.

King, L.S. and P. Agre. 1996. Pathophysiology of the aquaporin water channels. *Annu. Rev. Physiol.* 58:619–648.

Kozono, D., X. Ding, I. Iwasaki, X. Meng, Y. Kamagata, P. Agre and Y. Kitagawa. 2003. Functional expression and characterization of an archaeal aquaporin: AqpM from *Methanothermobacter marburgensis*. *J. Biol. Chem.* 278:10649–10656.

Kruse, E., N. Uehlein and R. Kaldenhoff. 2006. The aquaporins. *Genome Biol.* 7:206.

Maeshima, M. and F. Ishikawa. 2008. ER membrane aquaporins in plants. *Pflügers Archiv.—Eur. J. Physiol.* 456:709–716.

Marchissio, M.J., D.E.A. Francés, C.E. Carnovale and R.A. Marinelli. 2012. Mitochondrial aquaporin-8 knockdown in human hepatoma HepG2 cells causes ROS-induced mitochondrial depolarization and loss of viability. *Toxicol. Appl. Pharmacol.* 264:246–254.

Maurel, C., L. Verdoucq, D.-T. Luu and V. Santoni. 2008. Plant aquaporins: Membrane channels with multiple integrated functions. *Annu. Rev. Plant Biol.* 59:595–624.

Murata, K., K. Mitsuoka, T. Hirai, T. Walz, P. Agre, J.B. Heymann, A. Engel and Y. Fujiyoshi. 2000. Structural determinants of water permeation through aquaporin-1. *Nature* 407:599–605.

Park, J.H. and J.M.H. Saier. 1996. Phylogenetic characterization of the MIP family of transmembrane channel proteins. *J. Membr. Biol.* 153:171–180.

Perez Di Giorgio, J., G. Soto, K. Alleva, C. Jozefkowicz, G. Amodeo, J. Muschietti and N. Ayub. 2014. Prediction of aquaporin function by integrating evolutionary and functional analyses. *J. Membr. Biol.* 247:107–125.

Pieńkowska, J., E. Kosicka, M. Wojtkowska, H. Kmita and A. Lesicki. 2014. Molecular identification of first putative aquaporins in snails. *J. Membr. Biol.* 247:239–252.

Preston, G.M., T.P. Carroll, W.B. Guggino and P. Agre. 1992. Appearance of water channels in *Xenopus* oocytes expressing red cell CHIP28 protein. *Science* 256:385–387.

Sies, H. 2014. Role of metabolic H_2O_2 generation: Redox signaling and oxidative stress. *J. Biol. Chem.* 289:8735–8741.

Soria, L.R., E. Fanelli, N. Altamura, M. Svelto, R.A. Marinelli and G. Calamita. 2010. Aquaporin-8-facilitated mitochondrial ammonia transport. *Biochem. Biophys. Res. Commun.* 393:217–221.

Soto, G., K. Alleva, G. Amodeo, J. Muschietti and N.D. Ayub. 2012. New insight into the evolution of aquaporins from flowering plants and vertebrates: Orthologous identification and functional transfer is possible. *Gene* 503:165–176.

Sui, H., B.-G. Han, J.K. Lee, P. Walian and B.K. Jap. 2001. Structural basis of water-specific transport through the AQP1 water channel. *Nature* 414:872–878.

Verma, R.K., N.D. Prabh and R. Sankararamakrishnan. 2014. New subfamilies of major intrinsic proteins in fungi suggest novel transport properties in fungal channels: Implications for the host-fungal interactions. *BMC Evol. Biol.* 14:173–173.

Zardoya, R. 2005. Phylogeny and evolution of the major intrinsic protein family. *Biol. Cell.* 97:397–414.

Zardoya, R., X. Ding, Y. Kitagawa and M.J. Chrispeels. 2002. Origin of plant glycerol transporters by horizontal gene transfer and functional recruitment. *Proc. Natl. Acad. Sci. U.S.A.* 99:14893–14896.

Zardoya, R. and S. Villalba. 2001. A phylogenetic framework for the aquaporin family in eukaryotes. *J. Mol. Evol.* 52:391–404.

CHAPTER 3

Aquaporin Structure and Selectivity

Ardeschir Vahedi-Faridi and Andreas Engel

CONTENTS

Abstract	33
3.1 Introduction	34
3.2 Structure of Human AQPs	37
3.2.1 Sequence Homology	37
3.2.2 Ternary Structure and Fold	38
3.2.3 Pore Structure and Specificity	39
3.2.4 Surface Structure of AQPs Involved in Membrane Junctions	45
3.3 Unsolved Riddles	47
3.3.1 Anion Transport	47
3.3.2 Gas Transport	47
3.4 Conclusions	48
Acknowledgement	48
References	48

ABSTRACT

PROGRESS IN THE STRUCTURE determination of aquaporins has led to a deep under-standing of water and solute permeation by these small integral membrane proteins. Atomic structures have allowed the water permeation and exclusion of protons to be monitored by molecular dynamics simulations. Such numerical experiments have provided a framework for assessing the water, solute and gas permeation by site-directed mutations. In spite of this, further structural and molecular dynamics analyses are required to elucidate the basis for regulation as well as for gas permeation, processes that are still to be deciphered. Here, we discuss the structures of human aquaporins and compare their pore architectures, which provide the basis for the channel's specificity.

3.1 INTRODUCTION

Water flows through the membranes of all living cells either by diffusion through the lipid bilayer or by permeation through a highly specific pore that accommodates water but neither ions nor protons (H_3O^+). Whereas the former process requires an activation energy of $E_a > 10$ kcal/mol, the latter occurs with an activation energy of $E_a < 5$ kcal/mol. These observations suggested the existence of membrane channels with unique structural properties. Aquaporin (AQP)-1, the first member of this ancient membrane protein family was identified by its expression in *Xenopus* oocytes (Preston et al. 1992). Since then, thousands of putative AQP sequences have been found, several hundreds have been verified, and many have been expressed and functionally and structurally studied. Early biophysical characterizations and sequence analysis of the major intrinsic protein (now known as AQP0) (Gorin et al. 1984) of lens fiber cell membranes and AQP1 (Preston and Agre 1991) indicated that these channels consist of six α-helical membrane-spanning segments and connecting loops of variable length. The intriguing sequence homology between the first and second half makes an early gene duplication event likely.

These channel proteins facilitate the passive permeation of water (AQPs) and small, uncharged solutes (aquaglyceroporins, GLPs) (Heymann and Engel 1999; 2000; Zardoya 2005). Because both AQPs and GLPs allow water to pass, AQP refers in general to both types of AQPs. They are present in all kingdoms of life, demonstrating their central role in maintaining normal physiology of all organisms. Few AQP genes are present in unicellular organisms, such as archaea, bacteria, or yeast, encoding water channels and glycerol facilitators. More AQP genes are found in the genomes of multicellular organisms, an amazing variety in plants: *Arabidopsis thaliana* contains 38 putative AQP genes (Maurel 2007). The human body expresses 13 AQPs with specific organ, tissue and cellular localization (Table 3.1) (Verkman 2011; Day et al. 2014).

Because each half exhibits three transmembrane domains, the two halves must integrate in the membrane in opposite orientations, a basic structural property of the AQP membrane channels that relates to their internal pseudo-twofold symmetry. Both long loops between helices 2 (H2) and H3 (loop B) and between H5 and H6 (loop E) bear the highly conserved Asn-Pro-Ala (NPA) motif, which is the hallmark of AQP sequences (Figure 3.1). Site-directed mutagenesis experiments on these loops led to the hourglass model, which predicted loops B and E to meet in the center of the membrane to form the channel (Jung et al. 1994).

Advances in electron crystallography led to the first atomic model of AQP1 (Murata et al. 2000). The structure of the bacterial GLP, GlpF, was solved by x-ray crystallography at the same time (Fu et al. 2000). The subsequent x-ray structure of AQP1 at 2.2 Å resolution confirmed the model established by electron crystallography and showed the position of water molecules in the pore (Sui et al. 2001). Since then, the structures of several AQPs and GLPs have been solved, either by electron or by x-ray crystallography, the latter contributing the bulk of structural information (for human AQPs, see Table 3.2). Electron crystallography has given structural information on the lipid–protein interactions for the lens water channel AQP0 at 1.9 Å resolution and has provided insight into the adhesive function of AQP0 (Gonen et al. 2004a; Gonen et al. 2005), as well as AQP4 (Tani et al., 2009).

TABLE 3.1   Human Aquaporins: Tissue-Specific Location, Permeability, Size and Pore Motif

	Location	Permeability	Molecular Weight (Da)/Length	Pore Motif
AQP0	Eye	H_2O (low)	28,122/263	NPA (68–70)– NPA (184–186)
AQP1	Red blood cells, kidney, brain, lung	H_2O (high), CO_2	28,526/269	NPA (76–78)– NPA (192–194)
AQP2	Kidney	H_2O (high)	28,837/271	NPA (68–70)– NPA (184–186)
AQP3	Kidney, epithelial cells	H_2O (high), glycerol (high), NH_3 (low), urea (low)	31,544/292	NPA (83–85)– NPA (215–217)
AQP4	Kidney, brain, muscle, stomach, lung	H_2O (high)	34,830/323	NPA (97–99)– NPA (213–215)
AQP5	Stomach, pancreas, lung, glands, eye, ear	H_2O (high)	28,292/265	NPA (69–10)– NPA (185–187)
AQP6	Kidney	H_2O (high), anions (high)	29,370/282	NPA (82–84)– NPA (196–198)
AQP7	Kidney, fat cells, testis, sperm	H_2O (high), glycerol (high), NH_3 (low), urea (low)	37,232/342	NAA (94–96)– NPS (226–228)
AQP8	Pancreas, testis, liver, kidney, airways, heart, glands	H_2O (high), H_2O_2 (high), NH_3 (high)	27,381/261	NPA (92–94)– NPA (210–212)
AQP9	Liver, leukocytes, brain, testis	H_2O (low), glycerol (high), NH_3 (high), urea (high)	31,431/295	NPA (84–86)– NPA (216–218)
AQP10	Epithelial cells	H_2O (low), glycerol (high), urea (high)	31,763/301	NPA (82–84)– NPA (214–216)
AQP11	Kidney, liver, testis, brain	???	30,203/271	NPC (99–101)– NPA (216–218)
AQP12	Pancreas	???	31,475/295	NPT (81–83)– NPA (200–202)

Progress in structure determination (Gonen and Walz 2006) and molecular dynamic studies (de Groot and Grubmuller 2005) provided the basis to understand the selectivity of these channels for the permeation of water or small solutes and the strict exclusion of ions including protons. A number of other permeants than water and glycerol, including CO_2, NO, H_2O_2, NH_3, $As(OH)_3$, $Sb(OH)_3$, and even $Si(OH)_4$, have been shown to pass specific AQPs (Wu and Beitz 2007; Geyer et al. 2013). In addition, these structures have given the framework for assessing the water and solute permeation in great detail by site-directed mutations (Beitz et al. 2004; Beitz et al. 2006).

In mammals, AQP1, AQP2, AQP4, AQP5 and AQP8 function primarily as bidirectional water-selective pores. In addition to water, AQP8 also facilitates passage of ammonia (Saparov et al. 2007) and hydrogen peroxide (Bienert et al. 2007). AQP3, AQP7, AQP9 and AQP10 allow both water and glycerol to pass, while AQP9 appears to transport other small polar solutes, including amino acids, sugars and even arsenite (Wu and Beitz 2007). Involved in pH regulation of intracellular compartments, AQP6 has been found to operate both as water and as chloride channel (Yasui et al. 1999a). AQP11 and AQP12 are the most

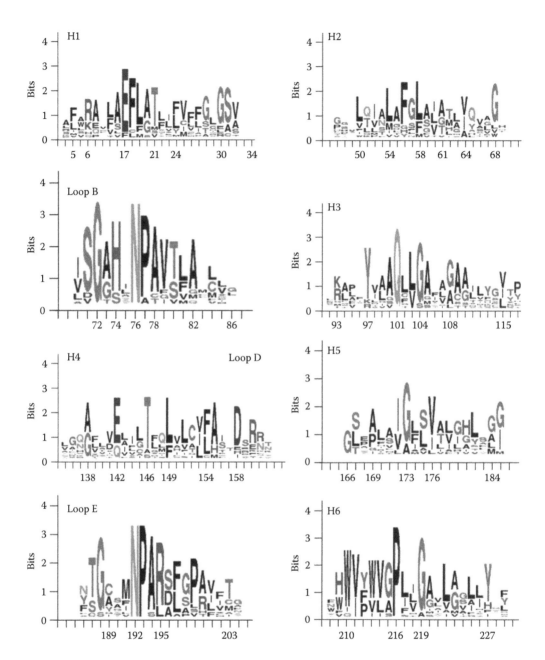

FIGURE 3.1 **(See colour insert.)** Sequence logos (Schneider and Stephens 1990) visualize the conservation of residues at particular positions in the sequences of 13 human aquaporins. The sequence alignment executed by Clustal W (Larkin et al. 2007) was converted to sequence logos using the 'WebLogo' facility *weblogo.berkeley.edu/logo.cgi*. Logos are displayed with the residue numbers for AQP1. The scale gives the certainty of finding a particular amino acid type at each position. Whereas loops B, D and E and all the transmembrane helices are highly conserved termini and loops A and C exhibit a higher variability (not shown). The sequence logos of 13 human AQPs compare favourably to an alignment of AQP sequences representing 46 phylogenic clusters (Heymann and Engel 2000), suggesting human AQPs to include the major aquaporin types.

TABLE 3.2   Human Aquaporins: Structures

	PDB	Resolution (Å)(Residues)	Method	References
AQP0	1YMG	2.24 (6–239)	X-ray (open pore)	Harries et al. (2004)
	2B6P	2.40 (2–263)	X-ray (open pore)	Gonen et al. (2005)
	2C32	7.01 (6–239)	X-ray	Palanivelu et al. (2006)
	1SOR	2.90 (5–239)	E-diff (closed pore)	Gonen et al. (2004a)
	2B6O	1.90 (5–239)	E-diff (closed pore)	Gonen et al. (2005)
AQP1	1FQY	3.80 (8–233)	E-diff	Murata et al. (2000)
	1H6I	3.54 (9–233)	E-diff	de Groot et al. (2001)
	1IH5	3.70 (9–232)	E-diff	Ren et al. (2001)
	4CSK	3.28 (3–235)	X-ray	Ruiz Carrillo et al. (2014)
	1J4N	2.20 (1–249)	X-ray	Sui et al. (2001)
AQP2	4NEF	2.75 (2–241)	X-ray	Frick et al. (2014)
	4OJ2	3.05 (5–257)	X-ray (S256A)	unpublished
AQP4	3GD8	1.80 (32–254)	X-ray	Ho et al. (2009)
	2D57	3.20 (31–254)	E-diff	Hiroaki et al. (2006)
	2ZZ9	2.80 (30–253)	E-diff (S180D)	Tani et al. (2009)
	3IYZ	10.0 (30–253)	E-diff (S180D)	Mitsuma et al. (2010)
AQP5	3D9S	2.00 (1–245)	X-ray	Horsefield et al. (2008)

distantly related paralogs (Morishita et al. 2005). They may both be intracellular AQPs as well, but their function is still unclear (Ishibashi et al. 2014).

Some mammalian AQPs are involved in the formation of junctional structures (Engel et al. 2008). AQP0 plays a major role as an adhesion molecule in the stacking of lens fiber cells. Loss of this function causes loss of stacking order and thus a turbidity of the lens, commonly known as cataract (Shiels et al. 2001). Freeze-fracture electron microscopy of lens tissue has revealed the presence of orthogonal arrays, which belong to the 11–13 nm 'thin lens junctions' (Costello et al. 1989) that assemble after proteolytic cleavage of AQP0. Orthogonal arrays of AQP4 have also been observed by freeze-fracture analysis of mouse kidney, skeletal muscle, and brain tissue, but their function has remained elusive (Verbavatz et al. 1997). The adhesive property of AQP4 arrays has been recognized, and the tongue-into-groove packing arrangement of the two layers has been structurally resolved at atomic resolution (Hiroaki et al. 2006; Tani et al. 2009).

This chapter elucidates the structural framework of human AQPs and compares their pore architectures that confer the specificity for solute permeation. Sequence analyses reveal homology also in the N- and C-termini, which may be related to targeting AQPs to specific cellular location or to gating. These properties are discussed elsewhere in this book (Maria Gourdon, Susanna Törnroth-Horsefield).

3.2 STRUCTURE OF HUMAN AQPs

3.2.1 Sequence Homology

The large number of AQP sequences available is an excellent source of information to compare the structure and function of AQPs. It is helpful for analyzing AQPs whose structure is not available yet. In an early work, a phylogenetic analysis was used to define type sequences to avoid extreme over-representation of some subfamilies, and as a measure of the quality of multiple sequence alignment (Heymann and Engel 1999). Eight conserved segments emerge from this and a subsequent analysis (Heymann and Engel 2000) that define the core architecture of six transmembrane helices and two functional loops, B and E, projecting into the plane of the membrane. Hydrophobic and conservation periodicity, as well as correlated mutations across the alignment allowed the assignment and orientation of the helices in the bilayer to be predicted before the atomic structure of AQP1 became available.

Figure 3.1 displays the conservation of residues within all human AQPs using sequence logos (Schneider and Stephens 1990). The eight conserved segments shown concern six transmembrane helices, H1–H6, and two long loops (loop B and loop E), the latter representing the most conserved features of AQPs. Loop D is also conserved, while loops A and C are more variable. The similarity to the more complete previous analysis is striking. It suggests the human AQPs to represent all major AQP subfamilies.

3.2.2 Ternary Structure and Fold

All AQPs form tetramers with each monomer functioning as an independent pore (Figure 3.2a and b). Tetramers are held together by extensive interactions between the monomers (Figure 3.2b), whose surfaces also exhibit a fraction of polar residues that do not pack in the lipid bilayer (Figure 3.2c).

Six transmembrane helices arranged in a right-handed helical bundle provide the framework for the pore (Figure 3.2a). Conserved residues (Figure 3.1) contribute to their tight interactions. The hallmark of the AQP fold consists of loop B connecting H2 and H3, and loop E connecting H5 and H6, each loop containing a short α-helical segment after the n-terminal NPA motif, called HB and HE. Helices H1–H2–HB–H3 and helices H4–H5–HE–H6 share a considerable sequence homology arising from an ancient gene duplication event (Figure 3.1; Preston and Agre 1991; Heymann and Engel 2000). Because the two halves insert in opposite orientations in the lipid bilayer, helices H1–H2–HB–H3 relate to helices H4–H5–HE–H6 by a quasi-twofold symmetry in the plane of the membrane. Loops B and E meet in the middle of the membrane, where the prolines belonging to the highly conserved NPA motifs stack by van der Waals interactions to form a platform from which the helices HB and HE emanate toward the cytosolic and extracellular surfaces (Figure 3.2a). Loop C comprising about 25 residues connects H3 and H4, that is the first and second half of the AQP, stretching across the extracellular surface of the protein. Figure 3.3 illustrates the structural homology of the AQP core and the variability of C-termini.

AQP1 tetramers are held together by extensive interactions between the monomers. Transmembrane helices 1 and 2 from one monomer engage in a left-handed coiled-coil interaction with helices 4 and 5 of a neighbouring monomer. Helices from neighbouring monomers also interact outside the membrane: helix 1 interacts with helix 5 of the adjacent monomer at the extracellular surface and helix 2 interacts with helix 4 of the adjacent monomer at the cytoplasmic surface. Figure 3.2b documents the tightness of these intermolecular contacts. Interactions between loops also contribute to tetramer stability, prominently with the four loops A surrounding the fourfold axis of the tetramer on the extracellular surface (Murata et al. 2000; Sui et al. 2001).

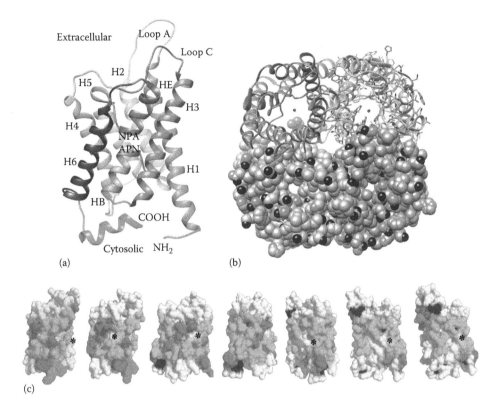

FIGURE 3.2 **(See colour insert.)** Fold and tertiary structure of aquaporins. (a) The color-coded bovine AQP1 monomer (PDB ID: 1J4N) reveals the AQP fold, with H1, H2, HB and H3 forming the first half of the protein and H4, H5, HE and H6 the second half. The two half helices HB and HE emanate outward from the platform formed by the prolines of the NPA motifs in the center of the pore. (b) Aquaporins and aquaglyceroporins exist as tetramers, which feature four independent pores (two pores are marked by asterisks). Tight packing of the monomers into a tetramer is indicated by two monomers rendered by spheres in Chimera (Pettersen et al. 2004). (c) The surface of an AQP1 monomer shows mainly hydrophobic (yellow) and aromatic (green) residues. However, polar residues (white/grey, highlighted by asterisks on their right side) are interspersed between hydrophobic residues and must be buried in the interfaces between protomers.

3.2.3 Pore Structure and Specificity

Overall, the conduction pores of AQPs are roughly 25 Å long (see Figure 3.5) and exhibit two sites interacting strongly with water, the constriction and the NPA motif. Pore helices HB and HE contain highly conserved residues that line one surface of the water pore. Conserved residues originating from helices 2 and 5 and the C-terminal halves of helices 1 and 4 form the remaining surface of the pore. Permeating molecules are coordinated to the channel through a combination of backbone carbonyl and amino acid side chain interactions, the amphipathic nature of the channel complementing the chemical nature of the conducted molecules. The narrowest constriction is located close to the extracellular pore mouth. In water-specific AQPs, it is approximately 2.8 Å in diameter, that is identical to that of a water molecule, and about 3.4 Å in GLPs matching the diameter of a

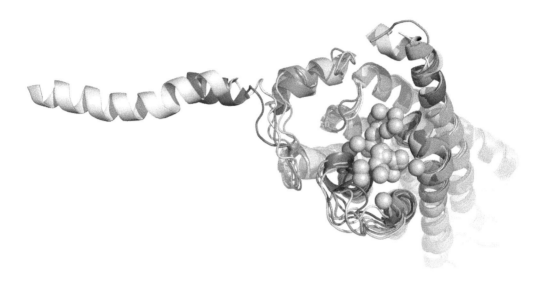

FIGURE 3.3 **(See colour insert.)** Structural superposition of all deposited human aquaporin structures (AQP0, green and blue; AQP1, pink; AQP2, grey and orange; AQP4, salmon and turquoise; AQP5, yellow), showing the different orientations of C-termini.

carbon hydroxyl group of polyols such as glycerol (Figure 3.4). This constriction, referred to as aromatic residue/arginine (ar/R) constriction or selectivity filter (SF), is formed by four residues, that is Phe56, His180, Cys189, and Arg195 in AQP1 or Trp48, Gly191, Phe200, and Arg205 in GlpF. The positively charged arginine is to some extent involved in proton exclusion because exchange to a valine residue results in a small but measurable proton leakage (Beitz et al. 2006; Li et al., 2011). A histidine is typical for water-specific AQPs, which together with the highly conserved arginine provides a hydrophilic edge in

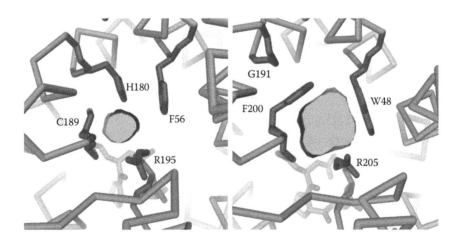

FIGURE 3.4 **(See colour insert.)** Comparison of the AQP1 and GlpF pores. The principal constriction is distinctly smaller in aquaporins (left) than in aquaglyceroporins (right), yet the arginine is highly conserved in both channels. Aromatic residues are more prominent in aquaglyceroporins, and they serve as a greasy slide to facilitate passage of glycerol and other polyols.

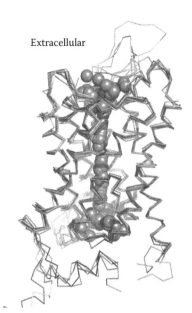

Extracellular

FIGURE 3.5 **(See colour insert.)** Structural superposition of all deposited human aquaporin structures (AQP0, AQP1, AQP2, AQP4 and AQP5), showing the distribution of the channel waters along the aquaporin pores. Rigid-body alignments were performed with the CCP4 program suite. Only coordinates with fitted pore waters were used, from both high-resolution electron and x-ray diffraction crystallography (PDB codes: AQP0 [2B6O, 1YMG], AQP1 [1J4N], AQP2 [4OJ2], AQP4 [2ZZ9, 3GD8], AQP5 [3D9S]).

juxtaposition to an aromatic residue. Cys189 is the site for blockage of water permeation through AQP1 by $HgCl_2$ (Preston et al. 1993; Zhang et al. 1993). Its side chain extends into the pore from the extracellular side of the constriction (Figure 3.4, left panel) indicating that $HgCl_2$ physically blocks the aqueous pathway (Murata et al. 2000).

Six water molecules were found to form a single file through the pore of AQP1 (Sui et al. 2001). Since then, further high-resolution structures of human AQPs became available that reveal the pore waters. Coordinate alignments of these structures reveal that solvents occupy similar positions, with slight variations as a result of differences in the positions of interacting side chains (Figure 3.5). As this figure shows, the water distribution follows the shape of the AQP pores funneling out at the extracellular and the cytosolic side. In general, waters that are observed in closely similar positions would be expected to play a structural role. However, for aquaporins they depict specific hydrogen bonding sites that position the waters in the channel in distinct orientations as they move through the pore.

Figure 3.6a shows a well-ordered water-mediated hydrogen bond network of pore waters in the proximity of the NPA motifs for the high-resolution structures of AQP0 and AQP4. AQP4 depicts a network of seven continuous waters connected via hydrogen bonding. This network is flanked by hydrophobic residues (see Figure 3.6 from top to bottom) PHE-77, PHE-48, ILE-81, ILE-96, ILE-193, VAL-85, VAL-174 and VAL-100 and stabilized via hydrogen bonding to residues ALA-210 (SF), HIS-201 (SF), ARG-216 (SF), ASN-213 (NPA) and ASN-97 (NPA), HIS-95, GLY-94 and GLY-93. Structural modulations and selected

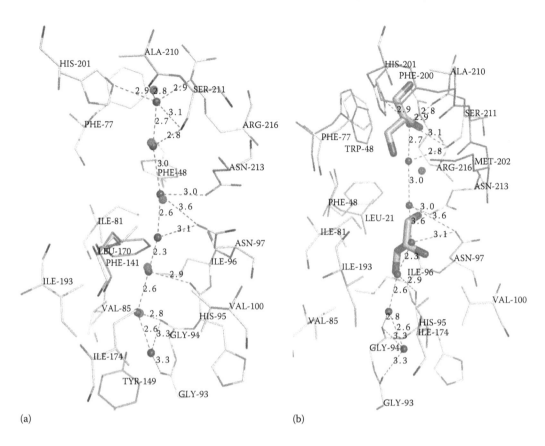

(a)   (b)

FIGURE 3.6   **(See colour insert.)** The network of ordered waters in the AQP pore center depends on specific differences in the amino acid environment. (a) Structural superposition of AQP0 and AQP4. The channel side chains of AQP4 are drawn in grey, and its pore waters are depicted as red spheres. AQP0 pore waters are shown as blue spheres. (b) Structural superposition of AQP4 (grey) and GlpF (salmon) (PDB code: 1FX8) highlights differences in the amino acid composition of the center pore region. AQP4 pore waters are depicted as red spheres. Glycerol molecules in the GlpF pore are shown in grey/red and the interspersed water as blue sphere.

disruption of this continuous pore water network may thus result in the lowering of the overall diffusion rate in AQPs. Superposition of the central pore waters from the high-resolution structure of AQP0, which has a lower water permeability than AQP4 shows that five of the seven network water positions are occupied likewise, while positions 4 and 7 (Figure 3.6a, top to bottom) are empty, thus resulting in a non-continuous water network for AQP0. Both of these positions are sterically excluded: at position 4 an ILE to PHE substitution (PHE-141 in blue) is taking up space at the center of the two NPA motifs and for position 7 the close proximity of TYR-149 inhibits the water occupation at this location. Interestingly, the high resolution structure of Aqy1 (sole water channel in *Pichia pastoris*), revealed 4 water molecules at the SF region too close to each other to be occupied at the same time, giving rise to the hypothesis that water moves through the pore in pairs at a van der Waals distance of 3 angstroms (Kosinska Eriksson et al., 2013).

A structural superposition of AQP4 with GlpF (Figure 3.6b) gives additional insight into the selective permeability of the glycerol-conducting channel GlpF (Fu et al. 2000). The comparison identifies the amphipathic selectivity filter formed by TRP-48 and PHE-200, positioned at a right angle to each other, forming a hydrophobic wedge around the glycerol channel. These additional van der Waals interactions formed with the glycerol backbone are not present in the AQP4 channel, thus favouring water transport. At the location of the second glycerol near the center of the pore, PHE-48 is replaced with the much smaller hydrophobic residue LEU-21, thus increasing the effective pore diameter in the constriction region. It is interesting to note that PHE-48 in AQP4 corresponds to PHE-24 in AQP1; the sequence logos in Figure 3.1 show that the prominent residues at this position are predominantly either PHE (water pores) or LEU (glycerol pores). The following residues are found at this position in AQP3 (LEU), AQP9 (LEU), AQP10 (LEU) and AQP11 (ALA), indicating their similarity to GlpF.

In GlpF, and essentially in all other GLPs, the ar/R region is more hydrophobic than that of AQP1 due to the lack of histidine and substitution of the cysteine by a second aromatic residue. The ladder of aromatic residues forms a 'greasy slide' that allows GlpF to efficiently conduct glycerol, small linear polyols and urea but makes GlpF (and GLPs in general) less efficient water channels (Stroud et al. 2003).

A less narrow constriction is located at the center of the pore in the NPA region, but the interaction between water molecules and the protein is nevertheless prominent at this site. The two asparagines are the capping amino acids at the positive ends of helices HB and HE and act as hydrogen donors to the oxygen atoms of passing permeants. In addition, water that enters this region is re-oriented by the dipoles of the emanating half helices HB and HE, such that hydrogen bonds between neighbouring water molecules in the chain are disrupted. This mechanism was initially proposed to prevent the formation of a proton wire throughout the pore and represents a major energy barrier for proton conductance (Murata et al. 2000). Subsequent molecular dynamics simulations combined with quantum mechanical calculations of proton hopping probabilities demonstrated that protons are excluded from the central region of the channel by a strong free energy barrier, resulting from the dipole moments of HB and HE (de Groot and Grubmuller 2005). The recent 0.88 Å structure of the *P. pastoris* water channel, Aqy1, visualizes the H-bond donor interactions of NPA's asparagine residues to passing water molecules, and documents a polarized water–water H-bond network within the channel. The four too closely spaced SF water positions revealed imply strongly correlated water movements that break the connectivity of SF waters to other water molecules within the channel and prevent proton transport via a Grotthuss mechanism (Kosinska Eriksson et al., 2013). In GlpF, a similar free energy barrier blocks proton permeation as well. Selectivity by size may also be of relevance in the NPA constriction, because almost all the GLPs have two leucine residues opposite the two asparagines instead of a leucine and a phenylalanine in AQPs (Heymann and Engel 2000). This combination results in a somewhat larger pore diameter that is suitable for solutes larger than water.

The remaining part of the AQP pore is lined by hydrophobic residues that expose main-chain carbonyl oxygens to the pore surface. These oxygens distribute as a ladder along one side of the pore and serve as hydrogen bond acceptor sites to efficiently funnel small hydrogen bond donor molecules, such as water, urea or polyols, through the AQP channel. The formation of hydrogen bonds between the AQP protein and the permeant also compensates for the solvatation energy cost when a molecule enters from the bulk solution into the pore.

In spite of the progress in high resolution structure determination (Kosinska Eriksson et al., 2013) it is instructive to follow as an example a water molecule during its transit though the pore of human AQP1 (Figure 3.7) (de Groot and Grubmuller 2001). At the extracellular side (top), the pore is relatively wide and water molecules interact mainly with the A and C loops through LYS-36 and SER-123, respectively. The first strong interaction site is found in the (ar/R) constriction formed by the side chains of ARG-195, HIS-180, GLY-190 and PHE-56. Within this region, between loop E and helix E, the hydrophobic PHE-56 side chain orients the water molecules such as to enforce strong hydrogen bonds to ARG-195 and HIS-180. Further down the channel, the carbonyl groups of residues ILE-191, GLY-190, and CYS-189 interact with the water molecules in the pore. The next strong interaction site for water with the protein is in the middle of the channel, with both asparagines of the NPA motifs on one side and the hydrophobic side chains of PHE-24, VAL-176, and ILE-191

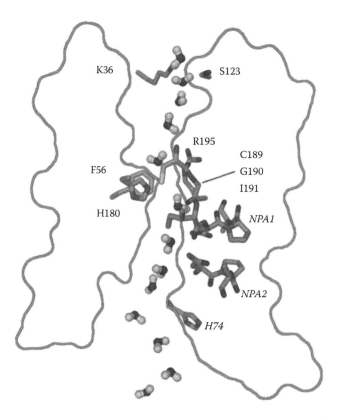

FIGURE 3.7 **(See colour insert.)** In AQP1, the pore has a length of about 25 Å and connects the extracellular with the cytosolic vestibules. Water molecules are from a snapshot of a molecular dynamics simulation (see de Groot and Grubmuller 2001) and reveal their tendency to orient in the electric field of the channel, with the hydrogen atoms preferentially oriented toward the extracellular (top halve of the pore) or the cytosolic surface (bottom halve of the pore).

on the other. At this site, water molecules rotate by 180° as the result of the electrostatic field produced by HB and HE. The pore widens toward the intracellular side (bottom), and water molecules interact only weakly with the pore. The main interaction sites are formed by the main-chain carbonyl groups of the residues following the second NPA motif, GLY-72 and ALA-73, and the side chain of HIS-74. It is remarkable that except for CYS-189 all these residues established by the molecular dynamics simulation before water molecules were resolved correspond to those depicted in Figure 3.6. Since the channel is rather symmetric in its nature, water permeation occurs in both directions, with the water flux following the osmotic gradient.

The pore structure dictates the permeation rate and the specificity. Changes in this structure resulting from changes in the pH or from other external factors will change the flux of water or solutes and is termed gating. Since the specificity of the pore is high, only small changes of a few or even a single residue are required to produce the gating phenomenon. Small changes of critical residues at the pore entrances may obstruct the pore as well. Therefore, high-resolution structures are required to identify gating mechanisms in AQPs.

3.2.4 Surface Structure of AQPs Involved in Membrane Junctions

The role of AQP0 in membrane junctions was recognized early. Reconstitution of purified AQP0 into liposomes showed that it caused the vesicles to cluster (Dunia et al. 1987). Moreover, orthogonal arrays observed by freeze-fracture electron microscopy and 'thin lens junctions' were investigated in the late 1970s (Kistler and Bullivant 1980). Structural clues on the adhesive properties of AQP0 emerged when 2D crystallization experiments resulted in single-layered and double-layered crystals, which were analyzed by electron crystallography (Hasler et al. 1998) and AFM (Fotiadis et al. 2000). Proteolytic cleavage appeared to induce AQP0-based junctions, because AQP0 exists as a full-length protein of 26 kDa in young fiber cells of the lens cortex, where junctions are rarely found. Older fiber cells buried deeper in the lens core exhibit more junctions than the cortex tissue, and some of the AQP0 in the lens core is proteolytically cleaved (e.g. Takemoto et al. 1986). This hypothesis was experimentally tested by in vitro proteolysis of purified protein as well as AQP0 2D arrays (Gonen et al. 2004a). When detergent-solubilized, full-length AQP0 (26 kDa) was treated with chymotrypsin, the resulting 22 kDa fragment eluted at a higher molecular weight from a sizing column than full-length AQP0, suggesting that cleaved AQP0 tetramers form pairs.

Double-layered 2D crystals produced with a mixture of full-length and proteolytically cleaved AQP0 isolated from the lens core made it possible to determine the structure of junctional AQP0 by electron crystallography, first at 3 Å resolution (Gonen et al. 2004b) and then at 1.9 Å resolution (Gonen et al. 2005). Besides a wealth of information on the protein–lipid interactions and the closed conformation of AQP0, the structure of the junction between extracellular AQP0 surfaces was elucidated. Contacts formed by residues in loop C involve a Pro–Pro motif (Pro109 and Pro110), which is part of a one-turn helix (helix HC), and residues Arg113 and Pro123. Although loop C is in a virtually identical conformation in junctional and non-junctional AQP0 determined by x-ray crystallography, cleavage of the cytoplasmic tails correlates with a rearrangement in loop A (Gonen et al. 2005). In the loop conformation of non-junctional AQP0, Pro38 is at some distance from the center of the tetramer. In addition, Trp34 lies above the pore and projects outward, blocking the approach of a second tetramer, and Arg33 is positioned in between two monomers. In contrast, loop A has reconfigured in the junctional AQP0 tetramer, positioning Pro38 so that it can form a rosette-like structure at the center of the tetramer and mediate a major junctional contact. The side chains of Arg33 and Trp34 also swap positions, so that Trp34 no longer interferes with the close approach of another tetramer. In the completed junction, all three residues interact with the corresponding residues from the opposing tetramer.

Similar to AQP0, AQP4 was found by freeze-fracture techniques to produce orthogonal arrays in murine kidneys, skeletal muscles and brain (Verbavatz et al. 1997). These arrays were not present in AQP4 knockout mice. Although a hypothetical role of AQP4 as osmosensor was proposed (Venero et al. 1999), AQP4 had not been implicated in cell adhesion, and the interactions between AQP4 tetramers in adjoining membranes are apparently weak. However, they are sufficiently significant to dictate the formation of double-layered 2D crystals when reconstituting the AQP4 isoform M23 but not the AQP4 isoform M1. The latter isoform has a longer N-terminus, which apparently influences the fine structure of the AQP4 extracellular surface. The 3.2 Å structure revealed that AQP4 tetramers in the two membranes interact with each other through their extracellular surfaces (Hiroaki et al. 2006). Rather than being exactly stacked, as in the case of AQP0, the AQP4 tetramers are shifted so that a tetramer in one membrane is at the center of four tetramers in the adjoining membrane. AQP4 features a short 3_{10} helix in extracellular loop C, helix HC, which contains the two residues Pro139 and Val142 that mediate the interactions between opposing tetramers. In the 2D crystals, each AQP4 tetramer interacts with four tetramers in the adjoining membrane so that an orthogonal array would significantly enhance AQP4-mediated adhesion. To understand whether the double-layered 2D crystals were just a result of the in vitro 2D crystallization of AQP4 or whether AQP4 could also form junctions in vivo, AQP4 was expressed in L-cells, a fibroblast cell line that does not express endogenous cell adhesion molecules. These cells showed some clustering, which was not observed when AQP1 was expressed (Hiroaki et al. 2006). Moreover, thin sections through the hypothalamus revealed large membrane junctions between glial lamellae with short stretches of separated membranes. Immuno-labelling studies localized AQP4 to the separated membranes as well as to the junctional regions (Hiroaki et al. 2006). Together, these results suggest a possible role of AQP4 in junction formation in vivo and provide an explanation for the involvement of AQP4 in osmoregulation (Venero et al. 1999). In the double-layered 2D crystals, the AQP4 molecules show a tight tongue-into-groove packing of their extracellular surfaces, which results in a partial blockage of the extracellular pore entrances. The formation of AQP4 junctions in vivo would thus potentially lead to a reduced water permeability of glial cell plasma membranes. Conversely, rapid water flow through the channels could drive the interacting membranes apart and thus resolve the junctions. As a result of the characteristics of AQP4, glial cells expressing a high ratio of AQP4M1 would produce small AQP4 arrays providing weak adhesion between membranes, which would easily separate and be sensitive to small water flows resulting from small osmotic differences. Vice versa, glial cells expressing a high ratio of AQP4M23 would produce large AQP4 arrays providing stronger adhesion between membranes that would withstand larger water flows associated with larger osmotic differences.

3.3 UNSOLVED RIDDLES

3.3.1 Anion Transport

AQP6 is located predominantly in intracellular membrane vesicles in multiple types of renal epithelia, where it colocalizes alongside H^+-ATPase (Yasui et al. 1999b). When expressed in *Xenopus* oocytes, AQP6 exhibits low basal water permeability, but when treated with the water channel blocker for AQP1, AQP2 and AQP5, Hg^{2+}, the water permeability of AQP6-expressing oocytes rapidly increases 10-fold and is accompanied by ion conductance. At pH values less than 5.5, anion conductance is rapidly and reversibly activated in AQP6-expressing oocytes, which suggests a role in acid secretion (Yasui et al. 1999a). Although AQP6 exhibits the critical cysteine (C189 in AQP1 Murata et al. 2000) and its sequence is closest to AQP2 and AQP5, the molecular mechanism related to AQP6's unusual features remains elusive. A major problem in solving this riddle is the lack of a system that allows AQP6 to be expressed in sufficient quantities.

3.3.2 Gas Transport

Much discussed experiments in *Xenopus* oocytes indicated a remarkable CO_2 uptake that was induced by expressing AQP1 (Nakhoul et al. 1998). In another contribution, a CO_2 permeability comparable to that of human AQP1 was reported for the tobacco plasma membrane AQP NtAQP1 (Uehlein et al. 2003). It was further shown that NtAQP1 overexpression increases membrane permeability for CO_2 and water and accelerates leaf growth. Electrophysiological work on oocytes expressing AQP1 demonstrated NH3 movement through AQP1 (Nakhoul and Lee Hamm, 2013). The permeation of ammonia/ammonium across AQP8 was demonstrated by planar bilayer experiments, verifying the exclusion of NH^{4+} or H^+ by lack of current under voltage clamp conditions (Saparov et al. 2007). The single-channel water permeability coefficient measured was more than twofold smaller than the single-channel ammonia permeability. This permeability ratio suggested that electrically silent ammonia transport may be a major function of AQP8.

These findings promoted molecular dynamics simulations that address the question of CO_2 permeation through human AQP1 (Hub and de Groot 2006). The free energy profiles derived from the simulations gave a barrier of ≈23 kJ/mol in the ar/R constriction region of the water pore, compared to a barrier of ≈4 kJ/mol found for a palmitoyl-oleoyl-phosphatidylethanolamine lipid bilayer membrane, which has served as model bilayer for the simulations. These simulations also included the cavity about the fourfold axis of the tetramer. Two major barriers for CO_2 can be identified, one of 12 ± 2 kJ/mol at z = 7.5 Å near the extracellular entrance to the central cavity. Four V50 residues of the four monomers located at this aperture force the CO_2 molecules to lose favourable interactions to neighbouring water molecules. The second barrier has a similar height and is at a site where four D48 residues surround the CO_2 molecule. Taken together, the free energy barrier for CO_2 permeation through the central cavity of AQP1 is significantly smaller than for the monomeric channel. A further and more detailed molecular dynamics analysis came to conclude that small hydrophobic solutes such as NO or CO_2 are expected to diffuse through the water channel of AQP1 at a very low rate, whereas they should do this at a 30 times higher rate through GlpF (Hub and de Groot 2008). Another MD analysis of human AQP1showed that CO_2 movement is feasible both through the monomeric water pores and the central pore, with the central pore's accounting for about as much CO_2 permeability as the four monomeric pores (Wang et al., 2007). The observation that $HgCl_2$, which inhibits CO_2 permeation (Cooper and Boron, 1998), argues that at least some CO_2 moves through the monomeric water pore. Preliminary work suggests that about half the channel-dependent flux of CO_2 moves through the four water pores, and about half through the central pore (Musa-Aziz et al., 2009).

Most recently Musa-Aziz, Geyer and coworkers compared water, CO_2 and NH_3 permeability for nine human AQPs expressed in Xenopus oocytes (Geyer et al., 2013), the first demonstration of gas selectivity by channels. These authors observed that AQP1 and 9 are permeable to all three molecules. They also confirmed that AQP6 lacks water permeability but showed that AQP6 is permeable to CO_2 and NH_3. In addition, these experiments demonstrated that AQP3, 7 and 8 exhibit water and NH_3 (but not CO_2) permeation, AQP0, AQP4-M23, and AQP5 show water and CO_2 (but not NH_3) permeation; and that AQP2 and AQP4-M1 have water but lack both CO_2 and NH_3 permeability. Because NH_3 and H_2O have similar electronic structures, it seems plausible that NH_3 permeates through an AQP via a monomeric water pore. However, it is not clear why NH_3 permeates some AQPs (1, 3, 6, 7, 8, 9) but not others, or why NH_3 permeates AQP6 but H_2O does not. If both the monomeric water pores and the central pore contribute to the CO_2 permeability of AQP1, it is not clear why some AQPs (2, 3, AQP4-M1, 7, 8) have no significant CO_2 permeability. To elucidate the gas transport by AQPs, site-directed mutations together with high-resolution structures and additional molecular dynamics simulations of the respective mutants would be required.

3.4 CONCLUSIONS

Structural studies have paved the avenue to a deep understanding of AQPs, small ancient proteins present in all kingdoms of life. They provide efficient transmembrane pathways for water, small uncharged solutes and gas molecules. Their specificity is related to the highly precise and conserved channel architecture. Because permeation of substrate molecules is passive, regulation of AQPs requires either gating or their controlled transport and integration to specific membrane compartments. These processes still need to be further investigated.

ACKNOWLEDGEMENT

This work has been supported by the Transcontinental EM Initiative for Membrane Protein Structure, funded by the NIH Protein Structure Initiative under Grant U54GM094598.

REFERENCES

Beitz E, Pavlovic-Djuranovic S, Yasui M, Agre P and Schultz JE (2004) Molecular dissection of water and glycerol permeability of the aquaglyceroporin from *Plasmodium falciparum* by mutational analysis. *Proceedings of the National Academy of Sciences of the United States of America* **101**:1153–1158.

Beitz E, Wu B, Holm LM, Schultz JE and Zeuthen T (2006) Point mutations in the aromatic/arginine region in aquaporin 1 allow passage of urea, glycerol, ammonia, and protons. *Proceedings of the National Academy of Sciences of the United States of America* **103**:269–274.

Bienert GP, Moller AL, Kristiansen KA, Schulz A, Moller IM, Schjoerring JK and Jahn TP (2007) Specific aquaporins facilitate the diffusion of hydrogen peroxide across membranes. *Journal of Biological Chemistry* **282**:1183–1192.

Blank ME and Ehmke H (2003) Aquaporin-1 and $HCO_3(-)$-Cl-transporter-mediated transport of CO_2 across the human erythrocyte membrane. *Journal of Physiology* **550**:419–429.

Cooper GJ and Boron WF (1998) Effect of PCMBS on CO_2 permeability of *Xenopus* oocytes expressing aquaporin 1 or its C189S mutant. *American Journal of physiology* **275**:C1481–1486.

Costello MJ, McIntosh TJ and Robertson JD (1989) Distribution of gap junctions and square array junctions in the mammalian lens. *Investigative Ophthalmology & Visual Science* **30**:975–989.

Day RE, Kitchen P, Owen DS, Bland C, Marshall L, Conner AC, Bill RM and Conner MT (2014) Human aquaporins: Regulators of transcellular water flow. *Biochimica et Biophysica Acta* **1840**:1492–1506.

de Groot BL, Engel A and Grubmuller H (2001) A refined structure of human aquaporin-1. *FEBS Letters* **504**:206–211.

de Groot BL and Grubmuller H (2001) Water permeation across biological membranes: Mechanism and dynamics of aquaporin-1 and GlpF. *Science* **294**:2353–2357.

de Groot BL and Grubmuller H (2005) The dynamics and energetics of water permeation and proton exclusion in aquaporins. *Current Opinion in Structural Biology* **15**:176–183.

Dunia I, Manenti S, Rousselet A and Benedetti EL (1987) Electron microscopic observations of reconstituted proteoliposomes with the purified major intrinsic membrane protein of eye lens fibers. *Journal of Cell Biology* **105**:1679–1689.

Engel A, Fujiyoshi Y, Gonen T and Walz T (2008) Junction-forming aquaporins. *Current Opinion in Structural Biology* **18**:229–235.

Fotiadis D, Hasler L, Muller DJ, Stahlberg H, Kistler J and Engel A (2000) Surface tongue-and-groove contours on lens MIP facilitate cell-to-cell adherence. *Journal of Molecular Biology* **300**:779–789.

Frick A, Eriksson UK, de Mattia F, Oberg F, Hedfalk K, Neutze R, de Grip WJ, Deen PM and Tornroth-Horsefield S (2014) X-ray structure of human aquaporin 2 and its implications for nephrogenic diabetes insipidus and trafficking. *Proceedings of the National Academy of Sciences of the United States of America* **111**:6305–6310.

Fu D, Libson A, Miercke LJ, Weitzman C, Nollert P, Krucinski J and Stroud RM (2000) Structure of a glycerol-conducting channel and the basis for its selectivity. *Science* **290**:481–486.

Geyer RR, Musa-Aziz R, Qin X and Boron WF (2013) Relative CO(2)/NH(3) selectivities of mammalian aquaporins 0–9. *American Journal of Physiology Cell Physiology* **304**:C985–C994.

Gonen T, Cheng Y, Kistler J and Walz T (2004a) Aquaporin-0 membrane junctions form upon proteolytic cleavage. *Journal of Molecular Biology* **342**:1337–1345.

Gonen T, Cheng Y, Sliz P, Hiroaki Y, Fujiyoshi Y, Harrison SC and Walz T (2005) Lipid–protein interactions in double-layered two-dimensional AQP0 crystals. *Nature* **438**:633–638.

Gonen T, Sliz P, Kistler J, Cheng Y and Walz T (2004b) Aquaporin-0 membrane junctions reveal the structure of a closed water pore. *Nature* **429**:193–197.

Gonen T and Walz T (2006) The structure of aquaporins. *Quarterly Reviews of Biophysics* **39**:361–396.

Gorin MB, Yancey SB, Cline J, Revel JP and Horwitz J (1984) The major intrinsic protein (MIP) of the bovine lens fiber membrane: Characterization and structure based on cDNA cloning. *Cell* **39**:49–59.

Harries WE, Akhavan D, Miercke LJ, Khademi S and Stroud RM (2004) The channel architecture of aquaporin 0 at a 2.2-A resolution. *Proceedings of the National Academy of Sciences of the United States of America* **101**:14045–14050.

Hasler L, Walz T, Tittmann P, Gross H, Kistler J and Engel A (1998) Purified lens major intrinsic protein (MIP) forms highly ordered tetragonal two-dimensional arrays by reconstitution. *Journal of Molecular Biology* **279**:855–864.

Heymann JB and Engel A (1999) Aquaporins: Phylogeny, structure, and physiology of water channels. *News in Physiological Sciences* **14**:187–193.

Heymann JB and Engel A (2000) Structural clues in the sequences of the aquaporins. *Journal of Molecular Biology* **295**:1039–1053.

Hiroaki Y, Tani K, Kamegawa A, Gyobu N, Nishikawa K, Suzuki H, Walz T et al. (2006) Implications of the aquaporin-4 structure on array formation and cell adhesion. *Journal of Molecular Biology* **355**:628–639.

Ho JD, Yeh R, Sandstrom A, Chorny I, Harries WE, Robbins RA, Miercke LJ and Stroud RM (2009) Crystal structure of human aquaporin 4 at 1.8 Å and its mechanism of conductance. *Proceedings of the National Academy of Sciences of the United States of America* **106**:7437–7442.

Horsefield R, Norden K, Fellert M, Backmark A, Tornroth-Horsefield S, Terwisscha van Scheltinga AC, Kvassman J, Kjellbom P, Johanson U and Neutze R (2008) High-resolution x-ray structure of human aquaporin 5. *Proceedings of the National Academy of Sciences of the United States of America* **105**:13327–13332.

Hub JS and de Groot BL (2006) Does CO_2 permeate through aquaporin-1? *Biophysical Journal* **91**:842–848.

Hub JS and de Groot BL (2008) Mechanism of selectivity in aquaporins and aquaglyceroporins. *Proceedings of the National Academy of Sciences of the United States of America* **105**:1198–1203.

Ishibashi K, Tanaka Y and Morishita Y (2014) The role of mammalian superaquaporins inside the cell. *Biochimica et Biophysica Acta* **1840**:1507–1512.

Jung J, Preston G, Smith B, Guggino W and Agre P (1994) Molecular structure of the water channel through aquaporin CHIP. The hourglass model. *Journal of Biological Chemistry* **269**:14648–14654.

Kistler J and Bullivant S (1980) Lens gap junctions and orthogonal arrays are unrelated. *FEBS Letters* **111**:73–78.

Kosinska Eriksson U, Fischer G, Friemann R, Enkavi G, Tajkhorshid E and Neutze R (2013) Subangstrom resolution x-ray structure details aquaporin-water interactions. *Science* **340**:1346–1349.

Larkin MA, Blackshields G, Brown NP, Chenna R, McGettigan PA, McWilliam H, Valentin F et al. (2007) Clustal W and clustal X version 2.0. *Bioinformatics* **23**:2947–2948.

Li H, Chen H, Steinbronn C, Wu B, Beitz E, Zeuthen T and Voth GA (2011) Enhancement of proton conductance by mutations of the selectivity filter of aquaporin-1. *Journal of Molecular Biology* **407**:607–620.

Maurel C (2007) Plant aquaporins: Novel functions and regulation properties. *FEBS Letters* **581**:2227–2236.

Mitsuma T, Tani K, Hiroaki Y, Kamegawa A, Suzuki H, Hibino H, Kurachi Y and Fujiyoshi Y (2010) Influence of the cytoplasmic domains of aquaporin-4 on water conduction and array formation. *Journal of Molecular Biology* **402**:669–681.

Morishita Y, Matsuzaki T, Hara-chikuma M, Andoo A, Shimono M, Matsuki A, Kobayashi K et al. (2005) Disruption of aquaporin-11 produces polycystic kidneys following vacuolization of the proximal tubule. *Molecular and Cellular Biology* **25**:7770–7779.

Murata K, Mitsuoka K, Hirai T, Walz T, Agre P, Heymann BJ, Engel A and Fujiyoshi Y (2000) Structural determinants of water permeation through aquaporin-1. *Nature* **407**:599–605.

Musa-Aziz R, Chen LM, Pelletier MF and Boron WF (2009) Relative CO_2/NH_3 selectivities of AQP1, AQP4, AQP5, AmtB, and RhAG. *Proceedings of the National Academy of Sciences of the United States of America* **106**:5406–5411.

Nakhoul NL, Davis BA, Romero MF and Boron WF (1998) Effect of expressing the water channel aquaporin-1 on the CO_2 permeability of *Xenopus* oocytes. *American Journal of Physiology* **274**:C543–C548.

Nakhoul NL and Lee Hamm L (2013) Characteristics of mammalian Rh glycoproteins (SLC42 transporters) and their role in acid-base transport. *Molecular Aspects of Medicine* **34**:629–637.

Palanivelu DV, Kozono DE, Engel A, Suda K, Lustig A, Agre P and Schirmer T (2006) Co-axial association of recombinant eye lens aquaporin-0 observed in loosely packed 3D crystals. *Journal of Molecular Biology* **355**:605–611.

Pettersen EF, Goddard TD, Huang CC, Couch GS, Greenblatt DM, Meng EC and Ferrin TE (2004) UCSF Chimera – A visualization system for exploratory research and analysis. *Journal of Computational Chemistry* **25**:1605–1612.

Prasad GV, Coury LA, Finn F and Zeidel ML (1998) Reconstituted aquaporin 1 water channels transport CO_2 across membranes. *Journal of Biological Chemistry* **273**:33123–33126.

Preston GM and Agre P (1991) Isolation of the cDNA for erythrocyte integral membrane protein of 28 kilodaltons: Member of an ancient channel family. *Proceedings of the National Academy of Sciences of the United States of America* **88**:11110–11114.

Preston GM, Jung JS, Guggino WB and Agre P (1993) The mercury-sensitive residue at cysteine 189 in the CHIP28 water channel. *Journal of Biological Chemistry* **268**:17–20.

Preston GM, Piazza Carroll T, Guggino WB and Agre P (1992) Appearance of water channels in *Xenopus* oocytes expressing red cell CHIP28 protein. *Science* **256**:385–387.

Ren G, Reddy VS, Cheng A, Melnyk P and Mitra AK (2001) Visualization of a water-selective pore by electron crystallography in vitreous ice. *Proceedings of the National Academy of Sciences of the United States of America* **98**:1398–1403.

Ruiz Carrillo D, To Yiu Ying J, Darwis D, Soon CH, Cornvik T, Torres J and Lescar J (2014) Crystallization and preliminary crystallographic analysis of human aquaporin 1 at a resolution of 3.28 Å. *Acta Crystallographica Section F: Structural Biology Communications* **70**:1657–1663.

Saparov SM, Liu K, Agre P and Pohl P (2007) Fast and selective ammonia transport by aquaporin-8. *Journal of Biological Chemistry* **282**:5296–5301.

Schneider TD and Stephens RM (1990) Sequence logos: A new way to display consensus sequences. *Nucleic Acids Research* **18**:6097–6100.

Shiels A, Bassnett S, Varadaraj K, Mathias R, Al-Ghoul K, Kuszak J, Donoviel D, Lilleberg S, Friedrich G and Zambrowicz B (2001) Optical dysfunction of the crystalline lens in aquaporin-0-deficient mice. *Physiological Genomics* **7**:179–186.

Stroud RM, Miercke LJ, O'Connell J, Khademi S, Lee JK, Remis J, Harries W, Robles Y and Akhavan D (2003) Glycerol facilitator GlpF and the associated aquaporin family of channels. *Current Opinion in Structural Biology* **13**:424–431.

Sui H, Han BG, Lee JK, Walian P and Jap BK (2001) Structural basis of water-specific transport through the AQP1 water channel. *Nature* **414**:872–878.

Takemoto L, Takehana M and Horwitz J (1986) Covalent changes in MIP26K during aging of the human lens membrane. *Investigative Ophthalmology & Visual Science* **27**:443–446.

Tani K, Mitsuma T, Hiroaki Y, Kamegawa A, Nishikawa K, Tanimura Y and Fujiyoshi Y (2009) Mechanism of aquaporin-4's fast and highly selective water conduction and proton exclusion. *Journal of Molecular Biology* **389**:694–706.

Uehlein N, Lovisolo C, Siefritz F and Kaldenhoff R (2003) The tobacco aquaporin NtAQP1 is a membrane CO_2 pore with physiological functions. *Nature* **425**:734–737.

Venero JL, Vizuete ML, Ilundain AA, Machado A, Echevarria M and Cano J (1999) Detailed localization of aquaporin-4 messenger RNA in the CNS: Preferential expression in periventricular organs. *Neuroscience* **94**:239–250.

Verbavatz JM, Ma T, Gobin R and Verkman AS (1997) Absence of orthogonal arrays in kidney, brain and muscle from transgenic knockout mice lacking water channel aquaporin-4. *Journal of Cell Science* **110 (Pt 22)**:2855–2860.

Verkman AS (2011) Aquaporins at a glance. *Journal of Cell Science* **124**:2107–2112.

Wang Y, Cohen J, Boron WF, Schulten K and Tajkhorshid E (2007) Exploring gas permeability of cellular membranes and membrane channels with molecular dynamics. *Journal of Structural Biology* **157**:534–544.

Wu B and Beitz E (2007) Aquaporins with selectivity for unconventional permeants. *Cellular and Molecular Life Sciences* **64**:2413–2421.

Yasui M, Hazama A, Kwon TH, Nielsen S, Guggino WB and Agre P (1999a) Rapid gating and anion permeability of an intracellular aquaporin. *Nature* **402**:184–187.

Yasui M, Kwon TH, Knepper MA, Nielsen S and Agre P (1999b) Aquaporin-6: An intracellular vesicle water channel protein in renal epithelia. *Proceedings of the National Academy of Sciences of the United States of America* **96**:5808–5813.

Zardoya R (2005) Phylogeny and evolution of the major intrinsic protein family. *Biology of the Cell* **97**:397–414.

Zhang R, van Hoek AN, Biwersi J and Verkman AS (1993) A point mutation at cysteine 189 blocks the water permeability of rat kidney water channel CHIP28k. *Biochemistry* **32**:2938–2941.

CHAPTER 4

Regulation of Eukaryotic Aquaporins

Maria Nyblom and Susanna Törnroth-Horsefield

CONTENTS

Abstract		53
4.1	Introduction	54
4.2	AQP Gating	55
	4.2.1 Gating of Plant AQPs	56
	4.2.1.1 PIP Gating Mechanism	56
	4.2.2 Gating of Yeast AQPs	58
	4.2.2.1 Gating Mechanism of Aqy1	59
	4.2.3 Gating of Mammalian AQPs	59
	4.2.3.1 Gating of AQP0	60
	4.2.3.2 Are Other Mammalian AQPs Gated?	61
4.3	AQP Trafficking	62
	4.3.1 Trafficking of Mammalian AQPs	62
	4.3.1.1 Trafficking of AQP2	62
	4.3.1.2 Trafficking of Other Mammalian AQPs	66
	4.3.2 Trafficking of Plant AQPs	67
4.4	Common Structural Themes in Regulated AQPs	68
	4.4.1 AQP Gating Involves a Cytoplasmic Constriction Site	68
	4.4.2 N-Termini Adopt Two Conformations	68
	4.4.3 Binding of Ca^{2+}	70
4.5	Conclusion	70
References		71

ABSTRACT

MEMBRANE-BOUND WATER CHANNELS known as aquaporins (AQPs) facilitate water transport across biological membranes along osmotic gradients. Since all living cells depend on their ability to maintain water homeostasis, this must be tightly regulated. In eukaryotes, this is achieved by gating, which involves a conformational change of the protein, thereby physically blocking water transport, or by trafficking in which AQPs are shuttled between intracellular storage sites and the plasma membrane.

Gating is common amongst plant AQPs in response to environmental stress and has been shown to be triggered by phosphorylation, pH and binding of divalent cations. Gating has been demonstrated for yeast AQPs for which it is believed to confer protection against osmotic shock and rapid freezing. In mammals, AQP regulation is mainly achieved through trafficking. Thirteen AQPs have been identified in humans, the majority of which are regulated by trafficking in response to a wide range of stimuli. The far best character-ized trafficking mechanism is that of AQP2 in the kidney collecting duct where it plays a key role in urine concentration. AQP2 trafficking is controlled by the pituitary hormone vasopressin that stimulates phosphorylation of the AQP2 C-terminus, triggering translo-cation of AQP2 from intracellular storage vesicles to the apical membrane. Defective traf-ficking of human AQPs can lead to several disease states, for example nephrogenic diabetes insipidus (AQP2) and Sjögren's syndrome (AQP5).

In this chapter, we give an overview of what is known about the regulation of eukaryotic AQPs, focusing particularly on structure–function relationships. We discuss the physi-ological role of AQP regulation, specific regulatory mechanisms and reoccurring themes in both gating and trafficking.

4.1 INTRODUCTION

All living cells depend on their ability to maintain water homeostasis. Cell membranes have an intrinsic water permeability that depends on the composition of the lipid bilayer, allowing for the equilibration of osmotic gradients in a matter of seconds to minutes. In many instances, however, a much higher rate of water flow across the membrane is needed. This is achieved through the action of aquaporins (AQPs), membrane-bound water chan-nels that facilitate water flow across membranes along osmotic gradients, allowing for up to 100-fold increase in water transport rate compared to diffusion through the lipid bilayer alone (Haines 1994).

Structural data have revealed that AQPs share a common structure illustrated in Figure 4.1a. AQPs are homotetramers with each protomer comprising six transmembrane helices surrounding a narrow channel through which water is transported in a single file. Loops B and E fold into the membrane from opposite sides, forming two half-membrane spanning helices which each harbours a copy of the AQP signature asparagine–proline–alanine (NPA) motif at their N-terminal ends. Toward the extracellular side, the chan-nel narrows at the aromatic/arginine region (ar/R-region), forming a selectivity filter that allows only water to pass through water-specific AQPs or, in the case of aquaglyceroporins (AQGPs), also permits the transport of other small solutes such as glycerol and urea. The two half helices focus their positive dipole moment at the NPA motifs, creating an electro-static barrier that, together with the water-bonding properties of the ar/R-region, efficiently excludes ion and proton transport (de Groot et al. 2003; Kosinska Eriksson et al. 2013).

An important aspect of AQP-mediated water transport is that, in contrast to water diffu-sion across the lipid bilayer, it can be regulated in response to cellular signals or environmen-tal changes. This is particularly common in higher eukaryotes such as plants and mammals

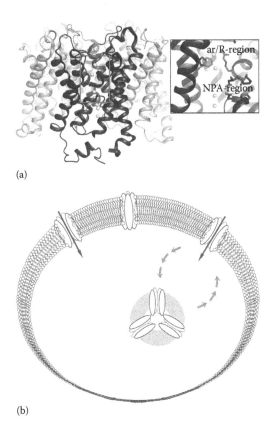

(a)

(b)

FIGURE 4.1 (a) Structure of human AQP5 (PDB entry 3D9S) illustrating the common aquaporin fold. One monomer is shown in black with water molecules in the water-conducting channel shown as light-grey spheres. (b) Schematic of the two aquaporin regulatory mechanisms. Gating is illustrated as an aquaporin in open and closed conformations, whereas trafficking is shown as recycling between a storage vesicle (grey circle) and the cell membrane. Black arrows indicate water transport.

where a tight control of water transport in different tissues is of fundamental physiological importance, but is also seen in microbes where it is believed to confer protection against osmotic shock and rapid freezing (Tornroth-Horsefield et al. 2010). In addition to regulation at the transcriptional/translational level, eukaryotes have evolved to regulate their AQPs post-translationally. Figure 4.1b shows the two main regulatory mechanisms: (1) controlling water transport through individual AQPs *via* conformational changes, so-called gating, or (2) altering the number of AQPs present in a certain membrane, a process known as trafficking. The aim of this chapter is to give an overview of what is currently known about regulation of eukaryotic AQPs by both mechanisms. With a focus on structure–function relationships, we discuss the physiological role of AQP regulation, specific regulatory mechanisms and reoccurring structural themes common for both gating and trafficking.

4.2 AQP GATING

Gating is a common regulatory mechanism for membrane-bound channels. It involves conformational changes that upon a certain trigger allow the channel to switch between an open and a closed state, thereby permitting or blocking transport of molecules across the membrane. Within the AQP family, gating is known to be an important regulatory mechanism for plant AQPs, playing a key role in whole plant cell water balance (Chaumont and Tyerman 2014). Gating also regulates water flow through yeast AQPs and is believed to be important for protection against freezing and osmotic shock (Ahmadpour et al. 2014). Amongst mammalian AQPs, gating has been shown to regulate the water permeability of AQP0 in a pH- and Ca^{2+}-dependent manner (Nemeth-Cahalan et al. 2004) but otherwise remains to be conclusively shown.

4.2.1 Gating of Plant AQPs

Plants depend on a tight regulation of water movement across cellular membranes and tissues. As such, they express an unusually large array of AQP isoforms; the genomes of *Arabidopsis thaliana*, *Zea mays* and *Oryza sativa* contain more than 30 different AQP genes compared to only 13 in humans. In higher plants, these are typically divided into four subgroups, differing in sequence and subcellular localization, which are as follows: plasma membrane intrinsic proteins (PIPs), tonoplast intrinsic proteins (TIPs), NOD26-like intrinsic proteins and small and basic intrinsic proteins (Li et al. 2014).

In addition, land-living plants have evolved to rapidly adapt to fluctuations in water supply through AQP gating. This is best characterized in the PIP isoforms in which gating is triggered by phosphorylation in response to drought (Johansson et al. 1998; Van Wilder et al. 2008), by pH in response to flooding (Tournaire-Roux et al. 2003; Alleva et al. 2006; Verdoucq et al. 2008) and by divalent cations (Alleva et al. 2006; Verdoucq et al. 2008). Phosphorylation and pH have also been shown to trigger gating of members of the TIP subfamily (Maurel et al. 1995; Leitao et al. 2012); however, this does not seem to be as conserved as for the PIPs.

Plant AQPs have also been suggested to be gated in a mechanosensitive manner. Mechanical stimuli in the form of increased pressure were shown to affect water transport in the green algae *Chara corallina* (Ye et al. 2004), young corn roots (Wan et al. 2004) and a tonoplast AQP from grapevine (Leitao et al. 2014). For the spinach AQP SoPIP2;1, an increase in proteoliposome water transport in response to mercury treatment was suggested to be a result of increased membrane stiffness due to mercury binding to the lipids (Frick et al. 2013a). In all these cases, the change in water transport was suggested to occur through a conformational change of the protein.

4.2.1.1 PIP Gating Mechanism

A conserved structural mechanism for PIP gating emerged from the x-ray structure of SoPIP2;1 in open and closed conformations (Tornroth-Horsefield et al. 2006). Figure 4.2a highlights the most important residues involved in this mechanism as outlined in the succeeding text. The key element is a conformational change of intracellular loop D that, in the closed conformation, occludes the channel from the cytoplasmic side through the insertion of Leu197 into the opening. This creates a hydrophobic barrier that hinders water transport through the channel. In the open state, loop D is displaced, removing Leu197, thereby allowing water to pass through the channel. The closed state is stabilized through a

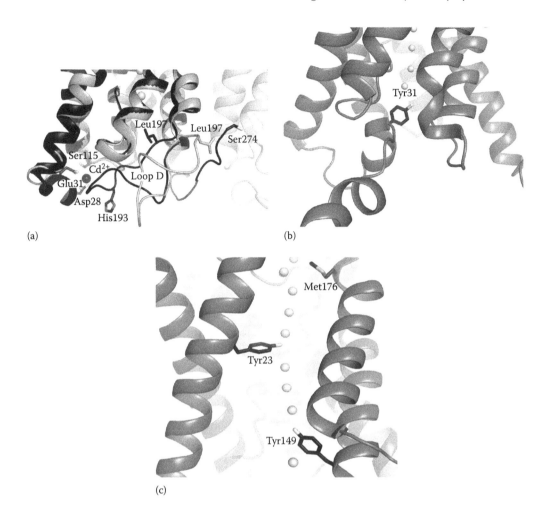

FIGURE 4.2 Structural details of gated aquaporins. Water molecules are shown as light-grey spheres. (a) Overlay of the open (grey, PDB entry 2B5F) and closed (black, PDB entry 1Z98) conformations of SoPIP2;1 illustrating how gating involves a conformational change of loop D. In the closed state, Leu197 (black) plugs the opening, whereas in the open state (grey), it is significantly displaced. Cd^{2+} (dark-grey sphere) in the structure is replaced by Ca^{2+} in vivo. Residues involved in gating by pH and phosphorylation are labelled. (b) Crystal structure of *P. pastoris* Aqy1 (PDB entry 2W2E) showing how the channel entrance is blocked by Tyr31. (c) Crystal structure of bovine AQP0 (PDB entry 1YMG) showing how the side chains of Met179, Tyr23 and Tyr149 constrict the water-conducting channel. At Tyr149, the channel is too narrow to allow water to pass.

network of interactions between loop D, Arg118 in loop B and a divalent cation-binding site at the N-terminus, occupied by Cd^{2+} in the structure but presumed to bind Ca^{2+} in vivo. All the involved residues, including the Cd^{2+} ligands Asp28 and Glu31, Arg118 and the whole of loop D, are fully conserved in all PIPs, suggesting that this is a common gating mechanism.

During conditions of drought, PIPs close due to dephosphorylation of one or two conserved serine residues. This involves Ser115 in loop B (fully conserved in all PIPs) and Ser274 at the C-terminus (conserved in the PIP2 subgroup). In their dephosphorylated form, both serine residues participate in interactions which stabilized the closed conformation of loop D. Phosphorylation breaks these interactions, pushing the equilibrium towards the open channel. Specifically, phosphorylation of Ser115 causes a structural perturbation of the Ca^{2+}-binding site, thus disrupting the network that anchors loop D onto the N-terminus. The second serine, Ser274, interacts with residues at the C-terminal end of loop D in a neighbouring protomer within the tetramer, thus stabilizing the closed state. In the phosphorylated state, these interactions are broken and the C-terminus moves away, thus removing a steric clash that was previously hindering the channel opening (Tornroth-Horsefield et al. 2006; Nyblom et al. 2009).

Gating by pH occurs in response to a rapid drop in cytosolic pH due to anoxia during flooding and has been attributed to protonation of a single histidine in loop D (Tournaire-Roux et al. 2003). Upon protonation, this histidine (His193 in SoPIP2;1) is suggested to flip toward the Ca^{2+}-binding site, making interactions that stabilize the closed state (Tornroth-Horsefield et al. 2006). Mutation of the Ca^{2+} ligands to alanines is known to cause reduced pH sensitivity in PIPs from *A. thaliana*, supporting a link between the Ca^{2+}-binding site and pH-induced gating (Verdoucq et al. 2008). In the absence of Ca^{2+}, the N-terminus which harbours the Ca^{2+}-binding site moves away to a significant distance, preventing an interaction with His193. Instead, the protonated His193 interacts with the backbone of loop B, allowing for an alternative way of stabilizing loop D in the closed state when Ca^{2+} is not present (Frick et al. 2013b).

The ability of Ca^{2+} to inhibit water transport through PIPs was first demonstrated in *A. thaliana* suspension cells (Gerbeau et al. 2002) and later shown to arise from a direct gating effect on the cytoplasmic side of the protein (Alleva et al. 2006). As described earlier, Ca^{2+} binds at the N-terminus and plays a key role in anchoring loop D in the closed conformation (Tornroth-Horsefield et al. 2006). Mutation of the two residues serving as ligands abolishes Ca^{2+}-dependent gating, supporting that this is in fact the site responsible for this effect (Verdoucq et al. 2008). In SoPIP2;1, a second divalent cation-binding site has been identified between loop D and the C-terminus (Frick et al. 2013a). This is also likely a Ca^{2+} site and could play a role in stabilizing the C-terminus in order to favour its interaction with loop D in the neighbouring protomer. The involvement of two Ca^{2+} sites in PIP gating is supported by the bi-phasic dose–response curve seen for Ca^{2+} inhibition of water transport in *Beta vulgaris* storage roots (Alleva et al. 2006). This leads to the conclusion that PIPs contain a high affinity site in the nM range and a lower affinity site in the μM range, likely corresponding to the N-terminal and C-terminal sites, respectively.

4.2.2 Gating of Yeast AQPs

Small unicellular organisms such as yeast have a large surface-to-volume ratio, and in most conditions, water diffusion across the lipid bilayer alone is sufficient for water homeostasis (Tanghe et al. 2006). However, during certain conditions of stress, a faster transmembrane water transport may be needed. Yeast expressing AQPs have higher tolerance against freezing as these facilitate a rapid flow of water out of the cell, thereby preventing intracellular ice formation. Furthermore, in *Saccharomyces cerevisiae*, AQGPs play important roles in the regulation of cellular osmolytes following osmotic shock (Tamas et al. 1999).

Gating as a regulatory mechanism for yeast AQPs was first demonstrated for the AQGP Fps1p from *S. cerevisiae* (Tamas et al. 2003). By reducing glycerol export through Fps1p during hyperosmotic stress, the yeast cells are able to retain glycerol in order to maintain intracellular osmolarity. Gating of Fps1p has been shown to depend on domains in the N-terminus and C-terminus, both of which are significantly longer than their counterparts in AQPs from other organisms. The presence of an extended N-terminus is a characteristic frequently encountered among water-specific AQPs as well as AQGPs in yeast. In addition, phosphorylation of a threonine residue within the N-terminal regulatory domain affects the transport activity of Fps1p (Thorsen et al. 2006). The water-specific Aqy1 from *Pichia pastoris* is also gated, providing increased tolerance to rapid freezing (Fischer et al. 2009). As with Fps1p, gating of Aqy1 involves the extended N-terminus and has been suggested to be regulated by phosphorylation, in this case, of a serine residue in loop B.

4.2.2.1 Gating Mechanism of Aqy1

A mechanism for how yeast AQPs may be gated emerged from the high-resolution structure of Aqy1 from *P. pastoris* (Fischer et al. 2009). The structure revealed that the extended N-terminus forms a tightly wound helical bundle which occludes the channel entrance on the cytoplasmic side through the insertion of a tyrosine residue (Tyr31) (Figure 4.2b). A similar bundle formation can be found in the mechanosensitive ion channel MscL from *Mycobacterium tuberculosis*, for which membrane tension induces channel opening and a rapid efflux of solutes (Chang et al. 1998). Molecular dynamic simulations show that mechanosensitive gating is indeed a plausible regulatory mechanism for Aqy1. This also explains why measurements in spheroplasts give a lower water transport rate than in the smaller proteoliposomes as the higher membrane curvature could trigger channel opening (Fischer et al. 2009). Membrane tension has also been suggested to regulate water transport through Aqy1 and Aqy2 from *S. cerevisiae* (Soveral et al. 2008). The high degree of conservation regarding the presence of an extended N-terminus and key residues therein suggests that this may well be a common regulatory mechanism for yeast AQPs.

In addition to gating by mechanosensitivity, phosphorylation has been suggested to regulate water transport through *P. pastoris* Aqy1. This involves a serine residue in loop B (Ser107) that participates in a hydrogen-bonding network with Tyr31 at the channel entrance. Molecular dynamics simulations have revealed that phosphorylation of this residue causes structural perturbations in this region similar to those induced by mechanical stress, leading to channel opening. It thus seems that phosphorylation and membrane-mediated mechanical stress trigger the same opening mechanism, where mechanosensitive gating offers a release valve for sudden osmotic changes, while phosphorylation may fine-tune water transport under more normal conditions (Fischer et al. 2009).

4.2.3 Gating of Mammalian AQPs

In contrast to plant AQPs for which gating is well established, gating as a general regulatory mechanism of mammalian AQPs is a controversial issue. Several mammalian AQPs have been suggested to be gated, including AQP0 (Nemeth-Cahalan et al. 2004), AQP2 (Eto et al. 2010), AQP4 (Zelenina et al. 2002; Gunnarson et al. 2008; Song and Gunnarson 2012) and AQP5 (Janosi and Ceccarelli 2013). However, the only mammalian AQP for which gating has been convincingly demonstrated is AQP0, the main AQP in lens fiber cells.

4.2.3.1 Gating of AQP0

AQP0 is exclusively found in the plasma membrane of eye lens fiber cells where it serves a dual function; it regulates water permeability across the fiber cell plasma membrane and it promotes cell-to-cell adhesion by participating in gap junctions. To keep AQP0 relatively insensitive to rapid osmotic changes, for example from tears, it has evolved to become a poor water transporter with a water permeability 15–45 times lower than that of AQP1. Instead, the fiber cell compensates by expressing a very large amount of AQP0 channels (>60% of total membrane proteins), ensuring a uniform response to osmotic changes throughout the lens (Harries et al. 2004). AQP0 has been proposed to close upon junction formation following proteolytic cleavage of the C-terminus (Gonen et al. 2005). In addition, Ca^{2+} and pH affect the water permeability of AQP0, leading to a twofold–threefold increase in water transport rate (Nemeth-Cahalan et al. 2004). In all these cases, the modulating effect has been suggested to occur *via* direct gating of the channel.

Structural studies of sheep and bovine AQP0 show that the water-conducting channel is narrower than in other mammalian AQPs (Gonen et al. 2004, 2005; Harries et al. 2004). As seen in Figure 4.2c, two tyrosine side chains in particular restrict the channel diameter: Tyr23, which break the continuous water chain seen in other AQP structures, and Tyr149, which creates an additional constriction site at the cytoplasmic channel entrance. This constriction site is at the same position along the channel as those found in gated plant and yeast AQPs (Tornroth-Horsefield et al. 2006; Fischer et al. 2009). In its junction form, conformational changes of side chains lining the AQP0 water-conducting channel cause this to narrow even further, most notably at Met176 located towards to extracellular side (Gonen et al. 2005) (Figure 4.2c). How these conformational changes are relayed by protein–protein interactions during junction formation remains to be shown.

Several studies have shown that a decrease in external pH increases the water transport through AQP0, with a doubled transport rate at pH 6.5 compared to pH 10.5 (Nemeth-Cahalan et al. 2004). The pH sensitivity has been suggested to involve His40 in extracellular loop A as well as Tyr149 of the cytoplasmic constriction site. However, AQP0 structures determined at pH 6 and pH 10.5 show no conformational difference for these or any other residue that might play a role in pH gating. The structural mechanism behind pH gating of AQP0 is therefore still unknown. It has been suggested that this occurs through a cooperative mechanism where allosteric changes within the entire tetramer is responsible for the increased water transport of AQP0 at lower pH (Nemeth-Cahalan et al. 2013).

Regulation of AQP0 water transport by Ca^{2+} is mediated through an interaction between calmodulin (CaM) in its Ca^{2+}-bound form and the AQP0 C-terminus (Reichow and Gonen 2008). This interaction occurs at the cytoplasmic side of the membrane where one CaM molecule binds two AQP0 C-terminal helices simultaneously, resulting in channel closure (Reichow et al. 2013). Rather than sterically hindering water transport itself, CaM allosterically modulates AQP0 water permeability by stabilizing the closed state at the cytoplasmic constriction site, with Tyr149 serving as a dynamic channel gate. The affinity between CaM and AQP0 has been shown to depend on the phosphorylation status of serine residues in the C-terminus of AQP0, suggesting that phosphorylation might play a role in CaM-mediated regulation (Reichow and Gonen 2008; Rose et al. 2008).

4.2.3.2 Are Other Mammalian AQPs Gated?

While there are convincing evidence that supports gating of AQP0, gating as a regulatory mechanism of other mammalian AQPs is still under debate. In rat AQP4, the main water channel in the brain, phosphorylation of a Ser180 close to the cytoplasmic entrance of the channel has been suggested to trigger channel closure by interacting with the C-terminus (Zelenina et al. 2002). However, mutating this serine to aspartate in order to mimic phosphorylation does not induce any structural changes or confer any difference in water transport rate when compared to the wild-type AQP4 (Mitsuma et al. 2010). Molecular dynamics simulations further indicate that, although an interaction between phosphorylated Ser180 and the C-terminus can occur, this does not cause channel closure (Sachdeva and Singh 2014). Gating of AQP4 has also been suggested to be triggered by phosphorylation of Ser111 in loop B, leading to opening of the channel (Gunnarson et al. 2008; Song and Gunnarson 2012). Again, structural information disputes this, as the crystal structure of rat AQP4 is clearly open despite the lack of phosphorylation of this residue (Ho et al. 2009). Moreover, neither functional nor molecular dynamics simulation studies show a difference in AQP4 water permeability upon phosphorylation of Ser111 (Assentoft et al. 2013).

Direct gating has also been proposed to regulate AQP2 found in the kidney collecting duct. AQP2 plays a key role in urine concentration, and its regulation by trafficking in response to vasopressin is well characterized (see Section 4.3.1). A crucial event in AQP2 trafficking is the phosphorylation of Ser256 at the C-terminus. Interestingly, phosphorylation of this residue has also been shown to enhance the water permeability of AQP2 in proteoliposomes, implying a role in channel gating (Eto et al. 2010). Functional studies in other systems contradict this, showing no effect on AQP2 water permeability upon phosphorylation of Ser256 or any other known phosphorylation site (Lande et al. 1996; Kamsteeg et al. 2000; Moeller et al. 2009b). Molecular dynamics simulations have further indicated that human AQP5 could be gated at a cytoplasmic constriction site, possibly involving phosphorylation as a trigger (Janosi and Ceccarelli 2013). In the absence of supporting structural and functional data, this, however, remains highly speculative.

Similarly as for plant AQPs, studies show that water transport through mammalian AQPs may be affected by mechanical stimuli, an effect that could be important for cell volume regulation. A combined experimental and modelling approach showed that oocytes expressing human AQP1 have reduced water permeability when membrane tension was increased (Ozu et al. 2013). The membrane properties have also been shown to affect human AQP4, increased membrane stiffness through the addition of cholesterol and reduced water transport (Tong et al. 2012). More studies are needed to establish whether this indeed involves gating of mammalian AQPs in a mechanosensitive manner.

All the studies mentioned earlier concern gating of water-specific AQPs. Amongst mammalian AQGPs, gating as a regulatory mechanism has only been suggested for AQP3. Transport through human AQP3 is reduced by external acidification (Zeuthen and

Klaerke 1999; Zelenina et al. 2003) as well as divalent metal ions such as Ni^{2+} and Cu^{2+} (Zelenina et al. 2003, 2004). Whether this reflects gating in a physiologically relevant manner remains to be determined. To summarize, there is not yet enough evidence to support gating as a general mechanism of mammalian AQPs.

4.3 AQP TRAFFICKING

The ability to modify the expression of membrane proteins at the cell surface is an important property of eukaryotic cells. It involves the exocytotic translocation of proteins from intracellular compartments to cell surface membranes on a given signal. When no longer needed, protein molecules are again internalized, stored for another round of redistribution, or targeted for degradation. This is known as trafficking and has been shown to regulate membrane proteins involved in many essential physiological processes, including nerve signal transmission (AMPAR, NMDAR, GABAAR), glucose uptake in response to insulin (GLUT4), secretion of gastric acid (H^+/K^+ ATPase) and urine concentration (AQP2) (Chieregatti and Meldolesi 2005).

Trafficking is an important regulatory mechanism of AQPs. By altering AQP abundance, cells are able to rapidly modify the water permeability of specific cellular membranes in a matter of minutes. This is particularly common in mammals where AQP trafficking is essential for controlling water homeostasis in several tissues (Conner et al. 2013). Trafficking of AQPs has also been reported to occur in plants, where it has been shown to provide protection against abiotic as well as biotic stress (Luu and Maurel 2013).

4.3.1 Trafficking of Mammalian AQPs

In mammals, 13 different AQP isoforms have been identified, which are expressed in a wide range of tissues in a tissue-dependent manner. A majority of these are regulated by trafficking in response to distinct cellular signals. Dysfunctional AQP trafficking plays a role in several human disease states such as nephrogenic diabetes insipidus (NDI) (Moeller et al. 2013), Sjögren's syndrome (Horsefield et al. 2008) and cholestatic liver disease (Lehmann et al. 2008).

Although the specific trigger varies, trafficking of mammalian AQPs shares several features suggesting that there might well be a common mechanism for translocation to the plasma membrane. Such a mechanism could involve the following: a cellular signal such as a hormone or change in tonicity triggers a signalling cascade that most often results in kinase-mediated phosphorylation of specific residues in the AQP molecule. This causes AQP-containing vesicles to move along the microtubule network and fuse with the plasma membrane (Conner et al. 2013). Table 4.1 summarizes what is currently known regarding the triggers and mediators that are involved in trafficking of mammalian AQPs.

4.3.1.1 Trafficking of AQP2

The by far best characterized AQP in terms of its regulation by trafficking is AQP2, the key ingredients of which is summarized in Figure 4.3a. AQP2 is found in the principal cells of the kidney collecting duct where it is essential for concentration of urine. By regulating the amount of AQP2 present in the apical membrane, the body is able to fine-tune the urine

TABLE 4.1 Trafficking of Mammalian Aquaporins

AQP	Trigger	Mediator	Residue	PM	Experimental System	References
AQP0	–	PKC	S235	+	RK3 epithelial cells	Golestaneh et al. (2008)
AQP1	Secretin	cAMP		+	Cholangiocytes	Marinelli et al. (1999)
	Hypotonicity	PKC	T157/T239	+	HEK293	Conner et al. (2012)
	Hypertonicity	–		+	Human neutrophils	Loitto et al. (2007)
AQP2	Vasopressin	PKA	S256	+	Kidney tissue	Katsura et al. (1997)
	Hypertonicity	–		+	Collecting duct cells	Hasler et al. (2005)
AQP3	Adrenaline	–		+	Caco-2 cells	Yasui et al. (2008)
	Isoprenaline	–		+	Adipocytes	Rodriguez et al. (2011)
	Hypotonicity	–		+	Keratinocytes	Garcia et al. (2011)
	Hypertonicity	–		+	MDCK cells	Matsuzaki et al. (2001)
AQP4	Histamine	PKA		–	Human gastric cells	Carmosino et al. (2007)
	Vasopressin	PKC	S180	–	Xenopus oocytes	Moeller et al. (2009a)
	Glutamate	PKC	S111	+	Human astrocytes	Gunnarson et al. (2008)
	Hypertonicity	–		+	Rat astrocytes	Arima et al. (2003)
AQP5	Adrenaline	–		+	Rat parotid cells	Ishikawa et al. (1999)
	Acetylcholine	PKG		+	Rat parotid cells	Ishikawa et al. (2002)
	Vasoactive intestinal polypeptide	PKA		+	Duodenum	Parvin et al. (2005)
	Lipopolysaccharide	–		+	Lung epithelial cells	Ohinata et al. (2005)
	Hypertonicity	–		+	Lung epithelial cells	Hoffert et al. (2008)
	–	PKA	S152	–	MDCK cells	Karabasil et al. (2009)
	–	PKA		+	Lung epithelial cells	Yang et al. (2003)
AQP6	–	PKA		no	MDCK cells	Beitz et al. (2006)
AQP7	Isoprenaline	–		+	Human adipocytes	Rodriguez et al. (2011)
AQP8	–	cAMP		+	Hepatocytes	Garcia et al. (2011)
	Hypotonicity	–		+	Amnion epithelial cells	Qi et al. (2009)
AQP9	Hypertonicity	–		+	Rat astrocytes	Arima et al. (2003)
	–	PKC	S11	+	Human neutrophils	Karlsson et al. (2011)

The table summarizes known triggers and mediators for trafficking of mammalian AQPs. Specific residues are indicated if known. An increase and decrease in plasma membrane localization is indicated with + and –, respectively.

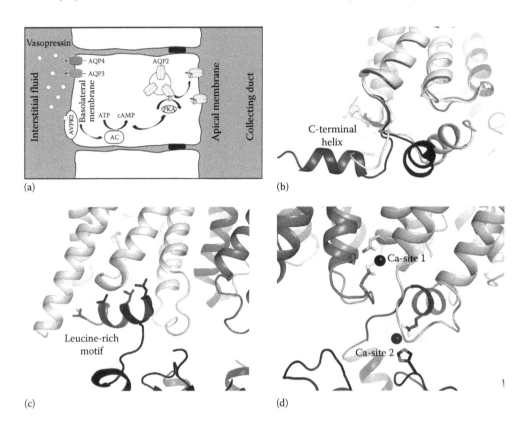

FIGURE 4.3 Structural insights into AQP2 trafficking. (a) A schematic diagram describing how vasopressin mediates trafficking of AQP2 to the apical membrane of the collecting duct principal cells. Binding of vasopressin (small grey spheres) to the vasopressin receptor (AVPR2) in the basolateral membrane causes an increase in the production of cAMP by adenylyl cyclase that stimulates phosphorylation of AQP2 by PKA. This triggers the fusion of AQP2-containing intracellular storage vesicles with the apical membrane, leading to an increased water uptake from urine. The absorbed water leaves the cell via AQP3 and AQP4 in the basolateral membrane. (b) Overlay of four monomers from the crystal structure of human AQP2 (PDB entry 4NEF) showing the conformational flexibility of the C-terminal helix. (c) Interactions within an AQP2 crystal illustrate how a leucine-rich motif at the C-terminal helix binds in the cytoplasmic vestibule of another AQP2 molecule. This interaction mimics a physiological interaction between AQP2 and the regulatory protein LIP5. (d) Two Cd^{2+} ions (dark grey spheres) are found per tetramer in the AQP2 structure representing Ca^{2+} sites in vivo (Ca site 1 and Ca site 2). Ligating residues are highlighted in stick representation.

volume output to meet current demands. This regulation is tightly controlled by the antidiuretic hormone arginine vasopressin (AVP). Upon dehydration, AVP is released from the pituitary gland and binds the vasopressin receptor (V2R) in the basolateral membrane. This causes an increase in intracellular cAMP, leading to PKA-mediated phosphorylation of the AQP2 C-terminus and the subsequent fusion of AQP2-containing storage vesicles with the apical membrane. As a result, the water permeability of the apical membrane is increased and water is reabsorbed from the urine. When the water balance has been restored, AQP2 is internalized and stored for another round of vasopressin-mediated shuttling to the apical membrane or targeted for degradation (van Balkom et al. 2002). Failure to recruit AQP2 to the apical membrane leads to NDI, a water balance disorder in which patients lack the ability to concentrate urine. This results in urine volumes up to 20 L/day and can lead to severe dehydration. Congenital NDI occurs as a result of mutations in either the vasopressin receptor or AQP2 itself. More than 50 NDI-causing mutations have been identified in AQP2, most of which cause endoplasmic reticulum (ER) retention (Moeller et al. 2013).

Multiple post-translational modification sites at the C-terminus govern AQP2 trafficking. The most important one is Ser256, the phosphorylation of which is critical for targeting to the apical membrane (Katsura et al. 1997). Additional phosphorylation at Ser264 and 269 is also triggered by AVP, and all three sites are phosphorylated in AQP2 found in the apical membrane (Fenton et al. 2008; Hoffert et al. 2008). Further post-translational modifications through ubiquitination of Lys270 is involved in targeting AQP2 for internalization and degradation (Kamsteeg et al. 2006). Another important factor involved in AQP2 trafficking is protein–protein interactions. A number of proteins have been shown to interact with AQP2 on its way to and from the apical membrane, including Spa1, actin, heat shock protein 70 (Hsc70) and the lysosomal trafficking regulator interacting protein LIP5. In most cases, this interaction has been suggested to be mediated by the AQP2 C-terminus (Frick et al. 2014).

4.3.1.1.1 Structural Insights into AQP2 Trafficking   Recently, the crystal structure of human AQP2 gave the first structural insights into how AQP2 is regulated by trafficking (Frick et al. 2014). The structure revealed a unique flexibility of the C-terminus with a short C-terminal helix adopting four different conformations within the tetramer (Figure 4.3b). Neither of these conformations has been observed previously in mammalian AQPs in which this C-terminal helix always occupies the same position along the cytoplasmic interface. The flexibility arises from a hinge region formed by two consecutive prolines (only one proline is present in other mammalian AQPs) and is likely to be important for AQP2 trafficking. Figure 4.3c illustrates how in one protomer the C-terminal helix interacts with a symmetry-related AQP2 molecule within the crystal through a hydrophobic motif consisting of four leucine residues. This interaction likely mimics the protein–protein interactions during AQP2 trafficking, specifically the interaction between AQP2 and LIP5, a lysosomal protein known to bind proteins with leucine-rich regions (van Balkom et al. 2009; Skalicky et al. 2012). LIP5 is involved in targeting AQP2 to multivesicular bodies, leading to subsequent degradation. In AQP0, the analogous motif interacts with calmodulin (Reichow et al. 2013), suggesting that exposed hydrophobic residues on the C-terminal helix are a common motif for protein–protein interactions in mammalian AQPs.

A surprising finding in the structure of AQP2 was the presence of two divalent cations per tetramer. The positions of these ions are shown in Figure 4.3d. As with the plant AQP SoPIP2;1, these were occupied by Cd^{2+} in the structure but are most likely Ca^{2+} in vivo (Tornroth-Horsefield et al. 2006). Both Ca^{2+}-binding sites are located on the cytoplasmic side and seem to be important for protein–protein interactions involving the C-terminus. The first Ca^{2+} binds between two protomers and causes rearrangement of loop D, thereby creating binding pocket in which the C-terminal helix from the symmetry-related molecule binds. The second Ca^{2+} binds at the C-terminus, positioning the C-terminal helix for this interaction. A direct role for Ca^{2+} in AQP2 trafficking is an unexplored topic. It is, however, known that binding of AVP to the vasopressin receptor does not only cause an increase in cAMP but also in intracellular Ca^{2+} concentration, and indications that Ca^{2+} release from ryanodine-sensitive stores are important for AQP2 targeting to the apical membrane (Noda and Sasaki 2006).

4.3.1.2 Trafficking of Other Mammalian AQPs

In comparison to AQP2, trafficking of other mammalian AQPs is less well known; however, the knowledge is constantly increasing. The current status of the field is summarized in Table 4.1 with the major points highlighted in the succeeding text.

In cholangiocytes, trafficking of AQP1 can be triggered by secretin, a hormone that stimulates ductal bile formation. Similarly as for AQP2, secretin binds its receptor in the basolateral membrane, causing an increase in intracellular cAMP. This triggers AQP1 translocation to the apical membrane, resulting in a higher water permeability and increased bile secretion (Marinelli et al. 1999). AQP1 can also be rapidly translocated to the plasma membrane in response to hypotonicity. This is dependent on PKC phosphorylation as well as Ca^{2+} influx through transient receptor potential channels and is believed to be important for regulatory volume decrease following cell swelling (Conner et al. 2012).

Hypotonicity also affects the cellular distribution of AQP8 in amniotic epithelial cells, suggesting a role in intramembranous water transport and balance in the amniotic fluid (Qi et al. 2009). In hepatocytes, AQP8 is mainly stored intracellularly but can target the plasma membrane in response to elevated cAMP levels. Interestingly, AQP8 lacks consensus sites for cAMP-dependent kinases, suggesting an alternative route for cAMP-mediated regulation (Koyama et al. 1998).

In brain astrocytes, membrane localization of AQP4 is a prerequisite for the formation of brain edema following stroke or traumatic brain injury (Papadopoulos and Verkman 2007). Hence, controlling AQP4 trafficking is an attractive route for treating this severe disease state. Both histamine and vasopressin stimulate the removal of AQP4 from the plasma membrane by increasing its internalization rate (Carmosino et al. 2007; Moeller et al. 2009a). In the case of vasopressin, internalization is triggered by phosphorylation of Ser180 by PKC, while phosphorylation of a second serine at position 111 seems to counteract this. The same phosphorylation sites were suggested to be involved in gating of AQP4 as discussed in Section 4.2.3.2; however, this has later been refuted.

Another example of a mammalian AQP regulated by trafficking is AQP5, mainly expressed in secretory glands, lungs and airways. Dysfunctional trafficking of AQP5 leads to Sjögren's syndrome, a disease manifesting itself as dry eyes and mouth. Several different molecules have been linked to AQP5 trafficking, including acetylcholine and adrenaline (Ishikawa et al. 1998). It has further been shown that AQP5 can be translocated to the plasma membrane in a PKA-dependent manner (Yang et al. 2003; Parvin et al. 2005), and a

PKA consensus site in cytoplasmic loop D has been shown to be phosphorylated in human bronchial epithelial cells in vivo (Woo et al. 2008). However, whether AQP5 trafficking is directly regulated by phosphorylation of this or any of the other PKA-sites present in the C-terminus remains to be unambiguously demonstrated.

Dysfunctional regulation of AQP0 in the eye lens has been linked to the formation of cataract. As described in Section 4.2.3.1, AQP0 is regulated by gating. In addition, putative phosphorylation sites at the C-terminus has been implicated in targeting of AQP0 to the plasma membrane (Ball et al. 2003). In particular, phosphorylation of Ser235 by PKC is proposed to induce AQP0 translocation (Golestaneh et al. 2008). This is particularly interesting as phosphorylation of this residue reduces the binding affinity of CaM, suggesting a role in gating. It thus seems that phosphorylation of Ser235 may serve a dual purpose in increasing water permeability of lens plasma membrane by altering the water transport rate of individual water channels as well as their abundance in the membrane.

Trafficking also regulates the mammalian AQGPs AQP3, AQP7 and AQP9, which are involved in lipid and glycerol metabolism. The adrenergic receptor agonist isoprenaline stimulates the translocation of both AQP3 and AQP7 in adipocytes (Rodriguez et al. 2011). A similar effect is seen in a colon epithelial cell line (Caco-2) where adrenaline triggers plasma membrane localization of AQP3 involving a PKC-dependent pathway (Yasui et al. 2008). PKC-mediated phosphorylation has been suggested to regulate basolateral membrane targeting as well as internalization of AQP3, using two distinct sites. PKC has also been proposed to phosphorylate AQP9, causing its translocation to the plasma membrane in response to a hyperosmotic trigger (Loitto et al. 2007; Karlsson et al. 2011).

Finally, AQP6 is an interesting exception as it is primarily located within the cell where it mainly transports anions (Yasui et al. 1999). Continuous surface expression of AQP6 led to apoptosis wherefore studies have focused on the mechanism behind its intracellular retention. For this purpose, a sequence at the N-terminus has been shown to be crucial as well as sufficient for maintaining AQP6 within the cytoplasm (Beitz et al. 2006).

4.3.2 Trafficking of Plant AQPs

The first evidence for stimuli-induced trafficking of plant AQPs came from studies of the ice plant *Mesembryanthemum crystallinum* (Vera-Estrella et al. 2004). It was shown that osmotic stress through the hypertonic treatment with mannitol induced the redistribution of McTIP1;2 from the vacuole to intracellular vesicles. This redistribution was dependent on protein glycosylation. The subcellular localization of plant AQPs is also altered upon salt stress. In roots from *A. thaliana*, high salt exposure causes an increased internalization of PIPs from the plasma membrane leading to lower membrane water permeability (Boursiac et al. 2008). The same effect was seen when the cells were exposed to salicylic acid (SA), a signalling molecule involved in plant defense and the response against various abiotic stresses. PIP internalization could also be induced by the exogenous application of H_2O_2, while treatment with catalase as a reactive oxygen species (ROS) scavenger abolished both the redistribution induced by both salt and SA. This strongly suggests that stress-induced trafficking of PIPs is mediated by ROS. The subcellular localization of AtPIP2;1 is further affected by phosphorylation of a serine residue in the C-terminus. Phosphorylation of this residue has been shown to be necessary for correct plasma membrane targeting of AtPIP2;1 under normal conditions as well as its intracellular accumulation under salt stress (Prak et al. 2008).

Plasma membrane expression of PIPs can also be controlled through oligomerization between isoforms. Studies in maize have shown that members of the PIP1 subgroup must interact with PIP2s and form mixed tetramers in order to be targeted to the plasma membrane. Such PIP1-PIP2 tetramers have been identified in maize as well as in *Mimosa pudica* (Zelazny et al. 2007). The formation of mixed tetramers also affects their water permeability. When expressed in oocytes, the water transport rate of mixed PIP1–PIP2 tetramers is lower than that of PIP2 homotetramers (Fetter et al. 2004). This provides the cells with an alternative way of regulating plasma membrane water permeability by forming different PIP1–PIP2 combinations in specific cell types and/or in response to environmental stimuli. However, further experiments are needed to elucidate when and how mixed tetramers are formed as well as the physiological relevance for such a regulatory mechanism.

4.4 COMMON STRUCTURAL THEMES IN REGULATED AQPs

Structural data from eukaryotic AQPs show that in addition to the conserved AQP fold and shared passive mechanism for water transport, there are also common structural themes governing their regulation. This does not only include AQPs regulated by the same mechanism but also the reoccurrence of structural principles amongst AQPs regulated by both gating and trafficking. These are further discussed in the succeeding text.

4.4.1 AQP Gating Involves a Cytoplasmic Constriction Site

AQP gating involves the formation of a second constriction site at the cytoplasmic channel entrance, considerably narrower than the ar/R-region, thereby efficiently blocking water transport. Channel closure may be achieved by larger conformational changes as in SoPIP2;1 and Aqy1 or small side-chain movements as in AQP0 (Gonen et al. 2004; Harries et al. 2004; Tornroth-Horsefield et al. 2006; Fischer et al. 2009). In all cases, a specific residue is inserted into the channel, tyrosine in the case of Aqy1 and AQP0 and leucine in the case of SoPIP2;1. Figure 4.4a shows how these residues, despite being well separated in sequence (N-terminus for Aqy1, loop B for AQP0 and loop D for SoPIP2;1), overlap perfectly structurally. This is an elegant example of convergent evolution, whereby AQPs from three major eukaryotic divisions (plants, animals and yeast) have independently evolved to gate water transport through a structurally conserved mechanism.

4.4.2 N-Termini Adopt Two Conformations

Structural data for eukaryotic AQPs have shown that the N-terminus adopts two different conformations: (1) a short cytoplasmic N-terminal helix (as in SoPIP2;1) or (2) a full-turn extension of TM helix 1 into the cytoplasm (as in AQP1) (Sui et al. 2001; Tornroth-Horsefield et al. 2006). These two conformations are illustrated in Figure 4.4b. The first conformation allows for an interaction to form between a glutamate at the N-terminus and a serine and arginine in loop B, all of which are conserved in eukaryotic AQPs. In SoPIP2;1, the interaction between Glu31, Ser115 and Arg118 plays a key role in the gating mechanism

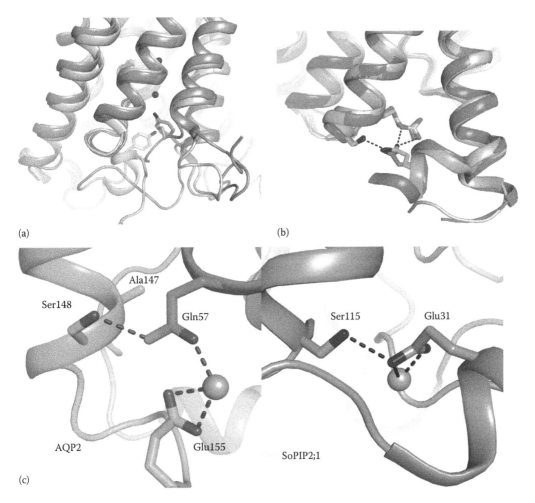

(a)

(b)

(c)

AQP2   Ala147   Ser148   Gln57   Ser115   Glu31   Glu155   SoPIP2;1

FIGURE 4.4 **(See colour insert.)** Common themes in aquaporin regulation. (a) Overlay of SoPIP2;1 (green, PDB entry 1Z98), yeast Aqy1 (yellow, PDB entry 2W2E) and bovine AQP0 (PDB entry 1YMG) showing how the position of the residue blocking the pore during gating is structurally conserved. (b) Overlay of bovine AQP1 (grey, PDB entry 1J4N), human AQP5 (blue, PDB entry 3D9S) and monomers A (light purple) and D (dark purple) of human AQP2 (PDB entry 4NEF) illustrating the two N-terminal conformations seen in aquaporins from higher eukaryotes. The glutamate–serine–arginine interaction motif found in AQP5 and AQP2 monomer A is highlighted. Hydrogen bonds are illustrated by dotted lines. (c) Comparison between Ca^{2+}-binding sites in AQP2 and SoPIP2;1. Cd^{2+} ions found in the two structures are shown as grey spheres and represents Ca^{2+} ions in vivo. In both AQP2 and SoPIP2;1, there is a hydrogen bond between a serine residue situated in a phosphorylation site and the ion ligand. In AQP2, Glu155 (a second ion ligand) as well as Ala147, a known nephrogenic diabetes insipidus–causing mutation site, is also highlighted.

by anchoring loop D in the closed conformation (Figure 4.2a). Phosphorylation of Ser115 breaks this interaction, causing the N-terminus to extend into the cytoplasm, thereby adopting the same conformation as in the constitutively open AQP1 (Nyblom et al. 2009). It thus seems that the gating mechanism of SoPIP2;1 involves a switch between the two N-terminal conformations.

The glutamate–serine–arginine motif is also present in AQPs that are regulated by trafficking (Figure 4.4b). In human AQP5, this triad of residues further interacts with the C-terminal helix, stabilizing its position across the cytoplasmic interface (Horsefield et al. 2008). It has been suggested that breaking this interaction could induce C-terminal structural changes that flag AQP5 for trafficking. Interestingly, the structure of human AQP2 displays both N-terminal conformations: the short N-terminal helix and glutamate–serine–arginine interaction in protomer A and the extension of TM1 in protomer D (Frick et al. 2014). The N-terminus of the remaining protomers is disordered beyond residue 5 and 6, respectively. This suggests that an N-terminal conformational switch could be part of the structural mechanism behind AQP2 trafficking. Indeed, the N-terminus is known to be important for vasopressin-mediated trafficking of AQP2. Taken together, the reoccurrence of these two N-terminal conformations amongst eukaryotic AQPs regulated by both gating and trafficking implies a common structural principle governing their regulation.

4.4.3 Binding of Ca^{2+}

An unexpected similarity between the structures of SoPIP2;1 and human AQP2 was the fact that they both contained calcium-binding sites (Tornroth-Horsefield et al. 2006; Frick et al. 2014). Whereas the role of Ca^{2+} in plant AQP gating is well established (see Section 4.2.1), its function in AQP2 trafficking remains speculative. Nevertheless, a comparison between SoPIP2;1 and AQP2 calcium-binding sites reveals common features which may have functional implications.

AQP2 contains two Ca^{2+}-binding sites (Figure 4.3d), one of which is located between protomers A and D, in close proximity to a casein kinase II consensus site at Ser148 and with a hydrogen bond formed between Ser148 and the Ca^{2+} ligand Gln58. As seen in Figure 4.4c, this is highly reminiscent of SoPIP2;1 where Ser115, a key residue in gating by phosphorylation, hydrogen bonds to the Ca^{2+} ligand Glu31. Upon phosphorylation, the interaction between Ser115 and Glu31 is broken, and the Ca^{2+}-binding site is perturbed, causing the channel to open. Should Ser148 of AQP2 be phosphorylated, this may also disrupt the Ca^{2+}-binding site, leading to further structural changes within the protein. It must be said that there is no evidence for phosphorylation of this site; however, mutation of Ser148 to aspartate in order to mimic phosphorylation causes ER retention of AQP2 in oocytes (van Balkom et al. 2002). Furthermore, mutation of the Ca^{2+} ligand Gln58 as well as Ala147, immediately adjacent to this site, has been identified in patients suffering from NDI (Mulders et al. 1997; Lin et al. 2002). This suggests that this region of the protein is particularly sensitive for structural disturbances, possibly because of the loss of the Ca^{2+}-binding site.

4.5 CONCLUSION

To summarize, since their discovery more than 20 years ago, a wealth of biochemical and structural information regarding AQP regulation has emerged. The aim of this chapter has been to summarize this knowledge and to bring the pieces together for a more complete understanding of these important cellular processes. Although many aspects of AQP regulation remain unknown, inconclusive or speculative, we begin to understand the underlying structural and cellular principles regarding how post-translational modifications govern gating as well as trafficking. One of the major challenges in this field is now to take a step away from individual AQPs to more complex scenarios, for example by studying their interactions with other proteins, structurally as well as functionally. A deep understanding of the underlying structural mechanisms for AQP regulation, whether it is by gating or trafficking, is a crucial step towards designing compounds that modulate AQP activity with the aim to use these as therapeutic agents.

REFERENCES

Ahmadpour, D., C. Geijer, M. J. Tamas, K. Lindkvist-Petersson and S. Hohmann. 2014. Yeast reveals unexpected roles and regulatory features of aquaporins and aquaglyceroporins. *Biochim Biophys Acta* 1840(5):1482–1491.

Alleva, K., C. M. Niemietz, M. Sutka, C. Maurel, M. Parisi, S. D. Tyerman and G. Amodeo. 2006. Plasma membrane of *Beta vulgaris* storage root shows high water channel activity regulated by cytoplasmic pH and a dual range of calcium concentrations. *J Exp Bot* 57(3):609–621.

Arima, H., N. Yamamoto, K. Sobue, F. Umenishi, T. Tada, H. Katsuya and K. Asai. 2003. Hyperosmolar mannitol simulates expression of aquaporins 4 and 9 through a p38 mitogen-activated protein kinase-dependent pathway in rat astrocytes. *J Biol Chem* 278(45):44525–44534.

Assentoft, M., S. Kaptan, R. A. Fenton, S. Z. Hua, B. L. de Groot and N. MacAulay. 2013. Phosphorylation of rat aquaporin-4 at Ser(111) is not required for channel gating. *Glia* 61(7):1101–1112.

Ball, L. E., M. Little, M. W. Nowak, D. L. Garland, R. K. Crouch and K. L. Schey. 2003. Water permeability of C-terminally truncated aquaporin 0(AQP0 1–243) observed in the aging human lens. *Invest Ophthalmol Vis Sci* 44(11):4820–4828.

Beitz, E., K. Liu, M. Ikeda, W. B. Guggino, P. Agre and M. Yasui. 2006. Determinants of AQP6 trafficking to intracellular sites versus the plasma membrane in transfected mammalian cells. *Biol Cell* 98(2):101–109.

Boursiac, Y., J. Boudet, O. Postaire, D. T. Luu, C. Tournaire-Roux and C. Maurel. 2008. Stimulus-induced downregulation of root water transport involves reactive oxygen species-activated cell signalling and plasma membrane intrinsic protein internalization. *Plant J* 56(2):207–218.

Carmosino, M., G. Procino, G. Tamma, R. Mannucci, M. Svelto and G. Valenti. 2007. Trafficking and phosphorylation dynamics of AQP4 in histamine-treated human gastric cells. *Biol Cell* 99(1):25–36.

Chang, G., R. H. Spencer, A. T. Lee, M. T. Barclay and D. C. Rees. 1998. Structure of the MscL homolog from *Mycobacterium tuberculosis*: A gated mechanosensitive ion channel. *Science* 282(5397):2220–2226.

Chaumont, F. and S. D. Tyerman. 2014. Aquaporins: Highly regulated channels controlling plant water relations. *Plant Physiol* 164(4):1600–1618.

Chieregatti, E. and J. Meldolesi. 2005. Regulated exocytosis: New organelles for non-secretory purposes. *Nat Rev Mol Cell Biol* 6(2):181–187.

Conner, A. C., R. M. Bill and M. T. Conner. 2013. An emerging consensus on aquaporin translocation as a regulatory mechanism. *Mol Membr Biol* 30(1):1–12.

Conner, M. T., A. C. Conner, C. E. Bland, L. H. Taylor, J. E. Brown, H. R. Parri and R. M. Bill. 2012. Rapid aquaporin translocation regulates cellular water flow: Mechanism of hypotonicity-induced subcellular localization of aquaporin 1 water channel. *J Biol Chem* 287(14):11516–11525.

de Groot, B. L., T. Frigato, V. Helms and H. Grubmuller. 2003. The mechanism of proton exclusion in the aquaporin-1 water channel. *J Mol Biol* 333(2):279–293.

Eto, K., Y. Noda, S. Horikawa, S. Uchida and S. Sasaki. 2010. Phosphorylation of aquaporin-2 regulates its water permeability. *J Biol Chem* 285(52):40777–40784.

Fenton, R. A., H. B. Moeller, J. D. Hoffert, M. J. Yu, S. Nielsen and M. A. Knepper. 2008. Acute regulation of aquaporin-2 phosphorylation at Ser-264 by vasopressin. *Proc Natl Acad Sci USA* 105(8):3134–3139.

Fetter, K., V. Van Wilder, M. Moshelion and F. Chaumont. 2004. Interactions between plasma membrane aquaporins modulate their water channel activity. *Plant Cell* 16(1):215–228.

Fischer, G., U. Kosinska-Eriksson, C. Aponte-Santamaria, M. Palmgren, C. Geijer, K. Hedfalk, S. Hohmann, B. L. de Groot, R. Neutze and K. Lindkvist-Petersson. 2009. Crystal structure of a yeast aquaporin at 1.15 angstrom reveals a novel gating mechanism. *PLoS Biol* 7(6):e1000130.

Frick, A., U. K. Eriksson, F. de Mattia, F. Oberg, K. Hedfalk, R. Neutze, W. J. de Grip, P. M. Deen and S. Tornroth-Horsefield. 2014. X-ray structure of human aquaporin 2 and its implications for nephrogenic diabetes insipidus and trafficking. *Proc Natl Acad Sci USA* 111(17):6305–6310.

Frick, A., M. Jarva, M. Ekvall, P. Uzdavinys, M. Nyblom and S. Tornroth-Horsefield. 2013a. Mercury increases water permeability of a plant aquaporin through a non-cysteine-related mechanism. *Biochem J* 454(3):491–499.

Frick, A., M. Jarva and S. Tornroth-Horsefield. 2013b. Structural basis for pH gating of plant aquaporins. *FEBS Lett* 587(7):989–993.

Garcia, N., C. Gondran, G. Menon, L. Mur, G. Oberto, Y. Guerif, C. Dal Farra and N. Domloge. 2011. Impact of AQP3 inducer treatment on cultured human keratinocytes, ex vivo human skin and volunteers. *Int J Cosmet Sci* 33(5):432–442.

Gerbeau, P., G. Amodeo, T. Henzler, V. Santoni, P. Ripoche and C. Maurel. 2002. The water permeability of Arabidopsis plasma membrane is regulated by divalent cations and pH. *Plant J* 30(1):71–81.

Golestaneh, N., J. Fan, P. Zelenka and A. B. Chepelinsky. 2008. PKC putative phosphorylation site Ser235 is required for MIP/AQP0 translocation to the plasma membrane. *Mol Vis* 14:1006–1014.

Gonen, T., Y. Cheng, P. Sliz, Y. Hiroaki, Y. Fujiyoshi, S. C. Harrison and T. Walz. 2005. Lipid–protein interactions in double-layered two-dimensional AQP0 crystals. *Nature* 438(7068):633–638.

Gonen, T., P. Sliz, J. Kistler, Y. Cheng and T. Walz. 2004. Aquaporin-0 membrane junctions reveal the structure of a closed water pore. *Nature* 429(6988):193–197.

Gunnarson, E., M. Zelenina, G. Axehult, Y. Song, A. Bondar, P. Krieger, H. Brismar, S. Zelenin and A. Aperia. 2008. Identification of a molecular target for glutamate regulation of astrocyte water permeability. *Glia* 56(6):587–596.

Haines, T. H. 1994. Water transport across biological membranes. *FEBS Lett* 346(1):115–122.

Harries, W. E., D. Akhavan, L. J. Miercke, S. Khademi and R. M. Stroud. 2004. The channel architecture of aquaporin 0 at a 2.2-A resolution. *Proc Natl Acad Sci USA* 101(39):14045–14050.

Hasler, U., M. Vinciguerra, A. Vandewalle, P. Y. Martin and E. Feraille. 2005. Dual effects of hypertonicity on aquaporin-2 expression in cultured renal collecting duct principal cells. *J Am Soc Nephrol* 16(6):1571–1582.

Ho, J. D., R. Yeh, A. Sandstrom, I. Chorny, W. E. Harries, R. A. Robbins, L. J. Miercke and R. M. Stroud. 2009. Crystal structure of human aquaporin 4 at 1.8 A and its mechanism of conductance. *Proc Natl Acad Sci USA* 106(18):7437–7442.

Hoffert, J. D., R. A. Fenton, H. B. Moeller, B. Simons, D. Tchapyjnikov, B. W. McDill, M. J. Yu, T. Pisitkun, F. Chen and M. A. Knepper. 2008. Vasopressin-stimulated increase in phosphorylation at Ser269 potentiates plasma membrane retention of aquaporin-2. *J Biol Chem* 283(36):24617–24627.

Horsefield, R., K. Norden, M. Fellert, A. Backmark, S. Tornroth-Horsefield, A. C. Terwissscha van Scheltinga, J. Kvassman, P. Kjellbom, U. Johanson and R. Neutze. 2008. High-resolution x-ray structure of human aquaporin 5. *Proc Natl Acad Sci USA* 105(36):13327–13332.

Ishikawa, Y., T. Eguchi, M. T. Skowronski and H. Ishida. 1998. Acetylcholine acts on M3 muscarinic receptors and induces the translocation of aquaporin5 water channel via cytosolic Ca^{2+} elevation in rat parotid glands. *Biochem Biophys Res Commun* 245(3):835–840.

Ishikawa, Y., H. Iida and H. Ishida. 2002. The muscarinic acetylcholine receptor-stimulated increase in aquaporin-5 levels in the apical plasma membrane in rat parotid acinar cells is coupled with activation of nitric oxide/cGMP signal transduction. *Mol Pharmacol* 61(6):1423–1434.

Ishikawa, Y., M. T. Skowronski, N. Inoue and H. Ishida. 1999. Alpha(1)-adrenoceptor-induced trafficking of aquaporin-5 to the apical plasma membrane of rat parotid cells. *Biochem Biophys Res Commun* 265(1):94–100.

Janosi, L. and M. Ceccarelli. 2013. The gating mechanism of the human aquaporin 5 revealed by molecular dynamics simulations. *PLoS One* 8(4):e59897.

Johansson, I., M. Karlsson, V. K. Shukla, M. J. Chrispeels, C. Larsson and P. Kjellbom. 1998. Water transport activity of the plasma membrane aquaporin PM28A is regulated by phosphorylation. *Plant Cell* 10(3):451–459.

Kamsteeg, E. J., I. Heijnen, C. H. van Os and P. M. Deen. 2000. The subcellular localization of an aquaporin-2 tetramer depends on the stoichiometry of phosphorylated and nonphosphorylated monomers. *J Cell Biol* 151(4):919–930.

Kamsteeg, E. J., G. Hendriks, M. Boone, I. B. Konings, V. Oorschot, P. van der Sluijs, J. Klumperman and P. M. Deen. 2006. Short-chain ubiquitination mediates the regulated endocytosis of the aquaporin-2 water channel. *Proc Natl Acad Sci USA* 103(48):18344–18349.

Karabasil, M. R., T. Hasegawa, A. Azlina, N. Purwanti, J. Purevjav, C. Yao, T. Akamatsu and K. Hosoi. 2009. Trafficking of GFP-AQP5 chimeric proteins conferred with unphosphorylated amino acids at their PKA-target motif((152)SRRTS) in MDCK-II cells. *J Med Invest* 56(1–2):55–63.

Karlsson, T., M. Glogauer, R. P. Ellen, V. M. Loitto, K. E. Magnusson and M. A. Magalhaes. 2011. Aquaporin 9 phosphorylation mediates membrane localization and neutrophil polarization. *J Leukoc Biol* 90(5):963–973.

Katsura, T., C. E. Gustafson, D. A. Ausiello and D. Brown. 1997. Protein kinase A phosphorylation is involved in regulated exocytosis of aquaporin-2 in transfected LLC-PK1 cells. *Am J Physiol* 272(6 Pt 2):F817–F822.

Kosinska Eriksson, U., G. Fischer, R. Friemann, G. Enkavi, E. Tajkhorshid and R. Neutze. 2013. Subangstrom resolution x-ray structure details aquaporin-water interactions. *Science* 340(6138):1346–1349.

Koyama, N., K. Ishibashi, M. Kuwahara, N. Inase, M. Ichioka, S. Sasaki and F. Marumo. 1998. Cloning and functional expression of human aquaporin8 cDNA and analysis of its gene. *Genomics* 54(1):169–172.

Lande, M. B., I. Jo, M. L. Zeidel, M. Somers and H. W. Harris, Jr. 1996. Phosphorylation of aquaporin-2 does not alter the membrane water permeability of rat papillary water channel-containing vesicles. *J Biol Chem* 271(10):5552–5557.

Lehmann, G. L., M. C. Larocca, L. R. Soria and R. A. Marinelli. 2008. Aquaporins: Their role in cholestatic liver disease. *World J Gastroenterol* 14(46):7059–7067.

Leitao, L., C. Prista, M. C. Loureiro-Dias, T. F. Moura and G. Soveral. 2014. The grapevine tonoplast aquaporin TIP2;1 is a pressure gated water channel. *Biochem Biophys Res Commun* 450(1):289–294.

Leitao, L., C. Prista, T. F. Moura, M. C. Loureiro-Dias and G. Soveral. 2012. Grapevine aquaporins: Gating of a tonoplast intrinsic protein(TIP2;1) by cytosolic pH. *PLoS One* 7(3):e33219.

Li, G., V. Santoni and C. Maurel. 2014. Plant aquaporins: Roles in plant physiology. *Biochim Biophys Acta* 1840(5):1574–1582.

Lin, S. H., D. G. Bichet, S. Sasaki, M. Kuwahara, M. F. Arthus, M. Lonergan and Y. F. Lin. 2002. Two novel aquaporin-2 mutations responsible for congenital nephrogenic diabetes insipidus in Chinese families. *J Clin Endocrinol Metab* 87(6):2694–2700.

Loitto, V. M., C. Huang, Y. J. Sigal and K. Jacobson. 2007. Filopodia are induced by aquaporin-9 expression. *Exp Cell Res* 313(7):1295–1306.

Luu, D. T. and C. Maurel. 2013. Aquaporin trafficking in plant cells: An emerging membrane-protein model. *Traffic* 14(6):629–635.

Marinelli, R. A., P. S. Tietz, L. D. Pham, L. Rueckert, P. Agre and N. F. LaRusso. 1999. Secretin induces the apical insertion of aquaporin-1 water channels in rat cholangiocytes. *Am J Physiol* 276(1 Pt 1):G280–G286.

Matsuzaki, T., T. Suzuki and K. Takata. 2001. Hypertonicity-induced expression of aquaporin 3 in MDCK cells. *Am J Physiol Cell Physiol* 281(1):C55–C63.

Maurel, C., R. T. Kado, J. Guern and M. J. Chrispeels. 1995. Phosphorylation regulates the water channel activity of the seed-specific aquaporin alpha-TIP. *EMBO J* 14(13):3028–3035.

Mitsuma, T., K. Tani, Y. Hiroaki, A. Kamegawa, H. Suzuki, H. Hibino, Y. Kurachi and Y. Fujiyoshi. 2010. Influence of the cytoplasmic domains of aquaporin-4 on water conduction and array formation. *J Mol Biol* 402(4):669–681.

Moeller, H. B., R. A. Fenton, T. Zeuthen and N. Macaulay. 2009a. Vasopressin-dependent short-term regulation of aquaporin 4 expressed in *Xenopus* oocytes. *Neuroscience* 164(4): 1674–1684.

Moeller, H. B., N. MacAulay, M. A. Knepper and R. A. Fenton. 2009b. Role of multiple phosphorylation sites in the COOH-terminal tail of aquaporin-2 for water transport: Evidence against channel gating. *Am J Physiol Renal Physiol* 296(3):F649–F657.

Moeller, H. B., S. Rittig and R. A. Fenton. 2013. Nephrogenic diabetes insipidus: Essential insights into the molecular background and potential therapies for treatment. *Endocr Rev* 34(2): 278–301.

Mulders, S. M., N. V. Knoers, A. F. Van Lieburg, L. A. Monnens, E. Leumann, E. Wuhl, E. Schober, J. P. Rijss, C. H. Van Os and P. M. Deen. 1997. New mutations in the AQP2 gene in nephrogenic diabetes insipidus resulting in functional but misrouted water channels. *J Am Soc Nephrol* 8(2):242–248.

Nemeth-Cahalan, K. L., D. M. Clemens and J. E. Hall. 2013. Regulation of AQP0 water permeability is enhanced by cooperativity. *J Gen Physiol* 141(3):287–295.

Nemeth-Cahalan, K. L., K. Kalman and J. E. Hall. 2004. Molecular basis of pH and Ca^{2+} regulation of aquaporin water permeability. *J Gen Physiol* 123(5):573–580.

Noda, Y. and S. Sasaki. 2006. Regulation of aquaporin-2 trafficking and its binding protein complex. *Biochim Biophys Acta* 1758(8):1117–1125.

Nyblom, M., A. Frick, Y. Wang, M. Ekvall, K. Hallgren, K. Hedfalk, R. Neutze, E. Tajkhorshid and S. Tornroth-Horsefield. 2009. Structural and functional analysis of SoPIP2;1 mutants adds insight into plant aquaporin gating. *J Mol Biol* 387(3):653–668.

Ohinata, A., K. Nagai, J. Nomura, K. Hashimoto, A. Hisatsune, T. Miyata and Y. Isohama. 2005. Lipopolysaccharide changes the subcellular distribution of aquaporin 5 and increases plasma membrane water permeability in mouse lung epithelial cells. *Biochem Biophys Res Commun* 326(3):521–526.

Ozu, M., R. A. Dorr, F. Gutierrez, M. T. Politi and R. Toria. 2013. Human AQP1 is a constitutively open channel that closes by a membrane-tension-mediated mechanism. *Biophys J* 104(1): 85–95.

Papadopoulos, M. C. and A. S. Verkman. 2007. Aquaporin-4 and brain edema. *Pediatr Nephrol* 22(6):778–784.

Parvin, M. N., S. Kurabuchi, K. Murdiastuti, C. Yao, C. Kosugi-Tanaka, T. Akamatsu, N. Kanamori and K. Hosoi. 2005. Subcellular redistribution of AQP5 by vasoactive intestinal polypeptide in the Brunner's gland of the rat duodenum. *Am J Physiol Gastrointest Liver Physiol* 288(6):G1283–G1291.

Prak, S., S. Hem, J. Boudet, G. Viennois, N. Sommerer, M. Rossignol, C. Maurel and V. Santoni. 2008. Multiple phosphorylations in the C-terminal tail of plant plasma membrane aquaporins: Role in subcellular trafficking of AtPIP2;1 in response to salt stress. *Mol Cell Proteomics* 7(6):1019–1030.

Qi, H., L. Li, W. Zong, B. J. Hyer and J. Huang. 2009. Expression of aquaporin 8 is diversely regulated by osmotic stress in amnion epithelial cells. *J Obstet Gynaecol Res* 35(6):1019–1025.

Reichow, S. L., D. M. Clemens, J. A. Freites, K. L. Nemeth-Cahalan, M. Heyden, D. J. Tobias, J. E. Hall and T. Gonen. 2013. Allosteric mechanism of water-channel gating by Ca^{2+}-calmodulin. *Nat Struct Mol Biol* 20(9):1085–1092.

Reichow, S. L. and T. Gonen. 2008. Noncanonical binding of calmodulin to aquaporin-0: Implications for channel regulation. *Structure* 16(9):1389–1398.

Rodriguez, A., V. Catalan, J. Gomez-Ambrosi, S. Garcia-Navarro, F. Rotellar, V. Valenti, C. Silva et al. 2011. Insulin- and leptin-mediated control of aquaglyceroporins in human adipocytes and hepatocytes is mediated via the PI3K/Akt/mTOR signaling cascade. *J Clin Endocrinol Metab* 96(4):E586–E597.

Rose, K. M., Z. Wang, G. N. Magrath, E. S. Hazard, J. D. Hildebrandt and K. L. Schey. 2008. Aquaporin 0-calmodulin interaction and the effect of aquaporin 0 phosphorylation. *Biochemistry* 47(1):339–347.

Sachdeva, R. and B. Singh. 2014. Phosphorylation of Ser-180 of rat aquaporin-4 shows marginal affect on regulation of water permeability: Molecular dynamics study. *J Biomol Struct Dyn* 32(4):555–566.

Skalicky, J. J., J. Arii, D. M. Wenzel, W. M. Stubblefield, A. Katsuyama, N. T. Uter, M. Bajorek, D. G. Myszka and W. I. Sundquist. 2012. Interactions of the human LIP5 regulatory protein with endosomal sorting complexes required for transport. *J Biol Chem* 287(52): 43910–43926.

Song, Y. and E. Gunnarson. 2012. Potassium dependent regulation of astrocyte water permeability is mediated by cAMP signaling. *PLoS One* 7(4):e34936.

Soveral, G., A. Madeira, M. C. Loureiro-Dias and T. F. Moura. 2008. Membrane tension regulates water transport in yeast. *Biochim Biophys Acta* 1778(11):2573–2579.

Sui, H., B. G. Han, J. K. Lee, P. Walian and B. K. Jap. 2001. Structural basis of water-specific transport through the AQP1 water channel. *Nature* 414(6866):872–878.

Tamas, M. J., S. Karlgren, R. M. Bill, K. Hedfalk, L. Allegri, M. Ferreira, J. M. Thevelein, J. Rydstrom, J. G. Mullins and S. Hohmann. 2003. A short regulatory domain restricts glycerol transport through yeast Fps1p. *J Biol Chem* 278(8):6337–6345.

Tamas, M. J., K. Luyten, F. C. Sutherland, A. Hernandez, J. Albertyn, H. Valadi, H. Li et al. 1999. Fps1p controls the accumulation and release of the compatible solute glycerol in yeast osmoregulation. *Mol Microbiol* 31(4):1087–1104.

Tanghe, A., P. Van Dijck and J. M. Thevelein. 2006. Why do microorganisms have aquaporins? *Trends Microbiol* 14(2):78–85.

Thorsen, M., Y. Di, C. Tangemo, M. Morillas, D. Ahmadpour, C. Van der Does, A. Wagner et al. 2006. The MAPK Hog1p modulates Fps1p-dependent arsenite uptake and tolerance in yeast. *Mol Biol Cell* 17(10):4400–4410.

Tong, J., M. M. Briggs and T. J. McIntosh. 2012. Water permeability of aquaporin-4 channel depends on bilayer composition, thickness, and elasticity. *Biophys J* 103(9):1899–1908.

Tornroth-Horsefield, S., K. Hedfalk, G. Fischer, K. Lindkvist-Petersson and R. Neutze. 2010. Structural insights into eukaryotic aquaporin regulation. *FEBS Lett* 584(12):2580–2588.

Tornroth-Horsefield, S., Y. Wang, K. Hedfalk, U. Johanson, M. Karlsson, E. Tajkhorshid, R. Neutze and P. Kjellbom. 2006. Structural mechanism of plant aquaporin gating. *Nature* 439(7077):688–694.

Tournaire-Roux, C., M. Sutka, H. Javot, E. Gout, P. Gerbeau, D. T. Luu, R. Bligny and C. Maurel. 2003. Cytosolic pH regulates root water transport during anoxic stress through gating of aquaporins. *Nature* 425(6956):393–397.

van Balkom, B. W., M. Boone, G. Hendriks, E. J. Kamsteeg, J. H. Robben, H. C. Stronks, A. van der Voorde, F. van Herp, P. van der Sluijs and P. M. Deen. 2009. LIP5 interacts with aquaporin 2 and facilitates its lysosomal degradation. *J Am Soc Nephrol* 20(5):990–1001.

van Balkom, B. W., P. J. Savelkoul, D. Markovich, E. Hofman, S. Nielsen, P. van der Sluijs and P. M. Deen. 2002. The role of putative phosphorylation sites in the targeting and shuttling of the aquaporin-2 water channel. *J Biol Chem* 277(44):41473–41479.

Van Wilder, V., U. Miecielica, H. Degand, R. Derua, E. Waelkens and F. Chaumont. 2008. Maize plasma membrane aquaporins belonging to the PIP1 and PIP2 subgroups are in vivo phosphorylated. *Plant Cell Physiol* 49(9):1364–1377.

Vera-Estrella, R., B. J. Barkla, H. J. Bohnert and O. Pantoja. 2004. Novel regulation of aquaporins during osmotic stress. *Plant Physiol* 135(4):2318–2329.

Verdoucq, L., A. Grondin and C. Maurel. 2008. Structure–function analysis of plant aquaporin AtPIP2;1 gating by divalent cations and protons. *Biochem J* 415(3):409–416.

Wan, X., E. Steudle and W. Hartung. 2004. Gating of water channels(aquaporins) in cortical cells of young corn roots by mechanical stimuli(pressure pulses): Effects of ABA and of HgCl2. *J Exp Bot* 55(396):411–422.

Woo, J., Y. K. Chae, S. J. Jang, M. S. Kim, J. H. Baek, J. C. Park, B. Trink et al. 2008. Membrane trafficking of AQP5 and cAMP dependent phosphorylation in bronchial epithelium. *Biochem Biophys Res Commun* 366(2):321–327.

Yang, F., J. D. Kawedia and A. G. Menon. 2003. Cyclic AMP regulates aquaporin 5 expression at both transcriptional and post-transcriptional levels through a protein kinase A pathway. *J Biol Chem* 278(34):32173–32180.

Yasui, H., M. Kubota, K. Iguchi, S. Usui, T. Kiho and K. Hira2008. Membrane trafficking of aquaporin 3 induced by epinephrine. *Biochem Biophys Res Commun* 373(4):613–617.

Yasui, M., A. Hazama, T. H. Kwon, S. Nielsen, W. B. Guggino and P. Agre. 1999. Rapid gating and anion permeability of an intracellular aquaporin. *Nature* 402(6758):184–187.

Ye, Q., B. Wiera and E. Steudle. 2004. A cohesion/tension mechanism explains the gating of water channels(aquaporins) in Chara internodes by high concentration. *J Exp Bot* 55(396):449–461.

Zelazny, E., J. W. Borst, M. Muylaert, H. Batoko, M. A. Hemminga and F. Chaumont. 2007. FRET imaging in living maize cells reveals that plasma membrane aquaporins interact to regulate their subcellular localization. *Proc Natl Acad Sci USA* 104(30):12359–12364.

Zelenina, M., A. A. Bondar, S. Zelenin and A. Aperia. 2003. Nickel and extracellular acidification inhibit the water permeability of human aquaporin-3 in lung epithelial cells. *J Biol Chem* 278(32):30037–30043.

Zelenina, M., S. Tritto, A. A. Bondar, S. Zelenin and A. Aperia. 2004. Copper inhibits the water and glycerol permeability of aquaporin-3. *J Biol Chem* 279(50):51939–51943.

Zelenina, M., S. Zelenin, A. A. Bondar, H. Brismar and A. Aperia. 2002. Water permeability of aquaporin-4 is decreased by protein kinase C and dopamine. *Am J Physiol Renal Physiol* 283(2):F309–F318.

Zeuthen, T. and D. A. Klaerke. 1999. Transport of water and glycerol in aquaporin 3 is gated by H(+). *J Biol Chem* 274(31):21631–21636.

Yeast Aquaporins and Aquaglyceroporins

A Matter of Lifestyle

Mikael Andersson and Stefan Hohmann

CONTENTS

Abstract	77
5.1 Introduction	78
5.2 Orthodox Aquaporins in *Saccharomyces cerevisiae*	80
5.2.1 Shared Roles of Both AQY1 and AQY2 in Freeze Tolerance	82
5.2.2 Aqy1 and Sporulation	84
5.2.3 Aqy2 and Colony Morphology	85
5.3 Aquaglyceroporins in *Saccharomyces cerevisiae*	85
5.3.1 Role of Fps1: Glycerol Transport and Osmoregulation	86
5.3.2 Role of Fps1: Metalloid Transport and Resistance	87
5.3.3 Role of Fps1: Acetic Acid Transport and Resistance	88
5.3.4 Fps1 Regulation	89
5.3.5 Fps1 Sequence	90
5.3.6 Yfl054: The Uncharacterized Protein	92
5.3.7 Yfl054 Sequence	93
5.4 Summary and Future Aspects	93
Acknowledgements	94
References	94

ABSTRACT

YEASTS AND FUNGI POSSESS different numbers of aquaporins and aquaglyceroporins that can be divided into different subgroups. Yeast aquaporins not only appear to confer an adaptive advantage to freeze–thaw cycles but also may participate in determining cell surface properties. Those properties are in turn important for substrate adhesion and invasion. The aquaglyceroporin Fps1 appears to be a yeast invention and plays a central role in osmoregulation. Although restricted to a confined group of yeasts, Fps1 has evolved a rather complex regulation pattern at the crossroad of different mitogen-activated protein kinases.

5.1 INTRODUCTION

As the cost of sequencing decreases and new more powerful bioinformatics tools are constantly being developed, the last decade has seen a sharp increase in available genome sequence data. Those data allow us to do more refined comparisons than ever before and can help to provide us with a glimpse of the actual workings of the different proteins within a cell along with the evolutionary history and pressures exerted on the organism.

The physiological purpose of aquaporins and aquaglyceroporins is apparent to all biologists and biochemists, as water is critical for life. Thus, the proper regulation of the cellular water content is an important part in the life cycle of all organisms. Many mammalian, protozoan, plant and yeast orthodox aquaporins and aquaglyceroporins can also transport other compounds not directly tied to the water balance of the cell. Examples of these compounds are nitrate (Ikeda et al. 2002), urea (Pavlovic-Djuranovic et al. 2003) and CO_2 (Flexas et al. 2006; Navarro-Ródenas et al. 2012; Mori et al. 2014) as well as toxic metalloids such as arsenite and antimonite (Wysocki et al. 2001; Liu et al. 2002). Orthodox aquaporins and aquaglyceroporins have acquired other important functions during evolution and, hence, there is a need to investigate these proteins not just because of their role in water transport. It appears that microbes use orthodox aquaporins and aquaglyceroporins for maintenance of cell homeostasis under adverse conditions, during developmental stages where efficient substrate flux is needed or in the acquisition of tolerance to toxic compounds (Figure 5.1). The common baker's yeast/brewer's yeast *Saccharomyces cerevisiae* is one of our most frequently used and domesticated microorganisms with the earliest record of use by humans dating back around 10,000 years (McGovern et al. 2004; Fay and Benavides 2005). While *S. cerevisiae* is still used in its historical context in bread baking and the production of alcoholic beverages, it is nowadays also used extensively as a model organism for research of cellular processes, genetics, evolution and biotechnological applications. Due to a mix of its historical use, ease of cultivation and its biology along with tools that enable powerful genetics, *S. cerevisiae* has become the most well-studied eukaryotic cell, with only 15% of its genes lacking an annotated function (Botstein and Fink 2011).

In nature, *S. cerevisiae* is commonly found on fruits, flowers and other sugar-rich environments (Mortimer and Polsinelli 1999), including oak and beech trees (Sampaio and Goncalves 2008; Boynton and Greig 2014). Yeast has also been shown to have a close relationship with some insects (Reuter et al. 2007; Coluccio et al. 2008; Neiman 2011; Stefaninia et al. 2012; Good et al. 2014), which facilitates dispersion (Reuter et al. 2007; Coluccio et al. 2008; Neiman 2011). *S. cerevisiae* is specialized to ferment sugars to ethanol, a trait thought to have arisen as a consequence of the whole-genome duplication 80–100 million years ago (Ihmels et al. 2005; Thomson et al. 2005; Piskur et al. 2006). This genome duplication event has affected the evolution and abundance of orthodox aquaporins in *S. cerevisiae* and closely related species. Only those species that have undergone genome duplication have more than one aquaporin, while the other species possess one or none orthodox aquaporins (Will et al. 2010).

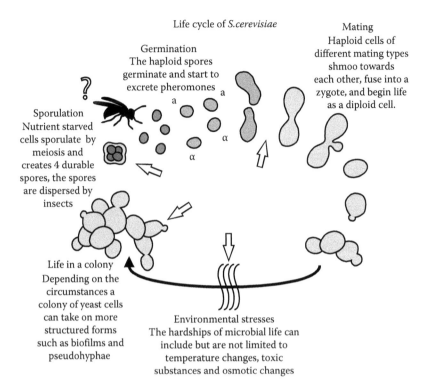

Life cycle of *S.cerevisiae*

Germination
The haploid spores germinate and start to excrete pheromones

Mating
Haploid cells of different mating types shmoo towards each other, fuse into a zygote, and begin life as a diploid cell.

Sporulation
Nutrient starved cells sporulate by meiosis and creates 4 durable spores, the spores are dispersed by insects

Life in a colony
Depending on the circumstances a colony of yeast cells can take on more structured forms such as biofilms and pseudohyphae

Environmental stresses
The hardships of microbial life can include but are not limited to temperature changes, toxic substances and osmotic changes

FIGURE 5.1 The life cycle of *Saccharomyces cerevisiae*. White arrows point to phases where aquaporins or aquaglyceroporins are important.

S. cerevisiae is widely regarded as a single-cell organism. However, wild-type strains show a wider variety of growth methods and can form multicellular structures and behaviours such as substrate invasive pseudohyphae (van Rijswijck et al. 2015), biofilms and structured or 'fluffy' colonies (Kuthan et al. 2003; Vopalenska et al. 2010; Vachova et al. 2011).

S. cerevisiae can reproduce sexually. It displays two mating types, MATa and MATα, which it can switch once every generation (Haber 2012). Yeast cells can grow vegetatively as both haploid MATa or MATα cells and diploid MATa/MATα cells. In the vegetative state, they grow by budding, where a daughter cell grows out of the mother as a small bud at the tip of the cell (Figure 5.1). MATa and MATα cells excrete pheromones that in conjunction with specific receptors are used to sense the presence of cells of the opposite mating type. When the MATa and MATα cells have sensed each other's presence, they will arrest their cell cycle in the G1 phase, start to grow toward each other, and eventually undergo fusion and karyogamy (Levin 2011; Haber 2012; Merlini et al. 2013) (Figure 5.1). Mating-type switching ensures that cells in nature commonly are diploid: any haploid cell will produce a mother and a daughter of opposite mating type, which then form a diploid. Under carbon and nitrogen, starvation diploids undergo meiosis and sporulation to form four haploid spores (Figure 5.1). The spores are low water content survival structures which are more durable than vegetative yeast cells (Neiman 2011). Each spore is enclosed by a plasma membrane and a rigid spore wall, with the plasma membrane and tough, sac-like cell wall derived from the mother cell serving as an ascus for the spore tetrad (Neiman 2011) (Figure 5.1).

The lifestyle and life cycle of *S. cerevisiae* involves many steps and circumstances where proper water homeostasis and thus osmoregulation are needed (Figure 5.1). The high-osmolarity glycerol pathway (HOG pathway) controls yeast osmoregulation and is very well studied. It is a conserved mitogen-activated protein kinase (MAPK) pathway similar to the human p38 signalling pathway, which is involved in inflammatory response and various cancers (Clark and Dean 2012; del Barco Barrantes and Nebreda 2012; Saito and Posas 2012). In addition to the HOG pathway with the Hog1/p38 MAPK, *S. cerevisiae* possesses three other MAPK: Fus3 and Kss1 (mating response), Kss1 (pseudohyphal development) and Slt2 (cell wall remodelling). Some of those pathways share components (Chen and Thorner 2007) and communicate with each other in manners that have not been fully elucidated (Tamás and Luyten 1999; Baltanas et al. 2013). The HOG pathway and the cell wall integrity (CWI) pathway maintain homeostasis toward the environment during developmental and environmental changes (Levin 2005; Saito and Posas 2012). The pheromone response (PR) pathway mediates the response to mating pheromones and controls subsequent developmental changes associated with mating (Dohlman and Thorner 2001; Merlini et al. 2013). The filamentous growth (FG) pathway is involved in the regulation of FG in response to nutrient limitation (Cullen and Sprague 2012).

S. cerevisiae contains four members of the major intrinsic protein (MIP) family of proteins. The two orthodox aquaporins, Aqy1 and Aqy2 (Soveral et al. 2010), are 88% similar and 81% identical and arose via whole-genome duplication. *S. cerevisiae* possesses two aquaglyceroporin genes, *FPS1* and *YFL054C*. In contrast to the two *AQY* genes, the two aquaglyceroporin paralogs share little identity; only one of them, Fps1, has a well-characterized function (Pettersson et al. 2005) (Figure 5.2).

In aquaporins, two asparagine–proline–alanine (NPA) sequence motifs located on the interhelical loops B and E are involved in the constriction of the channel (Fu et al. 2000; Nollert et al. 2001; de Groot et al. 2003). Another set of residues called the aromatic/arginine (Ar/R) selectivity filter also determines substrate specificity (Beitz et al. 2006; Wu and Beitz 2007; Hub and De Groot 2008; Mitani-Ueno et al. 2011; Azad et al. 2012). This selectivity filter consists of four residues that sit close to the constrictive channel formed by the NPA motifs in the 3D structure of the proteins (Figure 5.3a). The residues are located in transmembrane helix 2 (TM2) and 5 (TM5) and in loop E (LE1 and LE2) in proximity to the NPA motif; TM2 most often being an aromatic residue and LE2 being arginine (Hub and De Groot 2008).

5.2 ORTHODOX AQUAPORINS IN *SACCHAROMYCES CEREVISIAE*

Even through the two *S. cerevisiae* aquaporins Aqy1 (305 amino acids [AA]) and Aqy2 (289 AA) have been shown to function as orthodox aquaporins, studying their physiological roles has been troublesome. Most laboratory strains used, including the first one sequenced (Goffeau et al. 1997), lack functional versions of the Aqy1 and Aqy2 proteins

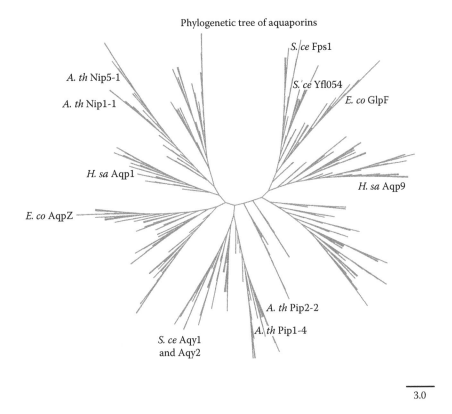

Phylogenetic tree of aquaporins

3.0

FIGURE 5.2 Neighbour joining phylogenetic tree based on the MIP motif of 800 random aquaporins. Aquaporins displayed are from the following organisms: *Saccharomyces cerevisiae*, *Homo sapiens*, *Arabidopsis thaliana*, and *Escherichia coli*.

(Bonhivers et al. 1998). Also, many *S. cerevisiae* wine strains and wild isolates lack functional versions of one or both of these aquaporins (Will et al. 2010). Lab strains usually have one out of two *AQY1* alleles, a functional *AQY1-1* or a non-functional *AQY1-2* that has the AA substitutions V121M and P255T (Bonhivers et al. 1998). The *AQY2* gene in lab strains has an internal 11bp deletion. This causes a frameshift and introduces a premature stop codon, resulting in a 149 AA long truncated non-functional protein (Laize et al. 2000; Carbrey et al. 2001a,b). Among lab strains, the common S288C has the non-functional *AQY1-2* allele, while the SK1 strain, which is frequently used in sporulation and meiosis research, has the functional *AQY1-1* allele (Sidoux-Walter et al. 2004). Σ1278b, a strain used for studying FG, has functional alleles of both *AQY1* and *AQY2* (Bonhivers et al. 1998; Laize et al. 1999, 2000; Meyrial et al. 2001). In Aqy-like proteins, both NPA motifs are highly conserved. The Ar/R selectivity filter is less conserved with only the main constituents of the aromatic TM2 (F98/F97 for Aqy1/Aqy2) and the LE2 (R233/R232 for Aqy1/Aqy2) arginine being well preserved (Figure 5.3b). The relative heterogeneity and lack of knowledge regarding Aqy regulation make it hard to see any preserved domains outside of the channel core.

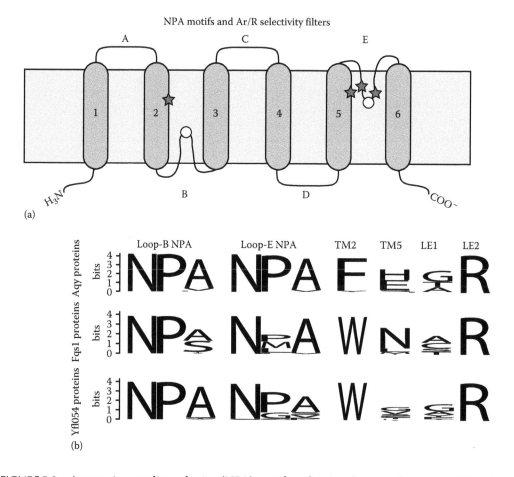

FIGURE 5.3 Asparagine–proline–alanine (NPA) motifs and Ar/R selectivity domains. (a) Topology map of a typical aquaporin. Transmembrane domains are numbered; loops are lettered. Dark stars and white dots represent approximate locations of Ar/R selectivity residues and NPA motifs, respectively. (b) Sequence logos represent the consensus sequence of the 168 (Aqy1/Aqy2), 30 (Fps1) or 41 (Yfl054) fungi proteins most similar to the respective protein. Sequence logos were produced with the WebLogo online tool (Schneider and Stephens 1990, Crooks et al. 2004).

5.2.1 Shared Roles of Both AQY1 and AQY2 in Freeze Tolerance

Although the strain variability with respect to functional aquaporins has long hampered the identification of physiologically relevant roles, adaptation to cold and freeze tolerance has repeatedly been reported to be associated with aquaporin function (Tanghe et al. 2002, 2004, 2005a,b; Soveral et al. 2006). Quantitative genetics demonstrated that 90% of the phenotypic variation in freeze tolerance between strains could be explained by the *AQY1* and *AQY2* loci, with *AQY2* representing approximately 66% of the phenotypic effect (Will et al. 2010). One of the more cold-resistant strains was also reported to have a 14-fold increase in *AQY2* expression, compared to other strains. Together with the previously reported role of *AQY2* in freeze tolerance (Tanghe et al. 2004) and the importance of yeast aquaporins in water flux under cold conditions (Soveral et al. 2006), these observations suggests that

Aqy2 and possibly Aqy1 might provide an evolutionary advantage for surviving freezing and cold temperatures (Will et al. 2010).

The impact on freeze tolerance by the two aquaporins seems to depend on growth conditions. In non-fermenting cells, the deletion of *AQY1*, but not *AQY2*, reduces freeze tolerance. The *aqy1Δ aqy2Δ* double-deletion strain exhibits freeze sensitivity similar to that of the *aqy1Δ* single mutant. In fermenting cells, however, both deletion of *AQY1* and *AQY2* individually reduce freeze tolerance, with *AQY2* deletion causing the biggest effect. The *aqy1Δ aqy2Δ* double-deletion strain is more freeze sensitive than either of the two single mutants in fermenting cells. The freeze sensitivity of this double-deletion mutant is rescued by heterologous overexpression of human *AQP1*, clearly demonstrating that it is the aquaporin water transport function of Aqy1 and Aqy2 that contributes to the freeze tolerance (Tanghe et al. 2002). Present ideas of the underlying mechanism of the aquaporin effect on freeze tolerance consider two factors. (1) A decrease of water conductivity of the plasma membrane has been observed as temperature decreases (Soveral et al. 2006). (2) The formation of ice crystals can rend and tear the protoplasm apart. The presence of aquaporins counteracts these two problems by making the plasma membrane more conductive to water even at low and freezing temperature. This increase in plasma membrane water conductivity facilitates a swift osmotically driven efflux of water from the cells when the surrounding water is freezing (Soveral et al. 2006). In turn, this water efflux diminishes the tendency for ice crystal formation by lowering the water content of the cell (Mazur 1970; Tanghe et al. 2002; Tanghe et al. 2004). Presumably, the water efflux is driven by the formation of ice outside of the cell, which in turn lowers the extracellular concentration of soluble water, driving the water out of the cell. This efflux also raises the intracellular concentrations of low freezing point solutes. The presence of aquaporins would make this water efflux much more efficient as it occurs at low temperatures (Ahmadpour et al. 2014). This hypothesis would also explain a previous observation that *S. cerevisiae* aquaporins only seem to be important for freeze tolerance to rapid freezing (Tanghe et al. 2004). During slower freezing conditions, the low water conductance of the plasma membrane would still be sufficient to prevent ice crystal formation (Ahmadpour et al. 2014).

Most freeze-tolerant strains isolated from nature were sampled from oak soil in northern more temperate climates, while the sensitive strains were sampled from warmer climates (Will et al. 2010). This indicate that *S. cerevisiae* strains that were originally isolated from non-temperate climates without regularly occurring freezing conditions more often lack functional *AQY* genes. It further appears that while aquaporins seem to promote freeze tolerance, they are detrimental to repeated cycles of high and low external osmolarity, which might explain why lab and industrial strains, which are exposed to such osmotic shifts, and strains from warmer climates lack functional aquaporins. In nature, the cells can encounter sugar-rich substrates of high osmolarity. Under such conditions, the loss of functional forms of both *AQY1* and *AQY2* give the cells a significant growth advantage over cells having functional forms of *AQY1* and *AQY2* (Will et al. 2010).

Compared to other *S. cerevisiae* strains, the freeze-resistant oak strains are thought to represent a more ancestral state. Those strains seem to be more closely related to the nearest relative of *S. cerevisiae*, *Saccharomyces paradoxus*. The loss of functional *AQY* genes in

S. cerevisiae is thought to be a relatively recent evolutionary event, possibly partly driven by domestication (Bonhivers et al. 1998; Laize et al. 2000; Carbrey et al. 2001b; Meyrial et al. 2001; Kvitek et al. 2008; Will et al. 2010). Loss of functional aquaporins is a result of non-neutral evolution in which the *AQY* paralog function has been lost in *S. cerevisiae* at multiple times during evolution. The forces that are driving this loss of aquaporin function in *S. cerevisiae* are probably the different selective pressures working on different subpopulations of *S. cerevisiae* (Will et al. 2010).

Since aquaporins seem to be detrimental under osmotic changes, cells that possess aquaporins should be able to control aquaporin activity. It appears that the Aqy2 protein is rapidly degraded upon hyperosmotic stress (Meyrial et al. 2001), and its expression diminished in a Hog1-dependent manner, presumably to hinder excess water loss from the cell to the hyperosmolar medium. Detailed mechanisms of these regulatory phenomena have not been reported.

5.2.2 Aqy1 and Sporulation

Aqy1 has been associated with sporulation (Sidoux-Walter et al. 2004; Karpel and Bisson 2006; Will et al. 2010). Sporulation is a complex process in the life cycle of *S. cerevisiae*, which requires a completed meiotic cycle preceding spore formation (Neiman 2011).

A study using the SK1 strain, which expresses a functional Aqy1, reported that the deletion of the *AQY1* gene led to a 38% reduction in spore fitness and that this likely depended on a mechanism that happens before the spores germinate, that is during spore formation, maturation or maintenance. The Aqy1 protein appeared to be located exclusively in the spore membrane, and it accumulated only after the spores had separated by cytoplasmic barriers (Sidoux-Walter et al. 2004). Other data suggested that *AQY1* was expressed earlier in the sporulation process (Chu et al. 1998; Primig et al. 2000), indicating that the Aqy1 protein expression might be controlled post-transcriptionally (Sidoux-Walter et al. 2004). The vast majority of Aqy1 protein appears to be degraded during spore germination, and deleting the *AQY1* gene did not seem to affect germination efficiency (Sidoux-Walter et al. 2004). It was concluded that Aqy1 affected spore maturation in *S. cerevisiae* (Sidoux-Walter et al. 2004).

Interestingly, *aqy1Δ* mutants of wild strains that express a functional allele of *AQY1* were compromised in spore formation. On the other hand, the *S. cerevisiae* strains that did not normally express a functional Aqy1 did not show any spore formation dependence on the *AQY1* gene, consistent with the data reporting that functional *AQY1* does not affect sporulation in wine yeasts (Karpel and Bisson 2006; Will et al. 2010). Hence, it appears that in strains adapted to an environment where Aqy1 contributes to freeze tolerance, Aqy1 function is involved in spore formation, while strains that opted against Aqy1 because of osmo-shift exposures found ways to sporulate without the involvement of Aqy1.

Exactly how Aqy1 exerts its function during sporulation is unknown. Aqy1 may facilitate the water efflux that is required to lower the water content in the spores. This idea is based on three facts. (1) The water content of the spore is heavily reduced and water is partially replaced by trehalose. (2) The Aqy1 protein is only detectable after the individual spores have separated from each other. (3) The thick, hard and durable spore wall may provide, together with increasing spore trehalose levels, the pressure to drive water out of the spores. Hence, the water content of spores is reduced because trehalose accumulates in the cell, while the rigid spore wall counteracts any volume change, and water is simply pushed out of the cell by turgor pressure. There is no direct experimental evidence supporting this idea (Ahmadpour et al. 2014).

5.2.3 Aqy2 and Colony Morphology

Aqy2 has been linked to the formation of the biofilm-like 'fluffy' colonies. These colonies are a three-dimensionally structured type of colony growth, rich in extracellular matrix. This specific phenotype is present in many wild isolates and makes the colonies appear as wrinkled patches on the surface of the substrate.

The *AQY2* gene is only expressed during logarithmic growth in nutrient-rich growth media and not during stationary phase (Meyrial et al. 2001). *AQY2* gene expression is controlled and regulated by three pathways that also control morphological developments. Upon hyperosmotic stress, *AQY2* expression and morphological developments are inhibited by the osmoregulatory MAPK Hog1. The protein kinase A pathway and FG pathway seem to have positive and negative effects, respectively, on *AQY2* gene expression. This pattern of regulation thus bears similarity to that of *FLO11*, whose product, a mucin-like cell wall glycoprotein, is known to be required for morphological development (Furukawa et al. 2009) such as formation of fluffy colonies (St'ovicek et al. 2010).

Aqy proteins have previously been associated with making the cells more hydrophobic, more adhesive to plastic and prone to clumping together (Carbrey et al. 2001b). Overexpression of functional *AQY2* increases plastic adhesion and agar invasion (Furukawa et al. 2009).

It has previously been reported that *AQY1* (not *AQY2*) and *FLO11* expression are tightly connected to the strength of the fluffy colony phenotype, with the most structured strains having the highest *AQY1* expression (St'ovicek et al. 2010). However, that study might actually report expression of *AQY2*; the technique used probably fails to distinguish between the two. In any case, aquaporins might influence the formation of fluffy colonies and cell surface properties by facilitating the hydration of cell wall glycoproteins and glucans. This could affect cell hydrophobicity and might enhance the exchange of nutrients and waste products within the colony and with its surrounding (Palkova and Vachova 2006; Ahmadpour et al. 2014). The ecology, biology and function of different growth modes in *S. cerevisiae* are relatively unexplored and, hence, the role of Aqy2 and Aqy1 in this context is still unclear and more research is needed.

5.3 AQUAGLYCEROPORINS IN *SACCHAROMYCES CEREVISIAE*

Fungal aquaglyceroporins can be divided into seven phylogenetic clusters: Fps1, Yfl054, α, β, γ1, γ2 and δ (Verma et al. 2014). Yeast aquaglyceroporins have traditionally been classified into Fps1-like proteins and Yfl054-like proteins, representing the two *S. cerevisiae* aquaglyceroporins (Pettersson et al. 2005). Fps1-like proteins only seem to exist among yeasts within the Ascomycetes, while the Yfl054 type of proteins seem to be present mostly in the filamentous Ascomycetes (Xu et al. 2013; Verma et al. 2014). The vast majority of the research regarding aquaglyceroporins in S. cerevisiae has been carried out on Fps1, simply because deletion or mutation of *FPS1* causes specific growth phenotypes.

Much like other members of the aquaporin and aquaglyceroporin family, Fps1 works as a homo-tetramer (Beese-Sims et al. 2011). Fps1 mediates the flux of glycerol, arsenite, antimonite, acetic acid and boric acid across the plasma membrane (Tamás and Luyten 1999; Wysocki et al. 2001; Nozawa et al. 2006), and it may also facilitate transport of acrolein, allyl alcohol, acetamide and other small amides (Shepherd and Piper 2010). Glycerol is the only known possible substrate reported for Yfl054, though the weak phenotype associated with this transport suggests that glycerol might not be its primary substrate (Oliveira 2003).

5.3.1 Role of Fps1: Glycerol Transport and Osmoregulation

Regulated glycerol efflux as part of the cellular osmoregulatory system appears to be the main function of Fps1 (Philips and Herskowitz 1997; Sutherland et al. 1997; Tamás and Luyten 1999; Hohmann 2002; Oliveira 2003; Saito and Posas 2012; Baltanas et al. 2013). Before Fps1 was discovered and characterized, it was thought that glycerol freely passes the plasma membrane lipid bilayer (Gancedo et al. 1968). It is now clear that very little, if there is any, glycerol passes the yeast plasma membrane freely (Oliveira 2003). During growth on glucose, glycerol is used as a compatible solute by S. cerevisiae (Blomberg and Adler 1989; Blomberg and Adler 1992). Glycerol is produced in S. cerevisiae from the glycolytic intermediate dihydroxyacetone phosphate in two steps catalyzed by glycerol-3-phosphate dehydrogenase (Gpd1 or Gpd2) and glycerol-3-phosphatase (Gpp1 or Gpp2). In the presence of fermentable carbon sources, glycerol is readily produced by yeast for osmoregulatory purposes. In the absence of fermentable carbon sources, glycerol can be taken up from the environment via a dedicated active glycerol import system, Stl1. Its expression is repressed by glucose and strongly upregulated under osmotic stress (Rep et al. 2000). Stl1 is unrelated to the MIP family and shows similarity to hexose transporters with 12 transmembrane domains. Stl1 mediates glycerol/H^+ symport and is required for growth with glycerol as a carbon source (Ferreira et al. 2005).

As previously mentioned, the HOG pathway controls the response to hyper-osmotic stress in S. cerevisiae. The Hog1 protein kinase is involved in the upregulation of expression of *GPD1*, *GPP2* and *STL1* and hence mediates an increased capacity to produce or take up glycerol (Saito and Posas 2012; Petelenz-Kurdziel et al. 2013; Babazadeh et al. 2014). Rapid closing of Fps1 ensures that glycerol produced by the cell does not leak out (Tamás and Luyten 1999). Under hypo-osmotic stress, glycerol instead needs to be transported out from the cell to even out the osmotic differences and alleviate the increase in turgor. Fps1 is thus rapidly opened upon hypo-osmotic stress to release glycerol. Cells lacking Fps1 instead accumulate glycerol and swell under hypo-osmotic stress (Luyten et al. 1995; Tamás and Luyten 1999; Beese et al. 2009). If the *fps1Δ* mutation is combined with other mutations that weaken the cell wall, hypo-osmotic stress becomes lethal and cells burst (Tamás and Luyten 1999). Consequently, treatments that weaken the cell wall such as calcofluor white and zymolyase also affect *fps1Δ* cells negatively (Beese et al. 2009). Other phenotypes of *fps1Δ* mutants likely coupled to the glycerol concentration and/or increase of turgor are heat stress sensitivity (Siderius et al. 2000; Beese et al. 2009) and altered membrane composition (Sutherland et al. 1997; Toh et al. 2001). Cells lacking Fps1 also suffer from a fitness disadvantage compared to cells that are able to export glycerol (Tamás and Luyten 1999; Beese et al. 2009). This fitness disadvantage is more pronounced when using a fermentative carbon under anaerobic conditions, as the cells then produce even more glycerol in order to re-oxidize NADH (Albers et al. 1996; Ansell et al. 1997; Björkqvist et al. 1997; Ahmadpour et al. 2014).

To maintain homeostasis, diverse pieces of information need to be integrated to a finely tuned functional response. This fine-tuning depends in large part on the four MAPK pathways present in *S. cerevisiae* (Posas et al. 1998). Mating is a good example of how the life cycle of *S. cerevisiae* involves the constant need for proper osmoregulation. Fps1 is required for mating, and Fps1 may be controlled in this situation by at least two MAPK, Hog1 and Slt2 (Philips and Herskowitz 1997; Baltanas et al. 2013). In cells grown under hyperosmotic conditions and treated with pheromone, Hog1 is phosphorylated and activated. This pheromone-dependent Hog1 activation requires the Slt2 MAPK and is caused by the release of glycerol into the medium via Fps1. This suggests that Slt2 may affect opening of Fps1 in response to an activated PR pathway. There is presently no direct experimental evidence that Slt2 phosphorylates Fps1 and the effect may also be indirect (Baltanas et al. 2013). The pheromone-dependent Hog1 activation also leads to an increased glycerol production in order for the cells to counteract the Fps1-mediated efflux of glycerol and maintain the intracellular glycerol concentration. This seemingly wasteful process of overproduction and Fps1-dependent leakage of glycerol probably allow cells to respond more rapidly to perturbations (Salvador 2008; Ahmadpour et al. 2014). A theory supported by the fact that cells overproducing glycerol via this mechanism recover twice as fast as cells lacking this mechanism (Baltanas et al. 2013).

Three non-mutually exclusive theories exist about the role of Fps1 during mating. (1) The increase in glycerol turnover allows cells to control turgor faster and more precisely. The remodelling of the cell wall and plasmogamy of the two haploid cells present significant osmotic challenges (Merlini et al. 2013). Thus, fast and precise control of turgor is a potentially important property during mating (Baltanas et al. 2013). (2) Glycerol might shape the pheromone gradient, and the release of glycerol through Fps1 could help the cell to fine-tune signal direction. (3) Cells subjected to mating pheromone accumulate reactive oxygen species (ROS) (Zhang et al. 2006). The high turnover of glycerol might allow cells to re-oxidize NADH via mitochondrial metabolism and reduce mitochondrial ROS production. The third option is strengthened by another observation; CWI mutants that cannot activate Slt2 under pheromone treatment have an elevated ROS concentration compared to wild-type controls (Zhang et al. 2006). This supports the notion that those cells are unable to activate Hog1 and thereby glycerol overproduction (Baltanas et al. 2013). Thus, the role of Fps1 in mating could eventually include more than osmoregulation.

5.3.2 Role of Fps1: Metalloid Transport and Resistance

Fps1 is permeable to toxic metalloids such as arsenite and antimonite (Wysocki et al. 2001; Thorsen et al. 2006; Beese et al. 2009; Maciaszczyk-Dziubinska et al. 2010; Lee et al. 2013).

The transport activity of the Fps1 protein is often assayed using arsenite because it produces a clear growth phenotype (Wysocki et al. 2001; Thorsen et al. 2006; Beese et al. 2009; Lee et al. 2013). Arsenite $As(OH)_3$ and antimonite $Sb(OH)_3$ are structurally similar to glycerol (Porquet and Filella 2007). Metalloids in nature pose a great threat to public health in several parts of the world. The toxicity coupled with the abundance in groundwater and soil results in contaminated drinking water and accumulation in staple food crops such as rice (Nordstrom 2002; Beyersmann and Hartwig 2008; Banerjee et al. 2013). Arsenic in particular is a common metalloid, which is toxic not only for humans but also for plants, animals and yeasts. Aquaporins provide one entry route for arsenic into these organisms (Tamás and Luyten 1999; Liu et al. 2002; Bienert et al. 2008; Isayenkov and Maathuis 2008; Maciaszczyk-Dziubinska et al. 2010). Therefore, organisms have evolved ways to protect themselves from these toxic metalloids, and in *S. cerevisiae*, Fps1 is involved in this process (Maciaszczyk-Dziubinska et al. 2010). In nature, arsenic occurs as trivalent arsenite As(III) and pentavalent arsenate As(V). As(III) enters cells through Fps1 and hexose transporters (Wysocki et al. 2001; Liu et al. 2004), and As(V) influx is mediated by phosphate transporters (Bun-ya et al. 1996). Once taken up, As(V) can be converted to As(III) by Acr2 (Mukhopadhyay and Rosen 1998) and can leave cells via Fps1 as $As(OH)_3$, while (Wysocki et al. 2001; Fu et al. 2009) $As(OH)_2O^-$ is actively pumped out of the cell by the Acr3 efflux pump (Wysocki and Bobrowicz 1997; Fu et al. 2009). The arsenite can also be sequestered by glutathione and pumped out by the ABC transporter Ycf1 (Wysocki and Bobrowicz 1997; Ghosh et al. 1999). *FPS1* deletion mutants display lower intracellular concentration of arsenite and are more resistant to arsenite (Wysocki et al. 2001).

Under metalloid stress, the HOG pathway is activated. It appears that Hog1 restricts arsenite flux through Fps1 (Thorsen et al. 2006). The transcription of the *FPS1* gene is also temporarily decreased (Wysocki et al. 2001), but the Fps1 protein itself is not degraded under these conditions. Counterintuitively, *FPS1* gene expression is tripled compared to non-arsenite-treated controls after 3 h of arsenite exposure. This overabundance of the Fps1 protein has a positive effect on the efflux of arsenite from the cell and arsenite-stressed cells that overexpress Fps1 display a lower intracellular arsenite concentration than cells that do not (Maciaszczyk-Dziubinska et al. 2010). The explanation for the phenotype is based on a difference between the ratios of $As(OH)_3$ and $As(OH)_2O^-$ between the cytoplasm and the extracellular environment. This difference could arise from the extrusion of glutathione that sequesters As(III) outside of the cell and thereby would lower its extracellular concentration (Maciaszczyk-Dziubinska et al. 2010; Thorsen et al. 2012). Thus, arsenite could flow via Fps1 following its concentration gradient. Hence, the cell would downregulate Fps1 to prevent arsenite influx, but once arsenite is present in the cell, it would upregulate Fps1 to facilitate arsenite efflux.

5.3.3 Role of Fps1: Acetic Acid Transport and Resistance

Fps1 also facilitates regulated transport of acetic acid (Mollapour and Piper 2007), which is toxic to *S. cerevisiae* (Teixeira et al. 2011). Acetic acid is a common compound in the *S. cerevisiae* environment as it is a by-product of sugar fermentation. Fps1 is the major entry route of the un-disassociated form of acetic acid into the cell, and deletion of *FPS1* renders cells resistant to acetic acid (Mollapour and Piper 2007). When subjected to acetic acid stress under conditions where acetic acid is mostly un-disassociated, the Fps1 protein is phosphorylated by the Hog1 MAP kinase, endocytosed and degraded in the vacuole (Mollapour and Piper 2007). It appears that acetic acid stress is the only known condition under which Fps1 is degraded following phosphorylation.

5.3.4 Fps1 Regulation

Fps1 is kept in its open state by interaction with the two cytosolic proteins Rgc1 and Rgc2 (also called Ask10). In the absence of Rgc1 and 2, it appears that Fps1 has a low basal activity because 20-fold overexpression of Fps1 suppresses hypo-osmosensitivity of an *rgc1Δ* *rgc2Δ* mutant (Beese et al. 2009). The two Rgc proteins are paralogs and are redundant; only the double mutant displays a full *fps1Δ* phenotype (Beese et al. 2009; Lee et al. 2013). The Rgc proteins bind Fps1 via a tripartite pleckstrin homology (PH) domain, which interacts with a partial PH domain on Fps1. The partial PH domain on Fps1 is located in the C-terminal cytosolic domain between residues 544 and 581 (Lee et al. 2013). Removal of the PH domain in Fps1 or the tripartite PH domain in Rgc2 abolishes most of the interaction between Fps1 and Rgc2 (Lee et al. 2013). A sequence between residues 611 and 614 on Fps1 is critical for Fps1 and Rgc protein interaction and its absence abolishes all Fps1 and Rgc2 interaction (Lee et al. 2013). Active Hog1 docks with Fps1 on the N-terminal cytosolic domain at residues 218 and 220. The docking site partially overlaps the known N-terminal regulatory site between residues 219 and 239. Absence of this N-terminal regulatory site renders Fps1 constitutively open and mediates sensitivity to hyperosmotic, arsenite and acetic acid stress (Tamas et al. 2003; Hedfalk et al. 2004; Mollapour and Piper 2007; Lee et al. 2013). When docked to Fps1, Hog1 appears to remove Rgc2 by phosphorylating Rgc2 on multiple sites, thereby facilitating closing of the Fps1 channel under hyperosmotic stress (Lee et al. 2013). Removal of the Hog1 docking site in Fps1 prevents Hog1 interaction with Fps1 and thus stabilizes the interaction with Rgc2 and Fps1, rendering Fps1 constitutively open (Lee et al. 2013).

Fps1 also has two potential MAPK phosphorylation sites, T231 and S537. Their exact roles remain unclear. T231 residue is located within the regulatory domain (residue 219–239) close to the first transmembrane domain within the N-terminal cytoplasmic domain of Fps1 (Tamas et al. 2003). T231 is phosphorylated by Hog1 under arsenite and acetic acid stress (Thorsen et al. 2006; Mollapour and Piper 2007). A T231A point renders Fps1 hyperactive and cells sensitive to As(III) and hyperosmotic stress (Thorsen et al. 2006). Under acetic acid stress, T231 phosphorylation targets the protein for ubiquitination, endocytosis and degradation in the vacuole (Mollapour and Piper 2007), rendering cells resistant to acetic acid stress. Deletion of *HOG1* does not completely abolish T231 phosphorylation under arsenite stress (Thorsen et al. 2006). It has also been shown that closing of Fps1 is essential for osmoadaptation in *hog1Δ* cells engineered to activate osmostress-induced genes via Fus3/Kss1, which is activated due to crosstalk upon hyperosmotic stress in *hog1Δ* cells (O'Rourke and Herskowitz 1998; Davenport et al. 1999; Rep et al. 2000; Babazadeh et al. 2014). This suggests that another protein kinase participates in T231 phosphorylation or that another kinase can partly take over in absence of Hog1.

The S537 site is located within a known regulatory domain on the C-terminal cytoplasmic domain of Fps1 just after the last transmembrane domain. This C-terminal domain (residue 535–546) is critical for cells to retain glycerol under hyperosmotic stress and tolerance to hyperosmotic stress (Hedfalk et al. 2004). The S537 residue is phosphorylated under acetic acid stress, possibly by activated Hog1 and/or other MAPK (Mollapour and Piper 2007). The T231A S537A double-point mutant of Fps1 cannot be phosphorylated in vivo under acetic acid stress and cannot be endocytosed and degraded (Mollapour and Piper 2007). Based on our unpublished data S537, it seems to be critical for the resistance to arsenite in cells overexpressing *FPS1*. This phenotype is apparently also dependent on the CWI MAPK Slt2, suggesting that Slt2 is a candidate regulator of Fps1 also under these conditions.

The TORC2-dependent protein kinase Ypk1 has recently been shown to keep Fps1 open by phosphorylating the previously uncharacterized Fps1 phosphorylation sites T147, S181, S185 and S570. In response to hyperosmotic stress this phosphorylation is inhibited and Fps1 is closed. Both of these events are Hog1 independent (Muir et al. 2015).

Taking together, Fps1 can be negatively regulated in at least two ways. (1) Hog1 binds to Fps1 and phosphorylates and thereby removes Rgc1 and Rgc2. This appears to happen in the case of temporary stresses such as osmostress, where the stimulus eventually disappears because of turgor adaptation (Mollapour and Piper 2007). (2) Hog1 and possibly other kinases phosphorylate Fps1, but subsequently, the Fps1 fate depends on the specific type of stress. Under acetic acid stress, Fps1 is phosphorylated, ubiquitinated, internalized and degraded. Under arsenite stress Hog1 instead seem to restrict the channel, while under osmotic stress Ypk1 stops phosphorylating Fps1 thus closing the channel. Hence, Hog1 is a negative regulator to Fps1 activity, and Rgc1 and Rgc2 are positive regulators. Ypk1 is a confirmed positive regulator under hyperosmotic stress while Slt2 is a candidate positive regulator under hypo-osmotic stress (Davenport et al. 1995) during mating and under arsenite stress (unpublished data). (Baltanas et al. 2013). Whether Slt2 mediates its apparent effects directly by phosphorylating Fps1 or perhaps via activating protein phosphatases that dephosphorylate Fps1 or the Rgc proteins remains to be established.

5.3.5 Fps1 Sequence

The normally highly conserved NPA motif present on the loop B and E of aquaporins is less well conserved in Fps1. This seem to be a general trend in fungal aquaglyceroporins, especially for the NPA motif present in the loop E (Verma et al. 2014). It is possible that the heterogeneity of the pore constriction could be compensated for by a more flexible pore structure (Bill et al. 2001). For the Ar/R selectivity filter, the characteristic aromatic TM2 residue (W330) and LE2 arginine residue (R483) are perfectly preserved in Fps1 proteins. The TM5 (N468) and LE1 (T477) residues are less preserved (Verma et al. 2014) (Figure 5.3b).

Known functional domains of Fps1 are commonly well conserved. In addition, there are some conserved sequence stretches presently lacking known function. A cluster of acidic residues centered around D75 seems to be somewhat conserved in acidity but not specific sequence (Pettersson et al. 2005). A domain centered around residue 109–118 in the

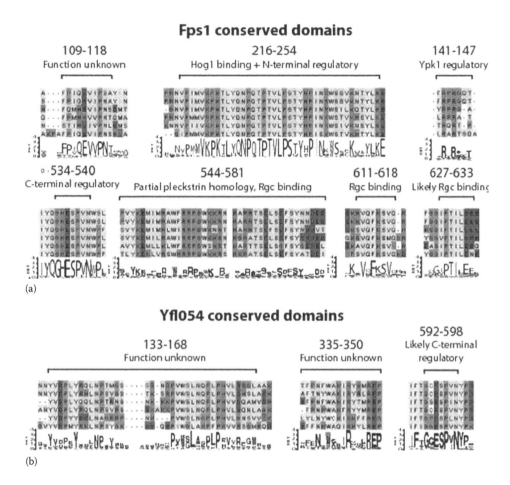

FIGURE 5.4 Conserved domains in yeast aquaglyceroporins. (a) Conserved domains in Fps1. (b) Conserved domains in Yfl054. BLASTP alignment shows the six best hits for each respective gene; sequence logos represent the consensus sequence of the 30 (Fps1) or 41 (Yfl054) fungi proteins most similar to the respective protein. Sequence logos were produced with the WebLogo online tool (Schneider and Stephens 1990, Crooks et al. 2004).

cytoplasmic N-terminal domain is highly conserved among all Fps1 proteins (Pettersson et al. 2005) (Figure 5.4a). A possible interaction partner for this domain could be the osmosensor Sho1, which is believed to bind preferentially to the N-terminal cytoplasmic domain (Lam et al. 2015). T147 and accompanying Ypk1 phospho acceptor motif is very conserved (Figure 4a), however, the two Ypk1 sites S181 and S185 is somewhat conserved but does not present a coherent double phosphorylation site in most Fps1 orthologs. The residues between 216 and 236 contain binding sites for Hog1 (Lee et al. 2013) and numerous residues critical for proper Fps1 regulation, including the T231 phosphorylation site. Overall, the sequence between residues P217 to the start of the first transmembrane domain at K254 is highly conserved among Fps1 proteins (Figure 5.4a).

The Fps1 C-terminal regulatory domain containing S537 is also present in a large number of other diverse fungal aquaporins, including Yfl054 (Figure 5.4a and b). This highly conserved domain appears to have the consensus sequence GGESPIN in fungal aquaglyceroporins. Fps1 and Yfl064 mostly share the consensus variant GHESPVN. The domain has been described earlier as HESPVN (Pettersson et al. 2005) and is important for the regulation of Fps1. Its removal leads to a partially regulated channel (Hedfalk et al. 2004), and the S537 residue is thought to affect the export function of Fps (unpublished data) (Ahmadpour et al. 2014). This domain is always located a few residues downstream of the last transmembrane domain (7 in Fps1, 6 for Yfl054). The MAPK phosphorylation target variants SP or TP (Tanoue and Nishida 2003) followed by a varying hydrophobic residue appear to be present in all cases. The very high conservation of this sequence stretch implies its importance for some basal property or regulation of the proteins. The partial PH domain (residue 544–581) involved in Rgc1/2 binding is not very preserved among Fps1 proteins neither in length nor sequence. However, two sites centered on W554 and S573 seem to be more conserved than other parts in this partial PH domain, the Ypk1 phosphorylation site S570 is also present here and is conserved among Fps1 orthologs, suggesting a possible link to Rgc1/2 binding (Figure 5.4a). The partial PH domain has been shown to be redundant (Lee et al. 2013), a fact that could explain the relatively low conservation. One of the sites critical for Rgc1 and Rgc2 binding is located between AA 611 and 614 (Lee et al. 2013) (Figure 5.4a). This specific sequence is well preserved among known Fps1 proteins. An additional preserved site is present further downstream around residue 627 and 632 (Figure 5.4a). Though not critical for Fps1 and Rgc1/2 interaction, this site has been associated with Rgc1/2 binding defects (Lee et al. 2013). Interestingly, several other yeasts Fps1 appears to have even more possible Ypk1 acceptor motifs than S.cerevisiae. Fps1proteins are also distinguished from other fungal aquaglyceroporins by having a long loop A linker domain between transmembrane domain 1 and 2.

5.3.6 Yfl054: The Uncharacterized Protein

Yfl054-type proteins are conserved over both Ascomycetes and Basidiomycetes but appear to be the most prevalent in filamentous Ascomycetes (Xu et al. 2013; Verma et al. 2014). Little phenotypic information associated with Yfl054 has been reported. The only known phenotype of *YFL054C* is a weak facilitation of glycerol diffusion in the presence of ethanol in yeast (Oliveira 2003; Mira et al. 2014). Expression of the *YFL054C* gene is upregulated during sporulation and entry into stationary phase (Gasch et al. 2000). Expression of the *YFL054C* ortholog in *Zygosaccharomyces bailii* is glucose repressed by the transcription factors Mig1 and Mig2 (Mira et al. 2014). Yfl054 has also been linked to biofilm formation; deletion of the *YFL054C* gene in the Σ1278b background significantly decreases the mRNA levels of the cell surface glycoprotein Flo11 which is essential for biofilm formation and invasive growth (Andersen et al. 2014). Green fluorescent protein (GFP) tagging and overexpression under the *MET25* promoter demonstrated that the protein localizes to the plasma membrane (unpublished data).

The species with the closest *Yfl054* orthologs are associated with brewing processes. Even more distantly related species such as *Schizosaccharomyces pombe* have a Yfl054 very similar to that of *S. cerevisiae* (46% identity and 59% similarity). Yfl054 and related proteins lack the conserved N-terminal Hog1 binding domain, the T231 phosphorylation site and the two

Rgc1 Rgc2 interacting domains present in the C-terminal cytoplasmic domain of Fps1. Taken together, this information suggests that Yfl054 is a plasma membrane protein, possibly performing its function during nutrient-limiting conditions where yeast tend to form biofilms and/or undergo invasive growth. The sequence differences in comparison to Fps1 indicate a different regulation. Yet, the great conservation of Yfl054 in such distant species such as *S. cerevisiae* and *S. pombe* may indicate a function conserved in evolution or horizontal transfer. Both species appear to coexist in the same environment (Odunfa 1985; Jespersen 2003).

5.3.7 Yfl054 Sequence

Yfl054-like proteins are more diverse (Verma et al. 2014), and conservation is more difficult to interpret due to the lack of known function. Like most fungal aquaporins, Yfl054 proteins do not have very conserved NPA motifs. The Ar/R selectivity filter has the typical aromatic (W388) and arginine (R541) residues but are otherwise highly variable in the TM5 and LE1 residues (Figure 5.3b). A moderately conserved domain of unknown function exists between Y133 and D168, with the residues L156 and P161 being exceptionally well conserved (Figure 5.4b). Parts of this domain has previously been reported as PXXSLXXPLPX (Pettersson et al. 2005). The stretch of 15 residues located just before the predicted TM1 between residues 332 and 346 is also well conserved. In contrast to the situation with Fps1, this domain does not contain any phosphorylation site (Figure 5.4b). As was mentioned earlier, the C-terminal GGESPVN domain is very well conserved in Yfl054 and most likely shares the same regulatory function as in Fps1 (Figure 5.4b). In conclusion, Yfl054-like proteins share extended N- and C-terminal cytoplasmic domains and several conserved sequence motifs, but their functional relevance remains unknown because the physiological role of Yfl064-like proteins has not yet been elucidated.

5.4 SUMMARY AND FUTURE ASPECTS

Orthodox aquaporins are important for specific events in yeast life cycle and in nature, such as resistance of cold and freezing and during developmental processes, such as spore maturation and formation of multicellular communities. Less is known about how these channels are regulated. *AQY* gene variability among yeast strains has highlighted a key aspect of biology in general, genetic variability between populations, evolution and ecology. The need to consider the natural habitat is vital to our understanding of processes that occur within the cell. For the aquaglyceroporin Fps1, a lot of the basic mechanisms regarding the function and regulation have now been established. Fps1 mediates regulated glycerol efflux. Its regulation depends on Hog1 and Slt2 to perform proper osmoregulation during various stages of the life cycle and within an osmotically shifting environment. But Fps1 also has protective roles to environmental hazards such as toxic metalloids and acetic acid, where Fps1 seems to be regulated differently depending on the type of stress. For a prolonged exposure to the toxic metalloid arsenite, Fps1 is upregulated and confers a net outflux of arsenite resulting in resistance to the toxic metalloid. In the case of acetic acid stress, Fps1 instead seems to be degraded to prevent the entry of the toxic substance into the cell. There are still numerous questions regarding the mechanistic effects of phosphorylation and the interplay of the regulatory components needed to modulate any relevant signal for an appropriate Fps1 channel response. Another interesting feature of Fps1 is its possible role in signal transduction. Fps1 appears to interact with the transmembrane scaffold protein Sho1 and effect downstream signalling of the HOG pathway (Lam et al. 2015). It remains to be shown if those effects are directly regulatory or due to the transport function of Fps1. It has only been theorized before that Fps1 might play a more active part in signalling, and it is likely that future investigations will expand this role of Fps1. The tight regulation of Fps1 by separate but interconnected and conserved signalling pathways offers a unique opportunity to study the interplay and output modulation of these pathways.

ACKNOWLEDGEMENTS

Work in our laboratory is supported by grants from the Swedish Research Council and the European Commission.

REFERENCES

Ahmadpour, D., C. Geijer, M. J. Tamas, K. Lindkvist-Petersson and S. Hohmann. 2014. Yeast reveals unexpected roles and regulatory features of aquaporins and aquaglyceroporins. *Biochim Biophys Acta* 1840(5):1482–1491.

Albers, E., C. Larsson, G. Lidén, C. Niklasson and L. Gustafsson. 1996. Influence of the nitrogen source on *Saccharomyces cerevisiae* anaerobic growth and product formation. *Appl Environ Microbiol* 62(9):3187–3195.

Andersen, K. S., R. Bojsen, L. G. R. Sørensen, M. W. Nielsen, M. Lisby, A. Folkesson and B. Regenberg. 2014. Genetic basis for *Saccharomyces cerevisiae* biofilm in liquid medium. *G3(Bethesda)* 4(9):1671–1680.

Ansell, R., K. Granath, S. Hohmann, J. M. Thevelein and L. Adler. 1997. The two isoenzymes for yeast NAD+-dependent glycerol 3-phosphate dehydrogenase encoded by *GPD1* and *GPD2* have distinct roles in osmoadaptation and redox regulation. *EMBO J* 16(9):2179–2187.

Azad, A. K., N. Yoshikawa, T. Ishikawa, Y. Sawa and H. Shibata. 2012. Substitution of a single amino acid residue in the aromatic/arginine selectivity filter alters the transport profiles of tonoplast aquaporin homologs. *Biochim Biophys Acta* 1818(1):1–11.

Babazadeh, R., T. Furukawa, S. Hohmann and K. Furukawa. 2014. Rewiring yeast osmostress signalling through the MAPK network reveals essential and non-essential roles of Hog1 in osmoadaptation. *Sci Rep* 4:4697.

Baltanas, R., A. Bush, A. Couto, L. Durrieu, S. Hohmann and A. Colman-Lerner. 2013. Pheromone-induced morphogenesis improves osmoadaptation capacity by activating the HOG MAPK pathway. *Sci Signal* 6(272):ra26.

Banerjee, M., N. Banerjee, P. Bhattacharjee, D. Mondal, P. R. Lythgoe, M. Martínez, J. Pan, D. A. Polya and A. K. Giri. 2013. High arsenic in rice is associated with elevated genotoxic effects in humans. *Sci Rep* 3:1–8.

Beese, S. E., T. Negishi and D. E. Levin. 2009. Identification of positive regulators of the yeast Fps1 glycerol channel. *PLoS Genet* 5(11):e1000738.

Beese-Sims, S. E., J. Lee and D. E. Levin. 2011. Yeast Fps1 glycerol facilitator functions as a homotetramer. *Yeast* 28(12):815–819.

Beitz, E., B. Wu, L. M. Holm, J. E. Schultz and T. Zeuthen. 2006. Point mutations in the aromatic/arginine region in aquaporin 1 allow passage of urea, glycerol, ammonia, and protons. *Proc Natl Acad Sci USA* 103(2):269–274.

Beyersmann, D. and A. Hartwig. 2008. Carcinogenic metal compounds: Recent insight into molecular and cellular mechanisms. *Arch Toxicol* 82(8):493–512.

Bienert, G. P., M. Thorsen, M. D. Schüssler, H. R. Nilsson, A. Wagner, M. J. Tamás and T. P. Jahn. 2008. A subgroup of plant aquaporins facilitate the bi-directional diffusion of As(OH)(3)and Sb(OH)(3)across membranes. *BMC Bio* 6:26–26.

Bill, R. M., K. Hedfalk, S. Karlgren, J. G. Mullins, J. Rydstrom and S. Hohmann. 2001. Analysis of the pore of the unusual major intrinsic protein channel, yeast Fps1p. *J Biol Chem* 276(39):36543–36549.

Björkqvist, S., R. Ansell, L. Adler and G. Lidén. 1997. Physiological response to anaerobicity of glycerol-3-phosphate dehydrogenase mutants of *Saccharomyces cerevisiae*. *Appl Environ Microbiol* 63(1):128–132.

Blomberg, A. and L. Adler. 1989. Roles of glycerol and glycerol-3-phosphate dehydrogenase(NAD+) in acquired osmotolerance of *Saccharomyces cerevisiae*. *J Bacteriol* 171(2):1087–1092.

Blomberg, A. and L. Adler. 1992. Physiology of osmotolerance in fungi. *Adv Microb Physiol* 33:145–212.

Bonhivers, M., J. M. Carbrey, S. J. Gould and P. Agre. 1998. Aquaporins in *Saccharomyces*: Genetic and functional distinctions between laboratory and wild-type strains. *J Biol Chem* 273(42):27565–27572.

Botstein, D. and G. R. Fink. 2011. Yeast: An experimental organism for 21st Century biology. *Genetics* 189(3):695–704.

Boynton, P. J. and D. Greig. 2014. The ecology and evolution of non-domesticated *Saccharomyces* species. *Yeast* 31(12):449–462.

Bun-ya, M., K. Shikata, S. Nakade, C. Yompakdee, S. Harashima and Y. Oshima. 1996. Two new genes, *PHO86* and *PHO87*, involved in inorganic phosphate uptake in *Saccharomyces cerevisiae*. *Curr Genet* 29(4):344–351.

Carbrey, J. M., M. Bonhivers, J. D. Boeke and P. Agre. 2001b. Aquaporins in *Saccharomyces*: Characterization of a second functional water channel protein. *Proc Natl Acad Sci USA* 98(3):1000–1005.

Carbrey, J. M., B. P. Cormack and P. Agre. 2001a. Aquaporin in *Candida*: Characterization of a functional water channel protein. *Yeast* 18(15):1391–6.

Chen, R. E. and J. Thorner. 2007. Function and regulation in MAPK signaling pathways: Lessons learned from the yeast *Saccharomyces cerevisiae*. *Biochim Biophys Acta* 1773(8):1311–1340.

Chu, S., J. DeRisi, M. Eisen, J. Mulholland, D. Botstein, P. O. Brown and I. Herskowitz. 1998. The transcriptional program of sporulation in budding yeast. *Science* 282(5389):699–705.

Clark, A. R. and J. L. E. Dean. 2012. The p38 MAPK pathway in rheumatoid arthritis: A sideways look. *Open Rheumatol J* 6:209–219.

Coluccio, A. E., R. K. Rodriguez, M. J. Kernan and A. M. Neiman. 2008. The yeast spore wall enables spores to survive passage through the digestive tract of *Drosophila*. *PLoS One* 3(8):e2873.

Crooks, G. E., Gary, H., John-Marc, C., and Steven, E. B. 2004. WebLogo: A Sequence Logo Generator. *Genome Research* no. 14(6):1188–1190.

Cullen, P. J. and G. F. Sprague, Jr. 2012. The regulation of filamentous growth in yeast. *Genetics* 190(1):23–49.

Davenport, K. D., K. E. Williams, B. D. Ullmann and M. C. Gustin. 1999. Activation of the *Saccharomyces cerevisiae* filamentation/invasion pathway by osmotic stress in high-osmolarity glycogen pathway mutants. *Genetics* 153(3):1091–1103.

Davenport, K. R., M. Sohaskey, Y. Kamada, D. E. Levin and M. C. Gustin. 1995. A second osmosensing signal transduction pathway in yeast hypotonic shock activates the PKC1 protein kinase-regulated cell integrity pathway. *J Biol Chem* 270(50):30157–30161.

de Groot, B. L., A. Engel and H. Grubmüller. 2003. The structure of the aquaporin-1 water channel: A comparison between cryo-electron microscopy and x-ray crystallography. *J Mol Biol* 325(3):485–493.

del Barco Barrantes, I. and A. R. Nebreda. 2012. Roles of p38 MAPKs in invasion and metastasis. *Biochem Soc Trans* 40(1):79–84.

Dohlman, H. G. and J. Thorner. 2001. Regulation of G protein-initiated signal transduction in yeast: Paradigms and principles. *Annu Rev Biochem* 70(1):703–754.

Fay, J. C. and J. A. Benavides. 2005. Evidence for domesticated and wild populations of *Saccharomyces cerevisiae*. *PLoS Genet* 1(1):66–71.

Ferreira, C., F. van Voorst, A. Martins, L. Neves, R. Oliveira, M. C. Kielland-Brandt, C. Lucas and A. Brandt. 2005. A member of the sugar transporter family, Stl1p Is the Glycerol/H(+) Symporter in *Saccharomyces cerevisiae*. *Mol Biol Cell* 16(4):2068–2076.

Flexas, J., M. Ribas-Carbo, D. T. Hanson, J. Bota, B. Otto, J. Cifre, N. McDowell, H. Medrano and R. Kaldenhoff. 2006. Tobacco aquaporin *NtAQP1* is involved in mesophyll conductance to CO_2 in vivo. *Plant J* 48(3):427–439.

Fu, D., A. Libson, L. J. W. Miercke, C. Weitzman, P. Nollert, J. Krucinski and R. M. Stroud. 2000. Structure of a glycerol-conducting channel and the basis for its selectivity. *Science* 290(5491):481–486.

Fu, H.-L., Y. Meng, E. Ordóñez, A. F. Villadangos, H. Bhattacharjee, J. A. Gil, L. M. Mateos and B. P. Rosen. 2009. Properties of arsenite efflux permeases(Acr3) from *Alkaliphilus metalliredigens* and *Corynebacterium glutamicum*. *J Biol Chem* 284(30):19887–19895.

Furukawa, K., F. Sidoux-Walter and S. Hohmann. 2009. Expression of the yeast aquaporin Aqy2 affects cell surface properties under the control of osmoregulatory and morphogenic signalling pathways. *Mol Microbiol* 74(5):1272–1286.

Gancedo, C., J. M. Gancedo and A. Sols. 1968. Glycerol metabolism in yeasts. *Eur J Biochem* 5(2):165–172.

Gasch, A. P., P. T. Spellman, C. M. Kao, O. Carmel-Harel, M. B. Eisen, G. Storz, D. Botstein and P. O. Brown. 2000. Genomic expression programs in the response of yeast cells to environmental changes. *Mol Biol Cell* 11(12):4241–4257.

Ghosh, M., J. Shen and B. P. Rosen. 1999. Pathways of As(III) detoxification in *Saccharomyces cerevisiae*. *Proc Natl Acad Sci USA* 96(9):5001–5006.

Goffeau, A. E. A., R. Aert, M. L. Agostini-Carbone, A. Ahmed, M. Aigle, L. Alberghina, K. Albermann, M. Albers, M. Aldea and D. Alexandraki. 1997. The yeast genome directory. *Nature* 387(6632):5–6.

Good, A. P., M.-P. L. Gauthier, R. L. Vannette and T. Fukami. 2014. Honey bees avoid nectar colonized by three bacterial species, but not by a yeast species, isolated from the bee gut. *PLoS One* 9(1):1–8.

Haber, J. E. 2012. Mating-type genes and MAT switching in *Saccharomyces cerevisiae*. *Genetics* 191(1):33–64.

Hedfalk, K., R. M. Bill, J. G. Mullins, S. Karlgren, C. Filipsson, J. Bergstrom, M. J. Tamas, J. Rydstrom and S. Hohmann. 2004. A regulatory domain in the C-terminal extension of the yeast glycerol channel Fps1p. *J Biol Chem* 279(15):14954–14960.

Hohmann, S. 2002. Osmotic stress signaling and osmoadaptation in yeasts. *Microbiol Mol Biol Rev* 66(2):300–372.

Hub, J. S. and B. L. De Groot. 2008. Mechanism of selectivity in aquaporins and aquaglyceroporins. *Proc Natl Acad Sci USA* 105(4):1198–1203.

Ihmels, J., S. Bergmann, M. Gerami-Nejad, I. Yanai, M. McClellan, J. Berman and N. Barkai. 2005. Rewiring of the yeast transcriptional network through the evolution of motif usage. *Science* 309:938–940.

Ikeda, M., E. Beitz, D. Kozono, W. B. Guggino, P. Agre and M. Yasui. 2002. Characterization of aquaporin-6 as a nitrate channel in mammalian cells. Requirement of pore-lining residue threonine 63. *J Biol Chem* 277(42):39873–39879.

Isayenkov, S. V. and F. J. M. Maathuis. 2008. The *Arabidopsis thaliana* aquaglyceroporin AtNIP7;1 is a pathway for arsenite uptake. *FEBS Lett* 582(11):1625–1628.

Jespersen, L. 2003. Occurrence and taxonomic characteristics of strains of *Saccharomyces cerevisiae* predominant in African indigenous fermented foods and beverages. *FEMS Yeast Res* 3(2):191–200.

Karpel, J. E. and L. F. Bisson. 2006. Aquaporins in *Saccharomyces cerevisiae* wine yeast. *FEMS Microbiol Lett* 257(1):117–123.

Kuthan, M., F. Devaux, B. Janderová, I. Slaninová, C. Jacq and Z. Palková. 2003. Domestication of wild *Saccharomyces cerevisiae* is accompanied by changes in gene expression and colony morphology. *Mol Microbiol* 47(3):745–754.

Kvitek, D. J., J. L. Will and A. P. Gasch. 2008. Variations in stress sensitivity and genomic expression in diverse *S. cerevisiae* isolates. *PLoS Genet* 4(10):e1000223.

Laize, V., R. Gobin, G. Rousselet, C. Badier, S. Hohmann, P. Ripoche and F. Tacnet. 1999. Molecular and functional study of *AQY1* from *Saccharomyces cerevisiae*: Role of the C-terminal domain. *Biochem Biophys Res Commun* 257:139–144.

Laize, V., F. Tacnet, P. Ripoche and S. Hohmann. 2000. Polymorphism of *Saccharomyces cerevisiae* aquaporins. *Yeast* 16(10):897–903.

Lam, M. H., J. Snider, M. Rehal, V. Wong, F. Aboualizadeh, L. Drecun, O. Wong et al. 2015. A comprehensive membrane interactome mapping of Sho1p reveals Fps1p as a novel key player in the regulation of the HOG pathway in *S. cerevisiae*. *J Mol Biol* 427(11):2088–2103.

Lee, J., W. Reiter, I. Dohnal, C. Gregori, S. Beese-Sims, K. Kuchler, G. Ammerer and D. E. Levin. 2013. MAPK Hog1 closes the *S. cerevisiae* glycerol channel Fps1 by phosphorylating and displacing its positive regulators. *Genes Dev* 27(23):2590–2601.

Levin, D. E. 2005. Cell wall integrity signaling in *Saccharomyces cerevisiae*. *Microbiol Mol Biol Rev* 69(2):262–291.

Levin, D. E. 2011. Regulation of cell wall biogenesis in *Saccharomyces cerevisiae*: The cell wall integrity signaling pathway. *Genetics* 189(4):1145–75.

Liu, Z., E. Boles and B. P. Rosen. 2004. Arsenic trioxide uptake by hexose permeases in *Saccharomyces cerevisiae*. *J Biol Chem* 279(17):17312–17318.

Liu, Z., J. Shen, J. M. Carbrey, R. Mukhopadhyay, P. Agre and B. P. Rosen. 2002. Arsenite transport by mammalian aquaglyceroporins AQP7 and AQP9. *Proc Natl Acad Sci USA* 99(9):6053–6058.

Luyten, K., J. Albertyn, W. F. Skibbe, B. A. Prior, J. Ramos, J. M. Thevelein and S. Hohmann. 1995. Fps1, a yeast member of the MIP family of channel proteins, is a facilitator for glycerol uptake and efflux and is inactive under osmotic stress. *EMBO J* 14(7):1360–1371.

Maciaszczyk-Dziubinska, E., I. Migdal, M. Migocka, T. Bocer and R. Wysocki. 2010. The yeast aquaglyceroporin Fps1p is a bidirectional arsenite channel. *FEBS Lett* 584(4):726–32.

Mazur, P. 1970. Cryobiology: The freezing of biological systems. *Science* 168(3934):939–949.

McGovern, P. E., J. Zhang, J. Tang, Z. Zhang, G. R. Hall, R. A. Moreau, A. Nunez et al. 2004. Fermented beverages of pre- and proto-historic China. *Proc Natl Acad Sci USA* 101(51):17593–17598.

Merlini, L., O. Dudin and S. G. Martin. 2013. Mate and fuse: How yeast cells do it. *Open Biol* 3(3):130008.

Meyrial, Â., V. Laize, Â. Gobin, P. Ripoche and S. Hohmann. 2001. Existence of a tightly regulated water channel in *Saccharomyces cerevisiae*. *Eur J Biochem* 268(2):334–343.

Mira, N. P., M. Munsterkotter, F. Dias-Valada, J. Santos, M. Palma, F. C. Roque, J. F. Guerreiro et al. 2014. The genome sequence of the highly acetic acid-tolerant *Zygosaccharomyces bailii*-derived interspecies hybrid strain ISA1307, isolated from a sparkling wine plant. *DNA Res* 21(3):299–313.

Mitani-Ueno, N., N. Yamaji, F.-J. Zhao and J. F. Ma. 2011. The aromatic/arginine selectivity filter of NIP aquaporins plays a critical role in substrate selectivity for silicon, boron, and arsenic. *J Exp Bot* 62(12):4391–4398.

Mollapour, M. and P. W. Piper. 2007. Hog1 mitogen-activated protein kinase phosphorylation targets the yeast Fps1 aquaglyceroporin for endocytosis, thereby rendering cells resistant to acetic acid. *Mol Biol Cell* 27(18):6446–6456.

Mori, I. C., J. Rhee, M. Shibasaka, S. Sasano, T. Kaneko, T. Horie and M. Katsuhara. 2014. CO_2 transport by PIP2 aquaporins of barley. *Plant Cell Physiol* 55(2):251–257.

Mortimer, R. and M. Polsinelli. 1999. On the origins of wine yeast. *Res Microbiol* 150(3):199–204.

Mukhopadhyay, R. and B. P. Rosen. 1998. *Saccharomyces cerevisiae ACR2* gene encodes an arsenate reductase. *FEMS Microbiol Lett* 168(1):127–136.

Navarro-Ródenas, A., J. M. Ruíz-Lozano, R. Kaldenhoff and A. Morte. 2012. The aquaporin *TcAQP1* of the desert truffle *Terfezia claveryi* is a membrane pore for water and CO_2 transport. *Mol Plant Microbe Interact* 25(2):259–266.

Neiman, A. M. 2011. Sporulation in the budding yeast *Saccharomyces cerevisiae*. *Genetics* 189(3):737–765.

Nollert, P., W. E. C. Harries, D. Fu, L. J. W. Miercke and R. M. Stroud. 2001. Atomic structure of a glycerol channel and implications for substrate permeation in aqua(glycero) porins. *FEBS Lett* 504(3):112–117.

Nordstrom, D. K. 2002. Worldwide occurrences of arsenic in ground water. *Science (Washington)* 296(5576):2143–2145.

Nozawa, A., J. Takano, M. Kobayashi, N. von Wiren and T. Fujiwara. 2006. Roles of *BOR1*, *DUR3*, and *FPS1* in boron transport and tolerance in *Saccharomyces cerevisiae*. *FEMS Microbiol Lett* 262(2):216–222.

Odunfa, S. A. 1985. African fermented foods. *Microbiol Ferment Foods* 2:155–191.

Oliveira, R. 2003. Fps1p channel is the mediator of the major part of glycerol passive diffusion in *Saccharomyces cerevisiae*: Artefacts and re-definitions. *Biochim Biophys Acta* 1613(1–2):57–71.

O'Rourke, S. M. and I. Herskowitz. 1998. The Hog1 MAPK prevents cross talk between the HOG and pheromone response MAPK pathways in *Saccharomyces cerevisiae*. *Genes Dev* 12(18):2874–2886.

Palkova, Z. and L. Vachova. 2006. Life within a community: Benefit to yeast long-term survival. *FEMS Microbiol Rev* 30(5):806–824.

Pavlovic-Djuranovic, S., J. E. Schultz and E. Beitz. 2003. A single aquaporin gene encodes a water/glycerol/urea facilitator in *Toxoplasma gondii* with similarity to plant tonoplast intrinsic proteins. *FEBS Lett* 555(3):500–504.

Petelenz-Kurdziel, E., C. Kuehn, B. Nordlander, D. Klein, K.-K. Hong, T. Jacobson, P. Dahl, J. Schaber, J. Nielsen, S. Hohmann and E. Klipp. 2013. Quantitative analysis of glycerol accumulation, glycolysis and growth under hyper osmotic stress. *PLoS Comput Biol* 9(6):e1003084.

Pettersson, N., C. Filipsson, E. Becit, L. Brive and S. Hohmann. 2005. Aquaporins in yeasts and filamentous fungi. *Biol Cell* 97:487–500.

Philips, J. and I. Herskowitz. 1997. Osmotic balance regulates cell fusion during mating in *Saccharomyces cerevisiae*. *J Cell Biol* 138(5):961–974.

Piskur, J., E. Rozpedowska, S. Polakova, A. Merico and C. Compag. 2006. How did *Saccharomyces* evolve to become a good brewer? *Trends Genet* 22(4):183–186.

Porquet, A. and M. Filella. 2007. Structural evidence of the similarity of Sb(OH)3 and As(OH)3 with glycerol: Implications for their uptake. *Chem Res Toxicol* 20(9):1269–1276.

Posas, F., M. Takekawa and H. Saito. 1998. Signal transduction by MAP kinase cascades in budding yeast. *Curr Opin Microbiol* 1(2):175–182.

Primig, M., R. M. Williams, E. A. Winzeler, G. G. Tevzadze, A. R. Conway, S. Y. Hwang, R. W. Davis and R. E. Esposito. 2000. The core meiotic transcriptome in budding yeasts. *Nat Genet* 26(4):415–423.

Rep, M., M. Krantz, J. M. Thevelein and S. Hohmann. 2000. The transcriptional response of *Saccharomyces cerevisiae* to osmotic shock Hot1p and Msn2p/Msn4p are required for the induction of subsets of high osmolarity glycerol pathway-dependent genes. *J Biol Chem* 275(12):8290–8300.

Reuter, M., G. Bell and D. Greig. 2007. Increased outbreeding in yeast in response to dispersal by an insect vector. *Curr Biol* 17(3):R81–R83.

Saito, H. and F. Posas. 2012. Response to hyperosmotic stress. *Genetics* 192(2):289–318.

Salvador, A. 2008. Uri Alon, *An Introduction to Systems Biology: Design Principles of Biological Circuits*, Chapman & Hall/CRC, London, ISBN 1584886420, GBP 30.99, 2007(320 pp.). *Math Biosci* 215(2):193–195.

Sampaio, J. P. and P. Goncalves. 2008. Natural populations of *Saccharomyces kudriavzevii* in Portugal are associated with oak bark and are sympatric with *S. cerevisiae* and *S. paradoxus*. *Appl Environ Microbiol* 74(7):2144–2152.

Schneider, T. D., and R. M. Stephens. 1990. Sequence logos: A new way to display consensus sequences. *Nucleic Acids Research* 18(20):6097–6100.

Shepherd, A. and P. W. Piper. 2010. The Fps1p aquaglyceroporin facilitates the use of small aliphatic amides as a nitrogen source by amidase-expressing yeasts. *FEMS Yeast Res* 10(5):527–534.

Siderius, M., O. Van Wuytswinkel, K. A. Reijenga, M. Kelders and W. H. Mager. 2000. The control of intracellular glycerol in *Saccharomyces cerevisiae* influences osmotic stress response and resistance to increased temperature. *Mol Microbiol* 36(6):1381–1390.

Sidoux-Walter, F., N. Pettersson and S. Hohmann. 2004. The *Saccharomyces cerevisiae* aquaporin Aqy1 is involved in sporulation. *Proc Natl Acad Sci USA* 101(50):17422–17427.

Soveral, G., C. Prista, T. F. Moura and M. C. Loureiro-Dias. 2010. Yeast water channels: An overview of orthodox aquaporins. *Biol Cell* 103(1):35–54.

Soveral, G., A. Veiga, M. C. Loureiro-Dias, A. Tanghe, P. Van Dijck and T. F. Moura. 2006. Water channels are important for osmotic adjustments of yeast cells at low temperature. *Microbiology* 152(Pt 5):1515–1521.

Stefaninia, I., L. Dapportob, J.-L. Legrasd, A. Calabrettaa, M. Di Paolag, C. De Filippoh, R. Violah, P. Caprettic, M. Polsinellib, S. Turillazzib and D. Cavalieri. 2012. Role of social wasps in *Saccharomyces cerevisiae* ecology and evolution. *Proc Natl Acad Sci USA* 109(33):13398–13403.

St'ovicek, V., L. Vachova, M. Kuthan and Z. Palkova. 2010. General factors important for the formation of structured biofilm-like yeast colonies. *Fungal Genet Biol* 47(12):1012–1022.

Sutherland, F. C., F. Lages, C. Lucas, K. Luyten, J. Albertyn, S. Hohmann, B. A. Prior and S. G. Kilian. 1997. Characteristics of Fps1-dependent and -independent glycerol transport in *Saccharomyces cerevisiae*. *J Bacteriol* 179(24):7790–7795.

Tamas, M. J., S. Karlgren, R. M. Bill, K. Hedfalk, L. Allegri, M. Ferreira, J. M. Thevelein, J. Rydstrom, J. G. Mullins and S. Hohmann. 2003. A short regulatory domain restricts glycerol transport through yeast Fps1p. *J Biol Chem* 278(8):6337–6345.

Tamás, M. J., K. Sutherland Luyten, F. C. Hernandez, A. Albertyn, J. Valadi, H. Li, H. Prior et al. 1999. Fps1p controls the accumulation and release of the compatible solute glycerol in yeast osmoregulation. *Mol Microbiol* 31(4):1087–1104.

Tanghe, A., J. M. Carbrey, P. Agre, J. M. Thevelein and P. Van Dijck. 2005a. Aquaporin expression and freeze tolerance in *Candida albicans*. *Appl Environ Microbiol* 71(10):6434–6437.

Tanghe, A., G. Kayingo, B. A. Prior, J. M. Thevelein and P. Van Dijck. 2005b. Heterologous aquaporin(AQY2-1) expression strongly enhances freeze tolerance of *Schizosaccharomyces pombe*. *J Mol Microbiol Biotechnol* 9(1):52–56.

Tanghe, A., P. Van Dijck, D. Colavizza and J. M. Thevelein. 2004. Aquaporin-mediated improvement of freeze tolerance of *Saccharomyces cerevisiae* is restricted to rapid freezing conditions. *Appl Environ Microbiol* 70(6):3377–3382.

Tanghe, A., P. Van Dijck, F. Dumortier, A. Teunissen, S. Hohmann and J. M. Thevelein. 2002. Aquaporin expression correlates with freeze tolerance in baker's yeast, and overexpression improves freeze tolerance in industrial strains. *Appl Environ Microbiol* 68(12):5981–5989.

Tanoue, T. and E. Nishida. 2003. Molecular recognitions in the MAP kinase cascades. *Cell Signal* 15(5):455–462.

Teixeira, M. C., N. P. Mira and I. Sá-Correia. 2011. A genome-wide perspective on the response and tolerance to food-relevant stresses in *Saccharomyces cerevisiae*. *Curr Opin Biotechnol* 22(2):150–156.

Thomson, J. M., E. A. Gaucher, M. F. Burgan, D. W. De Kee, T. Li, J. P. Aris and S. A. Benner. 2005. Resurrecting ancestral alcohol dehydrogenases from yeast. *Nat Genet* 37(6):630–635.

Thorsen, M., Y. Di, C. Tangemo, M. Morillas, D. Ahmadpour, C. Van der Does, A. Wagner et al. 2006. The MAPK Hog1p modulates Fps1p-dependent arsenite uptake and tolerance in yeast. *Mol Biol Cell* 17(10):4400–4410.

Thorsen, M., T. Jacobson, R. Vooijs, C. Navarrete, T. Bliek, H. Schat and M. J. Tamás. 2012. Glutathione serves an extracellular defence function to decrease arsenite accumulation and toxicity in yeast. *Mol Microbiol* 84(6):1177–1188.

Toh, T.-H., G. Kayingo, M. J. Merwe, S. G. Kilian, J. E. Hallsworth, S. Hohmann and B. A. Prior. 2001. Implications of *FPS1* deletion and membrane ergosterol content for glycerol efflux from *Saccharomyces cerevisiae*. *FEMS Yeast Res* 1(3):205–211.

Vachova, L., V. Stovicek, O. Hlavacek, O. Chernyavskiy, L. Stepanek, L. Kubinova and Z. Palkova. 2011. Flo11p, drug efflux pumps, and the extracellular matrix cooperate to form biofilm yeast colonies. *J Cell Biol* 194(5):679–687.

van Rijswijck, I. M., J. Dijksterhuis, J. C. Wolkers-Rooijackers, T. Abee and E. J. Smid. 2015. Nutrient limitation leads to penetrative growth into agar and affects aroma formation in *Pichia fabianii*, *P. kudriavzevii* and *Saccharomyces cerevisiae*. *Yeast* 32(1):89–101.

Verma, R. K., N. D. Prab and R. Sankararamakrishnan. 2014. New subfamilies of major intrinsic proteins in fungi suggest novel transport properties in fungal channels: Implications for the host-fungal interactions. *BMC Evol Biol* 14(173):1–16.

Vopalenska, I., V. St'ovicek, B. Janderova, L. Vachova and Z. Palkova. 2010. Role of distinct dimorphic transitions in territory colonizing and formation of yeast colony architecture. *Environ Microbiol* 12(1):264–277.

Will, J. L., H. S. Kim, J. Clarke, J. C. Painter, J. C. Fay and A. P. Gasch. 2010. Incipient balancing selection through adaptive loss of aquaporins in natural *Saccharomyces cerevisiae* populations. *PLoS Genet* 6(4):e1000893.

Wu, B. and E. Beitz. 2007. Aquaporins with selectivity for unconventional permeants. *Cell Mol Life Sci* 64(18):2413–2421.

Wysocki, R. and P. Bobrowicz. 1997. The *Saccharomyces cerevisiae* ACR3 gene encodes a putative membrane protein involved in arsenite transport. *J Biol Chem* 272(48):30061–30066.

Wysocki, R., C. C. Chéry, D. Wawrzycka, M. Van Hulle, R. Cornelis, J. M. Thevelein and M. J. Tamás. 2001. The glycerol channel Fps1p mediates the uptake of arsenite and antimonite in *Saccharomyces cerevisiae*. *Mol Microbiol* 40(6):1391–1401.

Xu, H., J. E. K. Cooke and J. J. Zwiazek. 2013. Phylogenetic analysis of fungal aquaporins provides insight into their possible role in water transport of mycorrhizal associations. *Botany* 91(8):495–504.

Zhang, N.-N., D. D. Dudgeon, S. Paliwal, A. Levchenko, E. Grote and K. W. Cunningham. 2006. Multiple signaling pathways regulate yeast cell death during the response to mating pheromones. *Mol Biol Cell* 17(8):3409–3422.

II

Aquaporins in Health and Disease

CHAPTER 6

Aquaporins in Health

Amaia Rodríguez, Leire Méndez-Giménez
and Gema Frühbeck

CONTENTS

Abstract   103
6.1 Introduction   104
6.2 Aquaporins   106
    6.2.1 AQP0: The Optical Function   106
    6.2.2 AQP1: The First Aquaporin   107
    6.2.3 AQP2: Vasopressin-Induced Aquaporin   108
    6.2.4 AQP4: Brain Function   108
    6.2.5 AQP5: Saliva, Tears and Pulmonary Secretion   109
    6.2.6 AQP6: Acid–Base Homeostasis   109
    6.2.7 AQP8: Digestive Fluid Secretion and Reproductive Function   110
6.3 Aquaglyceroporins   111
    6.3.1 AQP3: Skin Hydration and Cell Proliferation   112
    6.3.2 AQP7: Fat Metabolism and Insulin Secretion   112
    6.3.3 AQP9: Liver Gluconeogenesis and Cancer Treatment   114
    6.3.4 AQP10: Intestinal Water and Glycerol Absorption   115
6.4 Superaquaporins   115
    6.4.1 AQP11: Intravesicular Homeostasis and Oxidative Stress   116
    6.4.2 AQP12: Pancreatic Fluid Secretion   116
6.5 Concluding Remarks   116
Acknowledgements   117
References   117

ABSTRACT

AQUAPORINS (AQPs) ARE INTEGRAL plasma membrane proteins that facilitate a rapid transport of water, across biological membranes. Thirteen AQPs (AQP0–AQP12) have been identified so far in mammalian tissues. According to their structure and permeability properties, AQPs can be classified into three subgroups: AQPs (AQP0, AQP1, AQP2, AQP4, AQP5, AQP6 and AQP8), aquaglyceroporins (AQP3, AQP7, AQP9 and AQP10) and super-AQPs (AQP11 and AQP12). The functional importance of AQPs in mammalian pathophysiology has been extensively studied by analyzing the phenotype of transgenic knockout mice lacking these water channels. AQPs participate in many physiological and pathophysiological processes that include renal water absorption, neuro-homeostasis, tumour angiogenesis, fat metabolism, liver gluconeogenesis and reproduction. Until now functional studies of AQPs in humans are scarce and homozygous mutations in the genes encoding these water channels occasionally show divergent phenotypes to those observed in transgenic mice. Loss-of-function mutations in human AQPs cause congenital cataracts (AQP0), nephrogenic diabetes insipidus (AQP2), antibodies against the GIL blood group (AQP3) and impaired increase in circulating glycerol after exercise (AQP7). AQP single-nucleotide polymorphisms relevant in clinical medicine are beginning to be explored. Better understanding of the exact mechanisms and regulation of AQPs might be useful for designing potential drug targets against different metabolic disorders, such as stroke, glaucoma, brain edema, cancer and obesity.

6.1 INTRODUCTION

AQPs are channel-forming integral membrane proteins that allow the movement of water through cell membranes (King et al. 2004). To date, 13 members of the AQP family (AQP0–AQP12) have been identified in different mammalian tissues. Different names were assigned to these water channels at the moment of their cloning (e.g., MIP, CHIP28 or glycerol intrinsic protein [GLIP]) until the International Human Genome Nomenclature Committee adopted AQP as the referencing system. AQPs can be divided into three subgroups depending on their permeability and structure: AQPs, aquaglyceroporins and super-AQPs. AQPs (AQP0, AQP1, AQP2, AQP4, AQP5, AQP6 and AQP8) or *orthodox* AQPs are permeated only by water, while aquaglyceroporins (AQP3, AQP7, AQP9 and AQP10) also transport glycerol. In addition, some AQPs are also permeable to urea, arsenite, inorganic anions, gases (CO_2, NO, or O_2), H_2O_2 or some sugars (Yasui et al. 1999b; Liu et al. 2002; Bienert et al. 2007; Soria et al. 2010; Oberg et al. 2011). AQP8 facilitates ammonia and H_2O_2 transport very efficiently (Bienert et al. 2007; Soria et al. 2010). Based on these conductance properties, some authors have proposed that AQP8 might constitute a fourth branch of the AQP family termed ammoniaporin or peroxiporin (Liu et al. 2006; Soria et al. 2010; Marinelli and Marchissio 2013). Nonetheless, further studies are needed in order to associate the phylogenesis and the biophysical function of transport of this AQP.

AQPs assemble as tetramers in the plasma membrane, with each monomer forming a functional pore. The polypeptide chain of each AQP monomer is composed of six transmembrane α-helices and two highly conserved, hydrophobic Asn-Pro-Ala (NPA) consensus motifs, which overlap in the lipid bilayer forming a tridimensional *hourglass* structure that allows the movement of water through the pore (Agre et al. 1993; Jung et al. 1994b). In contrast to conventional AQPs, super-AQPs (AQP11 and AQP12) exhibit unique NPA boxes with unusual pore structure and functions (Ishibashi et al. 2000; Yakata et al. 2007; Calvanese et al. 2013). The functional importance of AQPs, aquaglyceroporins and super-AQPs in mammalian pathophysiology has been extensively studied by analyzing the phenotype of transgenic knockout mice lacking different AQPs (Table 6.1) (Wasson 2006). The impaired function of AQPs has been associated with several human diseases, such as congenital cataracts or nephrogenic diabetes insipidus. These studies have provided new

TABLE 6.1 Murine and Human Phenotypes of Aquaporin Deficiency

Name	Phenotype of AQP Knockout Mice	Phenotype of AQP Mutation in Humans	References
AQP0	Cataracts	Congenital cataracts	Shiels and Bassnett (1996), Berry et al. (2000), Francis et al. (2000), Geyer et al. (2006) and Yu et al. (2014)
AQP1	Polydipsia, defective proximal fluid reabsorption and impaired angiogenesis	Loss of Colton blood group, decreased urine-concentrating mechanism after water deprivation and reduced vascular pulmonary permeability	Preston et al. (1994), Smith et al. (1994), Ma et al. (1998) and King et al. (2001, 2002)
AQP2	Severe urine-concentrating defect	Nephrogenic diabetes insipidus	Deen et al. (1994), van Lieburg et al. (1994) and Rojek et al. (2006)
AQP3	Defective skin hydration, nephrogenic diabetes insipidus	Antibodies against GIL blood group, Ménière's disease and higher gallbladder cancer susceptibility	Ma et al. (2000, 2002), Roudier et al. (2002), Candreia et al. (2010) and Lesseur et al. (2012)
AQP4	Impaired vision, hearing and olfaction, reduced brain swelling, mild urine-concentrating defect	Not described	Ma et al. (1997a), Li and Verkman (2001), Li et al. (2002), Lu et al. (2008), Saadoun et al. (2008) and Tait et al. (2010)
AQP5	Impaired alveolar fluid clearance and saliva secretion, hyperresponsive bronchoconstriction	Nonepidermolytic palmoplantar keratoderma	Ma et al. (1999), Krane et al. (2001b), Nejsum et al. (2002) and Blaydon et al. (2013)
AQP6	Not described	Not described	Ma et al. (1996)
AQP7	Adult-onset obesity and insulin resistance, defective renal glycerol transport	Impaired increase of serum glycerol during exercise	Kondo et al. (2002), Maeda et al. (2004), Hara-Chikuma et al. (2005), Hibuse et al. (2005) and Sohara et al. (2005)
AQP8	Larger testes in males and increased follicles and number of embryos during pregnancy in females	Not described	Yang et al. (2005), Su et al. (2010) and Sha et al. (2011)
AQP9	Defective liver glycerol metabolism and delayed mortality after infection with the rodent malaria parasite, *Plasmodium berghei*	Not described	Liu et al. (2007) and Rojek et al. (2007)
AQP10	Murine *Aqp10*, a pseudogene	Not described	Morinaga et al. (2002)
AQP11	Polycystic kidney and liver	Not described	Morishita et al. (2005), Rojek et al. (2013) and Inoue et al. (2014)
AQP12	Increased susceptibility to caerulein-induced acute pancreatitis	Not described	Ohta et al. (2009)

insights into the underlying mechanisms of well-known human diseases, indicating that pharmacological modulation of water and/or glycerol transport targeting AQPs may provide novel opportunities for therapeutic interventions in several human disorders.

6.2 AQUAPORINS

AQPs are widely expressed in tissues implicated in high rates of active fluid transport (Figure 6.1). These channels play important functions in renal water absorption, lacrimation and aqueous dynamics in the eye, cerebrospinal fluid secretion, and generation of pulmonary secretions, among others.

6.2.1 AQP0: The Optical Function

AQP0 is the major intrinsic protein (original name MIP26) of lens fiber cells (Gorin et al. 1984). The crystalline lens constitutes an ocular component of high transparency and refractive index that is responsible for variable focusing of ray lights onto the photosensitive retina.

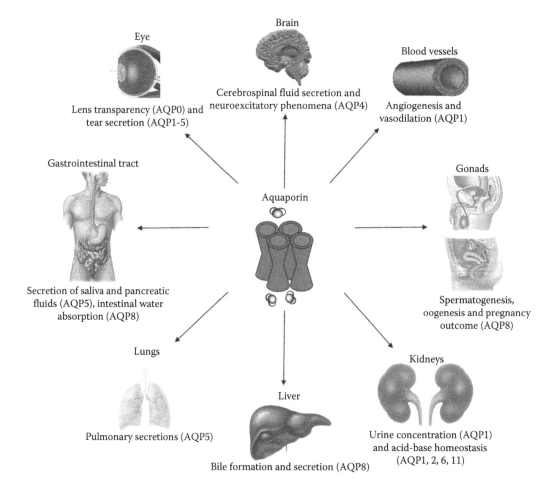

FIGURE 6.1 **(See colour insert.)** Role of aquaporins (AQPs) in the human body. The main AQPs are schematically represented indicating their location and function.

AQP0 is the most abundant protein in lens fiber cells, constituting approximately 50% of the total fiber cell membrane proteins. This AQP exerts a dual function in lens: as a water channel protein and as a cell–cell adhesion molecule (Mulders et al. 1995; Kumari et al. 2013). Homozygous or heterozygous mutations in murine *Aqp0* gene result in an autosomal-dominant bilateral lens cataract, highlighting the significance of this protein for optimal focusing and transparency of the crystalline lens (Shiels and Bassnett 1996; Shiels et al. 2001). AQP0 also exerts a key role in human optical function, since mutations in the *AQP0* gene result in autosomal-dominant congenital cataracts (Berry et al. 2000; Francis et al. 2000; Geyer et al. 2006; Yu et al. 2014). Furthermore, the post-translational glycation of AQP0 has been associated with age- and diabetes-related lens opacity (Swamy-Mruthinti 2001). These investigations highlight the significant role of AQP0 for lens transparency and homeostasis.

6.2.2 AQP1: The First Aquaporin

The identity of AQPs remained unknown until the discovery of a protein associated to the red cell Rh blood group antigens, the AQP1 (originally named channel-like integral protein of 28 kDa, CHIP28) (Agre et al. 1987; Preston and Agre 1991). AQP1 was described as a water-permeable membrane protein of red blood cells, contributing to the Colton blood group antigen, a minor blood group determinant (Preston et al. 1994; Smith et al. 1994). AQP1 is also strongly expressed in vascular endothelial cells and proliferating microvessels of human and rat malignant tumours, suggesting an important function of this AQP in tumoral angiogenesis (Saadoun et al. 2005). *Aqp1*-null mice show defective tumoral angiogenesis resulting from impaired endothelial cell migration, a key process of the angiogenic process accompanying tumour growth. AQP1-facilited cell migration appears to be relevant not only to angiogenesis but also to tumour spreading (Hu and Verkman 2006). In this sense, AQP1 expression increases the migration and metastatic potential in tumoral cells overexpressing this water channel. Nevertheless, although AQP1 is an excellent marker of microvasculature, its heterogeneous expression in different human tumors confirms that this water channel is not necessarily expressed in all neoplastic cells (Mobasheri et al. 2005). Interestingly, the presence of AQP1 in endothelial cells also allows the nitric oxide release to the extracellular space, a process required for endothelial-dependent relaxation (Herrera et al. 2006; Herrera and Garvin 2007).

The high expression of AQP1 at the apical and basolateral membranes of the proximal tubules suggested an essential role in the renal urine-concentrating mechanism (Deen et al. 1992; Sabolic et al. 1992). The generation of *Aqp1* knockout mice confirmed that AQP1 is required for the formation of concentrated urine by the kidney, since mice lacking the *Aqp1* gene have increased urine output (polyuria) and decreased urine-concentrating ability (Ma et al. 1998). After a 36 h water deprivation, *Aqp1*-deficient mice become profoundly dehydrated as a result of its urine-concentrating defect. Rare AQP1-null individuals have normal baseline renal function, as evidenced by normal glomerular filtration rates, free water clearance or lithium clearance (indices of proximal tubule function) (King et al. 2001). However, subjects with a homozygous mutation in the *AQP1* gene have a limited ability to maximally concentrate urine when water deprived. Thus, AQP1-null individuals are at risk of life-threatening clinical problems if they become dehydrated due to renal illness or environmental causes. Other phenotypical characteristics of rare AQP1-null subjects include a decreased pulmonary vascular permeability (King et al. 2002).

6.2.3 AQP2: Vasopressin-Induced Aquaporin

AQP2 was cloned from the apical membrane of collecting tubules in rat kidneys (originally named water channel–collecting duct protein, WCH-CD) (Fushimi et al. 1993). Renal water reabsorption and urine concentration are mandatory for most mammals to prevent water loss from the body. Pituitary anti-diuretic hormone vasopressin levels are elevated in response to dehydration and hypernatremia. Vasopressin induces the translocation of AQP2 from intracellular stores to the apical plasma membrane of the renal collecting tubules, rendering the cell permeable to water and, hence, favouring water reabsorption and urine concentration (Noda and Sasaki 2005; Frick et al. 2014). Defective AQP2 trafficking leads to nephrogenic diabetes insipidus, a water balance disorder characterized by large urine volumes leading to dehydration. *Aqp2*-deficient mice fail to thrive and die postnatally because of an excessive loss of extracellular fluid volume (Rojek et al. 2006). In contrast, the transgenic mice lacking AQP2 only in the renal collecting ducts survive to adulthood, but exhibit marked polyuria and severe urine-concentrating impairment in response to a 3 h water deprivation, indicating that AQP2 plays a key role in the regulation of body water balance. Several autosomal recessive mutations in the *AQP2* human gene have been shown to cause nephrogenic diabetes insipidus (Deen et al. 1994; van Lieburg et al. 1994).

6.2.4 AQP4: Brain Function

The cloning of AQP4 cDNA from rat brain and lungs was reported by two independent groups (Hasegawa et al. 1994; Jung et al. 1994a). AQP4 (originally defined as mercurial-insensitive water channel) constitutes the predominant water channel of the brain and has an important role in brain water homeostasis, although AQP4 also participates in other physiological processes, including a urine-concentrating mechanism (Ma et al. 1997b; Verkman et al. 2006). The expression of AQP4 in supportive cells next to excitable neurons, such as glial cells, retinal Müller cells, support cells in the inner ear and the olfactory epithelium, suggests an important role of this AQP in neuroexcitatory phenomena (Verkman 2012). In this sense, *Aqp4*-deficient mice show an impairment in vision, hearing and olfaction (Ma et al. 1997b; Li and Verkman 2001; Li et al. 2002; Lu et al. 2008). Growing evidence supports the involvement of AQP4 in other brain physiological functions, including the regulation of extracellular space volume, potassium buffering, cerebrospinal fluid circulation, interstitial fluid resorption, waste clearance, neuroinflammation, osmosensation, cell migration and Ca^{2+} signalling (Nagelhus and Ottersen 2014). An important role of AQP4 in the etiopathology of neuromyelitis optica, an inflammatory demyelinating disease that selectively affects optic nerves and spinal cord and is considered a severe variant of multiple sclerosis, has been also reported (Lennon et al. 2005). The defining feature of neuromyelitis optica is the development of serum IgG autoantibodies against extracellular epitopes of AQP4. The IgG binding to AQP4 in astrocytes triggers an inflammatory response leading to myelin loss and neurological deficits (Li et al. 2011). Based on the properties of AQP4 in the brain, it has been proposed that regulation of AQP4 might be useful as a novel therapeutic strategy not only for neuromyelitis optica but also against hydrocephalus, traumatic brain injury, epilepsy and stroke (King et al. 2004; Verkman 2012).

6.2.5 AQP5: Saliva, Tears and Pulmonary Secretion

AQP5 expression has been found in the apical membrane of serous acinar cells in salivary and lacrimal glands, type I alveolar and surface corneal epithelial cells (Raina et al. 1995). AQP5 plays an important role in glandular secretions of saliva and sweat, since mice lacking *Aqp5* gene exhibit a low production of hypertonic viscous saliva as well as a dramatic reduction of active sweat glands (Ma et al. 1999; Krane et al. 2001b; Nejsum et al. 2002). It has been recently described that mutations in *AQP5* gene causes autosomal-dominant diffuse nonepidermolytic palmoplantar keratoderma, a skin disorder that is characterized by the adoption of a white, spongy appearance of affected areas upon exposure to water (Blaydon et al. 2013). The presence of AQP5 in the acinar cells of lacrimal and salivary glands also opened up the hypothesis that abnormalities in AQP5 expression in these secretor glands may occur in patients with primary Sjögren's syndrome (PSS), an autoimmune disorder that is clinically characterized by dry eyes and mouth (Steinfeld et al. 2001; Tsubota et al. 2001). AQP5 misdistribution appears to be concomitant to the presence of inflammatory infiltrates and acinar destruction induced by the disease (Soyfoo et al. 2012). This hypothesis remains unclear, since other authors discarded a major role of AQP5 in the pathogenesis of the PSS because they did not observe changes in the distribution and expression of AQP5 in patients with PSS (Beroukas et al. 2001).

A strong expression of AQP5 has been found in intercalated ducts of the human pancreas, closely associated to the cystic fibrosis transmembrane conductance regulator, a marker of ductal electrolyte secretion, with both proteins declining with distance downstream from the intercalated duct (Burghardt et al. 2003). This tissue distribution suggests that AQP5 might play a role in coupling the flow of water to active electrolyte secretion in the terminal branches of the ductal tree.

The importance of AQP5 to human disease may also include disorders in the lungs and airways (Agre and Kozono 2003). This water channel has been shown to be expressed in alveolar type I and II cells, as well as in the tracheal and bronchial epithelium of mice (Raina et al. 1995; Krane et al. 2001a). *Aqp5*-deficient mice present bronchial hyperactivity after cholinergic stimulation as well as reduced airway inflammation and mucous hyperproduction during chronic allergic responses to house dust mite, suggesting a physiological role of AQP5 in modulating airway responsiveness and bronchoconstriction (Krane et al. 2001a; Shen et al. 2011). Interestingly, some forms of human asthma have been linked to chromosome 12q close to the site where the *AQP5* gene is located. In line with these observations, a reduced *Aqp5* gene expression has been detected in allergen and IL-13-induced mouse models of asthma (Krane et al. 2009).

6.2.6 AQP6: Acid–Base Homeostasis

AQP6 was cloned from the rat (WCH3) and human (hKID) kidney (Ma et al. 1996). AQP6 exhibits low water permeability and primarily transport anions (Yasui et al. 1999a; Liu et al. 2005). In glomeruli, AQP6 is present in the membrane vesicles within the podocytes' cell bodies and foot processes, suggesting a possible role in glomerular filtration (Yasui et al. 1999b). AQP6 also resides in the intracellular vesicles of acid-secreting α-intercalated cells of the collecting duct (Yasui et al. 1999b; Ohshiro et al. 2001). In particular, AQP6 localizes in vesicles containing V-type H^+-ATPase, a protein that participates in the secretion of acid into the urine. α-Intercalated cells respond to acidic–basic changes triggering the translocation of H^+-ATPase from the cytoplasmic vacuoles to the apical plasma membrane, where this ion pump secretes proton from the cells by using ATP supplied by the numerous mitochondria (Ohshiro et al. 2001). The anion permeability of AQP6 is increased fivefold by exposure to low pH and AQP6 expression in collecting ducts increases in response to chronic metabolic alkalosis or increased water intake (Promeneur et al. 2000). Taken together, AQP6 probably participates in the maintenance of acid–base homeostasis through urinary acid secretion regulation.

6.2.7 AQP8: Digestive Fluid Secretion and Reproductive Function

The cloning and functional analysis of AQP8 was reported by three independent research groups in 1997 (Ishibashi et al. 1997; Koyama et al. 1997; Ma et al. 1997a). AQP8 transcript expression has been found in different organs of the digestive system, such as salivary glands, pancreas, liver, gallbladder, small intestine and colon. Several possible functions have been proposed for AQP8, including secretion of saliva and pancreatic fluid, as well as intestinal fluid absorption/secretion (Yang et al. 2005; Portincasa and Calamita 2012). The presence of AQP8 in the hepatobiliary system might be important for bile formation and secretion (Masyuk and LaRusso 2006; Portincasa and Calamita 2012). Bile formation is initiated by hepatocytes and modified by secretory and absorptive processes in the epithelial cells of the intrahepatic ducts and gallbladder. In spite of its intracellular location in hepatocytes, under basal conditions, AQP8 is inserted into the plasma membrane to facilitate the transport of water together with AQP9 in response to hormonal stimuli, such as glucagon (Mazzone et al. 2006; Soria et al. 2009). This notion is supported by the fact that *AQP8* knockdown with interference RNA in human HepG2 hepatocytes shows a decline in the canalicular volume, suggesting an impaired basal canalicular water movement (Larocca et al. 2009). Moreover, based on the ability of AQP8 to transport ammonia and H_2O_2, recent reports have suggested its involvement in mitochondrial ammonia detoxification via ureagenesis (urea cycle), a way to metabolize amino acids and prevent the deleterious consequences of hyperammonemia and hepatic encephalopathy (Soria et al. 2010, 2013). In the gallbladder, AQP8 and AQP1 contribute to the water absorption and secretion required for bile formation and secretion (Calamita et al. 2005).

The abundance of AQP8 in male and female reproductive systems (testes and ovaries) has provided information for better understanding central processes that require water movement in the biology of reproduction (Ishibashi et al. 1997; McConnell et al. 2002). In the male reproductive tract, AQP8 is uniformly expressed in Sertoli cells, primary spermatocytes and elongated spermatids of the seminiferous tubules (Ishibashi et al. 1997; Calamita et al. 2001). AQP8 ontogeny and distribution in the male reproductive system indicate its involvement in the secretion of fluid to form the lumen of seminiferous tubules occurring during testes development and the fluid movements during spermatogenesis and sperm concentration as well as maturation (Yeung et al. 2010). On the other hand, several physi-ological processes need water movement in the female reproductive tract. Oocytogenesis or conversion of the oocyte into the mature ovum requires the formation and expansion of the fluid-filled antrum surrounding the cell. The water influx into ovarian antral fol-licles is mediated by AQP7, AQP8 and AQP9 expressed in the granulosa cells (McConnell et al. 2002). Transgenic female mice lacking the *Aqp8* gene show an increased number of mature follicles by reducing the apoptosis of granulosa cells, thus increasing their fertility (Su et al. 2010). In line with this potential role of AQP8 in fertility, several polymorphisms in the *AQP8* gene have been associated with the pathogenesis of polycystic ovary syndrome (Li et al. 2013). Besides its participation in the gametogenesis, AQP8 is also involved in the early stages of pregnancy (Richard et al. 2003). The implanting blastocyst expresses both AQP8 and AQP9, probably for fluid/solute transport during the embryo/placental develop-ment. In this sense, *Aqp8*-deficient pregnant mice have a significantly higher number of embryos as well as fetal weight, compared to wild-type controls, suggesting an important role in pregnancy outcome (Sha et al. 2011).

6.3 AQUAGLYCEROPORINS

Aquaglyceroporins (AQP3, AQP7, AQP9 and AQP10) are membrane pores permeable not only to water but also to small solutes, in particular, to glycerol, which is one of the main metabolites determining glycaemia (Reshef et al. 2003). Circulating glycerol results from lipolysis, diet-derived glycerol absorbed in the intestine and glycerol reabsorbed in the proximal tubules (Figure 6.2) (Echevarría et al. 1994; Kishida et al. 2000; Frühbeck 2005;

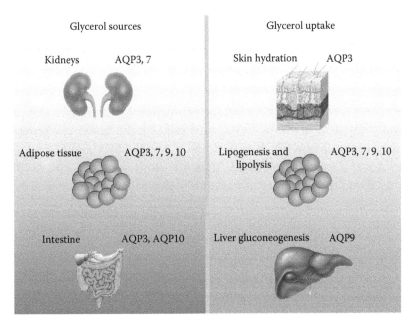

FIGURE 6.2 **(See colour insert.)** Participation of aquaglyceroporins in glycerol metabolism. The main tissues involved in glycerol absorption or efflux (*left panel*) and glycerol uptake (*right panel*) are represented.

Sohara et al. 2005). During fasting, hepatic glucose output embodies the main source of plasma glucose, and plasma glycerol becomes the major substrate for hepatic gluconeogenesis (Reshef et al. 2003; Rodríguez et al. 2011a). In addition, glycerol constitutes a key metabolite for lipid accumulation as the carbon backbone of triacylglycerols (TGs). Thus, the regulation of glycerol transport by aquaglyceroporins contributes to the control of fat accumulation and glucose homeostasis, among other biological functions.

6.3.1 AQP3: Skin Hydration and Cell Proliferation

AQP3 (originally named glycerol intrinsic protein) was initially cloned from rat kidney, being expressed in the apical and basolateral membranes of the proximal tubules (Echevarría et al. 1994; Ma et al. 1994). AQP3 plays an important role in the urine-concentrating mechanism, since *Aqp3*-deficient transgenic mice develop nephrogenic diabetes insipidus (Ma et al. 2000). Nonetheless, the main physiological roles of AQP3 include skin hydration and cell proliferation. AQP3 is expressed in the epidermal keratinocytes below the stratum corneum, the most superficial skin layer (Hara et al. 2002; Ma et al. 2002). Hydration of the stratum corneum constitutes an important determinant of skin appearance and physical properties, given that reduced hydration is found in aged skin and several skin disorders. Mice lacking AQP3 exhibit a reduced skin hydration and elasticity, delayed wound healing, as well as impaired retinoic acid-induced keratinocyte proliferation (Hara et al. 2002; Ma et al. 2002). Interestingly, topic or systemic glycerol administration or AQP3 re-expression in keratinocytes reverts the phenotype of skin abnormalities of *Aqp3*-deficient mice (Hara and Verkman 2003; Choudhary et al. 2014). In line with these observations, an abnormal epidermal AQP3 expression has been found in several skin pathologies, such as vitiligo, squamous cell carcinoma and psoriasis (Kim and Lee 2010; Voss et al. 2011; Lee et al. 2012). These findings suggest that AQP3 plays an important role in skin protection contributing to glycerol transport across the keratinocytes and provides a scientific rationale for the long-standing practice of including glycerol in cosmetic and medicinal preparations used to treat multiple skin disorders.

Although extremely rare, human homozygous mutations in the *AQP3* gene are associated with the development of antibodies against the GIL red blood cell group (Roudier et al. 2002), the onset of Menière's disease (Candreia et al. 2010) and higher susceptibility to bladder cancer (Lesseur et al. 2012). Thus, the clinical relevance of AQP3 remains to be disentangled and the discovery of novel physiological and pathological functions is warranted.

6.3.2 AQP7: Fat Metabolism and Insulin Secretion

The human *AQP7* gene, mapped to chromosome 9p13, was cloned from adipose tissue in 1997 (originally named AQPap) (Ishibashi et al. 1998). AQP7 has been considered for a long time the unique aquaglyceroporin in human adipose tissue, but AQP3, AQP9, AQP10 and most recently AQP11 represent novel additional pathways for glycerol transport in human adipocytes (Rodríguez et al. 2011b; Laforenza et al. 2013; Madeira et al. 2014). The glycerol channel AQP7 plays a pivotal role in adipose tissue enlargement and function as well as glucose homeostasis, since transgenic mice lacking the *Aqp7* gene have been shown to develop adult-onset obesity and insulin resistance (Maeda et al. 2004; Hara-Chikuma et al. 2005; Hibuse et al. 2005). The main reason for the adipocyte enlargement in *Aqp7* knockout mice is the progressive hypertrophy of fat cells, characterized by larger-sized lipid droplets (Hara-Chikuma et al. 2005; Hibuse et al. 2005). During periods of negative energy balance, such as fasting, starvation or long-term exercise, TGs stored in adipocytes are hydrolyzed by the enzyme adipocyte triglyceride lipase and hormone-sensitive lipase to glycerol and fatty acids, which are released into the bloodstream (Frühbeck et al. 2006, 2014). AQP7 facilitates the secretion of glycerol from adipocytes and a defective glycerol exit results in intracellular glycerol accumulation. Increased adipocyte glycerol concentrations would then increase TG biosynthesis, resulting in a progressive adipocyte hypertrophy (Hara-Chikuma et al. 2005). Nonetheless, human obesity is associated with a dysregulation of aquaglyceroporin expression in both fat depots (Marrades et al. 2006; Prudente et al. 2007; Catalán et al. 2008; Miranda et al. 2009; Rodríguez et al. 2011b). Visceral fat exhibits higher expression of AQP3 and AQP7, which might reflect an overall increase in lipolytic rate and glycerol release in this fat depot, while the repression of AQP7 in subcutaneous fat points to the promotion of an intracellular glycerol accumulation and a progressive adipocyte hypertrophy. Contrary to what is observed in *Aqp7*-deficient mice, lack of AQP7 protein in humans is not associated with diabetes or obesity. A single rare case of a human homozygous mutation in the coding region of the *AQP7* gene has been reported with the only apparent consequence of an impaired glycerol increase in response to exercise, reinforcing the role of this aquaglyceroporin in lipolysis (Kondo et al. 2002).

The increase in intracellular glycerol content in β-pancreatic cells has the potential to stimulate proinsulin mRNA and induce insulin secretion (Matsumura et al. 2007; Louchami et al. 2012). AQP7 expression has been reported in rat and murine pancreas as well as in the insulin-secreting pancreatic BRIN-BD11 cell line mediating the rapid entry of glycerol into β-cells (Matsumura et al. 2007; Best et al. 2009; Delporte et al. 2009). The raise in intracellular glycerol and the consequent GK activation stimulates pro-insulin biosynthesis and insulin secretion, probably through their participation in glycolysis and glycerol-3-phosphate shuttle in β-cells (Skelly et al. 2001; Matsumura et al. 2007). The increase in intracellular ATP/ADP ratio stimulated by these metabolic pathways induces the closure of ATP-sensitive K^+ channels, the β-cell membrane depolarization and the consequent opening of voltage-sensitive Ca^{2+} channels (Rodríguez et al. 2011a). The raise of cytosolic Ca^{2+} enhances the exocytosis of insulin-containing granules (Matsumura et al. 2007; Louchami et al. 2012). In this regard, *Aqp7*-deficient mice show elevated glycerol and GK activity in β-cells, causing an increase in islet TG levels, hyperinsulinemia and higher pancreatic insulin-1 and insulin-2 mRNA levels (Hibuse et al. 2005; Matsumura et al. 2007). Furthermore, *Aqp7*-null mice display a reduced β-cell mass and insulin content, suggesting a more efficient insulin biosynthesis and secretion (Matsumura et al. 2007). Further studies analyzing the regulation of AQP7 in human pancreas are needed to unravel its contribution to the development of type 2 diabetes.

6.3.3 AQP9: Liver Gluconeogenesis and Cancer Treatment

In 1998, a new AQP (AQP9) was isolated from rat liver, which was involved in the transport of water, glycerol and urea into the hepatocytes (Tsukaguchi et al. 1998). Although other glycerol channels are expressed in human liver, AQP9 is the most abundant aquaglyceroporin in human liver, where it plays a key control of hepatic gluconeogenesis and TG synthesis (Tsukaguchi et al. 1999; Rojek et al. 2007; Gena et al. 2013; Rodríguez et al. 2014). During fasting, plasma glycerol is introduced into hepatocytes by the liver-specific aquaglyceroporin AQP9, where it is converted into glycerol-3-phosphate by the GK enzymatic activity and used as a substrate for *de novo* synthesis of glucose or TG (Jelen et al. 2011; Calamita et al. 2012; Rodríguez et al. 2014). The coordinated regulation of adipose and hepatic aquaglyceroporins is extremely relevant to maintain whole-body glucose homeostasis (Kuriyama et al. 2002; Catalán et al. 2008; Rodríguez et al. 2011b). In rodents, insulin represses the expression of AQP7 and AQP9 through their negative insulin response elements in the promoter regions of these genes (Kuriyama et al. 2002; Rodríguez et al. 2006). Thus, under physiological conditions, insulin-mediated regulation of AQP7 and AQP9 may account for the increase or decrease of glycerol release from fat and gluconeogenesis in liver, in order to regulate the glucose production depending on the nutritional state (Méndez-Giménez et al. 2014). Nonetheless, the regulation of aquaglyceroporins in the human liver appears to be different to the control that takes place in mice. In human HepG2 hepatocytes, insulin increases AQP3 and AQP7 expression that might reflect the insulin-induced increase in TG content since both glycerol channels surround lipid droplets, while AQP9 is constitutively expressed in the plasma membrane and might contribute to glycerol influx (Rodríguez et al. 2011b). In human obesity and obesity-associated type 2 diabetes, non-alcoholic fatty liver diseases are associated with an up-regulation of AQP3, AQP7 and AQP9 glycerol channels in the adipose tissue leading to an increased glycerol efflux to the plasma as well as with a downregulation of AQP9 that might constitute a compensatory mechanism whereby the liver reduces the availability of intracellular glycerol as a gluconeogenic and lipogenic substrate (Catalán et al. 2008; Miranda et al. 2009; Rodríguez et al. 2011b; Rodríguez et al. 2014). In addition, an altered hepatic AQP9 expression has also been reported in hepatic inflammatory derangements, such as extrahepatic cholestasis as well as alcoholic and non-alcoholic steatohepatitis (Calamita et al. 2008; Potter et al. 2011; Rodríguez et al. 2014). In this regard, the hepatic AQP9 expression is decreased in parallel to the degree of lobular inflammation that might reflect an impaired hepatic function.

AQP9 constitutes the major route for arsenite uptake into mammalian cells, an observation potentially of large importance for understanding the action of arsenite as a human toxin and carcinogen, as well as its efficacy as a chemotherapeutic agent (Liu et al. 2002; Carbrey et al. 2009). The expression levels of AQP9 in cells obtained from patients with acute promyelocytic leukemia are closely correlated with the differential sensitivity to apoptosis induced by arsenic trioxide, the first line of treatment for acute promyelocytic leukemia (Hu et al. 2009; Iriyama et al. 2013). Moreover, the overexpression of AQP9 increases their sensitivity to arsenic trioxide in leukaemia cells and hepatocellular carcinoma and lung cancer cells (Bhattacharjee et al. 2004; Jablonski et al. 2007; Leung et al. 2007). Thus, the possibility of using pharmacological agents to increase the expression of the *AQP9* gene delivers the promise of new therapies for cancer treatment.

6.3.4 AQP10: Intestinal Water and Glycerol Absorption

AQP10 is an aquaglyceroporin abundantly expressed in human duodenum, jejunum and ileum, contributing to the absorption of water, glycerol and long sugar alcohols such as erythritol and xylitol (Hatakeyama et al. 2001; Mobasheri et al. 2004; Oberg et al. 2011). AQP10 is not expressed in mice, since the murine *Aqp10* gene has multiple structure defects leading to the production of non-functional proteins (Morinaga et al. 2002). For years, a paracellular pathway between epithelial cells has been proposed for water transport in the intestine. Since the identification of AQPs in the gastrointestinal tract, a transcellular pathway has been suggested for water movement across the absorptive epithelia via AQP10 (Hatakeyama et al. 2001). AQP10 induces water and glycerol uptake through a carrier-mediated mechanism, which is Na^+ independent and may be of facilitative type (Ishii et al. 2011). In this sense, AQP10 exerts an important role in glycerol absorption in the human small intestine, due to the fact that Western societies are used to consuming high-fat diets (Morinaga et al. 2002). Two splice variants of the human *AQP10* gene have been described in the human small intestine: AQP10v and AQP10 (Li et al. 2005). AQP10v is mainly expressed in the capillary endothelial cells of the small intestinal villi being possibly involved in the transport of water absorbed through the intestinal epithelium into blood (Hatakeyama et al. 2001). AQP10v is downregulated in villous epithelial cells of the human small intestine during acute cholera, suggesting a defensive mechanism of the host to reduce water permeability and, hence, limit the secretory response (Flach et al. 2007). On the other hand, AQP10 is localized in the cytoplasm of the gastro-entero-pancreatic (GEP) endocrine cells of human duodenum and jejunum (Li et al. 2005). In the small intestine, GEP cells secret several hormones, such as somatostatin, gastrin, glucagon or motilin. The cellular and subcellular localization of AQP10 variant suggests that this aquaglyceroporin participates in the secretion of polypeptidic hormones from GEP cells.

6.4 SUPERAQUAPORINS

AQP11 and AQP12 (originally named AQPX1 and AQPX2) are the most distantly related paralogs of the AQP family in humans (Ishibashi et al. 2014). Orthodox AQPs and aquaglyceroporins show two highly conserved NPA (Asn-Pro-Ala) sequence motifs, which overlap in the lipid layer forming a tridimensional *hourglass* structure that allows the movement of water and other small solutes through the pore (Agre et al. 1993). However, super-AQPs present an unusual sequence of the first NPA motif: Asn-Pro-Cys (NPC) in AQP11 and Asn-Pro-Thr (NPT) in AQP12 (Ishibashi et al. 2000; Yakata et al. 2007; Calvanese et al. 2013). In addition, super-AQPs localize on the membrane of intracellular organelles instead of the plasma membrane (Ishibashi 2006). Due to their particular structure and subcellular localization, AQP11 and AQP12 have been proposed as the third AQP subfamily, the super-AQPs. Paradoxically, the term super-AQPs suggests exceptional biological functions of AQP11 and AQP12, whereas the exact role of these AQP remains mostly elusive.

6.4.1 AQP11: Intravesicular Homeostasis and Oxidative Stress

AQP11 is highly expressed in testis while being moderately expressed in the kidney, liver, brain and adipose tissues (Gorelick et al. 2006; Madeira et al. 2014). To gain insight into the physiological role of AQP11, a transgenic mice lacking AQP11 was produced by Morishita et al. (2005). *Aqp11*-null mice die in the neonatal period due to renal failure with progressive vacuolization and cyst formation of the proximal tubule, leading to polycystic kidney development (Morishita et al. 2005; Tchekneva et al. 2008). An impaired glycosylation processing and aberrant membrane trafficking of polyscistin-1, a renal glycoprotein whose autosomal-dominant mutation leads to polycystic kidney disease, has been proposed as a key mechanism of cystogenesis in *Aqp11* knockout mice (Inoue et al. 2014). The vacuoles are also observed in hepatocytes close to the central vein as well as in the epithelium of intestinal villi where water is intensively absorbed (Morishita et al. 2005; Ishibashi et al. 2009). In particular, the specific hepatic deletion of AQP11 in mice results in disrupted rough endoplasmic reticulum (ER) homeostasis and increased sensitivity to ER injury upon metabolic challenge with amino acids (Rojek et al. 2013). Thus, AQP11 seems to play a relevant role in renal intravesicular homeostasis, which is essential for an adequate proximal tubular function, as well as in the control of ER stress, which has been involved in several chronic diseases including viral hepatitis, insulin resistance, NAFLD or liver ischemia.

6.4.2 AQP12: Pancreatic Fluid Secretion

AQP12 is specifically expressed in the acinar cells of the pancreas, but not in duct cells or islet cells (Itoh et al. 2005). AQP12 is located at intracellular sites and distributed in a granular, web-like pattern. The exocrine pancreas has the ability to secrete daily large amounts of fluid into the duodenum (1.0–2.5 L of juice containing digestive enzymes in humans). The intracellular localization of AQP12 in pancreatic acinar cells suggests a potential role of this super-AQP in the maturation and exocytosis of zymogen granules (Itoh et al. 2005). In this sense, *Aqp12* knockout mice exhibit an increased susceptibility to caerulein-induced acute pancreatitis, showing larger exocytic vesicles (vacuoles) in the pancreatic acini (Ohta et al. 2009). Hence, AQP12 might be involved in the proper secretion of pancreatic fluid following rapid and intense stimulation.

6.5 CONCLUDING REMARKS

The discovery of the AQP family has provided new insights into the molecular mechanisms underlying the onset of several human diseases. From a clinical point of view, the possibility of regulating AQP expression offers potentially different therapeutic approaches for a large number of pathologies. In this context, the design of small-molecule modulators of AQP expression/function may have clinical applications in the therapy of congenital cataracts, nephrogenic diabetes insipidus (AQP0 and AQP2 replacement, respectively), neuromyelitis optica (AQP4 modulators), obesity (AQP7 up-regulators) and tumour angiogenesis (AQP1 inhibitors), among others. Nonetheless, additional data, related to gene expression and protein stability, are needed to better establish a firm mechanistic basis for the involvement of AQPs in the ethiopathogenesis of these metabolic disorders. Undoubtedly, AQPs have broadened our understanding of the implications of water balance as well as water/glycerol transport in mammalian pathophysiology. Given the versatile functions of AQPs, additional and unexpected roles of these channels are sure to emerge in the coming years.

ACKNOWLEDGEMENTS

This work was supported by Fondo de Investigación Sanitaria-FEDER (PI12/00515 and PI13/01430) from the Spanish Instituto de Salud Carlos III, the Department of Health of the Gobierno de Navarra (61/2014), and the Plan de Investigación de la Universidad de Navarra (2011–2014). CIBER de Fisiopatología de la Obesidad y Nutrición (CIBERobn) is an initiative of the Instituto de Salud Carlos III, Spain.

REFERENCES

Agre, P. and Kozono, D. 2003. Aquaporin water channels: Molecular mechanisms for human diseases. *FEBS Lett* 555:72–78.

Agre, P., Preston, G. M., Smith, B. L. et al. 1993. Aquaporin CHIP: The archetypal molecular water channel. *Am J Physiol* 265:F463–F476.

Agre, P., Saboori, A. M., Asimos, A. and Smith, B. L. 1987. Purification and partial characterization of the Mr 30,000 integral membrane protein associated with the erythrocyte Rh(D) antigen. *J Biol Chem* 262:17497–17503.

Beroukas, D., Hiscock, J., Jonsson, R., Waterman, S. A. and Gordon, T. P. 2001. Subcellular distribution of aquaporin 5 in salivary glands in primary Sjogren's syndrome. *Lancet* 358:1875–1876.

Berry, V., Francis, P., Kaushal, S., Moore, A. and Bhattacharya, S. 2000. Missense mutations in MIP underlie autosomal dominant 'polymorphic' and lamellar cataracts linked to 12q. *Nat Genet* 25:15–17.

Best, L., Brown, P. D., Yates, A. P. et al. 2009. Contrasting effects of glycerol and urea transport on rat pancreatic b-cell function. *Cell Physiol Biochem* 23:255–264.

Bhattacharjee, H., Carbrey, J., Rosen, B. P. and Mukhopadhyay, R. 2004. Drug uptake and pharmacological modulation of drug sensitivity in leukemia by AQP9. *Biochem Biophys Res Commun* 322:836–841.

Bienert, G. P., Moller, A. L., Kristiansen, K. A. et al. 2007. Specific aquaporins facilitate the diffusion of hydrogen peroxide across membranes. *J Biol Chem* 282:1183–1192.

Blaydon, D. C., Lind, L. K., Plagnol, V. et al. 2013. Mutations in AQP5, encoding a water-channel protein, cause autosomal-dominant diffuse nonepidermolytic palmoplantar keratoderma. *Am J Hum Genet* 93:330–335.

Burghardt, B., Elkaer, M. L., Kwon, T. H. et al. 2003. Distribution of aquaporin water channels AQP1 and AQP5 in the ductal system of the human pancreas. *Gut* 52:1008–1016.

Calamita, G., Ferri, D., Bazzini, C. et al. 2005. Expression and subcellular localization of the AQP8 and AQP1 water channels in the mouse gall-bladder epithelium. *Biol Cell* 97:415–423.

Calamita, G., Ferri, D., Gena, P. et al. 2008. Altered expression and distribution of aquaporin-9 in the liver of rat with obstructive extrahepatic cholestasis. *Am J Physiol Gastrointest Liver Physiol* 295:G682–G690.

Calamita, G., Gena, P., Ferri, D. et al. 2012. Biophysical assessment of aquaporin-9 as principal facilitative pathway in mouse liver import of glucogenetic glycerol. *Biol Cell* 104:342–351.

Calamita, G., Mazzone, A., Bizzoca, A. and Svelto, M. 2001. Possible involvement of aquaporin-7 and -8 in rat testis development and spermatogenesis. *Biochem Biophys Res Commun* 288:619–625.

Calvanese, L., Pellegrini-Calace, M. and Oliva, R. 2013. In silico study of human aquaporin AQP11 and AQP12 channels. *Protein Sci* 22:455–466.

Candreia, C., Schmuziger, N. and Gurtler, N. 2010. Molecular analysis of aquaporin genes 1 to 4 in patients with Meniere's disease. *Cell Physiol Biochem* 26:787–792.

Carbrey, J. M., Song, L., Zhou, Y. et al. 2009. Reduced arsenic clearance and increased toxicity in aquaglyceroporin-9-null mice. *Proc Natl Acad Sci USA* 106:15956–15960.

Catalán, V., Gómez-Ambrosi, J., Pastor, C. et al. 2008. Influence of morbid obesity and insulin resistance on gene expression levels of AQP7 in visceral adipose tissue and AQP9 in liver. *Obes Surg* 18:695–701.

Choudhary, V., Olala, L. O., Qin, H. et al. 2015. Aquaporin-3 re-expression induces differentiation in a phospholipase D2-dependent manner in aquaporin-3-knockout mouse keratinocytes. *J Invest Dermatol* 135:499–507.

Deen, P. M., Dempster, J. A., Wieringa, B. and Van Os, C. H. 1992. Isolation of a cDNA for rat CHIP28 water channel: High mRNA expression in kidney cortex and inner medulla. *Biochem Biophys Res Commun* 188:1267–1273.

Deen, P. M., Verdijk, M. A., Knoers, N. V. et al. 1994. Requirement of human renal water channel aquaporin-2 for vasopressin-dependent concentration of urine. *Science* 264:92–95.

Delporte, C., Virreira, M., Crutzen, R. et al. 2009. Functional role of aquaglyceroporin 7 expression in the pancreatic b-cell line BRIN-BD11. *J Cell Physiol* 221:424–429.

Echevarría, M., Windhager, E. E., Tate, S. S. and Frindt, G. 1994. Cloning and expression of AQP3, a water channel from the medullary collecting duct of rat kidney. *Proc Natl Acad Sci USA* 91:10997–11001.

Flach, C. F., Qadri, F., Bhuiyan, T. R. et al. 2007. Differential expression of intestinal membrane transporters in cholera patients. *FEBS Lett* 581:3183–3188.

Francis, P., Berry, V., Bhattacharya, S. and Moore, A. 2000. Congenital progressive polymorphic cataract caused by a mutation in the major intrinsic protein of the lens, MIP (AQP0). *Br J Ophthalmol* 84:1376–1379.

Frick, A., Eriksson, U. K., de Mattia, F. et al. 2014. X-ray structure of human aquaporin 2 and its implications for nephrogenic diabetes insipidus and trafficking. *Proc Natl Acad Sci USA* 111:6305–6310.

Frühbeck, G. 2005. Obesity: Aquaporin enters the picture. *Nature* 438:436–437.

Frühbeck, G., Catalán, V., Gómez-Ambrosi, J. and Rodríguez, A. 2006. Aquaporin-7 and glycerol permeability as novel obesity drug-target pathways. *Trends Pharmacol Sci* 27:345–347.

Frühbeck, G., Méndez-Giménez, L., Fernández-Formoso, J. A., Fernández, S. and Rodríguez, A. 2014. Regulation of adipocyte lipolysis. *Nutr Res Rev* 27:63–93.

Fushimi, K., Uchida, S., Hara, Y. et al. 1993. Cloning and expression of apical membrane water channel of rat kidney collecting tubule. *Nature* 361:549–552.

Gena, P., Mastrodonato, M., Portincasa, P. et al. 2013. Liver glycerol permeability and aquaporin-9 are dysregulated in a murine model of non-alcoholic fatty liver disease. *PLoS One* 8:e78139.

Geyer, D. D., Spence, M. A., Johannes, M. et al. 2006. Novel single-base deletional mutation in major intrinsic protein (MIP) in autosomal dominant cataract. *Am J Ophthalmol* 141:761–763.

Gorelick, D. A., Praetorius, J., Tsunenari, T., Nielsen, S. and Agre, P. 2006. Aquaporin-11: A channel protein lacking apparent transport function expressed in brain. *BMC Biochem* 7:14.

Gorin, M. B., Yancey, S. B., Cline, J., Revel, J. P. and Horwitz, J. 1984. The major intrinsic protein (MIP) of the bovine lens fiber membrane: Characterization and structure based on cDNA cloning. *Cell* 39:49–59.

Hara, M., Ma, T. and Verkman, A. S. 2002. Selectively reduced glycerol in skin of aquaporin-3-deficient mice may account for impaired skin hydration, elasticity, and barrier recovery. *J Biol Chem* 277:46616–46621.

Hara, M. and Verkman, A. S. 2003. Glycerol replacement corrects defective skin hydration, elasticity, and barrier function in aquaporin-3-deficient mice. *Proc Natl Acad Sci USA* 100:7360–7365.

Hara-Chikuma, M., Sohara, E., Rai, T. et al. 2005. Progressive adipocyte hypertrophy in aquaporin-7-deficient mice: Adipocyte glycerol permeability as a novel regulator of fat accumulation. *J Biol Chem* 280:15493–15496.

Hasegawa, H., Ma, T., Skach, W., Matthay, M. A. and Verkman, A. S. 1994. Molecular cloning of a mercurial-insensitive water channel expressed in selected water-transporting tissues. *J Biol Chem* 269:5497–5500.

Hatakeyama, S., Yoshida, Y., Tani, T. et al. 2001. Cloning of a new aquaporin (AQP10) abundantly expressed in duodenum and jejunum. *Biochem Biophys Res Commun* 287:814–819.

Herrera, M. and Garvin, J. L. 2007. Novel role of AQP-1 in NO-dependent vasorelaxation. *Am J Physiol Renal Physiol* 292:F1443–F1451.

Herrera, M., Hong, N. J. and Garvin, J. L. 2006. Aquaporin-1 transports NO across cell membranes. *Hypertension* 48:157–164.

Hibuse, T., Maeda, N., Funahashi, T. et al. 2005. Aquaporin 7 deficiency is associated with development of obesity through activation of adipose glycerol kinase. *Proc Natl Acad Sci USA* 102:10993–10998.

Hu, J., Liu, Y. F., Wu, C. F. et al. 2009. Long-term efficacy and safety of all-trans retinoic acid/arsenic trioxide-based therapy in newly diagnosed acute promyelocytic leukemia. *Proc Natl Acad Sci USA* 106:3342–3347.

Hu, J. and Verkman, A. S. 2006. Increased migration and metastatic potential of tumor cells expressing aquaporin water channels. *FASEB J* 20:1892–1894.

Inoue, Y., Sohara, E., Kobayashi, K. et al. 2014. Aberrant glycosylation and localization of Polycystin-1 cause polycystic kidney in an AQP11 knockout model. *J Am Soc Nephrol* 25:2789–2799.

Iriyama, N., Yuan, B., Yoshino, Y. et al. 2013. Aquaporin 9, a promising predictor for the cytocidal effects of arsenic trioxide in acute promyelocytic leukemia cell lines and primary blasts. *Oncol Rep* 29:2362–2368.

Ishibashi, K. 2006. Aquaporin subfamily with unusual NPA boxes. *Biochim Biophys Acta* 1758:989–993.

Ishibashi, K., Koike, S., Kondo, S., Hara, S. and Tanaka, Y. 2009. The role of a group III AQP, AQP11 in intracellular organelle homeostasis. *J Med Invest* 56(Suppl.):312–317.

Ishibashi, K., Kuwahara, M., Gu, Y. et al. 1997. Cloning and functional expression of a new water channel abundantly expressed in the testis permeable to water, glycerol, and urea. *J Biol Chem* 272:20782–20786.

Ishibashi, K., Kuwahara, M., Kageyama, Y. et al. 2000. Molecular cloning of a new aquaporin superfamily in mammals: AQPX1 and AQPX2. In Holman S, Nielsen S eds. *Molecular Biology and Physiology of Water and Solute Transport*. New York: Kluwer Academic/Plenum Publishers, 2000, pp. 123–126.

Ishibashi, K., Tanaka, Y. and Morishita, Y. 2014. The role of mammalian superaquaporins inside the cell. *Biochim Biophys Acta* 1840:1507–1512.

Ishibashi, K., Yamauchi, K., Kageyama, Y. et al. 1998. Molecular characterization of human Aquaporin-7 gene and its chromosomal mapping. *Biochim Biophys Acta* 1399:62–66.

Ishii, M., Ohta, K., Katano, T. et al. 2011. Dual functional characteristic of human aquaporin 10 for solute transport. *Cell Physiol Biochem* 27:749–756.

Itoh, T., Rai, T., Kuwahara, M. et al. 2005. Identification of a novel aquaporin, AQP12, expressed in pancreatic acinar cells. *Biochem Biophys Res Commun* 330:832–838.

Jablonski, E. M., Mattocks, M. A., Sokolov, E. et al. 2007. Decreased aquaporin expression leads to increased resistance to apoptosis in hepatocellular carcinoma. *Cancer Lett* 250:36–46.

Jelen, S., Wacker, S., Aponte-Santamaria, C. et al. 2011. Aquaporin-9 protein is the primary route of hepatocyte glycerol uptake for glycerol gluconeogenesis in mice. *J Biol Chem* 286:44319–44325.

Jung, J. S., Bhat, R. V., Preston, G. M. et al. 1994a. Molecular characterization of an aquaporin cDNA from brain: Candidate osmoreceptor and regulator of water balance. *Proc Natl Acad Sci USA* 91:13052–13056.

Jung, J. S., Preston, G. M., Smith, B. L., Guggino, W. B. and Agre, P. 1994b. Molecular structure of the water channel through aquaporin CHIP. The hourglass model. *J Biol Chem* 269:14648–14654.

Kim, N. H. and Lee, A. Y. 2010. Reduced aquaporin3 expression and survival of keratinocytes in the depigmented epidermis of vitiligo. *J Invest Dermatol* 130:2231–2239.

King, L. S., Choi, M., Fernandez, P. C., Cartron, J. P. and Agre, P. 2001. Defective urinary-concentrating ability due to a complete deficiency of aquaporin-1. *N Engl J Med* 345:175–179.

King, L. S., Kozono, D. and Agre, P. 2004. From structure to disease: The evolving tale of aquaporin biology. *Nat Rev Mol Cell Biol* 5:687–698.

King, L. S., Nielsen, S., Agre, P. and Brown, R. H. 2002. Decreased pulmonary vascular permeability in aquaporin-1-null humans. *Proc Natl Acad Sci USA* 99:1059–1063.

Kishida, K., Kuriyama, H., Funahashi, T. et al. 2000. Aquaporin adipose, a putative glycerol channel in adipocytes. *J Biol Chem* 275:20896–20902.

Kondo, H., Shimomura, I., Kishida, K. et al. 2002. Human aquaporin adipose (AQPap) gene. Genomic structure, promoter analysis and functional mutation. *Eur J Biochem* 269:1814–1826.

Koyama, Y., Yamamoto, T., Kondo, D. et al. 1997. Molecular cloning of a new aquaporin from rat pancreas and liver. *J Biol Chem* 272:30329–30333.

Krane, C. M., Deng, B., Mutyam, V. et al. 2009. Altered regulation of aquaporin gene expression in allergen and IL-13-induced mouse models of asthma. *Cytokine* 46:111–118.

Krane, C. M., Fortner, C. N., Hand, A. R. et al. 2001a. Aquaporin 5-deficient mouse lungs are hyperresponsive to cholinergic stimulation. *Proc Natl Acad Sci USA* 98:14114–14119.

Krane, C. M., Melvin, J. E., Nguyen, H. V. et al. 2001b. Salivary acinar cells from aquaporin 5-deficient mice have decreased membrane water permeability and altered cell volume regulation. *J Biol Chem* 276:23413–23420.

Kumari, S. S., Gandhi, J., Mustehsan, M. H., Eren, S. and Varadaraj, K. 2013. Functional characterization of an AQP0 missense mutation, R33C, that causes dominant congenital lens cataract, reveals impaired cell-to-cell adhesion. *Exp Eye Res* 116:371–385.

Kuriyama, H., Shimomura, I., Kishida, K. et al. 2002. Coordinated regulation of fat-specific and liver-specific glycerol channels, aquaporin adipose and aquaporin 9. *Diabetes* 51:2915–2921.

Laforenza, U., Scaffino, M. F. and Gastaldi, G. 2013. Aquaporin-10 represents an alternative pathway for glycerol efflux from human adipocytes. *PLoS One* 8:e54474.

Larocca, M. C., Soria, L. R., Espelt, M. V., Lehmann, G. L. and Marinelli, R. A. 2009. Knockdown of hepatocyte aquaporin-8 by RNA interference induces defective bile canalicular water transport. *Am J Physiol Gastrointest Liver Physiol* 296:G93–G100.

Lee, Y., Je, Y. J., Lee, S. S. et al. 2012. Changes in transepidermal water loss and skin hydration according to expression of aquaporin-3 in psoriasis. *Ann Dermatol* 24:168–174.

Lennon, V. A., Kryzer, T. J., Pittock, S. J., Verkman, A. S. and Hinson, S. R. 2005. IgG marker of optic-spinal multiple sclerosis binds to the aquaporin-4 water channel. *J Exp Med* 202:473–477.

Lesseur, C., Gilbert-Diamond, D., Andrew, A. S. et al. 2012. A case–control study of polymorphisms in xenobiotic and arsenic metabolism genes and arsenic-related bladder cancer in New Hampshire. *Toxicol Lett* 210:100–106.

Leung, J., Pang, A., Yuen, W. H., Kwong, Y. L. and Tse, E. W. 2007. Relationship of expression of aquaglyceroporin 9 with arsenic uptake and sensitivity in leukemia cells. *Blood* 109:740–746.

Li, H., Kamiie, J., Morishita, Y. et al. 2005. Expression and localization of two isoforms of AQP10 in human small intestine. *Biol Cell* 97:823–829.

Li, J., Patil, R. V. and Verkman, A. S. 2002. Mildly abnormal retinal function in transgenic mice without Muller cell aquaporin-4 water channels. *Invest Ophthalmol Vis Sci* 43:573–579.

Li, J. and Verkman, A. S. 2001. Impaired hearing in mice lacking aquaporin-4 water channels. *J Biol Chem* 276:31233–31237.

Li, L., Zhang, H., Varrin-Doyer, M., Zamvil, S. S. and Verkman, A. S. 2011. Proinflammatory role of aquaporin-4 in autoimmune neuroinflammation. *FASEB J* 25:1556–1566.

Li, Y., Liu, H., Zhao, H. et al. 2013. Association of AQP8 in women with PCOS. *Reprod Biomed Online* 27:419–422.

Liu, K., Kozono, D., Kato, Y. et al. 2005. Conversion of aquaporin 6 from an anion channel to a water-selective channel by a single amino acid substitution. *Proc Natl Acad Sci USA* 102:2192–2197.

Liu, K., Nagase, H., Huang, C. G., Calamita, G. and Agre, P. 2006. Purification and functional characterization of aquaporin-8. *Biol Cell* 98:153–161.

Liu, Y., Promeneur, D., Rojek, A. et al. 2007. Aquaporin 9 is the major pathway for glycerol uptake by mouse erythrocytes, with implications for malarial virulence. *Proc Natl Acad Sci USA* 104:12560–12564.

Liu, Z., Shen, J., Carbrey, J. M. et al. 2002. Arsenite transport by mammalian aquaglyceroporins AQP7 and AQP9. *Proc Natl Acad Sci USA* 99:6053–6058.

Louchami, K., Best, L., Brown, P. et al. 2012. A new role for aquaporin 7 in insulin secretion. *Cell Physiol Biochem* 29:65–74.

Lu, D. C., Zhang, H., Zador, Z. and Verkman, A. S. 2008. Impaired olfaction in mice lacking aquaporin-4 water channels. *FASEB J* 22:3216–3223.

Ma, T., Frigeri, A., Hasegawa, H. and Verkman, A. S. 1994. Cloning of a water channel homolog expressed in brain meningeal cells and kidney collecting duct that functions as a stilbene-sensitive glycerol transporter. *J Biol Chem* 269:21845–21849.

Ma, T., Hara, M., Sougrat, R., Verbavatz, J. M. and Verkman, A. S. 2002. Impaired stratum corneum hydration in mice lacking epidermal water channel aquaporin-3. *J Biol Chem* 277:17147–17153.

Ma, T., Song, Y., Gillespie, A. et al. 1999. Defective secretion of saliva in transgenic mice lacking aquaporin-5 water channels. *J Biol Chem* 274:20071–20074.

Ma, T., Song, Y., Yang, B. et al. 2000. Nephrogenic diabetes insipidus in mice lacking aquaporin-3 water channels. *Proc Natl Acad Sci USA* 97:4386–4391.

Ma, T., Yang, B., Gillespie, A. et al. 1997a. Generation and phenotype of a transgenic knockout mouse lacking the mercurial-insensitive water channel aquaporin-4. *J Clin Invest* 100:957–962.

Ma, T., Yang, B., Kuo, W. L. and Verkman, A. S. 1996. cDNA cloning and gene structure of a novel water channel expressed exclusively in human kidney: Evidence for a gene cluster of aquaporins at chromosome locus 12q13. *Genomics* 35:543–550.

Ma, T., Yang, B., Gillespie, A. et al. 1998. Severely impaired urinary concentrating ability in transgenic mice lacking aquaporin-1 water channels. *J Biol Chem* 273:4296–4299.

Ma, T., Yang, B. and Verkman, A. S. 1997b. Cloning of a novel water and urea-permeable aquaporin from mouse expressed strongly in colon, placenta, liver, and heart. *Biochem Biophys Res Commun* 240:324–328.

Madeira, A., Fernandez-Veledo, S., Camps, M. et al. 2014. Human aquaporin-11 is a water and glycerol channel and localizes in the vicinity of lipid droplets in human adipocytes. *Obesity* 22:2010–2017.

Maeda, N., Funahashi, T., Hibuse, T. et al. 2004. Adaptation to fasting by glycerol transport through aquaporin 7 in adipose tissue. *Proc Natl Acad Sci USA* 101:17801–17806.

Marinelli, R. A. and Marchissio, M. J. 2013. Mitochondrial aquaporin-8: A functional peroxiporin? *Antioxid Redox Signal* 19:896.

Marrades, M. P., Milagro, F. I., Martínez, J. A. and Moreno-Aliaga, M. J. 2006. Differential expression of aquaporin 7 in adipose tissue of lean and obese high fat consumers. *Biochem Biophys Res Commun* 339:785–789.

Masyuk, A. I. and LaRusso, N. F. 2006. Aquaporins in the hepatobiliary system. *Hepatology* 43:S75–S81.

Matsumura, K., Chang, B. H., Fujimiya, M. et al. 2007. Aquaporin 7 is a b-cell protein and regulator of intraislet glycerol content and glycerol kinase activity, b-cell mass, and insulin production and secretion. *Mol Cell Biol* 27:6026–6037.

Mazzone, A., Tietz, P., Jefferson, J., Pagano, R. and LaRusso, N. F. 2006. Isolation and characterization of lipid microdomains from apical and basolateral plasma membranes of rat hepatocytes. *Hepatology* 43:287–296.

McConnell, N. A., Yunus, R. S., Gross, S. A. et al. 2002. Water permeability of an ovarian antral follicle is predominantly transcellular and mediated by aquaporins. *Endocrinology* 143:2905–2912.

Méndez-Giménez, L., Rodríguez, A., Balaguer, I. and Frühbeck, G. 2014. Role of aquaglyceroporins and caveolins in energy and metabolic homeostasis. *Mol Cell Endocrinol* 397:78–92.

Miranda, M., Ceperuelo-Mallafré, V., Lecube, A. et al. 2009. Gene expression of paired abdominal adipose AQP7 and liver AQP9 in patients with morbid obesity: Relationship with glucose abnormalities. *Metabolism* 58:1762–1768.

Mobasheri, A., Airley, R., Hewitt, S. M. and Marples, D. 2005. Heterogeneous expression of the aquaporin 1 (AQP1) water channel in tumors of the prostate, breast, ovary, colon and lung: A study using high density multiple human tumor tissue microarrays. *Int J Oncol* 26:1149–1158.

Mobasheri, A., Shakibaei, M. and Marples, D. 2004. Immunohistochemical localization of aquaporin 10 in the apical membranes of the human ileum: A potential pathway for luminal water and small solute absorption. *Histochem Cell Biol* 121:463–471.

Morinaga, T., Nakakoshi, M., Hirao, A., Imai, M. and Ishibashi, K. 2002. Mouse aquaporin 10 gene (AQP10) is a pseudogene. *Biochem Biophys Res Commun* 294:630–634.

Morishita, Y., Matsuzaki, T., Hara-chikuma, M. et al. 2005. Disruption of aquaporin-11 produces polycystic kidneys following vacuolization of the proximal tubule. *Mol Cell Biol* 25:7770–7779.

Mulders, S. M., Preston, G. M., Deen, P. M. et al. 1995. Water channel properties of major intrinsic protein of lens. *J Biol Chem* 270:9010–9016.

Nagelhus, E. A. and Ottersen, O. P. 2014. Physiological roles of aquaporin-4 in brain. *Physiol Rev* 93:1543–1562.

Nejsum, L. N., Kwon, T. H., Jensen, U. B. et al. 2002. Functional requirement of aquaporin-5 in plasma membranes of sweat glands. *Proc Natl Acad Sci USA* 99:511–516.

Noda, Y. and Sasaki, S. 2005. Trafficking mechanism of water channel aquaporin-2. *Biol Cell* 97:885–892.

Oberg, F., Sjohamn, J., Fischer, G. et al. 2011. Glycosylation increases the thermostability of human aquaporin 10 protein. *J Biol Chem* 286:31915–31923.

Ohshiro, K., Yaoita, E., Yoshida, Y. et al. 2001. Expression and immunolocalization of AQP6 in intercalated cells of the rat kidney collecting duct. *Arch Histol Cytol* 64:329–338.

Ohta, E., Itoh, T., Nemoto, T. et al. 2009. Pancreas-specific aquaporin 12 null mice showed increased susceptibility to caerulein-induced acute pancreatitis. *Am J Physiol Cell Physiol* 297:C1368–C1378.

Portincasa, P. and Calamita, G. 2012. Water channel proteins in bile formation and flow in health and disease: When immiscible becomes miscible. *Mol Aspects Med* 33:651–664.

Potter, J. J., Koteish, A., Hamilton, J. et al. 2011. Effects of acetaldehyde on hepatocyte glycerol uptake and cell size: Implication of aquaporin 9. *Alcohol Clin Exp Res* 35:939–945.

Preston, G. M. and Agre, P. 1991. Isolation of the cDNA for erythrocyte integral membrane protein of 28 kilodaltons: Member of an ancient channel family. *Proc Natl Acad Sci USA* 88:11110–11114.

Preston, G. M., Smith, B. L., Zeidel, M. L., Moulds, J. J. and Agre, P. 1994. Mutations in aquaporin-1 in phenotypically normal humans without functional CHIP water channels. *Science* 265:1585–1587.

Promeneur, D., Kwon, T. H., Yasui, M. et al. 2000. Regulation of AQP6 mRNA and protein expression in rats in response to altered acid-base or water balance. *Am J Physiol Renal Physiol* 279:F1014–F1026.

Prudente, S., Flex, E., Morini, E. et al. 2007. A functional variant of the adipocyte glycerol channel aquaporin 7 gene is associated with obesity and related metabolic abnormalities. *Diabetes* 56:1468–1474.

Raina, S., Preston, G. M., Guggino, W. B. and Agre, P. 1995. Molecular cloning and character-ization of an aquaporin cDNA from salivary, lacrimal, and respiratory tissues. *J Biol Chem* 270:1908–1912.

Reshef, L., Olswang, Y., Cassuto, H. et al. 2003. Glyceroneogenesis and the triglyceride/fatty acid cycle. *J Biol Chem* 278:30413–30416.

Richard, C., Gao, J., Brown, N. and Reese, J. 2003. Aquaporin water channel genes are differentially expressed and regulated by ovarian steroids during the periimplantation period in the mouse. *Endocrinology* 144:1533–1541.

Rodríguez, A., Catalán, V., Gómez-Ambrosi, J. and Frühbeck, G. 2006. Role of aquaporin-7 in the pathophysiological control of fat accumulation in mice. *FEBS Lett* 580:4771–4776.

Rodríguez, A., Catalán, V., Gómez-Ambrosi, J. and Frühbeck, G. 2011a. Aquaglyceroporins serve as metabolic gateways in adiposity and insulin resistance control. *Cell Cycle* 10:1548–1556.

Rodríguez, A., Catalán, V., Gómez-Ambrosi, J. et al. 2011b. Insulin- and leptin-mediated control of aquaglyceroporins in human adipocytes and hepatocytes is mediated via the PI3K/Akt/mTOR signaling cascade. *J Clin Endocrinol Metab* 96:E586–E597.

Rodríguez, A., Gena, P., Méndez-Giménez, L. et al. 2014. Reduced hepatic aquaporin-9 and glycerol permeability are related to insulin resistance in non-alcoholic fatty liver disease. *Int J Obes* 38:1213–1220.

Rojek, A., Fuchtbauer, E. M., Fuchtbauer, A. et al. 2013. Liver-specific aquaporin 11 knockout mice show rapid vacuolization of the rough endoplasmic reticulum in periportal hepatocytes after feeding amino acids. *Am J Physiol Gastrointest Liver Physiol* 304:G501–G515.

Rojek, A., Fuchtbauer, E. M., Kwon, T. H., Frokiaer, J. and Nielsen, S. 2006. Severe urinary concen-trating defect in renal collecting duct-selective AQP2 conditional-knockout mice. *Proc Natl Acad Sci USA* 103:6037–6042.

Rojek, A. M., Skowronski, M. T., Fuchtbauer, E. M. et al. 2007. Defective glycerol metabolism in aquaporin 9 (AQP9) knockout mice. *Proc Natl Acad Sci USA* 104:3609–3614.

Roudier, N., Ripoche, P., Gane, P. et al. 2002. AQP3 deficiency in humans and the molecular basis of a novel blood group system, GIL. *J Biol Chem* 277:45854–45859.

Saadoun, S., Bell, B. A., Verkman, A. S. and Papadopoulos, M. C. 2008. Greatly improved neu-rological outcome after spinal cord compression injury in AQP4-deficient mice. *Brain* 131:1087–1098.

Saadoun, S., Papadopoulos, M. C., Hara-Chikuma, M. and Verkman, A. S. 2005. Impairment of angiogenesis and cell migration by targeted aquaporin-1 gene disruption. *Nature* 434:786–792.

Sabolic, I., Valenti, G., Verbavatz, J. M. et al. 1992. Localization of the CHIP28 water channel in rat kidney. *Am J Physiol* 263:C1225–C1233.

Sha, X. Y., Xiong, Z. F., Liu, H. S., Zheng, Z. and Ma, T. H. 2011. Pregnant phenotype in aquaporin 8-deficient mice. *Acta Pharmacol Sin* 32:840–844.

Shen, Y., Wang, Y., Chen, Z. et al. 2011. Role of aquaporin 5 in antigen-induced airway inflamma-tion and mucous hyperproduction in mice. *J Cell Mol Med* 15:1355–1363.

Shiels, A. and Bassnett, S. 1996. Mutations in the founder of the MIP gene family underlie cataract development in the mouse. *Nat Genet* 12:212–215.

Shiels, A., Bassnett, S., Varadaraj, K. et al. 2001. Optical dysfunction of the crystalline lens in aquaporin-0-deficient mice. *Physiol Genomics* 7:179–186.

Skelly, R. H., Wicksteed, B., Antinozzi, P. A. and Rhodes, C. J. 2001. Glycerol-stimulated proinsu-lin biosynthesis in isolated pancreatic rat islets via adenoviral-induced expression of glycerol kinase is mediated via mitochondrial metabolism. *Diabetes* 50:1791–1798.

Smith, B. L., Preston, G. M., Spring, F. A., Anstee, D. J. and Agre, P. 1994. Human red cell aquaporin CHIP. I. Molecular characterization of ABH and Colton blood group antigens. *J Clin Invest* 94:1043–1049.

Sohara, E., Rai, T., Miyazaki, J. et al. 2005. Defective water and glycerol transport in the proximal tubules of AQP7 knockout mice. *Am J Physiol Renal Physiol* 289:F1195–F1200.

Soria, L. R., Fanelli, E., Altamura, N. et al. 2010. Aquaporin-8-facilitated mitochondrial ammonia transport. *Biochem Biophys Res Commun* 393:217–221.

Soria, L. R., Gradilone, S. A., Larocca, M. C. and Marinelli, R. A. 2009. Glucagon induces the gene expression of aquaporin-8 but not that of aquaporin-9 water channels in the rat hepatocyte. *Am J Physiol Regul Integr Comp Physiol* 296:R1274–R1281.

Soria, L. R., Marrone, J., Calamita, G. and Marinelli, R. A. 2013. Ammonia detoxification via ureagenesis in rat hepatocytes involves mitochondrial aquaporin-8 channels. *Hepatology* 57:2061–2071.

Soyfoo, M. S., Konno, A., Bolaky, N. et al. 2012. Link between inflammation and aquaporin-5 distribution in submandibular gland in Sjogren's syndrome? *Oral Dis* 18:568–574.

Steinfeld, S., Cogan, E., King, L. S. et al. 2001. Abnormal distribution of aquaporin-5 water channel protein in salivary glands from Sjogren's syndrome patients. *Lab Invest* 81:143–148.

Su, W., Qiao, Y., Yi, F. et al. 2010. Increased female fertility in aquaporin 8-deficient mice. *IUBMB Life* 62:852–857.

Swamy-Mruthinti, S. 2001. Glycation decreases calmodulin binding to lens transmembrane protein, MIP. *Biochim Biophys Acta* 1536:64–72.

Tait, M. J., Saadoun, S., Bell, B. A., Verkman, A. S. and Papadopoulos, M. C. 2010. Increased brain edema in aqp4-null mice in an experimental model of subarachnoid hemorrhage. *Neuroscience* 167:60–67.

Tchekneva, E. E., Khuchua, Z., Davis, L. S. et al. 2008. Single amino acid substitution in aquaporin 11 causes renal failure. *J Am Soc Nephrol* 19:1955–1964.

Tsubota, K., Hirai, S., King, L. S., Agre, P. and Ishida, N. 2001. Defective cellular trafficking of lacrimal gland aquaporin-5 in Sjogren's syndrome. *Lancet* 357:688–689.

Tsukaguchi, H., Shayakul, C., Berger, U. V. et al. 1998. Molecular characterization of a broad selectivity neutral solute channel. *J Biol Chem* 273:24737–24743.

Tsukaguchi, H., Weremowicz, S., Morton, C. C. and Hediger, M. A. 1999. Functional and molecular characterization of the human neutral solute channel aquaporin-9. *Am J Physiol* 277:F685–F696.

van Lieburg, A. F., Verdijk, M. A., Knoers, V. V. et al. 1994. Patients with autosomal nephrogenic diabetes insipidus homozygous for mutations in the aquaporin 2 water-channel gene. *Am J Hum Genet* 55:648–652.

Verkman, A. S. 2012. Aquaporins in clinical medicine. *Annu Rev Med* 63:303–316.

Verkman, A. S., Binder, D. K., Bloch, O., Auguste, K. and Papadopoulos, M. C. 2006. Three distinct roles of aquaporin-4 in brain function revealed by knockout mice. *Biochim Biophys Acta* 1758:1085–1093.

Voss, K. E., Bollag, R. J., Fussell, N. et al. 2011. Abnormal aquaporin-3 protein expression in hyperproliferative skin disorders. *Arch Dermatol Res* 303:591–600.

Wasson, K. 2006. Phenotypes of aquaporin mutants in genetically altered mice. *Comp Med* 56:96–104.

Yakata, K., Hiroaki, Y., Ishibashi, K. et al. 2007. Aquaporin-11 containing a divergent NPA motif has normal water channel activity. *Biochim Biophys Acta* 1768:688–693.

Yang, B., Song, Y., Zhao, D. and Verkman, A. S. 2005. Phenotype analysis of aquaporin-8 null mice. *Am J Physiol Cell Physiol* 288:C1161–C1170.

Yasui, M., Hazama, A., Kwon, T. H. et al. 1999a. Rapid gating and anion permeability of an intracellular aquaporin. *Nature* 402:184–187.

Yasui, M., Kwon, T. H., Knepper, M. A., Nielsen, S. and Agre, P. 1999b. Aquaporin-6: An intracellular vesicle water channel protein in renal epithelia. *Proc Natl Acad Sci USA* 96:5808–5813.

Yeung, C. H., Callies, C., Tuttelmann, F., Kliesch, S. and Cooper, T. G. 2010. Aquaporins in the human testis and spermatozoa – Identification, involvement in sperm volume regulation and clinical relevance. *Int J Androl* 33:629–641.

Yu, Y., Chen, P., Li, J. et al. 2014. A novel MIP gene mutation associated with autosomal dominant congenital cataracts in a Chinese family. *BMC Med Genet* 15:6.

CHAPTER 7

Renal Aquaporins

Role in Water Balance Disorders*

Tae-Hwan Kwon and Søren Nielsen

CONTENTS

Abstract   125
7.1   Introduction   126
7.2   Expression and Function of Aquaporins in the Kidney   127
7.3   Vasopressin Regulation of Kidney Aquaporins   130
7.4   Dysregulation of Renal Aquaporins in Water Balance Disorders   133
    7.4.1   Urinary Concentrating Defects   134
        7.4.1.1   Inherited Forms of Diabetes Insipidus   134
        7.4.1.2   Acquired Forms of NDI   135
        7.4.1.3   Lithium-Induced NDI   135
        7.4.1.4   Electrolytes Abnormality (Hypokalemia and Hypercalcemia)   139
        7.4.1.5   Ureteral Obstruction   140
        7.4.1.6   Acute and Chronic Renal Failure   142
7.5   Conditions Associated with Water Retention   143
    7.5.1   Congestive Heart Failure   143
    7.5.2   Hepatic Cirrhosis   144
    7.5.3   Experimental Nephrotic Syndrome   145
    7.5.4   SIADH and Vasopressin Escape   145
References   145

ABSTRACT

THE DISCOVERY OF AQUAPORIN (AQP) membrane water channels by Peter Agre and co-workers answered a long-standing biophysical question of how water crosses biological membranes and provided insight, at the molecular level, into the fundamental physiology of water balance regulation and the pathophysiology of water balance disorders. In the kidney, AQP1–AQP4, AQP6–AQP8 and AQP11 are expressed and multiple studies

---

* This chapter is an update of previous detailed reviews (Nielsen et al. 2002, 2007; Rojek et al. 2008; Kwon et al. 2009; Kwon et al. 2013).

have underscored the essential roles of AQP1–AQP4 in renal regulation of body water balance. Vasopressin regulates acutely the water permeability of the kidney collecting duct by regulation of AQP2 trafficking from intracellular vesicles to the apical plasma membrane. This involves complex signalling mechanisms. In addition, long-term regulation of AQP2 and AQP3 expression act in concert with acute regulation of AQP2 to tightly control collecting duct water reabsorption and hence body water balance. Importantly, mutations or dysregulation of renal AQPs is involved in water balance disorders. Mutations in AQP2 lead to nephrogenic diabetes insipidus (NDI). Downregulation of AQP2 expression and/or dysregulation of AQP2 trafficking leads to urinary concentrating defects seen in acute kidney injury, drug-induced polyuria (e.g. lithium-induced NDI), compulsive water drinking (polydipsia) and diseases associated with electrolyte disorders (hypokalemia and hypercalcemia). Conversely, upregulation and enhanced apical trafficking of AQP2 is involved in diseases associated with water retention such as congestive heart failure.

7.1 INTRODUCTION

Kidneys plays a key role in body water and electrolyte homeostasis. Thus, comprehending the mechanisms of water transport is essential to understanding mammalian kidney physiology and water balance. Because of its importance to human health, water permeability has been particularly well characterized in the mammalian kidney (Knepper and Burg 1983; Knepper et al, 2015). Approximately, 180 L/day of glomerular filtrate is generated in an average adult human, more than 90% of this is constitutively reabsorbed by the highly water-permeable proximal tubules and descending thin limbs in the loop of Henle. The ascending thin limbs and thick limbs are relatively impermeable to water and empty into renal distal tubules and ultimately into the collecting ducts. The collecting ducts are important clinically in water balance disorders, because they are the chief site of regulated water reabsorption. Basal epithelial water permeability in collecting duct principal cells is low, but the water permeability can become exceedingly high when stimulated with arginine vasopressin (AVP, also known as antidiuretic hormone). In this regard, the toad urinary bladder behaves like the collecting duct, and it has served as an important model of vasopressin-regulated water permeability. Stimulation of this epithelium with vasopressin produces an increase in water permeability in the apical membrane, which coincides with the redistribution of intracellular particles to the cell surface (Kachadorian et al. 1975, 1977; Wade and Kachadorian 1988). These particles were believed to contain water channels. The discovery of AQP1 by Agre and colleagues (Preston and Agre 1991; Preston et al. 1992; Smith and Agre 1991) explained the long-standing biophysical question of how water specifically crosses biological membranes and these studies led to the identification of a whole new family of membrane proteins, the AQP water channels. At present, at least eight AQPs are expressed at distinct sites in the kidney, and four members of this family (AQP1–AQP4) have been demonstrated to play pivotal roles in the physiology and pathophysiology for renal regulation of body water balance. In the present review, we will focus on the regulation of renal AQPs and in particular how the regulation of AQP2 takes place. In addition, a number of inherited and acquired conditions characterized by urinary concentration defects as well as common diseases associated with severe water retention are discussed in relation to the role of AQPs in the regulation and dysregulation of renal water transport.

7.2 EXPRESSION AND FUNCTION OF AQUAPORINS IN THE KIDNEY

Thirteen mammalian AQPs are now known (Agre et al. 2002; Nielsen et al. 2002, 2008b; Rojek et al. 2008), and they can be classified into three major subtypes, which are mainly determined by their transport capabilities: (1) the classical AQPs (AQP1, AQP2, AQP4 and AQP5), which are known to transport only water and thus serve essential roles in transcellular water transport; (2) aquaglyceroporins (AQP3, AQP7, AQP9 and AQP10), which are permeated by small uncharged molecules in addition to water; and (3) unorthodox AQPs (AQP6, AQP8, AQP11 and AQP12), whose function is currently being elucidated. Of the known AQPs, eight (AQP1, AQP2, AQP3, AQP4, AQP6, AQP7, AQP8 and AQP11) are expressed in the mammalian kidney (Table 7.1).

AQP1 protein (Agre et al. 1993; Preston and Agre 1991; Preston et al. 1992) is highly expressed in proximal tubules and descending thin limbs of the kidney (Nielsen et al. 1993c; Sabolic et al. 1992). AQP1 is also present in capillary endothelia (Nielsen et al. 1993b, 1995b), including the renal vasa recta. In addition, AQP1 is present in multiple water-permeable epithelia in the body (Bondy et al. 1993; Brown et al. 1993a; Gresz et al. 2001;

TABLE 7.1   Aquaporins in the Kidney

AQP	Localization (renal)	Subcellular Distribution	Regulation	Localization (extrarenal)
AQP1	S2, S3 segments of proximal tubules	Apical and basolateral plasma membranes	Glucocorticoids (peribronchiolar capillary endothelium)	Multiple tissues, including capillary endothelia, choroids plexus, ciliary and lens epithelium, ect.
AQP2	Collecting duct principal cells	Intracellular vesicles, apical plasma membrane	Vasopressin stimulates –short-term exocytosis –long-term biosynthesis	Epididymis
AQP3	Collecting duct principal cells	Basolateral plasma membrane	Vasopressin stimulates –long-term biosynthesis	Multiple tissues, including airway basal epithelia, conjunctiva, colon
AQP4	Collecting duct principal cells	Basolateral plasma membrane	Dopamine, protein kinase C	Multiple tissues, including central nervous system astroglia, ependyma, airway surface epithelia
AQP6	Collecting duct intercalated cells	Intracellular vesicles	Rapidly gated	Unknown
AQP7	S3 proximal tubules	Apical plasma membrane	Insulin (adipose tissue)	Multiple tissues, including adipose tissue, testis, and heart
AQP8	Proximal tubule, collecting duct cells	Intracellular domains	Unknown	Multiple tissues, including gastrointestinal tract, testis, and airways
AQP11	Proximal tubule	Intracellular domains	Unknown	Multiple tissues, including liver, testes and brain

Hasegawa et al. 1994a; Lai et al. 2001; Nielsen et al. 1993b; Stamer et al. 1994). The critical role of AQP1 in urinary concentration has been highlighted in studies using transgenic mice with knockout of the AQP1 gene (Ma et al. 1998). These mice were polyuric and had a reduced urinary concentrating capacity. The polyuria and impaired urinary concentration seen in AQP1 null mice could be explained by two mechanisms: impaired near isosmolar water reabsorption by proximal tubule and reduced medullary hypertonicity resulting from impaired countercurrent multiplication and exchange (Pallone et al. 2003; Verkman 2008). Consistent with this, isolated perfused proximal tubules and descending thin limbs had an 80% and 90% reduction in osmotic water permeability, respectively, illustrating an important role of AQP1 in water transport across these tubule segments (Schnermann et al. 1998). Moreover, these studies also emphasized the important role of transcellular rather than paracellular water transport in these tubule segments.

Soon after AQP1 was discovered to be a water channel, AQP2 was identified in renal collecting duct (Fushimi et al. 1993), where it is regulated by vasopressin and is involved in multiple clinical disorders (Kwon et al. 2001; Loonen et al. 2008; Nielsen et al. 2002, 2007; Schrier 2008). AQP2 is expressed in principal cells of the cortical, outer, and inner medullary collecting ducts and is abundant both in the apical plasma membrane and subapical vesicles (Nielsen et al. 1993a). AQP2 is the primary target for vasopressin regulation of collecting duct water permeability (Nielsen et al. 1995a; Sabolic et al. 1995; Yamamoto et al. 1995). This conclusion was established from studies showing a direct correlation between AQP2 expression and collecting duct water permeability in rats (Nielsen et al. 1995a), and in studies demonstrating that humans with mutations in the *AQP2* gene (Deen et al. 1994a; Loonen et al. 2008) or rats with 95% reduction in AQP2 expression have profound NDI (Christensen et al. 2006; Kwon et al. 2000; Marples et al. 1995a). As described later, body water balance is regulated both by short-term and long-term mechanisms, and it is now clear that AQP2 plays a fundamental role in both. Consistent with this, lack of functional AQP2 expression by generation of *AQP2* gene knockout mice (Rojek et al. 2006) produces a severe concentration defect, resulting in postnatal death. Morphological changes including renal medullary atrophy and dilation of the collecting ducts are also observed in these mice. Generation of AQP2−/− knockouts selectively in the collecting ducts, but not in the connecting tubule segments, rescues mice from the lethal phenotype (Rojek et al. 2006). However, body weight, urinary production, and the response to water deprivation are still severely impaired. This indicates an essential role of AQP2 in the renal tubular water reabsorption in both the connecting tubule segment and in the entire collecting duct.

AQP3 was identified in the kidney and other tissues and was found to be permeated by glycerol and water (Echevarria et al. 1994; Ishibashi et al. 1994; Ma et al. 1994). In the kidney, AQP3 is localized in the basolateral plasma membranes of connecting tubule cells, collecting duct principal cells, and inner medullary collecting duct cells (Coleman et al. 2000; Ecelbarger et al. 1995; Ishibashi et al. 1997b). AQP3 is expressed in the same cells as the vasopressin-regulated water channel AQP2 and is thought to mediate the basolateral exit of water that enters apically via AQP2. AQP3-deficient mice were shown to be severely polyuric (Ma et al. 2000), demonstrating that basolateral membrane water transport can also become a rate-limiting factor for water reabsorption. Moreover, urinary osmolality is reduced in AQP3−/− mice, and they fail to respond appropriately to 1-desamino-8-D-arginine vasopressin (dDAVP) stimulation, thus presenting with a urinary concentrating defect (Ma et al. 2000). Interestingly, AQP3 is not abundant in the cytoplasm, and there is no clear evidence for the short-term regulation of AQP3 by vasopressin-induced trafficking. However, AQP3 mRNA and protein expression are both upregulated by long-term stimulation of vasopressin which is seen during water deprivation or vasopressin infusion (Ecelbarger et al. 1995; Ishibashi et al. 1997b; Murillo-Carretero et al. 1999; Terris et al. 1996). In addition, sodium restriction or aldosterone infusion in normal and Brattleboro rats greatly increases the abundance of AQP3, while AQP3 abundance is markedly reduced during aldosterone deficiency, suggesting a direct effect of aldosterone on collecting duct AQP3 expression (Kwon et al. 2002).

AQP4, a $HgCl_2$-insensitive water channel, is most abundantly expressed in the brain and is present in the kidney collecting duct (Coleman et al. 2000; Frigeri et al. 1995; Hasegawa et al. 1994b; Terris et al. 1995). Although AQP3 and AQP4 are basolateral water channels, they are distributed differently along the collecting duct system, with the greatest abundance of AQP3 in the cortical and outer medullary collecting ducts and the greatest abundance of AQP4 in the inner medullary collecting duct (Terris et al. 1995). AQP4 is also found in basolateral membranes of proximal tubule S3 segments, although only in mice (Van Hoek et al. 2000). Using freeze fracture electron microscopy, orthogonal arrays of intramembrane particles was present in the basolateral membranes of the proximal tubule S3 segment in AQP4+/+, but not in AQP4−/− mice (Van Hoek et al. 2000). AQP4 appears to be regulated by PKC and dopamine, where stimulation by these factors decreases water permeability in AQP4-transfected cells (Zelenina et al. 2002). Targeted disruption of AQP4 in mice results in a 75% reduction in the osmotic water permeability of the inner medullary collecting duct (Chou et al. 1998). However, phenotypically the AQP4−/− mice appeared grossly normal, presenting with a very mild urinary concentrating defect (Ma et al. 1997). In double AQP3−/−/AQP4−/− knockout mice, urinary concentrating ability was slightly more impaired than that of the AQP3−/− mice (Ma et al. 2000), suggesting an important role of AQP3 in urinary concentration. It should also be noted that the localization of AQP2 in basolateral membranes in both the connecting tubule and inner medullary collecting duct raises the possibility that the observed effect is partly compensated by this mechanism (Christensen et al. 2003).

AQP6 was found to be localized in the subapical vesicles within intercalated cells of collecting duct, where it is coexpressed with the V-type H^+-ATPase (Yasui et al. 1999a,b). AQP6 appears functionally distinct from other known AQPs. Oocyte expression studies have revealed low water permeability of AQP6 during basal conditions, while in the presence of $HgCl_2$, a rapid increase in water permeability and ion conductance is observed (Yasui et al. 1999a). Additionally, reductions in pH (below 5.5) quickly and reversibly increase anion conductance and water permeability in AQP6-expressing oocytes (Yasui et al. 1999a). Subsequent studies have shown that the channel is permeable to halides with the highest permeability to NO_3^- (Ikeda et al. 2002), while the ionic selectivity becomes less specific after the addition of $HgCl_2$ (Hazama et al. 2002). Moreover, when Asn60, a residue unique in mammalian AQP6, is converted to glycine, a highly conserved residue in other mammalian AQPs, anionic permeability is abolished and osmotic water permeability is increased during basal conditions (Liu et al. 2005).

AQP7 is permeated by water and glycerol (Ishibashi et al. 1997a). AQP7 mRNA is abundantly expressed in the kidney (Ishibashi et al. 1997a; Kuriyama et al. 1997). Immunohistochemical studies in mouse and rat have localized AQP7 to the brush border of the proximal tubule S3 segment (Ishibashi et al. 2000; Nejsum et al. 2000), where AQP7 is colocalized with AQP1. AQP7 null mice have reduced water permeability in the proximal tubule brush border membrane (Sohara et al. 2005), whereas they do not exhibit a urinary concentrating defect or water balance abnormality since AQP1 is the major water transport pathway in the proximal tubule. AQP8 is a water channel found in intracellular domains of the proximal tubule and the collecting duct cells (Elkjaer et al. 2001); however, its function remains unclear. AQP11 is found in the cytoplasm of renal proximal tubule cells (Morishita et al. 2005). The exact function is not established, although deletion of the *AQP11* gene produces a severe phenotype with renal vacuolization and cyst formation (Morishita et al. 2005).

7.3 VASOPRESSIN REGULATION OF KIDNEY AQUAPORINS

AQP2 is expressed in principal cells of renal collecting duct, where it is the primary target for short-term regulation of collecting duct water permeability (Nielsen et al. 1995a; Sabolic et al. 1995; Yamamoto et al. 1995). AQP2 and AQP3 also are regulated either directly or indirectly by vasopressin through long-term effects that alter the abundance of these water channel proteins in collecting duct cells (DiGiovanni et al. 1994; Ecelbarger et al. 1995; Lankford et al. 1991; Nielsen et al. 1993a; Terris et al. 1996). Short-term regulation is the process by which vasopressin rapidly increases water permeability of principal cells by binding to vasopressin V2 receptors in the basolateral membranes, a response measurable within 5–30 min after increasing the peritubular vasopressin concentration (Kuwahara and Verkman 1989; Wall et al. 1992; Fenton and Moeller. 2008). It is believed that vasopressin, acting through a cyclic adenosine monophosphate (cAMP) cascade, causes intracellular AQP2 vesicles to fuse with the apical plasma membrane, which increases the number of water channels in the apical plasma membrane (Figure 7.1a through g). Long-term regulation of collecting duct water permeability is seen when circulating vasopressin levels are increased for 24 h or more, resulting in an increase in the maximal water permeability of the collecting duct epithelium (Lankford et al. 1991). This response is a consequence of an increase in the abundance of AQP2 water channels per cell in the collecting duct (DiGiovanni et al. 1994; Nielsen et al. 1993a), apparently due to increased transcription of the *AQP2* gene (Figure 7.1h).

Multiple studies with affinity-purified polyclonal antibodies to AQP2 have unequivocally established that AQP2 is specifically involved in the vasopressin-induced increases of renal collecting duct water permeability. Soon after the isolation of AQP2 cDNA and generation of specific antibodies (Fushimi et al. 1993), immunoperoxidase microscopy and immunoelectron microscopy clearly demonstrated that principal cells within renal collecting ducts contain abundant AQP2 in the apical plasma membranes and in subapical vesicles (DiGiovanni et al. 1994; Nielsen et al. 1993a). These studies strongly supported the 'shuttle hypothesis' originally proposed more than a decade earlier (Kachadorian et al. 1975). This hypothesis proposed that water channels can shuttle between an intracellular reservoir in subapical vesicles and the apical plasma membrane and that vasopressin alters water permeability by regulating the shuttling process (Wade et al. 1981). Shuttling of AQP2 was directly demonstrated in isolated perfused tubule studies (Nielsen et al. 1995a). In these studies, water permeability of isolated perfused collecting ducts was measured before or after stimulation with vasopressin, and the tubules were fixed directly for immunoelectron microscopic examination. Vasopressin stimulation resulted in a markedly decreased immunogold labelling of intracellular AQP2 accompanied by a fivefold increase in the appearance of AQP2 immunogold particles in the apical plasma membrane. This redistribution was associated with an increase in osmotic water permeability of similar magnitude. These findings were reproduced in vivo by injecting rats with vasopressin, which also caused

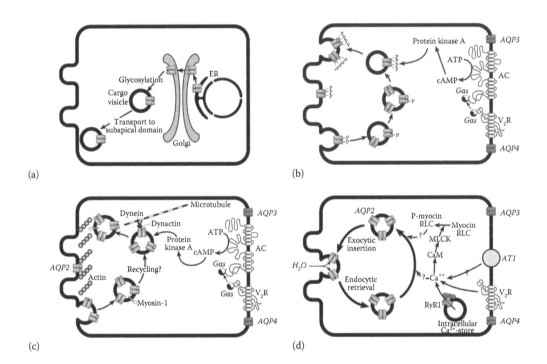

FIGURE 7.1 Regulation of aquaporin-2 (AQP2) trafficking (Panels a through h) and expression (Panel g) in collecting duct principal cells. (a) AQP2 is synthesized in the endoplasmic reticulum and transported via the Golgi apparatus and cargo vesicles to the subapical plasma membrane domain. (b) Vasopressin binding to basolateral G-protein-linked V2 receptor stimulates adenylyl cyclase; cyclic adenosine monophosphate (cAMP) activates protein kinase A to phosphorylate AQP2 in intracellular vesicles; phosphorylated AQP2 is exocytosed to the apical plasma membrane, resulting in increased apical membrane water permeability. (c) Overview of cytoskeletal elements, which may be involved in AQP2-trafficking. AQP2-containing vesicles may be transported along microtubules by dynein/dynactin. The cortical actin web may act as a barrier to fusion with the plasma membrane. (d) Intracellular calcium signalling and AQP2 trafficking. Increases in intracellular Ca^{2+} concentration may arise from stimulation of the V2 receptor. The existence and potential role of other receptors and pathways, for example VACM-1, in Ca^{2+} mobilization is still uncertain. The downstream targets of the calcium signal are unknown, and conflicting data exist on the importance of a rise in intracellular Ca^{2+} for the hydroosmotic response to vasopressin. *(Continued)*

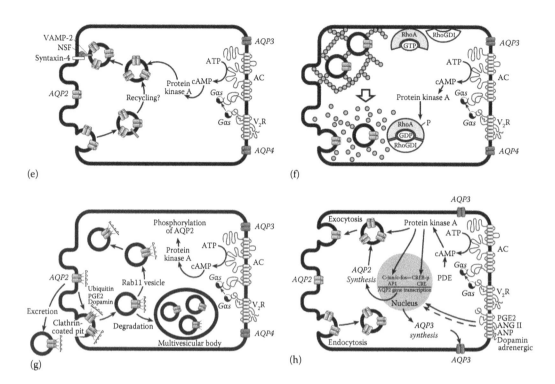

FIGURE 7.1 (*Continued*)   Regulation of aquaporin-2 (AQP2) trafficking (Panels a through h) and expression (Panel g) in collecting duct principal cells. (e) Vesicle-targeting receptors and AQP2 trafficking. AQP2 vesicles dock at the apical membrane by association of VAMP-2 with syntaxin-4 targets in the presence of N-ethylmalemide sensitive factor (NSF). The exact role of these remains to be established. (f) Changes in the actin cytoskeleton associated with AQP2 trafficking to the plasma membrane. Inactivation of RhoA by phosphorylation and increased formation of RhoA-RhoGDI complexes seem to control the dissociation of actin fibers seen after vasopressin stimulation. (g) Endocytic retrieval of AQP2 can be initiated by the removal of AVP or favoured by prostaglandin E2 (PGE2) or dopamine. Ubiquitination of AQP2 in the plasma membrane is proposed to play a role in the endocytotic events. Endocytosis of AQP2 occurs in clathrin-coated pits via a phosphatidylinositol-3-kinase-dependent mechanism. After membrane retrieval, AQP2 is transported to early endosomes, recycling vesicles or multivesicular bodies that may be targeted for proteasomal degradation. (h) Specifically, cAMP participates in the long-term regulation of AQP2 by increasing the levels of the catalytic subunit of PKA in the nuclei which is thought to phosphorylate transcription factors such as cyclic adenosine monophosphate response element binding protein and C-Jun/c-Fos. Binding of these factors are thought to increase gene transcription of AQP2 resulting in synthesis of AQP2 protein which in turn enters the regulated trafficking system. In parallel, AQP3 and AQP4 synthesis and trafficking to the basolateral plasma membrane takes place. Importantly, AQP2 regulation can be modified by a number of hormones including dopamine, atrial natriuretic peptide (ANP), PGE2 and adrenergic hormones. See text for details. GTP-binding protein (Gs), adenosine triphosphate (ATP), phosphorylation (P), vasopressin V2 receptors (V2R), adenylyl cyclase (AC), cAMP and protein kinase A (PKA). (Redrawn and modified from Nielsen et al. 2002.)

redistribution of AQP2 to the plasma membrane of collecting duct principal cells (Marples et al. 1995b; Sabolic et al. 1995). In contrast to the effect of vasopressin treatment, removal of vasopressin led to a reappearance of AQP2 in intracellular vesicles and a decline in osmotic water permeability in the isolated perfused collecting duct system (Nielsen et al. 1995a). Moreover, the offset response to vasopressin has been examined in vivo by acute treatment of rats with vasopressin-V_2-receptor antagonist (Christensen et al. 1998; Hayashi et al. 1994) or acute water loading (to reduce endogenous vasopressin levels) (Saito et al. 1997).

The long-term regulation of AQP2 occurs as a result of a vasopressin-induced increase in the total abundance of the AQP2 protein in collecting duct cells. These long-term actions are thought to be associated with regulatory processes at the transcriptional or post-transcriptional level. The half-life of the AQP2 protein could be increased by vasopressin. In cultured mpkCCD cells the half-life increased from 9 to 14 hours (Sandoval et al. 2013). Vasopressin enhances AQP2 protein abundance by altering its proteasomal degradation through a PKA- and p38-MAP kinase-dependent pathway (Nedvetsky et al. 2010). AQP2 is degraded in the proteasome and lysosome (Kamsteeg et al. 2006; Lee et al. 2011). The process of endocytosis and subsequent targeting to the proteasome and lysosome is thought to be regulated or ubiquitylation of the C-terminal tail of the AQP2 protein at lysine 270 (Kamsteeg et al. 2006). Identification of E3 ubiquitin-protein ligases specific to AQP2 degradation is currently under the investigation (Lee et al. 2011). Moreover, transcription of the AQP2 gene is markedly increased by vasopressin, resulting in increased cellular mRNA levels and increased AQP2 translation (Matsumura et al. 1997). Transcriptional regulation of AQP2 is thought to be a result of vasopressin-induced increase in intracellular cAMP levels with concomitant increases in activation of protein kinase A. In addition to the transcriptional regulation of AQP2, microRNA (miRNA) appears to play a role in the post-transcriptional regulation through inhibition of translation of target mRNA via translational regression of the RNA-induced silencing complex (Lee et al. 1993). A recent study specifically predicted miRNAs targeting AQP2 expression by in silico analysis, and two predicted AQP2-targeting miRNAs (miR-32 and miR-137) were further investigated to understand a novel molecular mechanism of AQP2 protein regulation (Kim et al. 2015). This study provides a novel insight on the regulation of AQP2 protein expression, at least in part, via RNA interference, i.e., AQP2-targting miRNAs. Other signal transduction pathways including prostaglandins, angiotensin II, aldosterone, PI3K/Akt pathways, cytoskeleton, intracellular Ca^{2+} concentration, and vesicle-targeting receptors have been described in previous studies and reviews (Nielsen et al. 2002; Kwon et al. 2005b; Kwon et al. 2013; Kwon et al. 2009; Moeller et al. 2012; Choi et al. 2015; Lee et al. 2007; Jung et al. 2015; Valenti et al. 2005).

7.4 DYSREGULATION OF RENAL AQUAPORINS IN WATER BALANCE DISORDERS

AQPs have been demonstrated to play a key role in the pathophysiology of a variety of water balance disorders. This section updates previous detailed reviews with regard to the critical role of AQP2 in water balance disorders (Kwon et al. 2001, 2009, 2013; Nielsen et al. 1999, 2000, 2002, 2007; Knepper et al. 2015).

7.4.1 Urinary Concentrating Defects

7.4.1.1 Inherited Forms of Diabetes Insipidus

The importance of AQP2 for urinary concentration was first demonstrated in a study by Deen et al. (1994a). Chromosome 12q13 harbors the human *AQP2* gene (GenBank accession number z29491) (Deen et al. 1994b; Sasaki et al. 1994). The *AQP2* gene comprises 4 exons distributed over ~5 kb of genomic DNA and three introns (Deen et al. 1994a,b; Sasaki et al. 1994). The 1.5 kb mRNA encodes a protein of 271 amino acids. Deen et al. (1994a) found mutated and non-functional AQP2 in patients with very severe NDI (non-X-linked NDI). The patient appeared to carry two-point mutations in the *AQP2* gene, one resulting in substitution of a cysteine for arginine 187 (R187C) in the third extracellular loop of the AQP2 and the other resulting in substitution of a proline for serine 216 (S216P) in the sixth transmembrane domain (Deen et al. 1994a). Subsequently, it was demonstrated that Brattleboro rats, which are vasopressin deficient and have extreme polyuria and therefore have central diabetes insipidus (DI), have reduced expression of AQP2 and very low AQP2 levels in the apical plasma membrane (DiGiovanni et al. 1994; Promeneur et al. 2000; Yamamoto et al. 1995).

There are two significant inherited forms of DI: central and nephrogenic. In central (or neurogenic) DI, normal vasopressin production is impaired. Central DI (CDI) is rarely hereditary in man, but usually occurring as a consequence of head trauma or diseases in the hypothalamus or pituitary gland. The Brattleboro rat provides an excellent model of this condition. These animals have a total or near-total lack of vasopressin production (Valtin and Schroeder 1964). Consequently, Brattleboro rats have substantially decreased expression levels of vasopressin-regulated AQP2 compared with the parent strain (Long Evans), and the AQP2 deficit was reversed by chronic vasopressin infusion, suggesting that patients lacking vasopressin are likely to have decreased AQP2 expression (DiGiovanni et al. 1994). The subsequent work showing that expression of AQP3 is also regulated by vasopressin implies that the expression levels of these water channels will also be decreased in patients with CDI. However, the most important denominator is the deficiency of AQP2 trafficking to the apical membrane. These deficits are likely to be the most important causes of the polyuria from which these patients of CDI suffer, which will be reversed by the desmopressin treatment. The second form of DI is called nephrogenic DI (NDI) and is caused by the inability of the kidney to respond to vasopressin stimulation. The most commonly hereditary cause (95% of the cases) is an X-linked disorder associated with mutations of the vasopressin V2 receptor making the collecting duct cells insensitive to vasopressin (Bichet 1996). The human gene that encodes for the V2 receptor (*AVPR2*) is located in chromosome region Xq28 and has 3 exons and 2 small introns (Birnbaumer et al. 1992; Seibold et al. 1992). The sequences of the cDNA predict a polypeptide of 371 amino acids with 7 transmembrane, 4 extracellular, and 4 cytoplasmic domains (Bichet 2008; Fujiwara and Bichet 2005; Mouillac et al. 1995). X-linked NDI is generally a rare disease in which the affected male patients do not concentrate their urine, whereas female individuals are not likely to be affected. However, heterozygous females can show variable degrees of polyuria because of skewed X-chromosome inactivation (Bichet 2008; Fujiwara and Bichet 2005). The incidence of X-linked NDI among male individuals was reported to be ~8.8 in 1,000,000 male live births in Quebec (Arthus et al. 2000). It is likely that this form of NDI will be associated with decreased expression of

AQP2, since the cells are unable to respond to circulating vasopressin. This will compound the lack of AQP2 trafficking. Consistent with this, urinary AQP2 levels are very low in patients with X-linked NDI (Deen et al. 1996; Kanno et al. 1995). However, since the amount of AQP2 in the urine appears to be determined largely by the response of the collecting duct cells to vasopressin (Wen et al. 1999) rather than their content of AQP2, the data must be interpreted with caution with respect to predicting AQP2 expression levels.

More rarely (5% of the cases), patients with hereditary NDI have mutations in the *AQP2* gene. Of these, more than 90% are reported with autosomal recessive NDI. Since these patients manifest a severe form of DI, the critical role of AQP2 in renal water conservation was established. To date, 39 mutations have been reported in *AQP2* gene, and among them, 32 mutations are involved in recessive NDI (Loonen et al. 2008). Nearly all the mutations in autosomal recessive NDI are found in the region encoding the AQP2 segment between the first and the last transmembrane domain. In contrast, seven mutations of *AQP2* gene have been described as autosomal dominant NDI, and all mutations in dominant NDI are found in the coding region of the C-terminal tail of AQP2, which has a critical role in AQP2 trafficking to the apical plasma membrane. Thus, AQP2 mutants in dominant NDI are sorted to other subcellular locations in the cells than wt-AQP2 (de Mattia et al. 2004; Kamsteeg et al. 2003; Kuwahara et al. 2001; Marr et al. 2002; Mulders et al. 1998; Procino et al. 2003). Patients with autosomal dominant mutations of *AQP2* gene are likely to demonstrate more mild clinical manifestations than recessive form (Kamsteeg et al. 2003; Kuwahara et al. 2001; Loonen et al. 2008; Marr et al. 2002): (1) in the dominant NDI, polyuria and polydipsia become apparent in the second half of the first year, whereas in recessive NDI, the symptoms are already present at birth, and (2) some patients with dominant NDI respond to dDAVP administration or dehydration by showing a transient increase in urine concentration.

7.4.1.2 Acquired Forms of NDI

In contrast to the rare inherited forms of DI (central and nephrogenic), acquired forms of NDI are much more common. A series of studies has been aimed at testing whether reduced expression and apical targeting of AQP2 might play a role in these polyuric conditions. For this purpose, several classic experimental protocols were used.

7.4.1.3 Lithium-Induced NDI

Lithium has been widely used for treating bipolar affective mood disorders (Timmer and Sands 1999). However, chronic lithium treatment also induce a decreased AQP2 and AQP3 expression in the collecting duct (Kwon et al. 2000; Marples et al. 1995a), resulting in a pronounced vasopressin-resistant polyuria and inability to concentrate urine (i.e. NDI) (Christensen et al. 1985; Kwon et al. 2000; Marples et al. 1995a). Lithium-induced polyuria is observed in approximately 20%–30% of lithium-treated patients (approximately 1 in 1000 of the population receives lithium treatment) and affected patients and experimental animals typically show a slow recovery of urinary concentrating ability when treatment is discontinued (Boton et al. 1987; Christensen et al. 2004). Lithium is almost exclusively excreted by the kidney (Radomski et al. 1950) and is reabsorbed by mechanisms similar to sodium. Approximately 75% of the filtered lithium is reabsorbed by the termination of the cortical thick ascending limb in the loop of Henle (see review Nielsen et al. 2008b). During conditions of restricted sodium intake, significant lithium reabsorption also occurs in the connecting tubule and collecting duct that is thought to be mediated by the amiloride-sensitive epithelial sodium channel ENaC. Thus, ENaC appears to play a central role in the development of lithium induced NDI (see review (Nielsen et al. 2008b)).

The progressive polyuria induced by chronic lithium treatment in rats is associated with a parallel downregulation of both total protein expression of AQP2 and AQP3. The AQP2 protein and mRNA expression in kidney cortex and medulla decreased to below ~30% of controls within 28 days of treatment (Christensen et al. 2004; Kwon et al. 2000; Laursen et al. 2004; Marples et al. 1995a; Nielsen et al. 2003). A recent in vivo study using the collecting duct–derived mpkCCDc14 cells indicated that decreased AQP2 protein expression is likely due to decreased AQP2 mRNA transcription while AQP2 protein stability was unchanged (Li et al. 2006). Immunohistochemistry has shown that the AQP2 downregulation in the kidney cortex predominantly occurs in the cortical collecting duct (CCD) while the connecting tubule appears to be less affected by lithium (Nielsen et al. 2006). Moreover, quantitative immunohistochemical analysis showed that lithium caused a cellular reorganization with a decreased fraction of AQP2- and AQP3-expressing principal cells relative to H$^+$-ATPase-expressing intercalated cells that may also play an important role for the lithium-induced polyuria (Christensen et al. 2004, 2006) (Figure 7.2). In addition

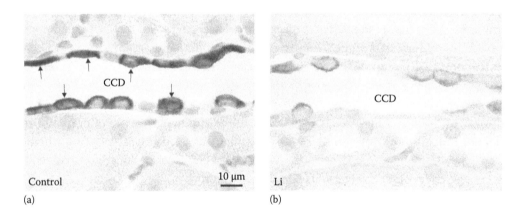

FIGURE 7.2 **(See colour insert.)** Downregulation of aquaporin-2 (AQP2) and cellular reorganization and composition in collecting duct of lithium-treated rats. Immunoperoxidase labelling of AQP2 in tissue section of kidney cortex of (a) control rats (Modified from Nielsen, J. et al., *Am. J. Physiol. Renal Physiol.*, 290, F438, 2006) and (b) rats treated with lithium for 4 weeks (Modified from Nielsen, J. et al., *Am. J. Physiol. Renal Physiol.*, 290, F438, 2006). In control rats, strong AQP2 labelling was present in the apical plasma membrane domain (arrows) and dispersed within the cytoplasm in the principal cells of the cortical collecting duct (CCD). In the lithium-treated rats, the AQP2 labelling in both the apical plasma membrane domain and cytoplasm of the principal cells of the CCD was drastically decreased (b) compared to control rats. To illustrate altered fraction of principal cells and intercalated cells, fluorescent double labelling was performed with polyclonal anti-AQP2 antibody (using a relatively high concentration, green) and monoclonal anti-H$^+$-ATPase antibody (red) in the proximal part of inner medulla, IM-1) from   *(Continued)*

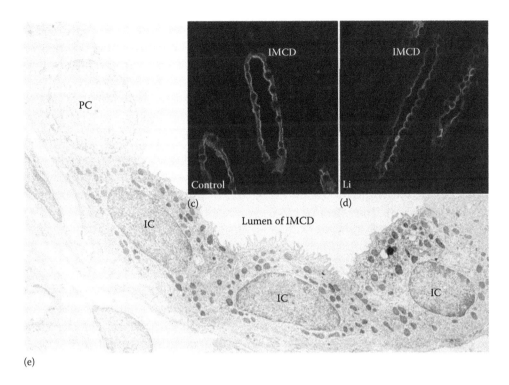

FIGURE 7.2 *(Continued)* **(See colour insert.)** downregulation of aquaporin-2 (AQP2) and cellular reorganization and composition in collecting duct of lithium-treated rats. Immunoperoxidase labelling of AQP2 in tissue section of kidney cortex of (c) control rats (Modified from Christensen, B.M. et al., *Am. J. Physiol. Cell Physiol.*, 286, C952, 2004) and (d) lithium-treated rats (Modified from Christensen, B.M. et al., *Am. J. Physiol. Cell Physiol.*, 286, C952, 2004). In both cortex (not shown) and IM-1, an increased density of H^+-ATPase positive cells was observed in (d) lithium-treated rats as compared to (c) controls. Immunoelectron microscopy of the inner medulla from lithium-treated rats shows three adjacent H^+-ATPase positive-intercalated cells which are highly unusual in the (e) normal rat kidney. (Modified from Kim, Y.H. et al., *Am. J. Physiol. Renal Physiol.*, 285, F1244, 2003.)

to decreased overall cellular AQP2 expression, quantitative immunoelectron microscopy demonstrated marked reduction of AQP2 in the apical plasma membrane as well as in the intracellular vesicles in rat inner medullary collecting duct principal cells (Marples et al. 1995a). Similarly, the AQP3 expression in the basolateral plasma membrane was reduced as demonstrated by immunoblotting and immunoelectron microscopy (Kwon et al. 2000). The significance of the decreased AQP2 and AQP3 in the collecting duct principal cells are underscored by the findings of similarly severe polyuria in the collecting duct–selective AQP2-deficient mice (Rojek et al. 2006) and AQP3 gene–deficient mice (Ma et al. 2000).

Several attempts have been made to find useful therapies for both CDI and NDI, and it is well established that thiazide and amiloride can have profound antidiuretic effects in patients with DI including lithium-induced NDI (Bedford et al. 2008; Crawford and Kennedy 1959). In rats with lithium-induced NDI, the antidiuretic effect of thiazide and amiloride has been associated with increased AQP2 expression (Bedford et al. 2008;

Kim et al. 2004a). Modulation of the renin–angiotensin–aldosterone system (RAAS) by captopril (an angiotensin-converting enzyme inhibitor) or spironolactone (a mineralocorticoid receptor blocker) also leads to decreased urine production in rats with DI (Henderson et al. 1979; Stamoutsos et al. 1981). We have shown that antidiuretic effect of spironolactone was associated with an increased apical AQP2 protein expression in the initial CCD in both rats with lithium-induced NDI and vasopressin-deficient Brattleboro rats (Nielsen et al. 2006). In the same study, we also showed that aldosterone infusion perturbed the polyuria in both lithium-treated rats and Brattleboro rats and that this is associated with decreased expression of AQP2 in the apical plasma membrane domain in the connecting tubule and initial collecting duct (while the AQP2 expression in the basolateral plasma membrane domain was increased). The effect of aldosterone and spironolactone observed in both lithium-treated rats could not be explained by altered plasma lithium concentrations. In another series of studies, we examined the effect of modulating RAAS in normal rats. In sodium-restricted rats, combined treatment with dDAVP and candesartan (an angiotensin II receptor antagonist) led to a blunted effect of dDAVP with decreased apical AQP2 targeting and increased urine production compared to rats treated with dDAVP alone (Kwon et al. 2005). On the other hand, combined treatment with dDAVP and aldosterone also blunted the effect of dDAVP with decreased apical (and increased basolateral) AQP2 and increased urine production compared to rats treated with dDAVP alone (de Seigneux et al. 2007). These results indicate that several pathways appear to modulate the response to vasopressin and aldosterone (and angiotensin II) with respect to AQP2 regulation both in normal and in polyuric rats (lithium-induced NDI and CDI). These studies indicate that angiotensin II and aldosterone receptors are likely to affect vasopressin signalling and that the effect of this may be more pronounced during states with altered vasopressin signalling (discussed in further detail in Nielsen et al. 2008b).

The mechanism responsible for the reduction in AQP2 expression is commonly thought to involve interference with the normal intracellular signalling of vasopressin (see review Nielsen et al. 2008b). The reduction in AQP2 and AQP3 expression may be induced by a lithium-dependent impairment in the production of cAMP by the inhibition of vasopressin-sensitive adenylate cyclase in collecting duct principal cells (Boton et al. 1987; Cogan and Abramow 1986; Cogan et al. 1987). The decreased cAMP production thus leads to a reduction in AQP2 and AQP3 expression as well as the inhibition of AQP2 targeting to the apical plasma membrane in response to lithium treatment. This is consistent with the presence of a cAMP-responsive element in the 5'-untranslated region of the AQP2 gene (Hozawa et al. 1996; Matsumura et al. 1997) and with the finding that mice with inherently low cAMP levels have decreased AQP2 expression (DI+/+ severe mouse) (Frokiaer et al. 1999). The vasopressin signalling is also known to be modulated by prostaglandin E2 (PGE2) by attenuating the anti-diuretic action of vasopressin on collecting duct water permeability by causing endocytic retrieval of AQP2 (see review Nielsen et al. 2008b). In lithium-induced NDI, urinary excretion of PGE2 is significantly increased in both humans and experimental animals (Hober et al. 1992; Kotnik et al. 2005; Sugawara et al. 1988) and COX-2 inhibition has been shown to ameliorate lithium-induced polyuria and PGE2 excretion in COX-1 null mice (Rao et al. 2005). Finally, in addition to vasopressin-dependent mechanisms, there is evidence that non-vasopressin-mediated effects may play in AQP2 (and AQP3) regulation during lithium-induced NDI (Marples et al. 1995a; Umenishi et al. 2006).

To further explore the broad and significant effects of lithium, we conducted combined proteomics and pathway analysis studies to identify novel candidate proteins and signalling mechanism affected by lithium (Nielsen et al. 2008a). Differential 2D gel electrophoresis combined with mass spectrometry identified 6 and 74 proteins with altered abundance compared to controls after 1 and 2 weeks of lithium treatment, respectively. Using bioinformatics analysis of the data indicated the protein changes were likely associated with changes in cellular functions affecting cell death, apoptosis, cell proliferation and morphology. Consistent with these results, follow-up studies revealed that several signalling pathways involved in these cellular functions, including the PKB/Akt-kinase and the mitogen-activated protein kinases including ERK, JNK and p38, were activated by lithium treatment. Activated PKB/Akt is a potential mediator of both JNK and P38 signalling pathways via activation of the apoptosis signal-regulating kinase 1, suggesting that these pathways may all be direct or indirect targets of lithium which leads to alteration in AQP2 and AQP3 expression.

The activation of PKB/Akt also indicated that the Wnt/beta-catenin (β-catenin) pathway, which regulates cellular proliferation, differentiation and apoptosis, could be affected. The glycogen synthase kinase type 3β (GSK3β, which is also known to be inhibited by lithium) (Klein and Melton 1996; Stambolic et al. 1996) plays a central role in this pathway and is inhibited by PKB/Akt. The GSK3β phosphorylation (causing inactivation) was increased consistent with the previous studies (Rao et al. 2005). In one of its many cellular functions, GSK3β functions as a negative regulator of the Wnt/beta-catenin (β-catenin) pathway by phosphorylating β-catenin targeting it for degradation. Consistent with the increased phosphorylation (and inactivation of GSK3β), β-catenin expression was increased and accumulated intracellularly after 2 weeks of lithium treatment compared with control rats (Nielsen et al. 2008a). When β-catenin is present intracellularly (during conditions with decreased GSK3β activation), it serves as a nuclear transcription factor activating T-cell factor (TCF)-dependent transcription (Novak and Dedhar 1999). Interestingly, mouse and rat AQP2 gene 5′ flanking regions contain the TCF consensus sites and altered Wnt/beta-catenin signalling could potentially play an important role in decreased AQP2 transcription in rats with lithium-induced NDI but potentially also in other conditions. Furthermore, TCF-dependent transcription has been demonstrated to regulate a number of proteins involved in cell cycle entry and may also play a role in the principal cell proliferation observed with lithium treatment.

In addition to the effects on the Wnt/beta-catenin pathway, lithium-induced inhibition of GSK3β has also been shown to increase PGE2 excretion (Rao et al. 2005). As discussed earlier, the increased PGE2 may counteract vasopressin actions by causing endocytic retrieval of AQP2, resulting in impaired urinary concentrating ability (Zelenina et al. 2000).

7.4.1.4 Electrolytes Abnormality (Hypokalemia and Hypercalcemia)

It has also been demonstrated that hypokalemia and hypercalcemia, which are relatively common electrolyte disorders, are well-known causes of acquired NDI. Rat models of these conditions are valuable tools to study the molecular defects, and it was shown that both conditions are associated with downregulation of AQP2 expression and apical targeting. In these two conditions, however, the downregulation of AQP2 was much more modest, as was the polyuria, further supporting a role of AQP2 expression for the urinary concentration (Marples et al. 1996b; Wang et al. 2002b). Extracellular calcium-sensing receptor (CaSR), originally cloned from the bovine parathyroid gland (Brown et al. 1993b), has also been localized at the kidney tubular segments including collecting duct (Brown et al. 1993b; Riccardi et al. 1996, 1998). CaSR is a G-protein-coupled receptor and recent studies demonstrated that CaSR is involved in the signal transduction pathways that link urinary calcium levels to AQP2 expression and apical targeting in the collecting duct principal cells (Bustamante et al. 2008; Procino et al. 2004). Hypercalciuria, commonly seen in a condition of hypercalcemia, could impair AQP2 targeting and thus reduce urinary concentrating ability possibly by a functional crosstalk in signalling transduction between vasopressin signalling and CaSR signalling. However, further studies are still required to understand the underlying pathogenetic mechanisms fully. A calcium-dependent calpain activation was also proposed to modulate AQP2 expression levels through AQP2 proteolysis in hypercalcemia (Puliyanda et al. 2003). In addition to the AQP2 downregulation, both conditions also have been shown to be associated with downregulation of Na-K-2Cl cotransporter (NKCC2) in the thick ascending limb (Elkjaer et al. 2002; Wang et al. 2002a). Expression of NKCC2 in the thick ascending limb is known to be regulated by dDAVP (Kim et al. 1999; Kwon et al. 1999b), and hence, this regulation could be significantly involved in the countercurrent multiplication system. Thus, downregulation of NKCC2 could reduce sodium and chloride reabsorption in the thick ascending limb and hence decreases medullary osmolality, also contributing to the polyuria and impaired urinary concentration in hypokalemia and hypercalcemia.

7.4.1.5 Ureteral Obstruction

A relatively common condition associated with long-term impaired urinary concentrating ability is obstruction of the urinary tract. Experimental bilateral obstruction of the ureters for 24 h was found to be associated with markedly reduced expression of AQP2, AQP3, AQP4 and AQP1 (Frokiaer et al. 1996; Li et al. 2001). In addition, BUO is associated with marked downregulation of key sodium transporters and urea transporters (Li et al. 2003). Following release of the obstruction, there is a marked polyuria during which period AQP2 and AQP3 levels remain downregulated up to 2 weeks after release providing an explanation at the molecular level for the observed post-obstructive polyuria. In a number of studies, BUO has been demonstrated to be associated with COX2 induction and cellular infiltration of the renal medulla (Cheng et al. 2004; Norregaard et al. 2005). Using specific COX2 inhibition to rats subjected to BUO, it was demonstrated that this treatment prevents downregulation of AQP2 and several sodium transporters located in the proximal tubule and mTAL (Cheng et al. 2004; Norregaard et al. 2005).

In contrast to BUO conditions, unilateral ureteral obstruction is not associated with changes in the absolute excretion of sodium and water since the non-obstructed kidney compensates for the reduced ability of the obstructed kidney to excrete solutes (Frokiaer et al. 1997; Li et al. 2001). These studies demonstrated a profound downregulation of AQP1, AQP2, AQP3, AQP4 and pAQP2 levels in the obstructed kidney, suggesting that local factors play a major role. Importantly, the role of PGs for the regulation of AQP2 in response to urinary tract obstruction has been highlighted in a number of publications (Norregaard et al. 2005, 2006). Interestingly, it was demonstrated that treatment with a specific COX-2 inhibitor prevented downregulation of AQP2 and reduced the post-obstructive polyuria indicating that COX-2 may be an important factor contributing to the impaired renal water and sodium handling in response to BUO (Norregaard et al. 2005). Moreover, the renin–angiotensin system is well known to be involved in the pathophysiological changes in renal function after obstruction of the ureter. In recent experiments, it was demonstrated that candesartan treatment from the onset of obstruction attenuated the reduction in GFR and prevented the reduced abundance of AQP2, NaPi2, and NKCC2 coinciding with a reduction in the post-obstructive polyuria (Jensen et al. 2006).

Congenital malformations of the kidney and urinary tract associated with ureteral obstruction account for a major proportion of renal insufficiency in infancy and childhood, but management of antenatally detected hydronephrosis is still debated (Shi et al. 2004b). To address this, the effect of neonatal partial unilateral ureter obstruction (PUUO) for 24 weeks was studied (Shi et al. 2004a,b). This resulted in a progressive decrease in RBF and a severe reduction in GFR in the obstructed kidney. The contralateral kidney counterbalanced the impairment of RBF and kidney growth. Obstruction was associated with severe hydronephrosis and obstructive nephropathy, shown as a marked reduction in total kidney protein content. These changes were associated with a decreased abundance of Na-K-ATPase, consistent with a significant natriuresis from the obstructed kidney. The abundance of AQP1, AQP2, and AQP3 was also reduced, consistent with the reduced GFR and solute-free water reabsorption. Importantly, the results demonstrated that the release after 4 weeks was associated with changes very similar to PUUO without the release of obstruction. In contrast, the release after 1 week of obstruction significantly attenuated the progressive reduction in RBF, and GFR was normal at 24 weeks of age. The development of hydronephrosis and obstructive nephropathy was prevented. Moreover, downregulation of renal Na-K-ATPase and AQP1 and AQP3 was prevented, consistent with attenuation of the natriuresis and decreased solute-free water reabsorption in the kidneys released 1 wk after onset of neonatal PUUO (Shi et al. 2004a,b). Moreover, it has recently been demonstrated that the treatment of neonatal rats subjected to PUUO at day 2 of life with candesartan prevented the reduction in RBF, GFR and dysregulation of AQP2 and Na,K-ATPase in response to congenital PUUO in rats, suggesting that AT1R blockade may protect the neonatally obstructed kidney against development of obstructive nephropathy (Topcu et al. 2007). These findings have been confirmed clinically by Valenti and co-workers, who demonstrated that children with severe unilateral hydronephrosis have reduced urinary AQP2 excretion levels (Murer et al. 2004). This suggests that urinary levels of AQP2 may be a useful biomarker for renal AQP2 levels. Importantly, decreased levels of urinary AQP2 excretion has also been demonstrated in various other significant clinical conditions with urinary concentrating defects, including nocturnal enuresis (Valenti et al. 2000, 2002).

7.4.1.6 Acute and Chronic Renal Failure

Acute (ARF) and chronic renal failure (CRF) are also important clinical conditions associated with polyuria and impaired urinary concentration. These are complex conditions, and in both cases, there is a wide range of glomerulotubular abnormalities contribute to the overall renal dysfunction. Ischemia and reperfusion (I/R)-induced experimental ARF in rats is a model that is widely used. In this model, there are structural alterations in renal tubule, in association with an impaired urinary concentration. ARF is complicated by defects of water and solute reabsorption in both the collecting duct and the proximal tubule (Hanley 1980; Tanner et al. 1973; Venkatachalam et al. 1978). Using an isolated tubule microperfusion model, water reabsorption in both the proximal tubule and the CCD was significantly impaired following ischemia (Hanley 1980) and no differences were found in either basal or vasopressin-induced cAMP levels in the outer or inner medulla in rats with ARF (Anderson et al. 1982). The results support the view that there are defects in the collecting duct water reabsorption. Consistent with these findings, it was demonstrated that AQP2 and AQP3 expression in the collecting duct as well as AQP1 expression in the proximal tubule are significantly decreased in ARF (Fernandez-Llama et al. 1999a; Kwon et al. 1999a). The decreased levels of AQPs were associated with impaired urinary concentration in rats with oliguric and non-oliguric ARF. Interestingly, reduced expression of AQP1, AQP2 and AQP3 and impaired urinary concentration ability were attenuated significantly by co-treatment with alpha-melanocyte-stimulating hormone (α-MSH), which is an anti-inflammatory cytokine that inhibits both neutrophil and nitric oxide pathways (Kwon et al. 1999a). It was also demonstrated that hemorrhagic shock-induced ARF is associated with decreased expression of collecting duct water channel AQP2 and AQP3 (Gong et al. 2003). And erythropoietin treatment (single or combined with α-MSH) in rats with I/R-induced ARF, which is known to prevent caspase-3, caspase-8 and caspase-9 activation in vivo and reduces apoptotic cell death (Sharples et al. 2004), prevents or reduces the urinary concentrating defects and the downregulation of AQP expression levels (Gong et al. 2004).

Patients with advanced CRF have urine which remains hypotonic to plasma despite the administration of supramaximal doses of vasopressin (Tannen et al. 1969). This vasopressin-resistant hyposthenuria specifically implies abnormalities in collecting duct water reabsorption in CRF patients. Previous studies demonstrated virtual absence of V_2 receptor mRNA in the inner medulla of CRF rat kidneys (Teitelbaum and McGuinness 1995), providing evidence for significant defects in the collecting duct water reabsorption in response to vasopressin. Consistent with these observations, AQP2 and AQP3 expression was downregulated in the collecting duct and decreased AQP2 expression was unchanged despite long-term dDAVP infusion in a rat model of 5/6 nephrectomy (Kwon et al. 1998).

7.5 CONDITIONS ASSOCIATED WITH WATER RETENTION

7.5.1 Congestive Heart Failure

Changes of hemodynamics and neurohormonal systems in congestive heart failure (CHF) result in renal sodium and water retention (Schrier RW, 1999). In particular, nonosmotic vasopressin release was demonstrated in patients with CHF (Szatalowicz VL et al, 1981). Two studies have examined the changes in renal AQP expression in rats with CHF induced by ligation of the left coronary artery to test if upregulation of AQP2 expression and targeting may play a role in the water retention in CHF (Nielsen et al. 1997; Xu et al. 1997). Both studies demonstrated that renal water retention in severe CHF in rats is associated with dysregulation of AQP2 in the renal collecting duct principal cells involving both an increase in the AQP2 expression and a marked redistribution of AQP2 to the apical plasma membrane (Nielsen et al. 1997; Xu et al. 1997). Rats with severe CHF had significantly elevated left ventricular end-diastolic pressures (LVEDPs) and had reduced plasma sodium concentrations (Nielsen et al. 1997). Immunoblotting revealed a threefold increase in AQP2 expression compared with sham-operated animals. These changes were associated with elevated LVEDP or hyponatremia, since animals with normal LVEDP and plasma sodium did not have increased AQP2 levels compared with sham-operated controls (Nielsen et al. 1997). Furthermore, this study showed an increased plasma membrane targeting, providing an explanation for the increased permeability of the collecting duct and an increase in water reabsorption. This may provide an explanation for excess free water retention in severe CHF and for the development of hyponatremia. In parallel, the other study showed upregulation of both AQP2 protein and AQP2 mRNA levels in kidney inner medulla and cortex in rats with CHF (Xu et al. 1997). These rats had significantly decreased cardiac output and, importantly, increased plasma vasopressin levels. Furthermore, in this study, administration of V2 antagonist OPC 31260 was associated with a significant increase in diuresis, a decrease in urine osmolality, a rise in plasma osmolality, and a significant reduction in AQP2 protein and AQP2 mRNA levels compared with untreated rats with CHF. Consistent with this, treatment of V2 receptor antagonist in human patients with heart failure is associated with a dose-related increase in water excretion and a decrease in urinary AQP2 excretion (Martin et al. 1999). Moreover, V2 receptor antagonist treatment (tolvaptan [OPC-41061]) in patients with CHF increases urine volume, decreases edema and normalizes serum sodium levels in patients with hyponatremia, compared with placebo-treated group (Gheorghiade et al. 2003). In addition to vasopressin-mediated water retention in patients with CHF, another pathophysiologic event occurring during CHF is associated with deleterious effects of angiotensin II and aldosterone (Schrier and Abraham. 1999). Several studies demonstrated that angiotensin II may exert an interaction with vasopressin in AQP2 trafficking and AQP2 protein expression (Kwon et al. 2005; Lee et al. 2007; Li et al. 2011). This finding was consistent with the findings in mice lacking angiotensin AT1A receptors in the collecting duct cells (Stegbauer et al. 2011). The study showed that levels of renal AQP2 protein after water deprivation were significantly lower in the collecting duct-specific AT1A receptor knockout mice compared with controls (Stegbauer et al. 2011), suggesting that AT1A receptors in the collecting duct cells modulate AQP2 levels and contribute to the concentration of urine.

7.5.2 Hepatic Cirrhosis

Hepatic cirrhosis is another chronic condition associated with water retention. It has been suggested that an important pathophysiological factor in the impaired ability to excrete water could be the increased levels of plasma vasopressin. However, unlike CHF, the changes in the expression of AQP2 protein levels vary considerably between different experimental models of hepatic cirrhosis. Several studies have examined the changes in renal AQP expression in rats with cirrhosis induced by common bile duct ligation (CBDL) (Fernandez-Llama et al. 1999b; Jonassen et al. 1998, 2000). The rats displayed impaired vasopressin-regulated water reabsorption despite normal plasma vasopressin levels. Consistent with this, semiquantitative immunoblotting showed a significant decrease in AQP2 expression in rats with hepatic cirrhosis (Fernandez-Llama et al. 1999b; Jonassen et al. 1998). In addition, the expression levels of AQP3 and AQP4 were downregulated in CBDL rats. This may predict a reduced water permeability of the collecting duct in this model (Fernandez-Llama et al. 1999b); hence, renal water reabsorption in the collecting duct is decreased in rats with compensated liver cirrhosis. In contrast, Fujita et al. (1995) demonstrated that hepatic cirrhosis induced by intra-peritoneal administration of carbon tetrachloride (CCl_4) was associated with a significant increase in both AQP2 protein levels and AQP2 mRNA expression. Interestingly, AQP2 mRNA expression correlated with the amount of ascites, suggesting that AQP2 may play a role in the abnormal water retention followed by the development of ascites in hepatic cirrhosis (Asahina et al. 1995). In a different model of CCl_4-induced cirrhosis, using CCl_4 inhalation, AQP2 expression was not increased (Fernandez-Llama et al. 2000). There was, however, evidence for increased trafficking of AQP2 to the plasma membrane, consistent with the presence of elevated levels of vasopressin in the plasma. Interestingly, there was a marked increase in AQP3 expression that is likely to be due to increased vasopressin levels. The pattern of increased AQP3 expression without upregulation of AQP2 is consistent with previous findings observed in the vasopressin escape (Ecelbarger et al. 1998), suggesting that the lack of increase in AQP2 expression could be a result of a normal compensatory response related to the escape phenomenon. Although the explanation for the differences between cirrhosis induced by CBDL and CCl_4 administration remains to be determined, it is well known that the dysregulation of body water balance depends on the severity of cirrhosis (Gines et al. 1998; Kim et al. 2005; Wood et al. 1988). CBDL results in a compensated cirrhosis characterized by peripheral vasodilation and increased cardiac output, whereas cirrhosis induced by 12 weeks of CCl_4 administration may be associated with the late state of decompensated liver cirrhosis characterized by sodium retention, edema and ascites (Gines et al. 1998; Kim et al. 2005; Levy and Wexler 1987). Thus the downregulation of AQP2 observed in milder forms of cirrhosis (i.e., in a compensated stage without water retention) may represent a compensatory mechanism to prevent the development of water retention. In contrast, the increased levels of vasopressin seen in severe 'non-compensated' cirrhosis with ascites may induce an inappropriate upregulation of AQP2 that would in turn participate in the development of water retention.

7.5.3 Experimental Nephrotic Syndrome

The nephrotic syndrome is characterized by extracellular volume expansion with excessive renal salt and water reabsorption. The underlying mechanisms of salt and water retention are poorly understood; however, they can be expected to be associated with dysregulation of solute transporters and water channels (Apostol et al. 1997; Kim et al. 2004b). In contrast to CHF and liver cirrhosis, a marked downregulation of AQP2 and AQP3 expression was demonstrated in rats with PAN-induced and adriamycin-induced nephrotic syndrome (Apostol et al. 1997; Fernandez-Llama et al. 1998a,b). The reduced expression of collecting duct water channels could represent a physiologically appropriate response to extracellular volume expansion. The signal transduction involved in this process is not clear, but circulating vasopressin levels are high in rats with PAN-induced nephrotic syndrome. Thus, the marked downregulation of AQP2 in experimental nephrotic syndrome may share similarities with the downregulation of AQP2 in water-loaded dDAVP-treated rats that escape from the action of vasopressin (Ecelbarger et al. 1997, 1998).

7.5.4 SIADH and Vasopressin Escape

Syndrome of inappropriate antidiuretic hormone secretion (SIADH) is associated with inappropriately high levels of circulating vasopressin, resulting in water retention and hyponatremia (Bartter and Schwartz 1967). In human patients, SIADH occurs most often with with neoplasia (notably small cell lung cancer), neurological diseases, and a variety of drugs including psychoactive drugs and chemotherapeutic agents (Baylis 2003; Sorensen et al. 1995). Experimental rat models of SIADH have demonstrated that AQP2 mRNA expression and AQP2 protein abundance are increased in the kidney collecting duct (Fujita et al. 1995; Ecelbarger et al. 1997). However, the degree of the hyponatremia is limited by a process that counteracts the water-retaining action of vasopressin, namely the process of 'vasopressin escape'. Vasopressin escape is characterized by an increase in urine volume with a decrease in urine osmolality independent of high circulating vasopressin levels. Studies of the vasopressin escape in rats have demonstrated that the escape is due to a suppression of dDAVP-stimulated AQP2 mRNA and protein levels by a process that affects AQP2 transcription rate (Ecelbarger et al. 1997). The process is associated with a fall in intracellular cAMP levels due to a decrease in coupling between the V2R and the adenylyl cyclases that it activates (Ecelbarger et al. 1998) and is associated with a downregulation of vasopressin V2R receptors (Saito T et al. 2001; Tian Y et al. 2000).

REFERENCES

Agre P, King LS, Yasui M et al. (2002) Aquaporin water channels – From atomic structure to clinical medicine. *J Physiol* 542:3–16.

Agre P, Preston GM, Smith BL et al. (1993) Aquaporin CHIP: The archetypal molecular water channel. *Am J Physiol* 265:F463–F476.

Anderson RJ, Gordon JA, Kim J et al. (1982) Renal concentration defect following nonoliguric acute renal failure in the rat. *Kidney Int* 21:583–591.

Apostol E, Ecelbarger CA, Terris J et al. (1997) Reduced renal medullary water channel expression in puromycin aminonucleoside – Induced nephrotic syndrome. *J Am Soc Nephrol* 8:15–24.

Arthus MF, Lonergan M, Crumley MJ et al. (2000) Report of 33 novel AVPR2 mutations and analysis of 117 families with X-linked nephrogenic diabetes insipidus. *J Am Soc Nephrol* 11:1044–1054.

Asahina Y, Izumi N, Enomoto N et al. (1995) Increased gene expression of water channel in cirrhotic rat kidneys. *Hepatology* 21:169–173.

Bajjalieh SM, Scheller RH (1995) The biochemistry of neurotransmitter secretion. *J Biol Chem* 270:1971–1974.

Bartter FC, Schwartz WB (1967) The syndrome of inappropriate secretion of antidiuretic hormone. *Am J Med* 42:790–806.

Baylis PH (2003) The syndrome of inappropriate antidiuretic hormone secretion. *Int J Biochem Cell Biol* 35(11):1495–1499.

Bedford JJ, Leader JP, Jing R et al. (2008) Amiloride restores renal medullary osmolytes in lithium-induced nephrogenic diabetes insipidus. *Am J Physiol Renal Physiol* 294:F812–F820.

Bichet DG (1996) Vasopressin receptors in health and disease. *Kidney Int* 49:1706–1711.

Bichet DG (2008) Vasopressin receptor mutations in nephrogenic diabetes insipidus. *Semin Nephrol* 28:245–251.

Birnbaumer M, Seibold A, Gilbert S et al. (1992) Molecular cloning of the receptor for human antidiuretic hormone. *Nature* 357:333–335.

Bondy C, Chin E, Smith BL et al. (1993) Developmental gene expression and tissue distribution of the CHIP28 water-channel protein. *Proc Natl Acad Sci USA* 90:4500–4504.

Boton R, Gaviria M, Batlle DC (1987) Prevalence, pathogenesis, and treatment of renal dysfunction associated with chronic lithium therapy. *Am J Kidney Dis* 10:329–345.

Brown D, Verbavatz JM, Valenti G et al. (1993a) Localization of the CHIP28 water channel in reabsorptive segments of the rat male reproductive tract. *Eur J Cell Biol* 61:264–273.

Brown EM, Gamba G, Riccardi D et al. (1993b) Cloning and characterization of an extracellular Ca(2+)-sensing receptor from bovine parathyroid. *Nature* 366:575–580.

Bustamante M, Hasler U, Leroy V et al. (2008) Calcium-sensing receptor attenuates AVP-induced aquaporin-2 expression via a calmodulin-dependent mechanism. *J Am Soc Nephrol* 19:109–116.

Cheng X, Zhang H, Lee HL et al. (2004) Cyclooxygenase-2 inhibitor preserves medullary aquaporin-2 expression and prevents polyuria after ureteral obstruction. *J Urol* 172:2387–2390.

Choi HJ, Jung HJ, Kwon TH. (2015) Extracellular pH affects phosphorylation and intracellular trafficking of AQP2 in inner medullary collecting duct cells. *Am J Physiol Renal Physiol* 308: F737–F748.

Chou CL, Ma T, Yang B et al. (1998) Fourfold reduction of water permeability in inner medullary collecting duct of aquaporin-4 knockout mice. *Am J Physiol* 274:C549–C554.

Christensen BM, Kim YH, Kwon TH et al. (2006) Lithium treatment induces a marked proliferation of primarily principal cells in rat kidney inner medullary collecting duct. *Am J Physiol Renal Physiol* 291:F39–F48.

Christensen BM, Marples D, Jensen UB et al. (1998) Acute effects of vasopressin V2-receptor antagonist on kidney AQP2 expression and subcellular distribution. *AJP – Renal Physiol* 275:F285–F297.

Christensen BM, Marples D, Kim YH et al. (2004) Changes in cellular composition of kidney collecting duct cells in rats with lithium-induced NDI. *Am J Physiol Cell Physiol* 286:C952–C964.

Christensen BM, Wang W, Frokiaer J et al. (2003) Axial heterogeneity in basolateral AQP2 localization in rat kidney: Effect of vasopressin. *Am J Physiol Renal Physiol* 284:F701–F717.

Christensen S, Kusano E, Yusufi AN et al. (1985) Pathogenesis of nephrogenic diabetes insipidus due to chronic administration of lithium in rats. *J Clin Invest* 75:1869–1879.

Cogan E, Abramow M (1986) Inhibition by lithium of the hydroosmotic action of vasopressin in the isolated perfused cortical collecting tubule of the rabbit. *J Clin Invest* 77:1507–1514.

Cogan E, Svoboda M, Abramow M (1987) Mechanisms of lithium-vasopressin interaction in rabbit cortical collecting tubule. *Am J Physiol* 252:F1080–F1087.

Coleman RA, Wu DC, Liu J et al. (2000) Expression of aquaporins in the renal connecting tubule. *Am J Physiol Renal Physiol* 279:F874–F883.

Crawford JD, Kennedy GC (1959) Chlorothiazide in diabetes insipidus. *Nature* 183:891–892.

de Mattia F, Savelkoul PJ, Bichet DG et al. (2004) A novel mechanism in recessive nephrogenic diabetes insipidus: Wild-type aquaporin-2 rescues the apical membrane expression of intracellularly retained AQP2-P262L. *Hum Mol Genet* 13:3045–3056.

de Seigneux S, Nielsen J, Olesen ET et al. (2007) Long-term aldosterone treatment induces decreased apical but increased basolateral expression of AQP2 in CCD of rat kidney. *Am J Physiol Renal Physiol* 293:F87–F99.

Deen PM, van Aubel RA, van Lieburg AF et al. (1996) Urinary content of aquaporin 1 and 2 in nephrogenic diabetes insipidus. *J Am Soc Nephrol* 7:836–841.

Deen PM, Verdijk MA, Knoers NV et al. (1994a) Requirement of human renal water channel aquaporin-2 for vasopressin-dependent concentration of urine. *Science* 264:92–95.

Deen PM, Weghuis DO, Sinke RJ et al. (1994b) Assignment of the human gene for the water channel of renal collecting duct Aquaporin 2 (AQP2) to chromosome 12 region q12→q13. *Cytogenet Cell Genet* 66:260–262.

DiGiovanni SR, Nielsen S, Christensen EI et al. (1994) Regulation of collecting duct water channel expression by vasopressin in Brattleboro rat. *Proc Natl Acad Sci USA* 91:8984–8988.

Ecelbarger CA, Chou CL, Lee AJ et al. (1998) Escape from vasopressin-induced antidiuresis: Role of vasopressin resistance of the collecting duct. *Am J Physiol* 274:F1161–F1166.

Ecelbarger CA, Nielsen S, Olson BR et al. (1997) Role of renal aquaporins in escape from vasopressin-induced antidiuresis in rat. *J Clin Invest* 99:1852–1863.

Ecelbarger CA, Terris J, Frindt G et al. (1995) Aquaporin-3 water channel localization and regulation in rat kidney. *Am J Physiol* 269:F663–F672.

Echevarria M, Windhager EE, Tate SS et al. (1994) Cloning and expression of AQP3, a water channel from the medullary collecting duct of rat kidney. *Proc Natl Acad Sci USA* 91:10997–11001.

Elkjaer ML, Kwon TH, Wang W et al. (2002) Altered expression of renal NHE3, TSC, BSC-1, and ENaC subunits in potassium-depleted rats. *Am J Physiol Renal Physiol* 283:F1376–F1388.

Elkjaer ML, Nejsum LN, Gresz V et al. (2001) Immunolocalization of aquaporin-8 in rat kidney, gastrointestinal tract, testis, and airways. *Am J Physiol Renal Physiol* 281:F1047–F1057.

Fenton RA, Moeller HB (2008) Recent discoveries in vasopressin-regulated aquaporin-2 trafficking. *Prog Brain Res* 170:571–579.

Fernandez-Llama P, Andrews P, Ecelbarger CA et al. (1998a) Concentrating defect in experimental nephrotic syndrome: Altered expression of aquaporins and thick ascending limb Na+ transporters. *Kidney Int* 54:170–179.

Fernandez-Llama P, Andrews P, Nielsen S et al. (1998b) Impaired aquaporin and urea transporter expression in rats with adriamycin-induced nephrotic syndrome. *Kidney Int* 53:1244–1253.

Fernandez-Llama P, Andrews P, Turner R et al. (1999a) Decreased abundance of collecting duct aquaporins in post-ischemic renal failure in rats. *J Am Soc Nephrol* 10:1658–1668.

Fernandez-Llama P, Jimenez W, Bosch-Marce M et al. (2000) Dysregulation of renal aquaporins and Na–Cl cotransporter in CCl4-induced cirrhosis. *Kidney Int* 58:216–228.

Fernandez-Llama P, Turner R, Dibona G et al. (1999b) Renal expression of aquaporins in liver cirrhosis induced by chronic common bile duct ligation in rats. *J Am Soc Nephrol* 10:1950–1957.

Franki N, Macaluso F, Schubert W et al. (1995) Water channel-carrying vesicles in the rat IMCD contain cellubrevin. *Am J Physiol* 269:C797–C801.

Frigeri A, Gropper MA, Turck CW et al. (1995) Immunolocalization of the mercurial-insensitive water channel and glycerol intrinsic protein in epithelial cell plasma membranes. *Proc Natl Acad Sci USA* 92:4328–4331.

Frokiaer J, Christensen BM, Marples D et al. (1997) Downregulation of aquaporin-2 parallels changes in renal water excretion in unilateral ureteral obstruction. *Am J Physiol* 273:F213–F223.

Frokiaer J, Marples D, Knepper MA et al. (1996) Bilateral ureteral obstruction downregulates expression of vasopressin-sensitive AQP-2 water channel in rat kidney. *Am J Physiol* 270:F657–F668.

Frokiaer J, Marples D, Valtin H et al. (1999) Low aquaporin-2 levels in polyuric DI+/+ severe mice with constitutively high cAMP-phosphodiesterase activity. *Am J Physiol* 276:F179–F190.

Fujita N, Ishikawa SE, Sasaki S et al. (1995) Role of water channel AQP-CD in water retention in SIADH and cirrhotic rats. *Am J Physiol* 269:F926–F931.

Fujiwara TM, Bichet DG (2005) Molecular biology of hereditary diabetes insipidus. *J Am Soc Nephrol* 16:2836–2846.

Fushimi K, Sasaki S, Marumo F (1997) Phosphorylation of serine 256 is required for cAMP-dependent regulatory exocytosis of the aquaporin-2 water channel. *J Biol Chem* 272:14800–14804.

Fushimi K, Uchida S, Hara Y et al. (1993) Cloning and expression of apical membrane water channel of rat kidney collecting tubule. *Nature* 361:549–552.

Gheorghiade M, Niazi I, Ouyang J et al. (2003) Vasopressin V2-receptor blockade with tolvaptan in patients with chronic heart failure: Results from a double-blind, randomized trial. *Circulation* 107:2690–2696.

Gines P, Berl T, Bernardi M et al. (1998) Hyponatremia in cirrhosis: From pathogenesis to treatment. *Hepatology* 28:851–864.

Gong H, Wang W, Kwon TH et al. (2003) Reduced renal expression of AQP2, p-AQP2 and AQP3 in haemorrhagic shock-induced acute renal failure. *Nephrol Dial Transplant* 18:2551–2559.

Gong H, Wang W, Kwon TH et al. (2004) EPO and alpha-MSH prevent ischemia/reperfusion-induced down-regulation of AQPs and sodium transporters in rat kidney. *Kidney Int* 66:683–695.

Gresz V, Kwon TH, Hurley PT et al. (2001) Identification and localization of aquaporin water channels in human salivary glands. *Am J Physiol Gastrointest Liver Physiol* 281:G247–G254.

Hanley MJ (1980) Isolated nephron segments in a rabbit model of ischemic acute renal failure. *Am J Physiol* 239:F17–F23.

Harris HW, Jr., Zeidel ML, Jo I et al. (1994) Characterization of purified endosomes containing the antidiuretic hormone-sensitive water channel from rat renal papilla. *J Biol Chem* 269:11993–12000.

Hasegawa H, Lian SC, Finkbeiner WE et al. (1994a) Extrarenal tissue distribution of CHIP28 water channels by in situ hybridization and antibody staining. *Am J Physiol* 266:C893–C903.

Hasegawa H, Ma T, Skach W et al. (1994b) Molecular cloning of a mercurial-insensitive water channel expressed in selected water-transporting tissues. *J Biol Chem* 269:5497–5500.

Hayashi M, Sasaki S, Tsuganezawa H et al. (1994) Expression and distribution of aquaporin of collecting duct are regulated by vasopressin V2 receptor in rat kidney. *J Clin Invest* 94:1778–1783.

Hazama A, Kozono D, Guggino WB et al. (2002) Ion permeation of AQP6 water channel protein. Single channel recordings after Hg^{2+} activation. *J Biol Chem* 277:29224–29230.

Henderson IW, McKeever A, Kenyon CJ (1979) Captopril (SQ 14225) depresses drinking and aldosterone in rats lacking vasopressin. *Nature* 281:569–570.

Hober C, Vantyghem MC, Racadot A et al. (1992) Normal hemodynamic and coagulation responses to 1-deamino-8-D-arginine vasopressin in a case of lithium-induced nephrogenic diabetes insipidus. Results of treatment by a prostaglandin synthesis inhibitor (indomethacin). *Horm Res* 37:190–195.

Hozawa S, Holtzman EJ, Ausiello DA (1996) cAMP motifs regulating transcription in the aquaporin 2 gene. *Am J Physiol* 270:C1695–C1702.

Ikeda M, Beitz E, Kozono D et al. (2002) Characterization of aquaporin-6 as a nitrate channel in mammalian cells. Requirement of pore-lining residue threonine 63. *J Biol Chem* 277:39873–39879.

Ishibashi K, Imai M, Sasaki S (2000) Cellular localization of aquaporin 7 in the rat kidney. *Exp Nephrol* 8:252–257.

Ishibashi K, Kuwahara M, Gu Y et al. (1997a) Cloning and functional expression of a new water channel abundantly expressed in the testis permeable to water, glycerol, and urea. *J Biol Chem* 272:20782–20786.

Ishibashi K, Sasaki S, Fushimi K et al. (1994) Molecular cloning and expression of a member of the aquaporin family with permeability to glycerol and urea in addition to water expressed at the basolateral membrane of kidney collecting duct cells. *Proc Natl Acad Sci USA* 91:6269–6273.

Ishibashi K, Sasaki S, Fushimi K et al. (1997b) Immunolocalization and effect of dehydration on AQP3, a basolateral water channel of kidney collecting ducts. *Am J Physiol* 272:F235–F241.

Jensen AM, Li C, Praetorius HA et al. (2006) Angiotensin II mediates downregulation of aquaporin water channels and key renal sodium transporters in response to urinary tract obstruction. *Am J Physiol Renal Physiol* 291:F1021–F1032.

Jonassen TE, Nielsen S, Christensen S et al. (1998) Decreased vasopressin-mediated renal water reabsorption in rats with compensated liver cirrhosis. *Am J Physiol* 275:F216–F225.

Jonassen TE, Promeneur D, Christensen S et al. (2000) Decreased vasopressin-mediated renal water reabsorption in rats with chronic aldosterone-receptor blockade. *Am J Physiol Renal Physiol* 278:F246–F256.

Jung HJ, Kim SY, Choi HJ, et al. (2015) Tankyrase-mediated beta-catenin activity regulates vaso-pressin-induced AQP2 expression in kidney collecting duct mpkCCDc14 cells. *Am J Physiol Renal Physiol* 308:F473–F486.

Kachadorian WA, Levine SD, Wade JB et al. (1977) Relationship of aggregated intramembranous particles to water permeability in vasopressin-treated toad urinary bladder. *J Clin Invest* 59:576–581.

Kachadorian WA, Wade JB, DiScala VA (1975) Vasopressin: Induced structural change in toad bladder luminal membrane. *Science* 190:67–69.

Kamsteeg EJ, Bichet DG, Konings IB et al. (2003) Reversed polarized delivery of an aquaporin-2 mutant causes dominant nephrogenic diabetes insipidus. *J Cell Biol* 163:1099–1109.

Kamsteeg EJ, Hendriks G, Boone M et al. (2006) Short-chain ubiquitination mediates the regulated endocytosis of the aquaporin-2 water channel. *Proc Natl Acad Sci USA* 103:18344–18349.

Kanno K, Sasaki S, Hirata Y et al. (1995) Urinary excretion of aquaporin-2 in patients with diabetes insipidus. *N Engl J Med* 332:1540–1545.

Kim JE, Jung HJ, Lee YJ, et al. (2015) Vasopressin-regulated miRNAs and AQP2-targeting miRNAs in kidney collecting duct cells. *Am J Physiol Renal Physiol* 308:F749–F764.

Kim GH, Ecelbarger CA, Mitchell C et al. (1999) Vasopressin increases Na-K-2Cl cotransporter expression in thick ascending limb of Henle's loop. *Am J Physiol* 276:F96–F103.

Kim GH, Lee JW, Oh YK et al. (2004a) Antidiuretic effect of hydrochlorothiazide in lithium-induced nephrogenic diabetes insipidus is associated with upregulation of aquaporin-2, Na–Cl co-transporter, and epithelial sodium channel. *J Am Soc Nephrol* 15:2836–2843.

Kim SW, Schou UK, Peters CD et al. (2005) Increased apical targeting of renal epithelial sodium channel subunits and decreased expression of type 2 11beta-hydroxysteroid dehydrogenase in rats with CCl4-induced decompensated liver cirrhosis. *J Am Soc Nephrol* 16:3196–3210.

Kim SW, Wang W, Nielsen J et al. (2004b) Increased expression and apical targeting of renal ENaC subunits in puromycin aminonucleoside-induced nephrotic syndrome in rats. *Am J Physiol Renal Physiol* 286:F922–F935.

Kim YH, Kwon TH, Christensen BM et al. (2003) Altered expression of renal acid–base transport-ers in rats with lithium-induced NDI. *Am J Physiol Renal Physiol* 285:F1244–F1257.

Klein PS, Melton DA (1996) A molecular mechanism for the effect of lithium on development. *Proc Natl Acad Sci USA* 93:8455–8459.

Knepper M, Burg M (1983) Organization of nephron function. *Am J Physiol* 244:F579–F589.

Knepper MA, Kwon TH, Nielsen S. (2015) Molecular physiology of water balance. *N Engl J Med* 372:1349–1358.

Kotnik P, Nielsen J, Kwon TH et al. (2005) Altered expression of COX-1, COX-2, and mPGES in rats with nephrogenic and central diabetes insipidus. *Am J Physiol Renal Physiol* 288:F1053–F1068.

Kuriyama H, Kawamoto S, Ishida N et al. (1997) Molecular cloning and expression of a novel human aquaporin from adipose tissue with glycerol permeability. *Biochem Biophys Res Commun* 241:53–58.

Kuwahara M, Iwai K, Ooeda T et al. (2001) Three families with autosomal dominant nephrogenic diabetes insipidus caused by aquaporin-2 mutations in the C-terminus. *Am J Hum Genet* 69:738–748.

Kuwahara M, Verkman AS (1989) Pre-steady-state analysis of the turn-on and turn-off of water permeability in the kidney collecting tubule. *J Membr Biol* 110:57–65.

Kwon TH, Nielsen J, Møller HB et al. (2009) Aquaporins in the kidney. *Handb Exp Pharmacol* 190:95–132.

Kwon TH, Frokiaer J, Fernandez-Llama P et al. (1999a) Reduced abundance of aquaporins in rats with bilateral ischemia-induced acute renal failure: Prevention by alpha-MSH. *Am J Physiol* 277:F413–F427.

Kwon TH, Frokiaer J, Fernandez-Llama P et al. (1999b) Altered expression of Na transporters NHE-3, NaPi-II, Na-K-ATPase, BSC-1, and TSC in CRF rat kidneys. *Am J Physiol* F257–F270.

Kwon TH, Frokiaer J, Knepper MA et al. (1998) Reduced AQP1, -2, and -3 levels in kidneys of rats with CRF induced by surgical reduction in renal mass. *Am J Physiol* 275:F724–F741.

Kwon TH, Hager H, Nejsum LN et al. (2001) Physiology and pathophysiology of renal aquaporins. *Semin Nephrol* 21:231–238.

Kwon TH, Laursen UH, Marples D et al. (2000) Altered expression of renal AQPs and Na(+) transporters in rats with lithium-induced NDI. *Am J Physiol Renal Physiol* 279:F552–F564.

Kwon TH, Nielsen J, Knepper MA et al. (2005) Angiotensin II AT1 receptor blockade decreases vasopressin-induced water reabsorption and AQP2 levels in NaCl-restricted rats. *Am J Physiol Renal Physiol* 288:F673–F684.

Kwon TH, Nielsen J, Masilamani S et al. (2002) Regulation of collecting duct AQP3 expression: Response to mineralocorticoid. *Am J Physiol Renal Physiol* 283:F1403–F1421.

Kwon TH, Frokiaer J, Nielsen S. (2013) Regulation of aquaporin-2 in the kidney: A molecular mechanism of body-water homeostasis. *Kidney Res Clin Pract* 32, 96–102.

Lai KN, Li FK, Lan HY et al. (2001) Expression of aquaporin-1 in human peritoneal mesothelial cells and its upregulation by glucose in vitro. *J Am Soc Nephrol* 12:1036–1045.

Lankford SP, Chou CL, Terada Y et al. (1991) Regulation of collecting duct water permeability independent of cAMP-mediated AVP response. *Am J Physiol* 261:F554–F566.

Laursen UH, Pihakaski-Maunsbach K, Kwon TH et al. (2004) Changes of rat kidney AQP2 and Na,K-ATPase mRNA expression in lithium-induced nephrogenic diabetes insipidus. *Nephron Exp Nephrol* 97:e1–e16.

Lee YJ, Song IK, Jang KJ et al. (2007) Increased AQP2 targeting in primary cultured IMCD cells in response to angiotensin II through AT1 receptor. *Am J Physiol Renal Physiol* 292:F340–F350.

Lee YJ, Lee JE, Choi HJ, et al. (2011) E3 ubiquitin-protein ligases in rat kidney collecting duct: response to vasopressin stimulation and withdrawal. *Am J Physiol Renal Physiol* 301: F883–F896.

Lee RC, Feinbaum RL, Ambros V. (1993) The C. elegans heterochronic gene lin-4 encodes small RNAs with antisense complementarity to lin-14. *Cell* 75:843–854.

Levy M, Wexler MJ (1987) Hepatic denervation alters first-phase urinary sodium excretion in dogs with cirrhosis. *Am J Physiol* 253:F664–F671.

Li C, Wang W, Kwon TH et al. (2001) Downregulation of AQP1, -2, and -3 after ureteral obstruction is associated with a long-term urine-concentrating defect. *Am J Physiol Renal Physiol* 281:F163–F171.

Li C, Wang W, Kwon TH et al. (2003) Altered expression of major renal Na transporters in rats with bilateral ureteral obstruction and release of obstruction. *Am J Physiol Renal Physiol* 285:F889–F901.

Li C, Wang W, Rivard CJ et al. (2011) Molecular mechanisms of angiotensin II stimulation on aquaporin-2 expression and trafficking. *Am J Physiol Renal Physiol* 300(5):F1255–1261.

Li Y, Shaw S, Kamsteeg EJ et al. (2006) Development of lithium-induced nephrogenic diabetes insipidus is dissociated from adenylyl cyclase activity. *J Am Soc Nephrol* 17:1063–1072.

Liu K, Kozono D, Kato Y et al. (2005) Conversion of aquaporin 6 from an anion channel to a water-selective channel by a single amino acid substitution. *Proc Natl Acad Sci USA* 102:2192–2197.

Loonen AJ, Knoers NV, van Os CH et al. (2008) Aquaporin 2 mutations in nephrogenic diabetes insipidus. *Semin Nephrol* 28:252–265.

Ma T, Frigeri A, Hasegawa H et al. (1994) Cloning of a water channel homolog expressed in brain meningeal cells and kidney collecting duct that functions as a stilbene-sensitive glycerol transporter. *J Biol Chem* 269:21845–21849.

Ma T, Song Y, Yang B et al. (2000) Nephrogenic diabetes insipidus in mice lacking aquaporin-3 water channels. *Proc Natl Acad Sci USA* 97:4386–4391.

Ma T, Yang B, Gillespie A et al. (1997) Generation and phenotype of a transgenic knockout mouse lacking the mercurial-insensitive water channel aquaporin-4. *J Clin Invest* 100:957–962.

Ma T, Yang B, Gillespie A et al. (1998) Severely impaired urinary concentrating ability in transgenic mice lacking aquaporin-1 water channels. *J Biol Chem* 273:4296–4299.

Marples D, Christensen S, Christensen EI et al. (1995a) Lithium-induced downregulation of aquaporin-2 water channel expression in rat kidney medulla. *J Clin Invest* 95:1838–1845.

Marples D, Frokiaer J, Dorup J et al. (1996b) Hypokalemia-induced downregulation of aquaporin-2 water channel expression in rat kidney medulla and cortex. *J Clin Invest* 97:1960–1968.

Marples D, Knepper MA, Christensen EI et al. (1995b) Redistribution of aquaporin-2 water channels induced by vasopressin in rat kidney inner medullary collecting duct. *Am J Physiol* 269:C655–C664.

Marr N, Bichet DG, Lonergan M et al. (2002) Heteroligomerization of an Aquaporin-2 mutant with wild-type Aquaporin-2 and their misrouting to late endosomes/lysosomes explains dominant nephrogenic diabetes insipidus. *Hum Mol Genet* 11:779–789.

Martin PY, Abraham WT, Lieming X et al. (1999) Selective V2-receptor vasopressin antagonism decreases urinary aquaporin-2 excretion in patients with chronic heart failure. *J Am Soc Nephrol* 10:2165–2170.

Matsumura Y, Uchida S, Rai T et al. (1997) Transcriptional regulation of aquaporin-2 water channel gene by cAMP. *J Am Soc Nephrol* 8:861–867.

Moeller HB, Fenton RA. (2012) Cell biology of vasopressin-regulated aquaporin-2 trafficking. *Pflugers Arch* 464:133–144.

Morishita Y, Matsuzaki T, Hara-chikuma M et al. (2005) Disruption of aquaporin-11 produces polycystic kidneys following vacuolization of the proximal tubule. *Mol Cell Biol* 25:7770–7779.

Mouillac B, Chini B, Balestre MN et al. (1995) The binding site of neuropeptide vasopressin V1a receptor. Evidence for a major localization within transmembrane regions. *J Biol Chem* 270:25771–25777.

Mulders SM, Bichet DG, Rijss JP et al. (1998) An aquaporin-2 water channel mutant which causes autosomal dominant nephrogenic diabetes insipidus is retained in the Golgi complex. *J Clin Invest* 102:57–66.

Murer L, Addabbo F, Carmosino M et al. (2004) Selective decrease in urinary aquaporin 2 and increase in prostaglandin E2 excretion is associated with postobstructive polyuria in human congenital hydronephrosis. *J Am Soc Nephrol* 15:2705–2712.

Murillo-Carretero MI, Ilundain AA, Echevarria M (1999) Regulation of aquaporin mRNA expression in rat kidney by water intake. *J Am Soc Nephrol* 10:696–703.

Nedvetsky PI, Tabor V, Tamma G, et al. (2010) Reciprocal regulation of aquaporin-2 abundance and degradation by protein kinase A and p38-MAP kinase. *J Am Soc Nephrol* 21:1645–1656.

Nejsum LN, Elkjaer M, Hager H et al. (2000) Localization of aquaporin-7 in rat and mouse kidney using RT-PCR, immunoblotting, and immunocytochemistry. *Biochem Biophys Res Commun* 277:164–170.

Nielsen J, Hoffert JD, Knepper MA et al. (2008a) Proteomic analysis of lithium-induced nephrogenic diabetes insipidus: Mechanisms for aquaporin 2 down-regulation and cellular proliferation. *Proc Natl Acad Sci USA* 105:3634–3639.

Nielsen J, Kwon TH, Christensen BM et al. (2008b) Dysregulation of renal aquaporins and epithelial sodium channel in lithium-induced nephrogenic diabetes insipidus. *Semin Nephrol* 28:227–244.

Nielsen J, Kwon TH, Praetorius J et al. (2003) Segment-specific ENaC downregulation in kidney of rats with lithium-induced NDI. *Am J Physiol Renal Physiol* 285:F1198–F1209.

Nielsen J, Kwon TH, Praetorius J et al. (2006) Aldosterone increases urine production and decreases apical AQP2 expression in rats with diabetes insipidus. *Am J Physiol Renal Physiol* 290:F438–F449.

Nielsen S, Chou CL, Marples D et al. (1995a) Vasopressin increases water permeability of kidney collecting duct by inducing translocation of aquaporin-CD water channels to plasma membrane. *Proc Natl Acad Sci USA* 92:1013–1017.

Nielsen S, DiGiovanni SR, Christensen EI et al. (1993a) Cellular and subcellular immunolocalization of vasopressin-regulated water channel in rat kidney. *Proc Natl Acad Sci USA* 90:11663–11667.

Nielsen S, Frokiaer J, Marples D et al. (2002) Aquaporins in the kidney: From molecules to medicine. *Physiol Rev* 82:205–244.

Nielsen S, Kwon TH, Christensen BM et al. (1999) Physiology and pathophysiology of renal aquaporins. *J Am Soc Nephrol* 10:647–663.

Nielsen S, Kwon TH, Dimke H et al (2007b) Aquaporin water channels in mammalian kidney. In: Alpern RJ, Hebert SC (eds.), *The Kidney*, 4th edn. Elsevier, San Diego, CA.

Nielsen S, Kwon TH, Frokiaer J et al. (2000) Key roles of renal aquaporins in water balance and water-balance disorders. *News Physiol Sci* 15:136–143.

Nielsen S, Kwon TH, Frokiaer J et al. (2007) Regulation and dysregulation of aquaporins in water balance disorders. *J Intern Med* 261:53–64.

Nielsen S, Pallone T, Smith BL et al. (1995b) Aquaporin-1 water channels in short and long loop descending thin limbs and in descending vasa recta in rat kidney. *Am J Physiol* 268:F1023–F1037.

Nielsen S, Smith BL, Christensen EI et al. (1993b) Distribution of the aquaporin CHIP in secretory and resorptive epithelia and capillary endothelia. *Proc Natl Acad Sci USA* 90:7275–7279.

Nielsen S, Smith BL, Christensen EI et al. (1993c) CHIP28 water channels are localized in constitutively water-permeable segments of the nephron. *J Cell Biol* 120:371–383.

Nielsen S, Terris J, Andersen D et al. (1997) Congestive heart failure in rats is associated with increased expression and targeting of aquaporin-2 water channel in collecting duct. *Proc Natl Acad Sci USA* 94:5450–5455.

Norregaard R, Jensen BL, Li C et al. (2005) COX-2 inhibition prevents downregulation of key renal water and sodium transport proteins in response to bilateral ureteral obstruction. *Am J Physiol Renal Physiol* 289:F322–F333.

Norregaard R, Jensen BL, Topcu SO et al. (2006) Cyclooxygenase type 2 is increased in obstructed rat and human ureter and contributes to pelvic pressure increase after obstruction. *Kidney Int* 70:872–881.

Novak A, Dedhar S (1999) Signaling through beta-catenin and Lef/Tcf. *Cell Mol Life Sci* 56:523–537.

Pallone TL, Turner MR, Edwards A et al. (2003) Countercurrent exchange in the renal medulla. *Am J Physiol Regul Integr Comp Physiol* 284:R1153–R1175.

Preston GM, Agre P (1991) Isolation of the cDNA for erythrocyte integral membrane protein of 28 kilodaltons: Member of an ancient channel family. *Proc Natl Acad Sci USA* 88:11110–11114.

Preston GM, Carroll TP, Guggino WB et al. (1992) Appearance of water channels in *Xenopus* oocytes expressing red cell CHIP28 protein. *Science* 256:385–387.

Procino G, Carmosino M, Marin O et al. (2003) Ser-256 phosphorylation dynamics of aquaporin 2 during maturation from the ER to the vesicular compartment in renal cells. *FASEB J* 17:1886–1888.

Procino G, Carmosino M, Tamma G et al. (2004) Extracellular calcium antagonizes forskolin-induced aquaporin 2 trafficking in collecting duct cells. *Kidney Int* 66:2245–2255.

Promeneur D, Kwon TH, Frokiaer J et al. (2000) Vasopressin V(2)-receptor-dependent regulation of AQP2 expression in Brattleboro rats. *Am J Physiol Renal Physiol* 279:F370–F382.

Puliyanda DP, Ward DT, Baum MA et al. (2003) Calpain-mediated AQP2 proteolysis in inner medullary collecting duct. *Biochem Biophys Res Commun* 303:52–58.

Radomski JL, Fuyathn, Nelson AA et al. (1950) The toxic effects, excretion and distribution of lithium chloride. *J Pharmacol Exp Ther* 100:429–444.

Rao R, Zhang MZ, Zhao M et al. (2005) Lithium treatment inhibits renal GSK-3 activity and promotes cyclooxygenase 2-dependent polyuria. *Am J Physiol Renal Physiol* 288:F642–F649.

Riccardi D, Hall AE, Chattopadhyay N et al. (1998) Localization of the extracellular Ca^{2+}/polyvalent cation-sensing protein in rat kidney. *Am J Physiol* 274:F611–F622.

Riccardi D, Lee WS, Lee K et al. (1996) Localization of the extracellular Ca(2+)-sensing receptor and PTH/PTHrP receptor in rat kidney. *Am J Physiol* 271:F951–F956.

Rojek A, Fuchtbauer EM, Kwon TH et al. (2006) Severe urinary concentrating defect in renal collecting duct-selective AQP2 conditional-knockout mice. *Proc Natl Acad Sci USA* 103:6037–6042.

Rojek A, Praetorius J, Frokiaer J et al. (2008) A current view of the mammalian aquaglyceroporins. *Annu Rev Physiol* 70:301–327.

Sabolic I, Katsura T, Verbavatz JM et al. (1995) The AQP2 water channel: Effect of vasopressin treatment, microtubule disruption, and distribution in neonatal rats. *J Membr Biol* 143:165–175.

Sabolic I, Valenti G, Verbavatz JM et al. (1992) Localization of the CHIP28 water channel in rat kidney. *Am J Physiol* 263:C1225–C1233.

Sandoval PC, Slentz DH, Pisitkun T et al. (2013) Proteome-wide measurement of protein half-lives and translation rates in vasopressin-sensitive collecting duct cells. *J Am Soc Nephrol* 24:1793–1805.

Saito T, Ishikawa SE, Sasaki S et al. (1997) Alteration in water channel AQP-2 by removal of AVP stimulation in collecting duct cells of dehydrated rats. *Am J Physiol* 272:F183–F191.

Saito T, Higashiyama M, Nagasaka S et al. (2001) Role of aquaporin-2 gene expression in hyponatremic rats with chronic vasopressin-induced antidiuresis. *Kidney Int* 60(4):1266–1276.

Sasaki S, Fushimi K, Saito H et al. (1994) Cloning, characterization, and chromosomal mapping of human aquaporin of collecting duct. *J Clin Invest* 93:1250–1256.

Schnermann J, Chou CL, Ma T et al. (1998) Defective proximal tubular fluid reabsorption in transgenic aquaporin-1 null mice. *Proc Natl Acad Sci USA* 95:9660–9664.

Schrier RW (2008) Vasopressin and aquaporin 2 in clinical disorders of water homeostasis. *Semin Nephrol* 28:289–296.

Schrier RW, Abraham WT (1999) Hormones and hemodynamics in heart failure. *N Engl J Med* 341:577–585.

Seibold A, Brabet P, Rosenthal W et al. (1992) Structure and chromosomal localization of the human antidiuretic hormone receptor gene. *Am J Hum Genet* 51:1078–1083.

Sharples EJ, Patel N, Brown P et al. (2004) Erythropoietin protects the kidney against the injury and dysfunction caused by ischemia-reperfusion. *J Am Soc Nephrol* 15:2115–2124.

Shi Y, Li C, Thomsen K et al. (2004a) Neonatal ureteral obstruction alters expression of renal sodium transporters and aquaporin water channels. *Kidney Int* 66:203–215.

Shi Y, Pedersen M, Li C et al. (2004b) Early release of neonatal ureteral obstruction preserves renal function. *Am J Physiol Renal Physiol* 286:F1087–F1099.

Smith BL, Agre P (1991) Erythrocyte Mr 28,000 transmembrane protein exists as a multisubunit oligomer similar to channel proteins. *J Biol Chem* 266:6407–6415.

Sohara E, Rai T, Miyazaki J et al. (2005) Defective water and glycerol transport in the proximal tubules of AQP7 knockout mice. *Am J Physiol Renal Physiol* 289:F1195–F1200.

Sorensen JB, Andersen MK, Hansen HH (1995) Syndrome of inappropriate secretion of antidiuretic hormone (SIADH) in malignant disease. *J Intern Med* 238(2):97–110.

Stambolic V, Ruel L, Woodgett JR (1996) Lithium inhibits glycogen synthase kinase-3 activity and mimics wingless signalling in intact cells. *Curr Biol* 6:1664–1668.

Stamer WD, Snyder RW, Smith BL et al. (1994) Localization of aquaporin CHIP in the human eye: Implications in the pathogenesis of glaucoma and other disorders of ocular fluid balance. *Invest Ophthalmol Vis Sci* 35:3867–3872.

Stamoutsos BA, Carpenter RG, Grossman SP (1981) Role of angiotensin-II in the polydipsia of diabetes insipidus in the Brattleboro rat. *Physiol Behav* 26:691–693.

Stegbauer J, Gurley SB, Sparks MA et al. (2011) AT1 receptors in the collecting duct directly modulate the concentration of urine. *J Am Soc Nephrol* 22(12):2237–2246.

Sugawara M, Hashimoto K, Ota Z (1988) Involvement of prostaglandinE2, cAMP, and vasopressin in lithium-induced polyuria. *Am J Physiol* 254:R863–R869.

Szatalowicz VL, Arnold PE, Chaimovitz C et al. (1981) Radioimmunoassay of plasma arginine vasopressin in hyponatremic patients with congestive heart failure. *N Engl J Med* 305:263–266.

Tannen RL, Regal EM, Dunn MJ et al. (1969) Vasopressin-resistant hyposthenuria in advanced chronic renal disease. *N Engl J Med* 280:1135–1141.

Tanner GA, Sloan KL, Sophasan S (1973) Effects of renal artery occlusion on kidney function in the rat. *Kidney Int* 4:377–389.

Teitelbaum I, McGuinness S (1995) Vasopressin resistance in chronic renal failure. Evidence for the role of decreased V2 receptor mRNA. *J Clin Invest* 96:378–385.

Terris J, Ecelbarger CA, Marples D et al. (1995) Distribution of aquaporin-4 water channel expression within rat kidney. *Am J Physiol* 269:F775–F785.

Terris J, Ecelbarger CA, Nielsen S et al. (1996) Long-term regulation of four renal aquaporins in rats. *Am J Physiol* 271:F414–F422.

Tian Y, Sandberg K, Murase T et al. (2000) Vasopressin V2 receptor binding is down-regulated during renal escape from vasopressin-induced antidiuresis. *Endocrinology* 141(1):307–314.

Timmer RT, Sands JM (1999) Lithium intoxication. *J Am Soc Nephrol* 10:666–674.

Topcu SO, Pedersen M, Norregaard R et al. (2007) Candesartan prevents long-term impairment of renal function in response to neonatal partial unilateral ureteral obstruction. *Am J Physiol Renal Physiol* 292:F736–F748.

Umenishi F, Narikiyo T, Vandewalle A et al. (2006) cAMP regulates vasopressin-induced AQP2 expression via protein kinase A-independent pathway. *Biochim Biophys Acta* 1758:1100–1105.

Valenti G, Laera A, Gouraud S et al. (2002) Low-calcium diet in hypercalciuric enuretic children restores AQP2 excretion and improves clinical symptoms. *Am J Physiol Renal Physiol* 283:F895–F903.

Valenti G, Laera A, Pace G et al. (2000) Urinary aquaporin 2 and calciuria correlate with the severity of enuresis in children. *J Am Soc Nephrol* 11:1873–1881.

Valenti G, Procino G, Tamma G, et al. (2005) Minireview: aquaporin 2 trafficking. *Endocrinology* 146:5063–5070.

Valtin H, Schroeder HA (1964) Familial hypothalamic diabetes insipidus in rats (Brattleboro strain). *Am J Physiol* 206:425–430.

Van Hoek AN, Ma T, Yang B et al. (2000) Aquaporin-4 is expressed in basolateral membranes of proximal tubule S3 segments in mouse kidney. *Am J Physiol Renal Physiol* 278:F310–F316.

Venkatachalam MA, Bernard DB, Donohoe JF et al. (1978) Ischemic damage and repair in the rat proximal tubule: Differences among the S1, S2, and S3 segments. *Kidney Int* 14:31–49.

Verkman AS (2008) Dissecting the roles of aquaporins in renal pathophysiology using transgenic mice. *Semin Nephrol* 28:217–226.

Wade JB, Kachadorian WA (1988) Cytochalasin B inhibition of toad bladder apical membrane responses to ADH. *Am J Physiol* 255:C526–C530.

Wade JB, Stetson DL, Lewis SA (1981) ADH action: Evidence for a membrane shuttle mechanism. *Ann N Y Acad Sci* 372:106–117.

Wall SM, Han JS, Chou CL et al. (1992) Kinetics of urea and water permeability activation by vasopressin in rat terminal IMCD. *Am J Physiol* 262:F989–F998.

Wang W, Kwon TH, Li C et al. (2002a) Reduced expression of Na-K-2Cl cotransporter in medullary TAL in vitamin D-induced hypercalcemia in rats. *Am J Physiol Renal Physiol* 282:F34–F44.

Wang W, Li C, Kwon TH et al. (2002b) AQP3, p-AQP2, and AQP2 expression is reduced in polyuric rats with hypercalcemia: Prevention by cAMP-PDE inhibitors. *Am J Physiol Renal Physiol* 283:F1313–F1325.

Wen H, Frokiaer J, Kwon TH et al. (1999) Urinary excretion of aquaporin-2 in rat is mediated by a vasopressin-dependent apical pathway. *J Am Soc Nephrol* 10:1416–1429.

Wood LJ, Massie D, McLean AJ et al. (1988) Renal sodium retention in cirrhosis: Tubular site and relation to hepatic dysfunction. *Hepatology* 8:831–836.

Xu DL, Martin PY, Ohara M et al. (1997) Upregulation of aquaporin-2 water channel expression in chronic heart failure rat. *J Clin Invest* 99:1500–1505.

Yamamoto T, Sasaki S, Fushimi K et al. (1995) Vasopressin increases AQP-CD water channel in apical membrane of collecting duct cells in Brattleboro rats. *Am J Physiol* 268:C1546–51.

Yasui M, Hazama A, Kwon TH et al. (1999a) Rapid gating and anion permeability of an intracellular aquaporin. *Nature* 402:184–187.

Yasui M, Kwon TH, Knepper MA et al. (1999b) Aquaporin-6: An intracellular vesicle water channel protein in renal epithelia. *Proc Natl Acad Sci USA* 96:5808–5813.

Zelenina M, Christensen BM, Palmer J et al. (2000) Prostaglandin E(2) interaction with AVP: Effects on AQP2 phosphorylation and distribution. *Am J Physiol Renal Physiol* 278:F388–F394.

Zelenina M, Zelenin S, Bondar AA et al. (2002) Water permeability of aquaporin-4 is decreased by protein kinase C and dopamine. *Am J Physiol Renal Physiol* 283:F309–F318.

CHAPTER 8

Vasopressin and the Regulation of Aquaporin-2 in Health and Disease

Giovanna Valenti and Grazia Tamma

CONTENTS

Abstract		157
8.1	Introduction	158
8.2	Short-Term Regulation of AQP2 by Vasopressin	159
	8.2.1 Trafficking Mechanism of AQP2	159
	8.2.1.1 Role of AKAPs on AQP2 Trafficking	160
	8.2.1.2 Cytoskeleton Dynamics in AQP2 Shuttle	161
	8.2.1.3 AQP2 Shuttle: Regulation of Exocytosis and Endocytosis	163
	8.2.1.4 Post-translational Modifications of AQP2: Phosphorylation, Ubiquitylation and Glutathionylation	166
8.3	Long-Term Regulation of AQP2 by Vasopressin	168
	8.3.1 AQP2 Synthesis	168
	8.3.2 Tonicity Modulation of AQP2 Expression	168
	8.3.3 Urinary Exosomes Excretion of AQP2	169
8.4	V2R Antagonist for the Treatment of Water Retaining Diseases	170
8.5	Conclusions	171
References		172

ABSTRACT

THE LOCALIZATION OF THE WATER channel aquaporin-2 (AQP2) is subjected to regulation by vasopressin. Vasopressin adjusts the amount of AQP2 in the plasma membrane by regulating its redistribution from intracellular vesicles into the plasma membrane allowing water entry into the cells and water exit through AQP3 and AQP4. This permits water reabsorption and urine concentration. Following binding of vasopressin to its V2R receptor, the rise in cAMP activates protein kinase A, which in turn phosphorylates AQP2 and thereby triggers the redistribution of AQP2. Several proteins participating in the control of cAMP-dependent AQP2 trafficking have been identified including SNAREs, annexin-2, hsc70, AKAPs and small GTPases of the Rho family proteins. Moreover, AQP2 has been found to be regulated by post-translational modifications (PTMs), such as ubiquitination and glutathionylation. Loss-of-function mutations of both V2R and AQP2 are associated with congenital nephrogenic diabetes insipidus characterized by a failure to concentrate urine. Conversely gain-of-function mutations of the V2R are associated with the nephrogenic syndrome of inappropriate antidiuresis characterized by positive water balance and hyponatremia. Vaptans, non-peptide vasopressin receptor antagonists, represent a new class of drugs developed for the treatment of euvolemic or hypervolemic hyponatremia. This chapter summarizes recent data elucidating molecular mechanisms underlying the trafficking of AQP2. The mechanism of action of vaptans and their current use in clinical practice is discussed.

8.1 INTRODUCTION

Maintenance of water balance is critically dependent on water intake, the sensation of thirst and the regulation of water excretion in the kidney, which is under the control of the antidiuretic hormone arginine vasopressin. Vasopressin is secreted into the circulation by the posterior pituitary gland, in response to an increase in serum osmolality or a decrease in blood volume. In the kidney, vasopressin binds to the V2 vasopressin receptor (V2R), which belongs to the superfamily of G protein–coupled receptors. V2Rs expressed in the basolateral membranes of collecting-duct cells in the last portion of the nephron activate the Gs protein, thus increasing the intracellular cAMP. The cAMP/protein kinase A (PKA) signal transduction cascade results in multiple phosphorylating events in the C-terminus of AQP2 of which phosphorylation at S256 is required for the vasopressin-regulated translocation of the AQP2-bearing vesicles toward the apical plasma membrane, thus increasing water luminal permeability. In addition to phosphorylation, AQP2 undergoes different regulated PTMs, such as ubiquitination and glutathionylation, which are likely to be fundamental for controlling AQP2 cellular localization, stability and function. Several proteins participating in the control of cAMP-dependent AQP2 trafficking have been identified including soluble NSF attachment protein receptors (SNAREs), annexin-2, hsc70, A kinase anchoring proteins (AKAPs) and small GTPases of the Rho family proteins controlling cytoskeletal dynamics.

In addition to regulate AQP2 trafficking, vasopressin also regulates the total amount of the water channel within the cell. Alterations in the AQP2 abundance as well as defects in vasopressin signalling in the renal collecting can seriously compromise the maintenance of water balance in the body.

Loss-of-function mutations of both V2R and AQP2 are associated with congenital nephrogenic diabetes insipidus (NDI), a rare genetic disorder, which can be quite severe in infants, characterized by a failure to concentrate urine (despite normal or elevated levels of vasopressin), polyuria, polydipsia and hypernatremia. More recently, it has been discovered that gain-of-function mutations of the V2R are associated with the nephrogenic syndrome of inappropriate antidiuresis (NSIAD). Inappropriate antidiuresis is the most common cause of hyponatremia, and it is characterized by the inability to excrete a free water load, inappropriately concentrated urine, hyponatremia, hypo-osmolality and natriuresis. The most common clinical entity is the idiopathic syndrome of inappropriate antidiuretic hormone secretion (SIADH), which is linked to hypersecretion of vasopressin. However, 10%–20% of patients with inappropriate antidiuresis display low or undetectable vasopressin circulating levels. The identification in these patients of gain-of- function mutations of the V2R led to the definition of NSIAD as a new clinical entity of inappropriate antidiuresis, which, in contrast to SIADH, is characterized by low or undetectable levels of vasopressin.

Vaptans represent a new class of drugs developed for the treatment of euvolemic or hypervolemic hyponatremia. These drugs are non-peptide vasopressin receptor antagonists causing water diuresis, namely aquaresis, thus increasing serum sodium.

This chapter highlights some of the new insights into vasopressin-dependent AQP2 regulation and function that have developed recently, with particular focus on the cell biological aspects of AQP2 regulation. Moreover, it will address the development and mechanism of action of vaptans and current use in clinical practice, as well as future perspectives.

8.2 SHORT-TERM REGULATION OF AQP2 BY VASOPRESSIN

The renal collecting duct water permeability is controlled by the hormone vasopressin through regulation of the water channel AQP2 in two processes: short- and long-term regulation. These processes are involved in modulating the trafficking of AQP2 at the apical plasma membrane and the total cellular abundance of AQP2 protein. The short-term regulation occurs over a period of minutes resulting in an increased trafficking of AQP2-bearing vesicles from an intracellular pool to the apical plasma membrane in response to vasopressin stimulus. The complex molecular signals controlling, at short-term, AQP2 trafficking in response to vasopressin include rapid changes of the activity of kinases, phosphatases, AKAPs, E3-ligases and regulation of cytoskeleton remodelling all important steps for exocytosis and endocytosis of AQP2.

The long-term regulation, instead, occurs over a period of hours to days due to vasopressin regulation of cellular AQP2 total abundance.

8.2.1 Trafficking Mechanism of AQP2

Pioneering studies on isolated toad bladder showed that vasopressin increases the unidirectional flux of water by increasing the permeability of the membrane (Hays and Leaf 1962). Measurements in isolated perfused collecting ducts also demonstrated a rapid increase in the osmotic water permeability in response to vasopressin (Grantham and Burg 1966). Vasopressin mainly targets its action on V2R localized at the basolateral membrane of renal principal cells. Binding of vasopressin to its specific V2R increases the intracellular concentration of cAMP via adenylyl cyclase type III and VI, resulting in the activation of PKA (Hoffert et al. 2005), although other kinases, such as Akt, Sgk and p38-MAPK, may play a role in response to vasopressin stimulation (Pisitkun et al. 2008; Nedvetsky et al. 2010; Rinschen et al. 2010). One target for PKA is AQP2 at serine-256, although AQP2 may be a substrate for other basophilic protein kinases such as Akt1 and protein kinase Cδ (Douglass et al. 2012). PKA-dependent phosphorylation of the water channel AQP2, at S256, is essential to promote the translocation of AQP2-bearing vesicles from an intracellular pool to the apical plasma membrane (Nielsen et al. 1993; Katsura et al. 1997). So far, PKA has been considered the main effector of cAMP; however, several studies clearly indicate that cAMP can also stimulate exocytosis independently of PKA action (Ozaki et al. 2000; Seino and Shibasaki 2005). cAMP activates exchange protein activated by cAMP (Epac) and PKA similarly (Christensen et al. 2003). Epac is a cAMP downstream effector expressed in different tissues (Kawasaki et al. 1998) and functions as a guanine nucleotide exchange factor for Rap1 and Ras (de Rooij et al. 1998; Kawasaki et al. 1998). Two isoforms (Epac1 and Epac2) have been described (Kawasaki et al. 1998), which are both expressed in renal tubules (Li et al. 2008). In collecting ducts, Epac1 is expressed in intercalated cells, whereas Epac2 is distributed in AQP2-expressing principal cells (Li et al. 2008). Stimulation of isolated renal tubules with a selective agonist of Epac mimics vasopressin-induced intracellular calcium release and AQP2 targeting at the apical plasma membrane independently of PKA signalling activation (Yip 2006). This study indicates that AQP2 exocytosis is controlled by PKA and Epac signalling pathways. However, in primary cultured inner medullar collecting duct (IMCD) cells, elevation of cAMP is sufficient to elicit AQP2 translocation to the plasma membrane, without any change in intracellular calcium level (Lorenz et al. 2003), likely suggesting that cultured IMCD cells have different features compared with native IMCD cells.

Besides cAMP elevation, vasopressin elicits a significant and transient rise in intracellular calcium (Star et al. 1988). Data from several laboratories highlighted that the V2R-mediated calcium release has an important role in the regulatory pathway modulating renal water reabsorption in response to vasopressin stimulation. Buffering of intracellular calcium with BAPTA prevented the rise in the osmotic water permeability coefficient (Pf) in response to vasopressin in IMCD (Chou et al. 2000; Yip 2002), suggesting that intracellular calcium mobilization is required for exocytotic insertion of AQP2. Moreover, the removal of extracellular free calcium in perfused IMCD did not prevent the initial calcium rise induced by vasopressin but inhibited the sustained oscillations (Yip 2002), which may be required to prolong the exocytotic activity induced by vasopressin. To date, the molecular mechanism controlling the vasopressin-dependent intracellular calcium release is far not completely clarified yet. Interestingly, in primary cultures of IMCD cells, ryanodine, a ryanodine receptor antagonist, prevents the translocation of AQP2 to the plasma membrane in response to vasopressin, suggesting a role for ryanodine-sensitive stores in the calcium-dependent AQP2 trafficking (Chou et al. 2000; Yip 2002).

8.2.1.1 Role of AKAPs on AQP2 Trafficking

AKAPs are scaffolding proteins playing a pivotal role in compartmentalization of cAMP/PKA signalling molecules (Beene and Scott 2007). Another novel feature of AKAPs is their function in the formation of signalling platform integrating cAMP signalling with other pathways as the ones mediated by Epac and extracellular signal-regulated kinase family (additional details related to this topic are further discussed in these papers: McConnachie et al. 2006; Beene and Scott 2007; Dodge-Kafka et al. 2008).

A correct compartmentalization of cAMP/PKA components by AKAPs is a prerequisite for vasopressin-dependent renal water reabsorption (Klussmann et al. 1999;

McSorley et al. 2006; Stefan et al. 2007). Incubation with Ht31 peptide, which prevents PKA-AKAP binding, impairs AQP2 fusion and insertion into the plasma membrane in IMCD cells (Klussmann et al. 1999,2000,2001a). A new splice variant of AKAP18, AKAP18delta, colocalizes with AQP2 and is expressed in immunoisolated AQP2-bearing vesicles (Henn et al. 2004). Additionally, AKAP18delta binds phosphodiesterases 4D (PDE4D), which degrades cAMP (Stefan et al. 2007). Vasopressin-dependent cAMP elevation results in PKA and AKAP18delta detachment (Henn et al. 2004) and translocation of AQP2 and PDE4D to the plasma membrane (Horner et al. 2012). Therefore, activation of PDE4D, via PKA-dependent phosphorylation, terminates PKA response by degrading cAMP locally and therefore decreasing the osmotic water permeability (Stefan et al. 2007).

8.2.1.2 Cytoskeleton Dynamics in AQP2 Shuttle

8.2.1.2.1 Microtubules Research over the past years have clarified that intracellular movement of vesicles requires a coordinate activity of cytoskeletal elements: microtubules and actin filaments (Desnos et al. 2007). Several studies have shown that cytoskeleton remodelling plays a key role in modulating AQP2 trafficking. Treatment with the microtubule-depolymerizing agents colchicine and nocodazole results in AQP2 vesicles scattering in the cytoplasm (Sabolic et al. 1995; Brown and Stow 1996) and impairs the vasopressin-dependent increase in water reabsorption (Phillips and Taylor 1989). Cold-induced depolymerization of microtubules increases the basolateral expression of AQP2 (Breton and Brown 1998), which is associated with AQP2 transcytosis (Yui et al. 2013).

In vivo and in vitro evidence clearly demonstrate the involvement of specific motor proteins in controlling the movement of vesicles along the microtubules (Allan and Schroer 1999). The minus end–directed motor protein, dynein, is involved in the regulation of vesicles movement toward the microtubule-organizing centre, while the plus end–directed motor proteins, kinesins, mediate the transport from the centre to the cell periphery. Interestingly, the motor proteins dynactin and dynein are expressed in the immunoisolated AQP2-bearing vesicles (Marples et al. 1998), likely suggesting a possible involvement of microtubule motor proteins in AQP2 trafficking. The expression of microtubule motor proteins has been confirmed by proteomic analysis of the protein pattern in AQP2 immunoisolated vesicles from renal IMCD (Barile et al. 2005). A study by Klussmann and co-workers gave some inputs on the role of microtubules in the regulation of AQP2 cellular distribution (Vossenkamper et al. 2007). The authors found that depolymerization of microtubules did not affect both cAMP-dependent translocation of AQP2 to the plasma membrane and endocytosis of AQP2, which was in line with the observations by Tajika (Tajika et al. 2005). Instead, microtubules play an important role in controlling perinuclear positioning of AQP2, co-expressed in Rab11-positive vesicles. Moreover, selective inhibition of dynein, by EHNA, resulted in AQP2 and Rab11 scattering in the entire cytoplasm, thus reproducing a similar effect to that induced by nocodazole (Vossenkamper et al. 2007). Together, these data suggest a role of microtubules and the motor protein dynein in the regulation of a proper AQP2 compartmentalization.

8.2.1.2.2 Actin   Dynamics of actin cytoskeleton are also known to be involved in controlling the cellular distribution of AQP2 (Valenti et al. 2005). Depolymerization of actin filaments has been described during vasopressin stimulation in toad bladder (Hays et al. 1993; Simon et al. 1993). Interestingly, in toad bladder epithelial cells treatment with cytochalasin D, a known actin-depolymerizing toxin, increases the fusion rate of aggrephores containing selective water channels, indicating that actin filaments might retard the fusion processes (Franki et al. 1992). Consistently, in renal collecting duct CD8 cells, treatment with okadaic acid, an inhibitor of type 1 and type 2A protein phosphatases, also decreases actin stress fibers and results in a significant increase in the osmotic water permeability (Valenti et al. 2000b). In this respect, it has been demonstrated that vasopressin-dependent F-actin depolymerization is dependent on AQP2 protein expression and that this is not dependent on the polarity of AQP2 membrane insertion (Yui et al. 2012). Together, these data indicate that actin cytoskeleton plays an important regulatory role in vasopressin-controlled AQP2 trafficking. Proteomic studies gave support to this view, showing that vasopressin-regulated actin network is potentially responsible for vasopressin-induced apical F-actin dynamics (Loo et al. 2013).

Our group deeply investigated the molecular signals integrating actin remodelling and AQP2 trafficking. Specifically, exposure of IMCD and CD8 cells to Clostridium toxin B, which inhibits proteins of Rho family (Rho, Rac and Cdc42) or Clostridium toxin C3, specifically inhibiting Rho, causes a partial depolymerization of actin filaments, an increase of the cells surface expression of AQP2 resulting in a significant increase of the osmotic water permeability, even in the absence of hormonal stimulation (Klussmann et al. 2001a,b; Tamma et al. 2001). Attenuation of RhoA activity is a physiological step during the signal transduction cascade activated by the hormonal stimulation, leading to a partial depolymerization of actin filaments, which facilitates the translocation of AQP2 to the apical plasma membrane (Tamma et al. 2003a). On another hand, Rho-induced actin polymerization decreases the cell surface expression of AQP2 indicating that RhoA exerts a bidirectional control in AQP2 cellular distribution by modulating actin remodelling (Tamma et al. 2003b, 2005a).

The ezrin–radixin–moesin (ERM) proteins cross-link actin filaments to membrane proteins, functioning as scaffolding proteins at the cell cortex (McClatchey and Fehon 2009). Importantly, ERM proteins are involved in controlling Rho activity, being upstream and downstream of Rho (Takahashi et al. 1997; Matsui et al. 1999; Mammoto et al. 2000). Inhibition of Rho kinase, by Y27632, decreases moesin phosphorylation at T558 and its interaction with actin filaments, thus facilitating AQP2 translocation to the plasma membrane (Tamma et al. 2005b). Together, these data provide a strong evidence for the participation of Rho signalling and the functional involvement of ERM proteins in vasopressin-dependent AQP2 trafficking.

Immunoaffinity chromatography studies, associated with 2D gels analysis, reveal that the C-terminus of AQP2 strongly binds β- and γ-isoforms of actin (Noda et al. 2004b). Phosphorylation of AQP2 at S256 increases the affinity of AQP2 to tropomyosin-5b (TM5b) and decreases its interaction with G-actin, thus inducing F-actin depolymerization (Noda et al. 2008). Moreover, proteomic studies showed that immunoisolated AQP2 vesicles from renal IMCD express several isoforms of the motor proteins myosin, which are important to promote AQP2 trafficking (Barile et al. 2005). In fact, vasopressin stimulation significantly increases the phosphorylation of calcium/calmodulin-dependent myosin light chain kinase whose inhibition impairs AQP2 shuttle (Chou et al. 2004). Instead, expression either of Rab11-FIP2 lacking the C2 domain, which cross-links myosin Vb to Rab11, or of a dominant-negative form of myosin Vb, which results in AQP2 accumulation in Rab11-positive cell compartment, thus preventing AQP2 trafficking (Nedvetsky et al. 2007).

Altogether these observations reveal that vasopressin-induced AQP2 trafficking requires a precise activation and configuration of actin and actin-binding proteins with important consequences for signalling events.

8.2.1.3 AQP2 Shuttle: Regulation of Exocytosis and Endocytosis

8.2.1.3.1 Exocytosis   It is well accepted that vasopressin increases the cell surface expression of AQP2, whether this increase is due to a decrease of endocytosis or an increase of the exocytosis remains unclear. In an elegant study, Dennis Brown has convincingly showed that vasopressin exerts its action by increasing the extent of exocytosis (Nunes et al. 2008). The soluble secreted yellow fluorescent protein (ssYFP), which is an exclusive marker for the secretory pathway, colocalizes with AQP2 when transfected in LLCPK1 cells and is secreted upon vasopressin stimulation. In contrast, the specific inhibitor of endocytosis, methyl-cyclodextrin, which accumulates AQP2 at the plasma membrane (Russo et al. 2006), had no effect on ssYFP secretion (Nunes et al. 2008), indicating that hormonal stimulation mainly activates the exocytotic pathway.

Temporal and spatial regulation of intracellular vesicles is modulated by small-GTPases activity. Rab (Rab stands for ras in the brain) proteins control many steps of membrane trafficking including vesicle formation, movement along cytoskeletal elements, proper cellular localization of cargo and membrane fusion. The expression of Rab3 and Rab5a in renal collecting duct epithelial cells was already demonstrated in 1995 (Liebenhoff and Rosenthal 1995). Further proteomic studies reveal that AQP2-bearing vesicles express many Rab isoforms (Rab1, Rab2, Rab4B, Rab5A-B-C, Rab7, Rab10, Rab11A-B, Rab14, Rab18, Rab21 and Rab25). However, the role of these important regulatory proteins, in AQP2 trafficking, remains mostly unidentified. To note, only the role of Rab11 has been clearly described (see in the preceding text) (Nedvetsky et al. 2007; Vossenkamper et al. 2007).

The molecular machinery for docking and fusion of vesicles to the apical membrane in epithelial cells and neurons is similar, and studies of synaptic vesicles fusion shed more light on the fusion mechanism in epithelial cell system as well. During membrane fusion, two separated lipid bilayers of membrane-bound cell compartment merge to create a unique layer. This complex mechanism consists of two phases and requires a precise coordination of several accessory proteins. During the first step, known as hemifusion, the two lipid bilayers attract each other to merge. Then, the fusion pore formation follows the merging (Chizmadzhev et al. 2000). Annexin-2 is highly expressed in renal tubules (Markoff and Gerke 2005) and plays a role in exocytosis and endocytosis, facilitating the preliminary phase of membrane fusion (Gerke and Moss 2002; Rescher and Gerke 2008). In renal collecting duct CD8 cells, stimulation with forskolin recruits annexin-2 in lipid raft (Tamma et al. 2008) together with AQP2 phosphorylated at serine 256 (Yu et al. 2008). In vitro and in vivo evidences indicate that annexin-2 is functionally involved in the fusion of highly purified AQP2 vesicles with plasma membrane since a synthetic peptide, inactivating annexin-2, impairs the fusion of AQP2-vesicles reducing the osmotic water permeability, in response to forskolin stimulation in intact cells (Tamma et al. 2008). These data clearly demonstrate that annexin-2 participates in fusion mechanism of AQP2.

The molecular mechanism for vesicles fusion was clarified in 1993 by the discoveries of SNARE (where NSF stands for *N*-ethyl-maleimide-sensitive fusion protein) in neuronal cells (Sollner et al. 1993a,b). The synaptic protein syntaxin, synaptosomal-associated protein (SNAP) and vesicle-associated membrane protein (VAMP, also known as synaptobrevin) were the first SNARE to be discovered and form the fusion machinery to bridge vesicle SNARE with the cognate partner membrane proteins allowing membrane merging (Rothman 1994).

Renal collecting duct and AQP2 immunoisolated vesicles express SNAP23, VAMP2, and syntaxin-3 and 4 (Mandon et al. 1996,1997; Inoue et al. 1998). Tetanus neurotoxin (TeNT), which cleaves synaptobrevin-like protein, decreases the cell surface expression of AQP2 in CD8 cells (Valenti et al. 2002). Selective protein silencing of VAMP2, VAMP3 and SNAP23 inhibits AQP2 insertion at the apical plasma membrane (Procino et al. 2008). Interestingly, it has been shown that knockout mice for VAMP8 develop hydronephrosis. In these animals, stimulation with forskolin or DDAVP fails in inducing AQP2 exocytosis, whereas VAMP3 null mice do not show any defect in urinary concentrating ability (Wang et al. 2009). Therefore, these findings indicate that VAMP8 is necessary and required for the fusion of AQP2-bearing vesicles, while, in vivo, the functional involvement of VAMP3 might be compensate by other SNARE proteins.

Besides SNARE proteins, other accessory proteins are known to be implicated in AQP2 shuttling as the heterotrimeric GTP-binding protein of the Gi family and signal-induced proliferation-associated protein 1 (SPA-1). Exposure of rabbit CD8 cells to synthetic peptide corresponding to the C-terminus of Gαi3 impairs the translocation of AQP2-bearing vesicles, likely suggesting that Gαi3 promotes AQP2 trafficking (Valenti et al. 1998), though the precise molecular mechanism involving Gαi3 on AQP2 trafficking is still not clear.

SPA-1, a GTPase for Rap1, interacts with AQP2 modulating its cellular distribution because in Madin-Darby canine kidney (MDCK) cells transfected with AQP2, the dominant-negative SPA-1 lacking Rap1 activity, inhibits the translocation of AQP2-bearing vesicles at the plasma membrane under hormonal stimulation. These observations are confirmed by in vivo studies showing that during water deprivation, AQP2 locates intracellularly in null mice for SPA-1 and at the luminal side in wild-type mice, as expected (Noda et al. 2004a).

8.2.1.3.2 Endocytosis   Endocytosis, the vesicle-mediated internalization of plasma membrane, involves the coordinate activity of a dynamic complex of proteins including clathrin, dynamin, hsc70, endophilin, amphiphysin, synaptojanin, epsin, adaptor protein-2 (AP-2) and flotillin (Grant and Donaldson 2009; Mettlen et al. 2009). The involvement of clathrin in AQP2 internalization is well described. Vasopressin stimulation is accompanied by coated pits formation in collecting duct luminal membranes (Brown and Orci 1983).

Further studies indicated that AQP2 is accumulated in clathrin-enriched membrane domains during vasopressin stimulation and washout. Indeed, expression of a dominant-negative form of dynamin (K44A), in LLC-PK1 cells, results in a significant increase of the cell surface expression of AQP2 (Sun et al. 2002). Dynamin is a GTPase controlling vesicle scission at the plasma membrane and the formation of clathrin-coated vesicles (De Camilli et al. 1995; Pucadyil and Schmid 2009). Therefore, in cells expressing the negative form of dynamin, the clathrin-dependent endocytosis is impaired causing AQP2 accumulation to the plasma membrane (Sun et al. 2002). A similar result is obtained by depleting membrane cholesterol by treating LLC-PK1 or MCD4 cells with methyl-β-cyclodextrin simvastatin or lovastatin, respectively (Russo et al. 2006; Procino et al. 2009; Li et al. 2011). These observations indicate that besides a vasopressin-regulated trafficking, AQP2 constitutively recycles between membrane and intracellular vesicles. Moreover, the cell surface expression of AQP2 increases by inhibiting the extent of endocytosis, as shown after vasopressin stimulation. Therefore, the possibility to develop new specific drugs, accumulating AQP2 at the plasma membrane by reducing endocytosis bypassing the altered V2R signalling in X-linked NDI, is an attractive challenge.

The heat shock protein 70 (HSP-70) and its cognate protein Hsc-70 are new players controlling AQP2 endocytosis. HSP-70 assists variegate processes in almost all cell compartments by controlling protein synthesis, folding, degradation and endocytosis (Bukau and Horwich 1998; Mayer and Bukau 1998). Hsc-70 is involved in the uncoating of clathrin-coated vesicles during endocytosis (Rothman and Schmid 1986; Ungewickell et al. 1995; Morgan et al. 2001). Several protein–protein interaction assays reveal that AQP2 interacts with several important components of the endocytotic machinery as AP-2, clathrin, dynamin and Hsp70/Hsc70. Indeed, while ATP reduces, AVP increases the direct interaction between Hsc-70 and AQP2. Disruption of Hsc-70 activity by infecting LLC-PK1-AQP2 cells with ATPase-deficient Hsc-70 causes AQP2 accumulation at the plasma membrane, indicating a functional role of this accessory protein on AQP2 internalization (Lu et al. 2007). LC-MS/MS analysis shows that Hsc70, HSP-70-1, HSP-70-2 and annexin-2 have a low affinity for a peptide reproducing the C-tail of AQP2 phosphorylated at S256 than to the un-phosphorylated form. In contrast, HSP-70-5 has a higher affinity for the phosphorylated peptide (Zwang et al. 2009). Though, the functional role of these accessory proteins remains to be established, it is important to bear in mind that AQP2 phosphorylation deeply influences the AQP2 binding ability to several signalling proteins. Recent studies have shown that myelin and lymphocyte-associated protein (MAL), also known as vesicle integral protein of 17 kDa (VIP17), binds AQP2 phosphorylated at S256 preferentially (Kamsteeg et al. 2007). MAL is a tetraspan membrane protein (it spans the membrane four times) and a detergent-resistant membrane-associated protein regulating apical sorting and targeting (Cheong et al. 1999; Marazuela et al. 2003); a crucial role of MAL for clathrin-dependent endocytosis has been shown as well. (Martin-Belmonte et al. 2003). In LLC-PK1-AQP2, internalization experiments reveal that MAL causes an increase of the apical expression of AQP2 by reducing the extent of endocytosis without affecting the ability of AQP2 to associate with detergent-resistant membrane domains (Kamsteeg et al. 2007).

8.2.1.4 Post-translational Modifications of AQP2: Phosphorylation, Ubiquitylation and Glutathionylation

PTMs of proteins represent the major level of regulation. In the past few years, evidence for extensive crosstalk between PTMs has accumulated (Hunter 2007). The combination of different PTMs on protein surfaces may generate a 'PTM code' that may initiate or prevent selective downstream events. Proteomics studies from Knepper's group have substantially improved our knowledge about the different types of PTMs on AQP2 (Barile et al. 2005; Hoffert et al. 2006), even though deciphering the AQP2 PTMs code is not complete and represents the next level of complexity and one of the biggest challenges for future research. At short term, vasopressin stimulation promotes the insertion of AQP2 at the plasma membrane and reduces the endocytosis process by modulating the extent of different PTMs of AQP2 such as phosphorylation, monoubiquitylation and polyubiquitylation (Kamsteeg et al. 2006; Nedvetsky et al. 2010).

8.2.1.4.1 Phosphorylation of AQP2

8.2.1.4.1.1 Serine 256   Vasopressin-induced translocation of AQP2-bearing vesicles to the plasma membrane requires PKA-dependent phosphorylation of AQP2 at S256. The crucial role of cAMP-dependent phosphorylation of AQP2 was first investigated by Kuwahara et al. (1995) in 1995. In this study, the authors have shown that oocytes expressing wild-type AQP2 respond to forskolin stimulation by increasing the osmotic water permeability (Pf). In contrast, oocytes expressing mutated forms of AQP2 at S256 are not sensitive to forskolin or cAMP treatment (Kuwahara et al. 1995). Interestingly, in 1997, two different groups have shown that vasopressin-dependent AQP2 trafficking is impaired in LLC-PK1 cells expressing AQP2-S256A, the mutant form mimicking the constitutively not phosphorylated AQP2 protein (Fushimi et al. 1997; Katsura et al. 1997). The effect of vasopressin on AQP2 phosphorylation has been further confirmed in rat renal papillae pre-labelled with ^{32}P (Nishimoto et al. 1999). In rat kidney, specific affinity-purified antibodies, recognizing AQP2 phosphorylated at S256 (AQP2-pS256), labelled both apical plasma membrane and intracellular vesicles. Treatment with a selective V2R antagonist significantly reduced the apical staining confirming the pivotal role of the hormone vasopressin in controlling the cellular distribution of AQP2 (Christensen et al. 2000). S256 phosphorylation is required for vasopressin-dependent AQP2 translocation to the apical plasma membrane (see in the preceding text) (Russo et al. 2006), while AQP2 internalization occurs independently of S256 de-phosphorylation (van Balkom et al. 2002; Nejsum et al. 2005).

Subsequently, phosphoproteomic studies revealed a more complex action of vasopressin on AQP2, showing that vasopressin stimulation increases S256, S264 and T269 while decreases S261 phosphorylation in the C-terminus (Hoffert et al. 2006). A time course study reveal that within few seconds the extent of S256 phosphorylation rapidly increases ($t_{1/2}$ = 41 s). This priming phosphorylation event is crucial in the regulation of AQP2 localization and results in the consequent phosphorylation of S269 ($t_{1/2}$ = 3.2 min), S264 ($t_{1/2}$ = 4.2 min) and S261 dephosphorylation ($t_{1/2}$ = 10.6 min) (Hoffert et al. 2008).

8.2.1.4.1.2 Serine 261 LLC-PK1 cells transfected with S261A and S261D mutant forms of AQP2 suggested that the phosphorylation state of AQP2 at Ser261 does not appreciably affect the trafficking of AQP2 (Lu et al. 2008). However, Deen's group demonstrated that Ser261 phosphorylation follows monoubiquitination at K270, endocytosis and stabilizes AQP2 ubiquitination and intracellular localization (Tamma et al. 2011). Furthermore, ATP and dopamine, which reduce the water permeability counteracting the action of vasopressin, cause AQP2 internalization and increased phosphorylation at S261 (Boone et al. 2011).

8.2.1.4.1.3 Serine 264 Serine 264 is located within the consensus site for PKC and casein kinase type 1 (Hoffert et al. 2006; Brown et al. 2008). Using a phosphospecific antibody, it has been shown that short-term vasopressin stimulation (30 min) causes a relocalization of AQP2-pS264 from predominantly intracellular vesicles, to both the basolateral and apical plasma membranes. After 60 min, part of the AQP2-pS264 was observed in clathrin-coated vesicles, early endosomal compartments, and recycling compartments, but not lysosomes (Fenton et al. 2008).

8.2.1.4.1.4 Serine 269 Quantitative mass spectrometry and immunoblotting analysis with phosphospecific antibodies showed that vasopressin increases the phosphorylation of AQP2 at Ser269 which is mainly located in the apical plasma membrane (Hoffert et al. 2008). This observation is confirmed analysing the cellular localization of the mutant protein mimicking the constitutive phosphorylated AQP2 at S269 (AQP2-S269D), indicating that S269 phosphorylation might be a strong apical retention signal (Hoffert et al. 2008; Moeller et al. 2009). The slower internalization corresponded with reduced interaction of S269D-mutated AQP2 with several proteins involved in endocytosis, including Hsp70, Hsc70, dynamin and clathrin heavy chain (Rice et al. 2012).

8.2.1.4.1.5 Ubiquitylation of AQP2 At short term, forskolin stimulation promotes K63-linked short-chain ubiquitylation of AQP2 at the apical plasma membrane (Kamsteeg et al. 2006). Interestingly, short-chain ubiquitylation increases during forskolin washout corresponding to a higher rate of AQP2 retrieval from the plasma membrane to cellular vesicles (Kamsteeg et al. 2006). Short-chain ubiquitylation of AQP2 has become recognized as a key signal for intracellular trafficking, endocytosis and degradation in lysosomes (Woelk et al. 2006). In situ mutagenesis experiments revealed that AQP2 is ubiquitylated, within the C-terminus, at lysine 270 (K270). The kinetic of endocytosis showed that AQP2-K270R mutant, defective in ubiquitylation, is internalized slower than wild-type AQP2 indicating that short-chain ubiquitylation increases AQP2 endocytosis and its localization to internal vesicles of multivesicular bodies (MVBs) (Kamsteeg et al. 2006). On the other forskolin, stimulation decreases AQP2 polyubiquitylation via p38-MAPK inhibition resulting in a significant decrease of S261 phosphorylation (Nedvetsky et al. 2010). These findings suggest a complex regulation of AQP2 trafficking and abundance via ubiquitylation because short-chain ubiquitylation of AQP2 at K270 may promote AQP2 endocytosis to desensitize the cAMP signalling cascade activated by the hormone vasopressin, which however reduces the extent of AQP2 polyubiquitylation to preserve cellular AQP2 abundance.

8.2.1.4.1.6 Glutathionylation of AQP2 Recent data from our group demonstrated for the first time that AQP2 is subjected to *S*-glutathionylation in native mammalian kidney and in renal cell culture and that this PTM is modulated by the oxidative stress (Tamma et al. 2014b). It is known that reactive oxygen species (ROS) and reactive nitrogen species can function as signalling components transducing extracellular or intracellular information and elaborating specific responses by promoting PTMs of thiol residues on target proteins (Pastore and Piemonte 2012). Topological analysis of AQP2 suggests that Cys75 and Cys79, on cytosolic B-loop, might be the target of *S*-glutathionylation. Indeed, previous data have shown that vasopressin stimulation increased *S*-glutathionylation of different proteins in mpkccd cells, indicating the involvement of ROS in vasopressin-activated signal transduction pathway (Sandoval et al. 2013). In addition, the significant increase in intracellular calcium level associated with vasopressin is reduced by inhibiting NADPH oxidases, the major source of ROS in the kidney (Ding et al. 2011). In turn, NADPH oxidase deficiency results in a relevant decrease in AQP2 mRNA synthesis (Feraille et al. 2014), indicating that NADPH oxidase-derived ROS may contribute to enhance AQP2 transcription. In this scenario, additional studies will be needed to better elucidate the physiological relevance of glutathionylation of AQP2.

8.3 LONG-TERM REGULATION OF AQP2 BY VASOPRESSIN

Long-term treatment with vasopressin increases the total abundance of cellular AQP2 (Nielsen et al. 1993). The total cellular AQP2 abundance is the resulted balance of protein synthesis and protein removal via degradation or exosomal secretion.

8.3.1 AQP2 Synthesis

AQP2 abundance depends on production of AQP2 mRNA or by direct regulation of translation. As stated earlier, vasopressin causes a significant increase in AQP2 mRNA in rat collecting ducts (Ecelbarger et al. 1997) indicating that the increase of AQP2 protein is correlated with an increase of AQP2 mRNA. However, in the past few years, it has become clear that alternative mechanisms controlling AQP2 abundance exist and AQP2 mRNA available for translation does not reflect the total AQP2 mRNA due to regulated sequestration of mRNAs and mRNA degradation via micro-RNA (miRNAs) (Alsaleh and Gottenberg 2014). Important information regarding transcription factors involved in the regulation of Aqp2 gene were obtained analysing its 5′-flanking regions. Several conserved binding motifs, including a cAMP-response element (CRE), GATA, Sp1, Ets, Hox, nuclear factor of activated T (NFAT), RXR sites, a Forkhead box, an AP1 binding site and a site for Kruppel-like factor (Klf) binding have been identified (Rai et al. 1997; Yu et al. 2009; Tchapyjnikov et al. 2010).

8.3.2 Tonicity Modulation of AQP2 Expression

Storm et al. (2003) in the inner medulla of rat kidney (IMCD cells) showed that the expression of AQP2 is regulated by external osmolality and solute composition, via the tonicity-responsive enhancer binding protein (TonEBP), which plays a key role in protecting renal cells from hypertonic stress by stimulating transcription of specific gene. Extensive studies on the expression of AQP2 have been performed in the mouse mpkCCD cell line (Bens et al. 1999). When grown on filters, the expression of AQP2 mRNA in mpkCCD cell is very low in the absence of vasopressin. Addition of vasopressin to the basolateral side significantly increased the endogenous AQP2 mRNA and protein levels (Hasler et al. 2002, 2005). In this cell model, AQP2 mRNA as well as AQP2 protein abundance, under iso-osmotic condition, decreased in cells deficient for TonEBP (Hasler et al. 2006), confirming the role of this motif in regulating AQP2 expression. Five members of the NFAT cells family of transcription factors were also propose to play a role in controlling AQP2 expression (Hasler et al. 2006). In particular, NFATc1 was found to enhance AQP2 promoter activity, since hypertonicity resulted in the nuclear translocation of NFATc1 enhancing AQP2 promoter activity (Li et al. 2007). Importantly, besides the TonEBP motif, AQP2 promoter has two novel segments (−283 to −252 and −157 to −126 bp) involved in the hypotonicity-induced AQP2 downregulation during vasopressin escape (Kortenoeven et al. 2011). In this context, however, it has to be underlined that tonicity-induced AQP2 gene expression is not associated neither to change in AQP2 protein abundance nor to altered AQP2 shuttle (Terris et al. 1996; Storm et al. 2003).

8.3.3 Urinary Exosomes Excretion of AQP2

Urinary excretion of intact AQP2 was firstly described in 1995 by Kanno et al. (1995). The molecular mechanism responsible of AQP2 delivery into the urine is not fully clarified yet, although it is well accepted that it occurs via exosome secretion (Pisitkun et al. 2004). Exosomes are the internal vesicles contained into MVBs (late endosomes) that are released to the lumen of the renal tubules (Pisitkun et al. 2006). The proteome of urinary exosomes has been published at a publicly accessible resource (Gonzales et al. 2009). Several groups are used to detect urinary AQP2 excretion as biomarker of renal physiology with regard AQP2-dependent urinary concentrating ability (Deen et al. 1996; Elliot et al. 1996; Valenti et al. 2000a; Procino et al. 2012; Tamma et al. 2014b). Specifically, an increase in AQP2 excretion is observed under vasopressin action as a result of AQP2 activation and translocation to the membrane, whereas a reduced AQP2 excretion reflects reduced activation of the water channel and reduced renal ability to conserve water.

However, the physiological signal promoting exosomes delivery into the urine is still not completely clarified, and therefore such data may be reliable only when urine collections are made under well-defined steady-state conditions. In particular, under steady-state conditions, an increase in urinary AQP2 excretion could due to an increased AQP2 synthesis or to a decrease in degradation. Both mechanisms depend on an increase of cellular abundance of AQP2, which could be reflective of vasopressin stimulation. For example, high urinary AQP2 has been correlated with the presence of arterial hypertension (Rocchetti et al. 2011). This is the reason why urinary AQP2 excretion can be considered a reliable biomarker of AQP2-dependent urinary concentrating ability.

8.4 V2R ANTAGONIST FOR THE TREATMENT OF WATER RETAINING DISEASES

Considering the role of renal water reabsorption in the osmotic and volume homeostasis, water balance disturbances can be due to factors directly affecting the osmobalance or, indirectly, to factors principally affecting the volume balance. Numerous studies reported the role of vasopressin and AQP2 in different water retaining diseases such as congestive heart failure, hepatic cirrhosis, nephritic syndrome and some types of hyponatremia (Schrier 2007).

Selective vasopressin antagonists would offer the opportunity to treat these disorders. Several peptidic vasopressin antagonists have been generated; however, because of their unexpected pharmacological property to act as not selective agonist, their use has been limited (Maric et al. 1998). By large screening analysis, a new class of non-peptide and orally available vasopressin antagonists, called vaptans, have been recently described (Izumi et al. 2014).

Mozavaptan (OPC-31260) is the first vasopressin antagonist active in humans. It binds V2R and not vascular V1 vasopressin receptor (Yamamura et al. 1992). Oral administration to health humans at a dose of 1 mg/kg increases the urinary output eight times, while the sodium urinary excretion does not change. In Japan, mozavaptan has been approved to treat hyponatremia associated with paraneoplastic disease (Wang et al. 2007) and in small cell gall bladder carcinoma exacerbated by the syndrome of inappropriate antidiuresis (Tamura and Takeuchi 2013). Another V1A and V2R antagonist, conivaptan or YM 087 has been proved to generate aquaresis, an electrolyte-sparing excretion of free water, which results in the correction of serum sodium concentration (Udelson et al. 2001). So far, intravenous conivaptan has been approved by the Food and Drug Administration (FDA) in the United States for the treatment of euvolemic hyponatremia. However, several side effects including thirst, headache, hypokalemia, polyuria, vomiting, diarrhea and phlebitis associated with mozavaptan and conivaptan administration have been described and studies for pediatric use are still in progress (Stefan et al. 2007).

Lixivaptan is also under clinical trials (Liamis et al. 2014). This compound is well tolerated at single and multiple total daily doses also in combination with furosemide, enalapril and spironolactone (Gerbes et al. 2003; Wong et al. 2003; Liamis et al. 2014). Most of the patients admitted in these trials had a diagnosis of cirrhosis which leads to dilutional hypervolemic hyponatremia. Endpoints include alteration in net fluid balance, free water clearance and serum osmolality. In the first study, 125 and 250 mg increased free water clearance, while 250 mg twice-daily doses increased plasma vasopressin and ameliorate plasma sodium concentrations compared to placebo over 4 days. However, 250 mg twice-daily doses caused strong thirst and significant dehydration (Wong et al. 2003). In another study, it has been shown that 100 mg is sufficient to normalize serum sodium concentration in 27% of patients, while 200 mg is sufficient in 50% patients. A 200 mg dose caused reduction in urine osmolality and body weight. Moreover, patients with SIADH were more responsive to lixivaptan than patients affected by liver cirrhosis though the causes of this insensitivity are still not clear (Gerbes et al. 2003).

Another selective V2R antagonist, satavaptan (SR121463) (Izumi et al. 2014), is under trial investigation (Soupart et al. 2006; Schrier 2007; Wang et al. 2007). Satavaptan is well tolerated and judged safe; SIADH patients, under fluid restriction, participated in the phase III of the clinical trial. At dose of 25–50 mg/day, this selective antagonist decreased urinary osmolality and corrected sodium concentration from 127 ± 5 to 140 ± 6 mmol/L, indicating that this compound might be useful to treat hyponatremic patients (Soupart et al. 2006).

Tolvaptan (OPC-41061) is a new and promising selective oral V2R antagonist (Schrier 2007; Wang et al. 2007). Two studies of tolvaptan effect in humans have been described (Gheorghiade et al. 2004; Konstam et al. 2007). The first trial included 254 patients with severe CHF and affected by hyponatremia in 28% of cases at baseline. Endpoints included, in a first stage, change in body weight while in further step urine sodium excretion, urine volume and urine osmolality are considered. All three doses admitted in this trial (30, 45, and 60 mg) were effective in inducing a significant body weight reduction, increased urine excretion, plasma sodium levels compared with placebo group and improved the typical CHF symptoms. As significant side effects, tolvaptan administration resulted in dry mouth, thirst and polyuria but no alterations in blood pressure, serum potassium concentrations or renal function were measured (Konstam et al. 2007). Similar results were obtained in a second double-blind trial with 319 CHF hospitalized patients receiving 30, 60 and 90 mg of tolvaptan. A 60-day tolvaptan treatment significantly reduced mortality in patients with renal disease or CHF treated (Gheorghiade et al. 2004). However, recent observations obtained by EVEREST trial with 4133 patients clearly indicate that long-term tolvaptan treatment does not reduce death risk, while short-term treatment showed multiple benefits in sense of reduction of body weight and improvement in dyspnoea and oedema (Konstam et al. 2007). Therefore, tolvaptan administration could be admit for short-term treatment in hospitalized CHF patients (additional details related to this topic are further described here (Lehrich and Greenberg 2008). Recent studies have demonstrated the efficacy of tolvaptan to affect AQP2 signalling. Specifically, tolvaptan abolished the vasopressin effects on AQP2 phosphorylation (Miranda et al. 2014), though the molecular basis of this effect is still not clarified.

8.5 CONCLUSIONS

The identification of the vasopressin-sensitive AQP2 water channel in principal cells of the kidney collecting duct, as essential channel for urine concentration and water balance, has opened an existing research field for understanding body water homeostasis.

In the last decade, significant progress has been made in understanding the molecular basis of AQP2 trafficking between intracellular vesicles and the cell surface. A variety of cell signalling are activated by vasopressin leading to regulation of AQP2 expression and trafficking, including hormone-induced AQP2 phosphorylation, interaction with SNAREs, GTP-binding proteins, cytoskeletal remodelling, interplay between AQP2 and its lipid environment. More recently, besides phosphorylation, the importance of PTMs of AQP2 such as mono- and polyubiquitylation and glutathionylation have emerged as key

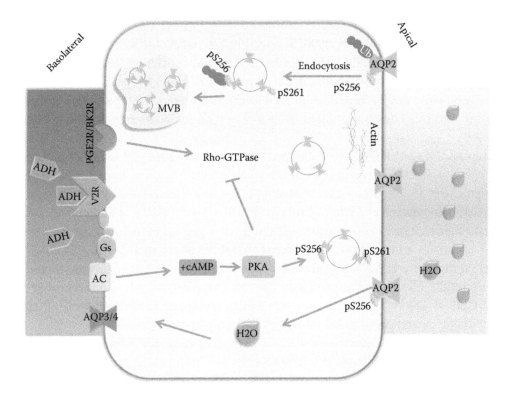

FIGURE 8.1 Schematic model of vasopressin-regulated AQP2 trafficking. In renal collecting duct, vasopressin promotes water reabsorption by binding to vasopressin V2 receptors (V2R). Activation of V2R stimulates adenylyl cyclase (AC) via the G protein Gs. The resulting increase in cAMP leads to the activation of PKA phosphorylating AQP2 at ser 256. Another PKA target is RhoA and RhoA phosphorylation decreases its activity, resulting in the depolymerization of F-actin and facilitating the insertion of AQP2 predominantly into the apical plasma membrane. Prostaglandin 2 (PGE2) and bradykinin (BK) counteract vasopressin response by activation of RhoA causing actin stabilization. The endocytic retrieval of AQP2 is controlled by post-translational modification (ubiquitination), which precedes AQP2 phosphorylation at ser-261 eventually leading to MVB sorting and degradation.

regulators of AQP2 trafficking and function (Figure 8.1). Vasopressin contributes to the regulation of both AQP2 trafficking and abundance in cells, which are defective in several water balance disorders. Recently, the availability of new, potent, orally active vasopressin receptor antagonists, the vaptans, has attracted attention as a possible therapy for water balance disorders characterized by water retention, although more studies are needed to better define their long-term safety and efficacy.

REFERENCES

Allan, V.J. and T.A. Schroer. 1999. Membrane motors. *Curr Opin Cell Biol* 11 (4):476–482.
Alsaleh, G. and J.E. Gottenberg. 2014. Characterization of microRNAs and their targets. *Methods Mol Biol* 1142:55–63.
Barile, M., T. Pisitkun, M.J. Yu et al. 2005. Large scale protein identification in intracellular aquaporin-2 vesicles from renal inner medullary collecting duct. *Mol Cell Proteomics* 4 (8):1095–1106.

Beene, D.L. and J.D. Scott. 2007. A-kinase anchoring proteins take shape. *Curr Opin Cell Biol* 19 (2):192–198.

Bens, M., V. Vallet, F. Cluzeaud et al. 1999. Corticosteroid-dependent sodium transport in a novel immortalized mouse collecting duct principal cell line. *J Am Soc Nephrol* 10 (5):923–934.

Boone, M., M.L. Kortenoeven, J.H. Robben, G. Tamma and P.M. Deen. 2011. Counteracting vasopressin-mediated water reabsorption by ATP, dopamine, and phorbol esters: Mechanisms of action. *Am J Physiol Renal Physiol* 300 (3):F761–F771.

Breton, S. and D. Brown. 1998. Cold-induced microtubule disruption and relocalization of membrane proteins in kidney epithelial cells. *J Am Soc Nephrol* 9 (2):155–166.

Brown, D., U. Hasler, P. Nunes, R. Bouley and H.A. Lu. 2008. Phosphorylation events and the modulation of aquaporin 2 cell surface expression. *Curr Opin Nephrol Hypertens* 17 (5):491–498.

Brown, D. and L. Orci. 1983. Vasopressin stimulates formation of coated pits in rat kidney collecting ducts. *Nature* 302 (5905):253–255.

Brown, D. and J.L. Stow. 1996. Protein trafficking and polarity in kidney epithelium: From cell biology to physiology. *Physiol Rev* 76 (1):245–297.

Bukau, B. and A.L. Horwich. 1998. The Hsp70 and Hsp60 chaperone machines. *Cell* 92 (3):351–366.

Cheong, K.H., D. Zacchetti, E.E. Schneeberger and K. Simons. 1999. VIP17/MAL, a lipid raft-associated protein, is involved in apical transport in MDCK cells. *Proc Natl Acad Sci USA* 96 (11):6241–6248.

Chizmadzhev, Y.A., P.I. Kuzmin, D.A. Kumenko, J. Zimmerberg and F.S. Cohen. 2000. Dynamics of fusion pores connecting membranes of different tensions. *Biophys J* 78 (5):2241–2256.

Chou, C.L., B.M. Christensen, S. Frische et al. 2004. Non-muscle myosin II and myosin light chain kinase are downstream targets for vasopressin signaling in the renal collecting duct. *J Biol Chem* 279 (47):49026–49035.

Chou, C.L., K.P. Yip, L. Michea et al. 2000. Regulation of aquaporin-2 trafficking by vasopressin in the renal collecting duct. Roles of ryanodine-sensitive Ca^{2+} stores and calmodulin. *J Biol Chem* 275 (47):36839–36846.

Christensen, A.E., F. Selheim, J. de Rooij et al. 2003. cAMP analog mapping of Epac1 and cAMP kinase. Discriminating analogs demonstrate that Epac and cAMP kinase act synergistically to promote PC-12 cell neurite extension. *J Biol Chem* 278 (37):35394–35402.

Christensen, B.M., M. Zelenina, A. Aperia and S. Nielsen. 2000. Localization and regulation of PKA-phosphorylated AQP2 in response to V(2)-receptor agonist/antagonist treatment. *Am J Physiol Renal Physiol* 278 (1):F29–F42.

De Camilli, P., K. Takei and P.S. McPherson. 1995. The function of dynamin in endocytosis. *Curr Opin Neurobiol* 5 (5):559–565.

Deen, P.M., R.A. van Aubel, A.F. van Lieburg and C.H. van Os. 1996. Urinary content of aquaporin 1 and 2 in nephrogenic diabetes insipidus. *J Am Soc Nephrol* 7 (6):836–841.

de Rooij, J., F.J. Zwartkruis, M.H. Verheijen et al. 1998. Epac is a Rap1 guanine-nucleotide-exchange factor directly activated by cyclic AMP. *Nature* 396 (6710):474–477.

Desnos, C., S. Huet and F. Darchen. 2007. 'Should I stay or should I go?': Myosin V function in organelle trafficking. *Biol Cell* 99 (8):411–423.

Ding, Y., A. Winters, M. Ding et al. 2011. Reactive oxygen species-mediated TRPC6 protein activation in vascular myocytes, a mechanism for vasoconstrictor-regulated vascular tone. *J Biol Chem* 286 (36):31799–31809.

Dodge-Kafka, K.L., A. Bauman and M.S. Kapiloff. 2008. A-kinase anchoring proteins as the basis for cAMP signaling. *Handb Exp Pharmacol* (186):3–14.

Douglass, J., R. Gunaratne, D. Bradford et al. 2012. Identifying protein kinase target preferences using mass spectrometry. *Am J Physiol Cell Physiol* 303 (7):C715–C727.

Ecelbarger, C.A., S. Nielsen, B.R. Olson et al. 1997. Role of renal aquaporins in escape from vasopressin-induced antidiuresis in rat. *J Clin Invest* 99 (8):1852–1863.

Elliot, S., P. Goldsmith, M. Knepper, M. Haughey and B. Olson. 1996. Urinary excretion of aquaporin-2 in humans: A potential marker of collecting duct responsiveness to vasopressin. *J Am Soc Nephrol* 7 (3):403–409.

Fenton, R.A., H.B. Moeller, J.D. Hoffert, M.J. Yu, S. Nielsen and M.A. Knepper. 2008. Acute regulation of aquaporin-2 phosphorylation at Ser-264 by vasopressin. *Proc Natl Acad Sci USA* 105 (8):3134–3139.

Feraille, E., E. Dizin, I. Roth et al. 2014. NADPH oxidase 4 deficiency reduces aquaporin-2 mRNA expression in cultured renal collecting duct principal cells via increased PDE3 and PDE4 activity. *PLoS One* 9 (1):e87239.

Franki, N., G. Ding, Y. Gao and R.M. Hays. 1992. Effect of cytochalasin D on the actin cytoskeleton of the toad bladder epithelial cell. *Am J Physiol* 263 (5 Part 1):C995–C1000.

Fushimi, K., S. Sasaki and F. Marumo. 1997. Phosphorylation of serine 256 is required for cAMP-dependent regulatory exocytosis of the aquaporin-2 water channel. *J Biol Chem* 272 (23):14800–14804.

Gerbes, A.L., V. Gulberg, P. Gines et al. 2003. Therapy of hyponatremia in cirrhosis with a vaso-pressin receptor antagonist: A randomized double-blind multicenter trial. *Gastroenterology* 124 (4):933–939.

Gerke, V. and S.E. Moss. 2002. Annexins: From structure to function. *Physiol Rev* 82 (2):331–371.

Gheorghiade, M., W.A. Gattis, C.M. O'Connor et al. 2004. Effects of tolvaptan, a vasopressin antag-onist, in patients hospitalized with worsening heart failure: A randomized controlled trial. *J Am Med Assoc* 291 (16):1963–1971.

Gonzales, P.A., T. Pisitkun, J.D. Hoffert, D. Tchapyjnikov et al 2009. Large-scale proteomics and phosphoproteomics of urinary exosomes. *J Am Soc Nephrol* 20(2):363–379.

Grant, B.D. and J.G. Donaldson. 2009. Pathways and mechanisms of endocytic recycling. *Nat Rev Mol Cell Biol* 10 (9):597–608.

Grantham, J.J. and M.B. Burg. 1966. Effect of vasopressin and cyclic AMP on permeability of iso-lated collecting tubules. *Am J Physiol* 211 (1):255–259.

Hasler, U., U.S. Jeon, J.A. Kim et al. 2006. Tonicity-responsive enhancer binding protein is an essential regulator of aquaporin-2 expression in renal collecting duct principal cells. *J Am Soc Nephrol* 17 (6):1521–1531.

Hasler, U., D. Mordasini, M. Bens et al. 2002. Long term regulation of aquaporin-2 expression in vasopressin-responsive renal collecting duct principal cells. *J Biol Chem* 277 (12):10379–10386.

Hasler, U., M. Vinciguerra, A. Vandewalle, P.Y. Martin and E. Feraille. 2005. Dual effects of hyper-tonicity on aquaporin-2 expression in cultured renal collecting duct principal cells. *J Am Soc Nephrol* 16 (6):1571–1582.

Hays, R.M., J. Condeelis, Y. Gao, H. Simon, G. Ding and N. Franki. 1993. The effect of vasopressin on the cytoskeleton of the epithelial cell. *Pediatr Nephrol* 7 (5):672–679.

Hays, R.M. and A. Leaf. 1962. Studies on the movement of water through the isolated toad bladder and its modification by vasopressin. *J Gen Physiol* 45:905–919.

Henn, V., B. Edemir, E. Stefan et al. 2004. Identification of a novel A-kinase anchoring protein 18 isoform and evidence for its role in the vasopressin-induced aquaporin-2 shuttle in renal principal cells. *J Biol Chem* 279 (25):26654–26665.

Hoffert, J.D., C.L. Chou, R.A. Fenton and M.A. Knepper. 2005. Calmodulin is required for vasopressin-stimulated increase in cyclic AMP production in inner medullary collecting duct. *J Biol Chem* 280 (14):13624–13630.

Hoffert, J.D., R.A. Fenton, H.B. Moeller et al. 2008. Vasopressin-stimulated increase in phos-phorylation at Ser269 potentiates plasma membrane retention of aquaporin-2. *J Biol Chem* 283 (36):24617–24627.

Hoffert, J.D., T. Pisitkun, G. Wang, R.F. Shen and M.A. Knepper. 2006. Quantitative phosphopro-teomics of vasopressin-sensitive renal cells: Regulation of aquaporin-2 phosphorylation at two sites. *Proc Natl Acad Sci USA* 103 (18):7159–7164.

Horner, A., F. Goetz, R. Tampe, E. Klussmann and P. Pohl. 2012. Mechanism for targeting the A-kinase anchoring protein AKAP18delta to the membrane. *J Biol Chem* 287 (51):42495–42501.

Hunter, T. 2007. The age of crosstalk: Phosphorylation, ubiquitination, and beyond. *Mol Cell* 28 (5):730–738.

Inoue, T., S. Nielsen, B. Mandon, J. Terris, B.K. Kishore and M.A. Knepper. 1998. SNAP-23 in rat kidney: Colocalization with aquaporin-2 in collecting duct vesicles. *Am J Physiol* 275 (5 Part 2):F752–F760.

Izumi, Y., K. Miura and H. Iwao. 2014. Therapeutic potential of vasopressin-receptor antagonists in heart failure. *J Pharmacol Sci* 124 (1):1–6.

Kamsteeg, E.J., A.S. Duffield, I.B. Konings et al. 2007. MAL decreases the internalization of the aquaporin-2 water channel. *Proc Natl Acad Sci USA* 104 (42):16696–16701.

Kamsteeg, E.J., G. Hendriks, M. Boone et al. 2006. Short-chain ubiquitination mediates the regulated endocytosis of the aquaporin-2 water channel. *Proc Natl Acad Sci USA* 103 (48):18344–18349.

Kanno, K., S. Sasaki, Y. Hirata et al. 1995. Urinary excretion of aquaporin-2 in patients with diabetes insipidus. *N Engl J Med* 332 (23):1540–1545.

Katsura, T., C.E. Gustafson, D.A. Ausiello and D. Brown. 1997. Protein kinase A phosphorylation is involved in regulated exocytosis of aquaporin-2 in transfected LLC-PK1 cells. *Am J Physiol* 272 (6 Part 2):F817–F822.

Kawasaki, H., G.M. Springett, N. Mochizuki et al. 1998. A family of cAMP-binding proteins that directly activate Rap1. *Science* 282 (5397):2275–2279.

Klussmann, E., B. Edemir, B. Pepperle et al. 2001a. Ht31: The first protein kinase A anchoring protein to integrate protein kinase A and Rho signaling. *FEBS Lett* 507 (3):264–268.

Klussmann, E., K. Maric and W. Rosenthal. 2000. The mechanisms of aquaporin control in the renal collecting duct. *Rev Physiol Biochem Pharmacol* 141:33–95.

Klussmann, E., K. Maric, B. Wiesner, M. Beyermann and W. Rosenthal. 1999. Protein kinase A anchoring proteins are required for vasopressin-mediated translocation of aquaporin-2 into cell membranes of renal principal cells. *J Biol Chem* 274 (8):4934–4938.

Klussmann, E., G. Tamma, D. Lorenz et al. 2001b. An inhibitory role of Rho in the vasopressin-mediated translocation of aquaporin-2 into cell membranes of renal principal cells. *J Biol Chem* 276 (23):20451–20457.

Konstam, M.A., M. Gheorghiade, J.C. Burnett, Jr. et al. 2007. Effects of oral tolvaptan in patients hospitalized for worsening heart failure: The EVEREST outcome trial. *JAMA* 297 (12):1319–1331.

Kortenoeven, M.L., M. van den Brand, J.F. Wetzels and P.M. Deen. 2011. Hypotonicity-induced reduction of aquaporin-2 transcription in mpkCCD cells is independent of the tonicity responsive element, vasopressin, and cAMP. *J Biol Chem* 286 (15):13002–13010.

Kuwahara, M., K. Fushimi, Y. Terada, L. Bai, F. Marumo and S. Sasaki. 1995. cAMP-dependent phosphorylation stimulates water permeability of aquaporin-collecting duct water channel protein expressed in *Xenopus* oocytes. *J Biol Chem* 270 (18):10384–10387.

Lehrich, R.W. and A. Greenberg. 2008. When is it appropriate to use vasopressin receptor antagonists? *J Am Soc Nephrol* 19 (6):1054–1058.

Li, S.Z., B.W. McDill, P.A. Kovach et al. 2007. Calcineurin-NFATc signaling pathway regulates AQP2 expression in response to calcium signals and osmotic stress. *Am J Physiol Cell Physiol* 292 (5):C1606–C1616.

Li, W., Y. Zhang, R. Bouley et al. 2011. Simvastatin enhances aquaporin-2 surface expression and urinary concentration in vasopressin-deficient Brattleboro rats through modulation of Rho GTPase. *Am J Physiol Renal Physiol* 301 (2):F309–F318.

Li, Y., I.B. Konings, J. Zhao, L.S. Price, E. de Heer and P.M. Deen. 2008. Renal expression of exchange protein directly activated by cAMP (Epac) 1 and 2. *Am J Physiol Renal Physiol* 295 (2):F525–F533.

Liamis, G., T.D. Filippatos and M.S. Elisaf. 2014. Treatment of hyponatremia: The role of lixivaptan. *Expert Rev Clin Pharmacol* 7 (4):431–441.

Liebenhoff, U. and W. Rosenthal. 1995. Identification of Rab3-, Rab5a- and synaptobrevin II-like proteins in a preparation of rat kidney vesicles containing the vasopressin-regulated water channel. *FEBS Lett* 365 (2–3):209–213.

Loo, C.S., C.W. Chen, P.J. Wang et al. 2013. Quantitative apical membrane proteomics reveals vasopressin-induced actin dynamics in collecting duct cells. *Proc Natl Acad Sci USA* 110 (42):17119–17124.

Lorenz, D., A. Krylov, D. Hahm et al. 2003. Cyclic AMP is sufficient for triggering the exocytic recruitment of aquaporin-2 in renal epithelial cells. *EMBO Rep* 4 (1):88–93.

Lu, H.A., T.X. Sun, T. Matsuzaki et al. 2007. Heat shock protein 70 interacts with aquaporin-2 and regulates its trafficking. *J Biol Chem* 282 (39):28721–28732.

Lu, H.J., T. Matsuzaki, R. Bouley, U. Hasler, Q.H. Qin and D. Brown. 2008. The phosphorylation state of serine 256 is dominant over that of serine 261 in the regulation of AQP2 trafficking in renal epithelial cells. *Am J Physiol Renal Physiol* 295 (1):F290–F294.

Mammoto, A., K. Takahashi, T. Sasaki and Y. Takai. 2000. Stimulation of Rho GDI release by ERM proteins. *Methods Enzymol* 325:91–101.

Mandon, B., C.L. Chou, S. Nielsen and M.A. Knepper. 1996. Syntaxin-4 is localized to the apical plasma membrane of rat renal collecting duct cells: Possible role in aquaporin-2 trafficking. *J Clin Invest* 98 (4):906–913.

Mandon, B., S. Nielsen, B.K. Kishore and M.A. Knepper. 1997. Expression of syntaxins in rat kidney. *Am J Physiol* 273 (5 Part 2):F718–F730.

Marazuela, M., A. Acevedo, M. Adrados, M.A. Garcia-Lopez and M.A. Alonso. 2003. Expression of MAL, an integral protein component of the machinery for raft-mediated pical transport, in human epithelia. *J Histochem Cytochem* 51 (5):665–674.

Maric, K., A. Oksche and W. Rosenthal. 1998. Aquaporin-2 expression in primary cultured rat inner medullary collecting duct cells. *Am J Physiol* 275 (5 Part 2):F796–F801.

Markoff, A. and V. Gerke. 2005. Expression and functions of annexins in the kidney. *Am J Physiol Renal Physiol* 289 (5):F949–F956.

Marples, D., T.A. Schroer, N. Ahrens, A. Taylor, M.A. Knepper and S. Nielsen. 1998. Dynein and dynactin colocalize with AQP2 water channels in intracellular vesicles from kidney collecting duct. *Am J Physiol* 274 (2 Part 2):F384–F394.

Martin-Belmonte, F., J.A. Martinez-Menarguez, J.F. Aranda, J. Ballesta, M.C. de Marco and M.A. Alonso. 2003. MAL regulates clathrin-mediated endocytosis at the apical surface of Madin-Darby canine kidney cells. *J Cell Biol* 163 (1):155–164.

Matsui, T., S. Yonemura and S. Tsukita. 1999. Activation of ERM proteins in vivo by Rho involves phosphatidyl-inositol 4-phosphate 5-kinase and not ROCK kinases. *Curr Biol* 9 (21):1259–1262.

Mayer, M.P. and B. Bukau. 1998. Hsp70 chaperone systems: Diversity of cellular functions and mechanism of action. *Biol Chem* 379 (3):261–268.

McClatchey, A.I. and R.G. Fehon. 2009. Merlin and the ERM proteins—Regulators of receptor distribution and signaling at the cell cortex. *Trends Cell Biol* 19 (5):198–206.

McConnachie, G., L.K. Langeberg and J.D. Scott. 2006. AKAP signaling complexes: Getting to the heart of the matter. *Trends Mol Med* 12 (7):317–323.

McSorley, T., E. Stefan, V. Henn et al. 2006. Spatial organisation of AKAP18 and PDE4 isoforms in renal collecting duct principal cells. *Eur J Cell Biol* 85 (7):673–678.

Mettlen, M., M. Stoeber, D. Loerke, C.N. Antonescu, G. Danuser and S.L. Schmid. 2009. Endocytic accessory proteins are functionally distinguished by their differential effects on the maturation of clathrin-coated pits. *Mol Biol Cell* 20 (14):3251–3260.

Miranda, C.A., J.W. Lee, C.L. Chou and M.A. Knepper. 2014. Tolvaptan as a tool in renal physiology. *Am J Physiol Renal Physiol* 306 (3):F359–F366.

Moeller, H.B., M.A. Knepper and R.A. Fenton. 2009. Serine 269 phosphorylated aquaporin-2 is targeted to the apical membrane of collecting duct principal cells. *Kidney Int* 75 (3):295–303.

Morgan, J.R., K. Prasad, S. Jin, G.J. Augustine and E.M. Lafer. 2001. Uncoating of clathrin-coated vesicles in presynaptic terminals: Roles for Hsc70 and auxilin. *Neuron* 32 (2):289–300.

Nedvetsky, P.I., E. Stefan, S. Frische et al. 2007. A Role of myosin Vb and Rab11-FIP2 in the aquaporin-2 shuttle. *Traffic* 8 (2):110–123.

Nedvetsky, P.I., V. Tabor, G. Tamma et al. 2010. Reciprocal regulation of aquaporin-2 abundance and degradation by protein kinase A and p38-MAP kinase. *J Am Soc Nephrol* 21 (10):1645–1656.

Nejsum, L.N., M. Zelenina, A. Aperia, J. Frokiaer and S. Nielsen. 2005. Bidirectional regulation of AQP2 trafficking and recycling: Involvement of AQP2-S256 phosphorylation. *Am J Physiol Renal Physiol* 288 (5):F930–F938.

Nielsen, S., S.R. DiGiovanni, E.I. Christensen, M.A. Knepper and H.W. Harris. 1993. Cellular and subcellular immunolocalization of vasopressin-regulated water channel in rat kidney. *Proc Natl Acad Sci USA* 90 (24):11663–11667.

Nishimoto, G., M. Zelenina, D. Li et al. 1999. Arginine vasopressin stimulates phosphorylation of aquaporin-2 in rat renal tissue. *Am J Physiol* 276 (2 Part 2):F254–F259.

Noda, Y., S. Horikawa, T. Furukawa et al. 2004a. Aquaporin-2 trafficking is regulated by PDZ-domain containing protein SPA-1. *FEBS Lett* 568 (1–3):139–145.

Noda, Y., S. Horikawa, E. Kanda et al. 2008. Reciprocal interaction with G-actin and tropomyosin is essential for aquaporin-2 trafficking. *J Cell Biol* 182 (3):587–601.

Noda, Y., S. Horikawa, Y. Katayama and S. Sasaki. 2004b. Water channel aquaporin-2 directly binds to actin. *Biochem Biophys Res Commun* 322 (3):740–745.

Nunes, P., U. Hasler, M. McKee, H.A. Lu, R. Bouley and D. Brown. 2008. A fluorimetry-based ssYFP secretion assay to monitor vasopressin-induced exocytosis in LLC-PK1 cells expressing aquaporin-2. *Am J Physiol Cell Physiol* 295 (6):C1476–C1487.

Ozaki, N., T. Shibasaki, Y. Kashima et al. 2000. cAMP-GEFII is a direct target of cAMP in regulated exocytosis. *Nat Cell Biol* 2 (11):805–811.

Pastore, A. and F. Piemonte. 2012. S-Glutathionylation signaling in cell biology: Progress and prospects. *Eur J Pharm Sci* 46 (5):279–292.

Phillips, M.E. and A. Taylor. 1989. Effect of nocodazole on the water permeability response to vasopressin in rabbit collecting tubules perfused in vitro. *J Physiol* 411:529–544.

Pisitkun, T., V. Jacob, S.M. Schleicher, C.L. Chou, M.J. Yu and M.A. Knepper. 2008. Akt and ERK1/2 pathways are components of the vasopressin signaling network in rat native IMCD. *Am J Physiol Renal Physiol* 295 (4):F1030–F1043.

Pisitkun, T., R. Johnstone and M.A. Knepper. 2006. Discovery of urinary biomarkers. *Mol Cell Proteomics* 5 (10):1760–1771.

Pisitkun, T., R.F. Shen and M.A. Knepper. 2004. Identification and proteomic profiling of exosomes in human urine. *Proc Natl Acad Sci USA* 101 (36):13368–13373.

Procino, G., C. Barbieri, M. Carmosino, F. Rizzo, G. Valenti and M. Svelto. 2009. Lovastatin-induced cholesterol depletion affects both apical sorting and endocytosis of aquaporin 2 in renal cells. *Am J Physiol Renal Physiol* 298(2):F266–F278.

Procino, G., C. Barbieri, G. Tamma et al. 2008. AQP2 exocytosis in the renal collecting duct—Involvement of SNARE isoforms and the regulatory role of Munc18b. *J Cell Sci* 121 (Part 12):2097–2106.

Procino, G., L. Mastrofrancesco, G. Tamma et al. 2012. Calcium-sensing receptor and aquaporin 2 interplay in hypercalciuria-associated renal concentrating defect in humans. An in vivo and in vitro study. *PLoS One* 7 (3):e33145.

Pucadyil, T.J. and S.L. Schmid. 2009. Conserved functions of membrane active GTPases in coated vesicle formation. *Science* 325 (5945):1217–1220.

Rai, T., S. Uchida, F. Marumo and S. Sasaki. 1997. Cloning of rat and mouse aquaporin-2 gene promoters and identification of a negative cis-regulatory element. *Am J Physiol* 273 (2 Part 2):F264–F273.

Rescher, U. and V. Gerke. 2008. S100A10/p11: Family, friends and functions. *Pflugers Arch* 455 (4):575–582.

Rice, W.L., Y. Zhang, Y. Chen, T. Matsuzaki, D. Brown and H.A. Lu. 2012. Differential, phosphorylation dependent trafficking of AQP2 in LLC-PK1 cells. *PLoS One* 7 (2):e32843.

Rinschen, M.M., M.J. Yu, G. Wang et al. 2010. Quantitative phosphoproteomic analysis reveals vasopressin V2-receptor-dependent signaling pathways in renal collecting duct cells. *Proc Natl Acad Sci USA* 107 (8):3882–3887.

Rocchetti, M.T., G. Tamma, D. Lasorsa et al. 2011. Altered urinary excretion of aquaporin 2 in IgA nephropathy. *Eur J Endocrinol* 165 (4):657–664.

Rothman, J.E. 1994. Intracellular membrane fusion. *Adv Second Messenger Phosphoprotein Res* 29:81–96.

Rothman, J.E. and S.L. Schmid. 1986. Enzymatic recycling of clathrin from coated vesicles. *Cell* 46 (1):5–9.

Russo, L.M., M. McKee and D. Brown. 2006. Methyl-beta-cyclodextrin induces vasopressin-independent apical accumulation of aquaporin-2 in the isolated, perfused rat kidney. *Am J Physiol Renal Physiol* 291 (1):F246–F253.

Sabolic, I., T. Katsura, J.M. Verbavatz and D. Brown. 1995. The AQP2 water channel: Effect of vasopressin treatment, microtubule disruption, and distribution in neonatal rats. *J Membr Biol* 143 (3):165–175.

Sandoval, P.C., D.H. Slentz, T. Pisitkun, F. Saeed, J.D. Hoffert and M.A. Knepper. 2013. Proteome-wide measurement of protein half-lives and translation rates in vasopressin-sensitive collecting duct cells. *J Am Soc Nephrol* 24 (11):1793–1805.

Schrier, R.W. 2007. The sea within us: Disorders of body water homeostasis. *Curr Opin Invest Drugs* 8 (4):304–311.

Seino, S. and T. Shibasaki. 2005. PKA-dependent and PKA-independent pathways for cAMP-regulated exocytosis. *Physiol Rev* 85 (4):1303–1342.

Simon, H., Y. Gao, N. Franki and R.M. Hays. 1993. Vasopressin depolymerizes apical F-actin in rat inner medullary collecting duct. *Am J Physiol* 265 (3 Part 1):C757–C762.

Sollner, T., M.K. Bennett, S.W. Whiteheart, R.H. Scheller and J.E. Rothman. 1993. A protein assembly-disassembly pathway in vitro that may correspond to sequential steps of synaptic vesicle docking, activation, and fusion. *Cell* 75 (3):409–418.

Sollner, T., S.W. Whiteheart, M. Brunner et al. 1993. SNAP receptors implicated in vesicle targeting and fusion. *Nature* 362 (6418):318–324.

Soupart, A., P. Gross, J.J. Legros et al. 2006. Successful long-term treatment of hyponatremia in syndrome of inappropriate antidiuretic hormone secretion with satavaptan (SR121463B), an orally active nonpeptide vasopressin V2-receptor antagonist. *Clin J Am Soc Nephrol* 1 (6):1154–1160.

Star, R.A., H. Nonoguchi, R. Balaban and M.A. Knepper. 1988. Calcium and cyclic adenosine monophosphate as second messengers for vasopressin in the rat inner medullary collecting duct. *J Clin Invest* 81 (6):1879–1888.

Stefan, E., B. Wiesner, G.S. Baillie et al. 2007. Compartmentalization of cAMP-dependent signaling by phosphodiesterase-4D is involved in the regulation of vasopressin-mediated water reabsorption in renal principal cells. *J Am Soc Nephrol* 18 (1):199–212.

Storm, R., E. Klussmann, A. Geelhaar, W. Rosenthal and K. Maric. 2003. Osmolality and solute composition are strong regulators of AQP2 expression in renal principal cells. *Am J Physiol Renal Physiol* 284 (1):F189–F198.

Sun, T.X., A. Van Hoek, Y. Huang, R. Bouley, M. McLaughlin and D. Brown. 2002. Aquaporin-2 localization in clathrin-coated pits: Inhibition of endocytosis by dominant-negative dynamin. *Am J Physiol Renal Physiol* 282 (6):F998–F1011.

Tajika, Y., T. Matsuzaki, T. Suzuki et al. 2005. Differential regulation of AQP2 trafficking in endosomes by microtubules and actin filaments. *Histochem Cell Biol* 124 (1):1–12.

Takahashi, K., T. Sasaki, A. Mammoto et al. 1997. Direct interaction of the Rho GDP dissociation inhibitor with ezrin/radixin/moesin initiates the activation of the Rho small G protein. *J Biol Chem* 272 (37):23371–23375.

Tamma, G., M. Carmosino, M. Svelto and G. Valenti. 2005a. Bradykinin signaling counteracts cAMP-elicited aquaporin 2 translocation in renal cells. *J Am Soc Nephrol* 16 (10):2881–2889.

Tamma, G., A. Di Mise, M. Ranieri et al. 2014a. A decrease in aquaporin 2 excretion is associated with bed rest induced high calciuria. *J Trans Med* 12:133.

Tamma, G., E. Klussmann, K. Maric et al. 2001. Rho inhibits cAMP-induced translocation of aquaporin-2 into the apical membrane of renal cells. *Am J Physiol Renal Physiol* 281 (6):F1092–F1101.

Tamma, G., E. Klussmann, J. Oehlke et al. 2005b. Actin remodeling requires ERM function to facilitate AQP2 apical targeting. *J Cell Sci* 118 (Part 16):3623–3630.

Tamma, G., E. Klussmann, G. Procino, M. Svelto, W. Rosenthal and G. Valenti. 2003a. cAMP-induced AQP2 translocation is associated with RhoA inhibition through RhoA phosphorylation and interaction with RhoGDI. *J Cell Sci* 116 (Part 8):1519–1525.

Tamma, G., G. Procino, M.G. Mola, M. Svelto and G. Valenti. 2008. Functional involvement of Annexin-2 in cAMP induced AQP2 trafficking. *Pflugers Arch* 456 (4):729–736.

Tamma, G., M. Ranieri, A. Di Mise, M. Centrone, M. Svelto and G. Valenti. 2014b. Glutathionylation of the aquaporin-2 water channel: A novel post-translational modification modulated by the oxidative stress. *J Biol Chem* 289 (40):27807–27813.

Tamma, G., J.H. Robben, C. Trimpert, M. Boone and P.M. Deen. 2011. Regulation of AQP2 localization by S256 and S261 phosphorylation and ubiquitination. *Am J Physiol Cell Physiol* 300 (3):C636–C646.

Tamma, G., B. Wiesner, J. Furkert et al. 2003b. The prostaglandin E2 analogue sulprostone antagonizes vasopressin-induced antidiuresis through activation of Rho. *J Cell Sci* 116 (Part 16):3285–3294.

Tamura, T. and K. Takeuchi. 2013. Small cell gall bladder carcinoma complicated by syndrome of inappropriate secretion of antidiuretic hormone (SIADH) treated with mozavaptan. *BMJ Case Rep* 2013.

Tchapyjnikov, D., Y. Li, T. Pisitkun, J.D. Hoffert, M.J. Yu and M.A. Knepper. 2010. Proteomic profiling of nuclei from native renal inner medullary collecting duct cells using LC-MS/MS. *Physiol Genomics* 40 (3):167–183.

Terris, J., C.A. Ecelbarger, S. Nielsen and M.A. Knepper. 1996. Long-term regulation of four renal aquaporins in rats. *Am J Physiol* 271 (2 Part 2):F414–F422.

Udelson, J.E., W.B. Smith, G.H. Hendrix et al. 2001. Acute hemodynamic effects of conivaptan, a dual V(1A) and V(2) vasopressin receptor antagonist, in patients with advanced heart failure. *Circulation* 104 (20):2417–2423.

Ungewickell, E., H. Ungewickell, S.E. Holstein et al. 1995. Role of auxilin in uncoating clathrin-coated vesicles. *Nature* 378 (6557):632–635.

Valenti, G., A. Laera, S. Gouraud et al. 2002. Low-calcium diet in hypercalciuric enuretic children restores AQP2 excretion and improves clinical symptoms. *Am J Physiol Renal Physiol* 283 (5):F895–F903.

Valenti, G., A. Laera, G. Pace et al. 2000a. Urinary aquaporin 2 and calciuria correlate with the severity of enuresis in children. *J Am Soc Nephrol* 11 (10):1873–1881.

Valenti, G., G. Procino, M. Carmosino et al. 2000b. The phosphatase inhibitor okadaic acid induces AQP2 translocation independently from AQP2 phosphorylation in renal collecting duct cells. *J Cell Sci* 113 (Part 11):1985–1992.

Valenti, G., G. Procino, U. Liebenhoff et al. 1998. A heterotrimeric G protein of the Gi family is required for cAMP-triggered trafficking of aquaporin 2 in kidney epithelial cells. *J Biol Chem* 273 (35):22627–22634.

Valenti, G., G. Procino, G. Tamma, M. Carmosino and M. Svelto. 2005. Minireview: Aquaporin 2 trafficking. *Endocrinology* 146 (12):5063–5070.

van Balkom, B.W., P.J. Savelkoul, D. Markovich et al. 2002. The role of putative phosphorylation sites in the targeting and shuttling of the aquaporin-2 water channel. *J Biol Chem* 277 (44):41473–41479.

Vossenkamper, A., P.I. Nedvetsky, B. Wiesner, J. Furkert, W. Rosenthal and E. Klussmann. 2007. Microtubules are needed for the perinuclear positioning of aquaporin-2 after its endocytic retrieval in renal principal cells. *Am J Physiol Cell Physiol* 293 (3):C1129–C1138.

Wang, C.C., C.P. Ng, H. Shi et al. 2009. A role of VAMP8/endobrevin in surface deployment of the water channel aquaporin 2. *Mol Cell Biol* 30(1):333–343.

Wang, W., C. Li, S.N. Summer, S. Falk and R.W. Schrier. 2007. Polyuria of thyrotoxicosis: Downregulation of aquaporin water channels and increased solute excretion. *Kidney Int* 72 (9):1088–1094.

Woelk, T., B. Oldrini, E. Maspero et al. 2006. Molecular mechanisms of coupled monoubiquitination. *Nat Cell Biol* 8 (11):1246–1254.

Wong, F., A.T. Blei, L.M. Blendis and P.J. Thuluvath. 2003. A vasopressin receptor antagonist (VPA-985) improves serum sodium concentration in patients with hyponatremia: A multicenter, randomized, placebo-controlled trial. *Hepatology* 37 (1):182–191.

Yamamura, Y., H. Ogawa, H. Yamashita et al. 1992. Characterization of a novel aquaretic agent, OPC-31260, as an orally effective, nonpeptide vasopressin V2 receptor antagonist. *Br J Pharmacol* 105 (4):787–791.

Yip, K.P. 2002. Coupling of vasopressin-induced intracellular Ca^{2+} mobilization and apical exocytosis in perfused rat kidney collecting duct. *J Physiol* 538 (Part 3):891–899.

Yip, K.P. 2006. Epac-mediated Ca(2+) mobilization and exocytosis in inner medullary collecting duct. *Am J Physiol Renal Physiol* 291 (4):F882–F890.

Yu, M.J., R.L. Miller, P. Uawithya et al. 2009. Systems-level analysis of cell-specific AQP2 gene expression in renal collecting duct. *Proc Natl Acad Sci USA* 106 (7):2441–2446.

Yu, M.J., T. Pisitkun, G. Wang et al. 2008. Large-scale quantitative LC-MS/MS analysis of detergent-resistant membrane proteins from rat renal collecting duct. *Am J Physiol Cell Physiol* 295 (3):C661–C678.

Yui, N., H.J. Lu, R. Bouley and D. Brown. 2012. AQP2 is necessary for vasopressin- and forskolin-mediated filamentous actin depolymerization in renal epithelial cells. *Biol Open* 1 (2):101–108.

Yui, N., H.A. Lu, Y. Chen, N. Nomura, R. Bouley and D. Brown. 2013. Basolateral targeting and microtubule-dependent transcytosis of the aquaporin-2 water channel. *Am J Physiol Cell Physiol* 304 (1):C38–C48.

Zwang, N.A., J.D. Hoffert, T. Pisitkun, H.B. Moeller, R.A. Fenton and M.A. Knepper. 2009. Identification of phosphorylation-dependent binding partners of aquaporin-2 using protein mass spectrometry. *J Proteome Res* 8 (3):1540–1554.

CHAPTER 9

Hepatobiliary, Salivary Glands and Pancreatic Aquaporins in Health and Disease

Giuseppe Calamita, Christine Delporte and Raúl A. Marinelli

CONTENTS

Abstract		182
9.1	Introduction	182
9.2	Aquaporins in Hepatobiliary Physiology	183
	9.2.1 Expression and Localization in Hepatocytes	183
	9.2.2 Roles in Bile Physiology	183
	9.2.2.1 Canalicular Bile	183
	9.2.2.2 Ductal Bile	185
	9.2.2.3 Gallbladder Bile	187
	9.2.3 Roles in Metabolic Homeostasis	187
	9.2.3.1 AQP9 in Liver Glycerol Metabolism	187
	9.2.3.2 Mitochondrial AQP8 in Ammonia Transport and Ureagenesis	189
	9.2.3.3 AQP8 and Glycogen Metabolism	189
	9.2.4 Hepatocyte mtAQP8 as Peroxiporin	190
9.3	Hepatobiliary Aquaporins in Disease	190
	9.3.1 Liver Cholestatic Disease	190
	9.3.1.1 Aquaporin Gene Therapy to Cholestasis	190
	9.3.2 Cholesterol Gallstone Disease	191
	9.3.3 Fatty Liver Disease and Insulin Resistance	191
	9.3.4 Implications in Other Hepatobiliary Clinical Disorders	192
9.4	Salivary Glands Aquaporins	192
	9.4.1 Expression and Localization	192
	9.4.2 Role in Saliva Secretion	192
	9.4.3 Implication in Xerostomic Conditions	194

9.5   Exocrine Pancreas Aquaporins   195
      9.5.1   Expression and Localization   195
      9.5.2   Role in Pancreatic Secretion   195
      9.5.3   Implications in Pancreatic Exocrine Diseases   195
9.6   Endocrine Pancreas Aquaporins   195
9.7   Conclusions and Future Perspectives   196
Acknowledgements   197
References   197

ABSTRACT

AQUAPORINS (AQPs) ARE CHANNEL proteins largely present in mammals where they facilitate the permeation of water and a variety of substrates across cellular membranes. AQPs exert pleiotropic roles in both health and disease. This chapter addresses the most recent acquisitions in terms of expression and regulation, as well as physiological and pathophysiological roles, of AQPs in the hepatobiliary tract, salivary glands and pancreas. A number of AQPs are found in liver, bile ducts and gallbladder where they are playing roles in bile formation, secretion and reabsorption. Liver AQPs are also implicated in energy homeostasis by acting in hepatic gluconeogenesis and fat metabolism and in important processes such as ammonia detoxification and mitochondrial release of hydrogen peroxide. Hepatobiliary AQPs are involved in clinical disorders including cholestasis, gallstone formation, insulin resistance, fatty liver disease, hepatic cirrhosis and hepatocarcinoma. Salivary and exocrine pancreas AQPs exert main roles in fluid secretion and contribute to the pathogenesis of xerostomia and pancreatic insufficiencies. Endocrine pancreas AQPs are suggested to be key regulators of intra-islet glycerol content as well as insulin production and secretion and to contribute to the pathogenesis of diabetes. This body of knowledge represents the mainstay of present and future research in a rapidly expanding field.

9.1 INTRODUCTION

Aquaporins (AQPs) are a family of channel proteins widely present in mammals where they facilitate the rapid movement of water and of small solutes and gases across biological membranes (Agre 2004). According to their transport properties and structure, the 13 mammalian AQPs are roughly subdivided into *orthodox AQPs*, homologues primarily permeable to water (AQP0, AQP1, AQP2, AQP4, AQP5, AQP6 and AQP8), *aquaglyceroporins*, AQPs conducting a series of small solutes, particularly glycerol, in addition to water (AQP3, AQP7, AQP9, AQP10), and *unorthodox* AQPs (AQP11 and AQP12), homologues whose transport properties remain to be fully established. AQP8 is also denoted as *ammoniaporin* or *peroxiporin*, based on its pronounced conductance to ammonia and hydrogen peroxide (besides water).

The expression and significance of AQPs is an object of intense investigation in several organs and tissues. Besides predicted functions, AQPs are being found to play unanticipated roles. This chapter will give a concise overview of the most recent updates on the regulation and function of AQPs in important districts such as the hepatobiliary tract, salivary glands (SG) and pancreas in health and disease.

9.2 AQUAPORINS IN HEPATOBILIARY PHYSIOLOGY

9.2.1 Expression and Localization in Hepatocytes

Hepatocytes, as polarized epithelial cells, are characterized by a basolateral membrane (BM), in contact with the sinusoidal blood, and an apical membrane (AP) directed toward bile canalicular lumen. Two AQPs are expressed at the protein level in rodent and human hepatocytes, that is AQP8 and AQP9 (for review, see Marinelli et al. 2011). N-glycosylated AQP8 is localized to pericanalicular vesicles and the canalicular plasma membrane of hepatocytes (Calamita et al. 2001; Elkjaer et al. 2001; García et al. 2001), while its non-glycosylated form is expressed in the inner mitochondrial membrane (Ferri et al. 2003; Calamita et al. 2005b). AQP8-mediated canalicular water permeability results from glucagon-induced AQP8 trafficking from intracellular vesicles to lipid rafts in canalicular plasma membranes (Gradilone et al. 2003; Tietz et al. 2005; Mazzone et al. 2006) through cAMP-PKA and PI3K signalling pathways (Gradilone et al. 2003) and likely microtubule-associated proteins (Gradilone et al. 2003, 2005). Glucagon also upregulates AQP8 protein expression through cAMP-PKA and PI3K pathways (Soria et al. 2009). Therefore, canalicular AQP8 modulates membrane water permeability and promotes osmotically driven water secretion (Marinelli et al. 2011; Portincasa and Calamita 2012). However, mitochondrial AQP8 does not appear to have major relevance in mediating the overall transport of water across mitochondrial membranes (Calamita et al. 2006; Gena et al. 2009). Based on the AQP8 permeability to ammonia (Saparov et al. 2007) and hydrogen peroxide (Bienert et al. 2007), recent evidence suggests its involvement in mitochondrial ammonia detoxification via ureagenesis (Soria et al. 2013) and in H_2O_2-dependent cellular signalling (Marchissio et al. 2012). AQP8 is also expressed at protein level in Kupffer cells during liver regeneration (Hung et al. 2012). AQP9 is specifically localized on the sinusoidal plasma membrane of hepatocytes (Elkjaer et al. 2000; Carbrey et al. 2003). As detailed in the succeeding text, hepatocyte AQP9 plays a key role in the uptake of glycerol (Jelen et al. 2011; Calamita et al. 2012) and the exit of urea (Jelen et al. 2012). AQP9 is thought to facilitate the basolateral movement of water as well as key metabolites with minimal osmotic perturbation. Hepatocyte expression or subcellular localization of AQP9 does not seem to be regulated by glucagon (Gradilone et al. 2003; Soria et al. 2009). AQP9 also permeates metalloids such as arsenite by providing a route for excretion of arsenic by the liver (Carbrey et al. 2009). Murine hepatocytes also express AQP11 protein, an AQP believed to be implicated in RER homeostasis and liver regeneration (see Ishibashi et al. 2014 for a review). The reported localization and suggested physiological and pathophysiological relevance of AQPs in hepatocyte are depicted in Table 9.1.

9.2.2 Roles in Bile Physiology

9.2.2.1 Canalicular Bile

Water transport across hepatocytes plays a significant role in bile production. Osmotically active substances, primarily bile salts, and other organic anions are actively transported into and concentrated within the canalicular lumen, resulting in the passive entry of water by osmosis. The excretion of bile salts via several transporters and exchangers is thought to be the major driving force for water movement from the sinusoidal blood to the bile

TABLE 9.1   Reported Localization and Suggested Physiological and Pathophysiological Relevance of Hepatobiliary AQPs

Hepatobiliary Compartment	Aquaporin	Cellular Location	Subcellular Location	Suggested Functional Involvement	Suggested Clinical Relevance
Liver	AQP8	Hepatocytes	APM; SAV; SER; IMM	Secretion of canalicular bile water; preservation of cytoplasm osmolarity during glycogen synthesis and degradation; mitochondrial ammonia detoxification and ureagenesis; mitochondrial H_2O_2 release	Cholestasis
	AQP9	Hepatocytes	BLM	Uptake of glycerol during starvation; import of water from sinusoidal blood; urea extrusion	Cholestasis; T2D; NAFLD
	AQP11	Hepatocytes	RER	RER homeostasis; liver regeneration	Cystic liver disease
Intrahepatic bile ducts	AQP1	Cholangiocytes	APM; SAV	Secretion and absorption of ductal bile water	Biliary cryptosporidiosis
	AQP4	Cholangiocytes	BLM	Secretion and absorption of ductal bile water	Undefined
Gallbladder	AQP1	Epithelial cells	APM; SAV; BLM	Gallbladder bile absorption/secretion (?)	Cholesterol gallstone disease
	AQP8	Epithelial cells	APM	Cystic bile absorption (?)	Cholesterol gallstone disease
Endothelia	AQP1	Portal sinusoids; PVP; BV	APM; BLM	Bile formation and flow	Liver cirrhosis

APM, Apical plasma membrane; BLM, Basolateral plasma membrane; BV, Blood vessels; IC, Intracellular location; NAFLD, Non-Alcoholic Fatty Liver Disease; PVP, Peribiliary vascular plexus; RER, Rough endoplasmic reticulum; SER, Smooth endoplasmic reticulum; SAV, Subapical membrane vesicles; T2D, Type 2 diabetes-

canaliculus (for review, see Boyer, 2013). The transcellular pathway accounts for most of the water entering the bile canaliculus, with minimal paracellular contribution (Marinelli et al. 2011; Boyer 2013). AQP8 plays a role in canalicular bile secretion (Huebert et al. 2002; Larocca et al. 2009) and is responsible for the rate-limiting water flow (Marinelli et al. 2003), whereas AQP9 allows basolateral water uptake (Marinelli et al. 2011). Moreover, hormones stimulating choleresis, such as endothelins and glucagon, increase canalicular AQP8 expression (Gradilone et al. 2003; Rodriguez et al. 2013). Thus, AQP8 would provide a molecular mechanism for the efficient coupling of osmotically active solutes and water transport during agonist-stimulated hepatocyte bile formation (Figure 9.1a).

9.2.2.2 Ductal Bile

Cholangiocytes, the epithelial cells that line the biliary tree, account for secretin-induced ductal bile secretion via a cAMP-dependent pathway (Boyer 2013) and activation of Cl$^-$ efflux via cystic fibrosis transmembrane conductance regulator (CFTR) driving the extrusion of HCO_3^- into the lumen via apical AE2 (i.e. the chloride/bicarbonate exchanger). Both HCO_3^- and Cl$^-$ provide the main driving force for the osmotic movement of water via apical AQP1 into the biliary lumen (Boyer 2013). AQP1 is expressed in human and rodent cholangiocytes (Nielsen et al. 1993; Marinelli et al. 1997) and plays a key role in osmotically driven apical water secretion during basal- and hormone-regulated ductal bile formation (for review, see Marinelli et al. 2011). AQP1 is also present in subapical membrane vesicles (Marinelli et al. 1999) and sometimes coexpressed with AE2 and CFTR (Tietz et al. 2003). Secretin is able to modulate the microtubule-dependent trafficking and exocytic insertion of these vesicles into the cholangiocyte AP (Marinelli et al. 1999; Tietz et al. 2003) suggesting the novel concept of a functional bile secretory unit.

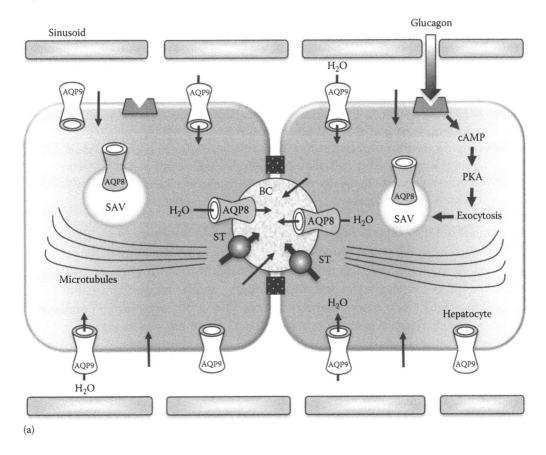

(a)

FIGURE 9.1 Proposed mechanisms of aquaporin-mediated water transport in hepatobiliary and gallbladder bile formation. (a) Hepatocytes. AQP8 facilitates the osmotic secretion of water into the bile canaliculus, whereas AQP9 contributes to the transport of water from the sinusoidal blood into the cell. Choleretic hormones, such as glucagon, can stimulate the microtubule-dependent canalicular targeting of AQP8-containing vesicles.   (*Continued*)

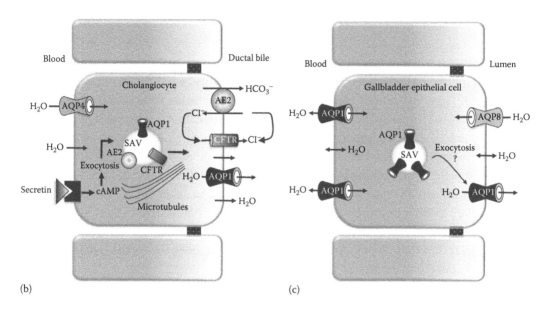

(b)   (c)

FIGURE 9.1 (*Continued*) Proposed mechanisms of aquaporin-mediated water transport in hepatobiliary and gallbladder bile formation. (b) Intrahepatic bile ducts cholangiocytes. Secretin, via cAMP, triggers the microtubule-dependent apical targeting and exocytic insertion of vesicles containing AQP1, CFTR Cl$^-$ channels and the Cl$^-$/HCO$_3^-$ exchanger AE2 into the apical membrane. The efflux of Cl$^-$ via cystic fibrosis transmembrane conductance regulator provides the luminal substrate to drive the extrusion of HCO$_3^-$ into the lumen via AE2. HCO$_3^-$ and Cl$^-$ ions provide the osmotic driving force for the movement of water from plasma (mainly via basolateral AQP4) to biliary lumen (via apical AQP1). (c) Gallbladder epithelial cells. AQP1 and AQP8 facilitate the osmotic absorption and secretion of water, respectively. BC, bile canaliculus; CFTR, cystic fibrosis transmembrane conductance regulator; SAV, subapical vesicle; ST, solute transporters.

Cholangiocytes also express AQP4 and AQP1 at the BM (Marinelli et al. 1999, 2000). AQP-facilitated water movement would allow the relative isosmolar status of the cell to be maintained during ductal bile formation and is consistent with the physical association between the BM of cholangiocytes and the peribiliary vascular plexus that surround bile ducts and from which bile water originates (Figure 9.1b). AQP1 expression in peribiliary vascular endothelia (Nielsen et al. 1993; Marinelli et al. 1999) suggests its important functional role in facilitating water transport from plasma to bile across cholangiocytes.

Surprisingly, cholangiocytes from *Aqp1*$^{-/-}$ knockout mice are not defective in water movement (Mennone et al. 2002). This suggests, in disagreement with the aforementioned findings in bile duct units, that AQP1 plays non-essential role in mouse ductal bile secretion. Deletion of a single AQP might not significantly affect cell water transport because other known or still unidentified AQPs could undergo compensatory upregulation. In fact, several AQPs are expressed in mouse cholangiocytes (Poling et al. 2014), including AQP8. Cholangiocyte AQP8 is largely localized on plasma membranes (Ueno et al. 2003) suggesting a likely role in ductal bile formation.

Intrahepatic bile ducts not only secrete but also absorb water. Studies in isolated rodent intrahepatic bile duct units demonstrated osmotically induced net water absorption

(Masyuk et al. 2002). The active absorption of sodium-coupled glucose (SGLT1) and bile salt apical sodium-dependent bile acid transporter (ASBT) cotransporters would induce osmotic water movement through AQP1 in the inward direction (Boyer 2013). Somatostatin, gastrin, and insulin, found to decrease cholangiocyte levels of cAMP (Boyer 2013), could inhibit the secretin-induced vesicular transport of AQP1, CFTR and AE2 to cholangiocyte AM, as well as the activation of CFTR with a consequent decrease in ductal bile secretion (Figure 9.1b). Under this condition, the absorption of water induced by glucose and bile salts would become relevant; this could explain why somatostatin can not only cause inhibition of ductal secretion but also promote net ductal water absorption.

9.2.2.3 Gallbladder Bile

Human and mouse gallbladder (GB) epithelial cells express AQP1 and AQP8 (for review, see Portincasa and Calamita 2012). AQP1 is localized at both the AP and BM of the epithelial cells lining the neck portion of human GB (Calamita et al. 2005a) and at the apical and BM and in subapical vesicles of the epithelial cells lining the neck and *corpus* portions of mouse GB (van Erpecum et al. 2006). Mouse GB *Aqp1* mRNA is slightly upregulated by leptin (Swartz-Basile et al. 2007). GB epithelium from different species expresses AQP8 at the plasma membrane and, at lesser extent, intracellularly (Calamita et al. 2005a). AQP8 and AQP1 are speculated to facilitate the transport of water through the AP, whereas basolateral AQP1 is to mediate the movement of water at the serosal side (Figure 9.1c). AQP1 might be translocated to the AP to secrete water as in the bile duct epithelium, a functional homologue of the GB epithelium, whereas apical AQP8 might account for the absorption of water from GB bile. Very high transepithelial osmotic water permeability was reported in intact mouse GB resulting from transcellular water movement through AQP1 (Li et al. 2009). However, AQP1 is not physiologically relevant to GB bile concentration and AQP8 does not functionally substitute for AQP1. This was not consistent with previous studies showing temporal association between decreased GB concentrating function and reduced AQP1 or AQP8 expression (van Erpecum et al. 2006), and leptin-deficient mice submitted to leptin replacement where leptin was reported to alter GB volume by mediating the AQP-mediated absorption/secretion of water (Goldblatt et al. 2002). Further work is warranted to clarify the question.

9.2.3 Roles in Metabolic Homeostasis

9.2.3.1 AQP9 in Liver Glycerol Metabolism

Glycerol is a source of glycerol-3-phosphate for triacylglycerols (TGs) synthesis and a substrate for hepatic gluconeogenesis during fasting (Reshef et al. 2003). The liver is responsible for 70%–90% of the whole-body glycerol metabolism. Glycerol utilization by hepatocytes is rate limited by AQP9 allowing glycerol uptake in hepatocytes in the first 24 h of starvation (Jelen et al. 2011; Calamita et al. 2012) (Figure 9.2). A mathematical model simulating the interplay among plasma glucose, circulating insulin, and liver AQP9 has been recently proposed based on Hill and step functions (D'Abbicco et al. 2015). The peroxisome proliferator-activated receptor α (PPARα) is central for the increased expression of AQP9 in fasted male (but not female) mice (Patsouris et al. 2004). The recent observation that hepatic AQP9 expression in male rats is reduced in response to PPARα agonist

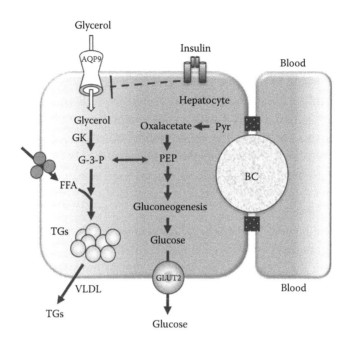

FIGURE 9.2   Working model of glycerol import by AQP9 in hepatocytes. Sinusoidal blood glycerol is imported by hepatocytes through AQP9. During fasting, the expression of liver AQP9 is negatively regulated by insulin. Glycerol is phosphorylated (G-3-P) by glycerol kinase and, depending on the metabolic state, is used to sustain gluconeogenesis or lipogenesis. BC, bile canaliculus; FFA, free fatty acids; GLUT2, glucose transporter 2; PEP, phosphoenolpyruvate; Pyr, pyruvate; TG, triacylglycerol; VLDL, very-low-density lipoprotein.

treatment was interpreted as a result suggesting that, in the fed state, PPARα activation directs glycerol into TG synthesis rather than into *de novo* synthesis of glucose (Lebeck et al. 2015). Disruption of the *Aqp9* gene in obese diabetic *db/db* mice reportedly diminishes plasma glucose concentrations between 10% and 40% (Rojek et al. 2007). In rodents, insulin represses hepatocyte *Aqp9* gene expression by acting on an insulin response element (Kuriyama et al. 2002), in accordance with AQP9 increases seen in insulin-resistant animal models (Carbrey et al. 2003). A gender-specific regulation is observed in rats where 17β-estradiol prevents increased hepatic AQP9 expression and glycerol import during starvation (Lebeck et al. 2012). In humans, hepatocytes also express AQP3 and AQP7 and, at low extent, AQP10. The regulation and expression of AQPs in the human liver appear to be different from the control that takes place in mice. In human, the expression of AQP9 is upregulated by insulin through the phosphatidylinositol 3-kinase/Akt/mammalian target of rapamycin (PI3K/Akt/mTOR) signalling cascade, whereas it is reduced by leptin (Rodriguez et al. 2011b) and AMP-activated protein kinase, via forkhead box a2 (Fox a2) (Yokoyama et al. 2011). Variable results have been observed regarding the hepatic expression profile of AQP9 in obese individuals with T2D and obese with no impairment of their glucose tolerance (Miranda et al. 2009; Rodriguez et al. 2011b). In terms of sexual dimorphism, obese women show lower hepatocyte glycerol permeability compared to obese men although the related expression levels of AQP9 do not appear to be significantly different (Rodriguez et al. 2014).

Aquaglyceroporins have been recently associated with caveolins as key integral membrane proteins involved in maintaining energy and metabolic homeostasis (Méndez-Giménez et al. 2014). While substantial is the evidence suggesting relevance for AQPs in metabolism and energy balance (Rodriguez et al. 2011a; Lebeck 2014), further work is warranted to fully elucidate their regulation.

9.2.3.2 Mitochondrial AQP8 in Ammonia Transport and Ureagenesis

AQP8 efficiently facilitates the membrane diffusional transport of ammonia in rat, mouse and human testis plasma membrane vesicles (Jahn et al. 2004; Holm et al. 2005; Liu et al. 2006; Yang et al. 2006; Saparov et al. 2007). Moreover, mitochondrial AQP8 (mtAQP8) is able to markedly increase ammonia transport across inner mitochondrial membranes (Soria et al. 2010).

Ammonia detoxification occurs mainly through conversion of ammonia to urea via the urea cycle in hepatocytes, a critical process for preventing hyperammonemia and hepatic encephalopathy and involving mitochondrial uptake of ammonia. No relevant function has been found for hepatocyte mtAQP8 in mitochondrial water transport (for review, see Gena et al. 2009), whereas a key role for mtAQP8 in ammonia detoxification via ureagenesis has been suggested based on recent experimental evidence (Soria et al. 2013). Basal- and glucagon-induced ureagenesis from ammonia were significantly reduced in hepatocytes undergoing mtAQP8 knockdown (Soria et al. 2013). Contrarily, mtAQP8 silencing induced no significant change in ureagenesis when glutamine or alanine, that is intramitochondrial nitrogen donors, was used (Soria et al. 2013). Additional support for an mtAQP8-mediated ammonia transport to supply urea cycle comes from in vivo studies in the rat where glucagon-induced ureagenesis was associated with upregulation of both hepatic mtAQP8 protein expression and diffusional ammonia permeability of inner mitochondrial membranes (Soria et al. 2013). Interestingly, in rats with hypothyroidism, a condition associated with increased hepatocyte urea synthesis, liver mtAQP8 is upregulated (Calamita et al. 2007). There is also evidence for the involvement of mtAQP8 in the molecular pathogenesis of defective hepatic ammonia detoxification in sepsis, since lipopolysaccharide-treated rats showed a downregulation in hepatic mtAQP8 expression and mitochondrial ammonia diffusion together with defective basal- and glucagon-stimulated ammonia-derived ureagenesis (Soria et al. 2014).

AQP9 has been recently reported to contribute to the efflux of urea, an ending metabolite, out of mouse hepatocytes where a urea transporter-like protein was hypothesized to have larger relevance in liver urea removal (Jelen et al. 2012).

9.2.3.3 AQP8 and Glycogen Metabolism

In hepatocytes, AQP8 immunoreactivity was also found at smooth endoplasmic reticulum (SER) membranes adjacent to glycogen granules (Ferri et al. 2003). AQP8 may also have a role in preserving cytoplasmic osmolality during glycogen synthesis and degradation by mediating the movement of water between SER lumen and cytoplasm. Work is necessary to evaluate this possibility.

9.2.4 Hepatocyte mtAQP8 as Peroxiporin

Hepatic mitochondria are important sources for the generation of reactive oxygen species (ROS), hydrogen peroxide (H_2O_2). H_2O_2 is normally released from hepatocyte mitochondria to participate in signal transduction pathways (Rigoulet et al. 2011). Transmembrane H_2O_2 transport is ensured by AQP8 (Miller et al. 2010) working as a *peroxiporin* (Sies 2014).

More recent studies in human hepatocarcinoma HepG2 cells suggest that mtAQP8 can facilitate the mitochondrial release of H_2O_2 (Marchissio et al. 2014). Moreover, *AQP8* knockdown caused ROS-induced mitochondrial depolarization via the mitochondrial permeability transition mechanism (Marchissio et al. 2012). Oxidant-induced mitochondrial dysfunction led to necrotic death of *AQP8*-knockdown HepG2 cells (Marchissio et al. 2014), a finding that might be relevant to therapeutic strategies against hepatoma cells. In non-hepatic cells, AQP8 has also been found to modulate NAD(P)H oxidases (Nox)-produced H_2O_2 transport through plasma membranes, whether hepatocyte canalicular AQP8 may be involved in such a mechanism has not been explored yet.

9.3 HEPATOBILIARY AQUAPORINS IN DISEASE

9.3.1 Liver Cholestatic Disease

Cholestasis is characterized by impairment of bile formation and is caused by several liver diseases. Chronic cholestasis can progress toward biliary cirrhosis and liver failure requiring liver transplantation (Lee and Boyer 2000). As AQPs play a physiological role in canalicular bile formation, defective canalicular AQP expression may lead to alterations of normal bile physiology. Defective expression of canalicular AQP8 protein was found to be present in experimental models of cholestasis (Carreras et al. 2003, 2007; Lehmann et al. 2008; Lehmann and Marinelli 2009) and could result from increased lysosomal and proteosomal protein degradation. The expression of sinusoidal AQP9 was also found to be downregulated in obstructive cholestasis (Calamita et al. 2008), suggesting that a reduced sinusoidal water uptake might also work as another contributing factor in obstructive cholestasis. These findings suggest that AQP are involved in the development of bile secretory dysfunction.

9.3.1.1 Aquaporin Gene Therapy to Cholestasis

The gene transfer of human *AQP1* via the adenoviral vector Ad*h*AQP1 has been successfully used to restore normal salivary flow to irradiated hypofunctional SG of experimental animals (Delporte et al. 1997) and humans (Baum et al. 2012). We showed that Ad*h*AQP1, administered by retrograde bile ductal infusion, induced hepatocyte canalicular *h*AQP1 expression and a concomitant increased in canalicular osmotic water permeability, bile flow and choleretic efficiency of endogenous bile salts (i.e. volume of bile/μmol of excreted bile salt) in an experimental model of estrogen-induced cholestasis (Carreras et al. 2007). These data suggest that the adenoviral transfer of *h*AQP1 gene to livers of estrogen-induced cholestatic rats improves bile flow by enhancing the AQP-mediated bile salt–induced canalicular water secretion (Marrone et al. 2014). This finding may have potential as novel treatment for certain liver cholestatic diseases.

9.3.2 Cholesterol Gallstone Disease

Gallstone disease is one of the most prevalent and costly digestive diseases in western countries, with a 10%–15% prevalence in adulthood (Portincasa et al. 2006). About 80% of the gallstones are cholesterol gallstones, while the remaining are pigment stones that contain less than 30% cholesterol (Portincasa et al. 2008). A decrement of the GB concentrating ability was associated with reduced expression of the GB AQP1 and AQP8 during lithogenesis obtained in C57L mice susceptible to diet-induced cholesterol gallstones (van Erpecum et al. 2006). Dysregulated AQP1 and AQP4 were found in the GB of obese leptin-deficient mice undergoing leptin replacement (Swartz-Basile et al. 2007). Work is in progress to verify if similar phenomena also occur in human cholesterol gallstone disease.

9.3.3 Fatty Liver Disease and Insulin Resistance

Non-alcoholic fatty liver disease (NAFLD), a worrisome health problem worldwide characterized by intrahepatic TG overaccumulation (Chalasani et al. 2012), is a common feature of metabolic syndrome being often associated with obesity, dyslipidemia and diabetes and mostly closely linked to insulin resistance. The mechanism of NAFLD pathogenesis is the object of intense investigation especially regarding complex systems ultimately resulting in excessive TG deposition in hepatocytes (Tiniakos et al. 2010). A role for AQP9 in controlling hepatic TG synthesis in NAFLD has been recently suggested after observing that both liver AQP9 and hepatocyte membrane glycerol permeability are reduced in mouse models of NAFLD (Gena et al. 2013) as well as in humans with obesity, insulin resistance and NAFLD (Rodriguez et al. 2014). AQP9 downregulation and hepatic reduction in glycerol permeability in insulin-resistant states have been interpreted as a compensatory mechanism, whereby the liver counteracts further TG accumulation within its parenchyma and reduces liver gluconeogenesis during NAFLD. However, this scenario should be contextualized on the pathophysiological profile and gender of the studied animal and human specimens since both the expression of AQP9 and hepatic uptake of glycerol resulted to have a different profile of regulation in *n*3 polyunsaturated fatty acid–depleted female rats (Portois et al. 2012), a model of metabolic syndrome displaying several features of the disease also including liver steatosis, and in rats fed with a high-fat diet (Cai et al. 2013). AQP9 increases in the early onset of steatosis, whereas it diminishes at a later stage of the disease, when consistent and excessive steatosis has occurred. Additional experimental information needs to be achieved before exhaustively assessing both the pathophysiological significance and regulation (i.e., the gender dimorphism found in morbid subjects) of AQP9 in a disease with multifactorial pathogenesis such as NAFLD. The interplay between liver AQP9 and fat AQP3 and AQP7 in NAFLD states accompanied with obesity and T2D also needs to fully be understood. Work is also warranted in investigating the potential selective manipulation of aquaglyceroporins and caveolins in liver and other organs with metabolic relevance in the treatment of NAFLD and other metabolic disorders (Méndez-Giménez et al. 2014).

9.3.4 Implications in Other Hepatobiliary Clinical Disorders

AQPs could also play a role in cystic liver disease (AQP1 andAQP11), primary biliary cirrhosis (AQP1), liver cirrhosis (AQP1), cryptosporidiosis (AQP1) and hepatocellular carcinoma (AQP9; Padma et al. 2009). The pathophysiological involvement of AQPs in such diseases has been reviewed in a recent report (Portincasa and Calamita 2012).

9.4 SALIVARY GLANDS AQUAPORINS

9.4.1 Expression and Localization

Mammalian SG comprise parotid glands, submandibular glands, sublingual glands and labial SG (LSG). SG contains acinar, ductal, endothelial and myoepithelial cells (Redman 1987). Table 9.2 summarizes the expression of AQPs in SG (for review, see Delporte and Steinfeld 2006; Larsen et al. 2011; Aure et al. 2014; Delporte, 2014).

9.4.2 Role in Saliva Secretion

The mechanisms accounting for final hypotonic saliva fluid secretion can be divided into two steps (Figure 9.3a; for details, see Lee et al. 2012; Delporte 2013, 2014). The first step occurs at the level of acinar cells and consists of the secretion of an isotonic fluid rich in

TABLE 9.2 Reported Localization of Salivary Glands and Pancreas Aquaporins

Aquaporin	Organism		
	Rat	**Mouse**	**Human**
AQP1	SMG and PG: EC	EXP: DC	SG: EC, MC
	EXP: DC, AC, EC		EXP: AC, CA, DC, EC
AQP3	ND	SMG and SLG	SG: AC
AQP4	SMG	ND	SG
	EXP		
AQP5	SMG: AC, DC	SMG, PG and SL: AC	SG: AC
	PG, AC	EXP: DC	
	EXP: CA, DC	ENP: BeC	EXP: DC
AQP6	PG: AC	ND	SG
AQP7	ENP: BeC	SMG and SLG: EC	SG
		ENP: BeC	
AQP8	SMG and PG: MC	SMG and SLG	EXP: AC
	EXP: AC	ENP: BeC	
AQP9	ND	SMG and SLG	ND
AQP11	ND	SMG and SLG	ND
AQP12	ND	EXP: AC	EXP

Note: Only mRNA has been detected when the cell type is not mentioned.

AC, Acinar cells; BeC, β-cells; CA, Centroacinar cells; DC, Ductal cells; EC, Endothelial cells; ENP, Endocrine pancreas; EXP, Exocrine pancreas: MC, Myoepithelial cells; ND, Not defined; SG, All salivary glands; SL, Sublingual glands; SMG, Submandibular glands; PG, Parotid glands.

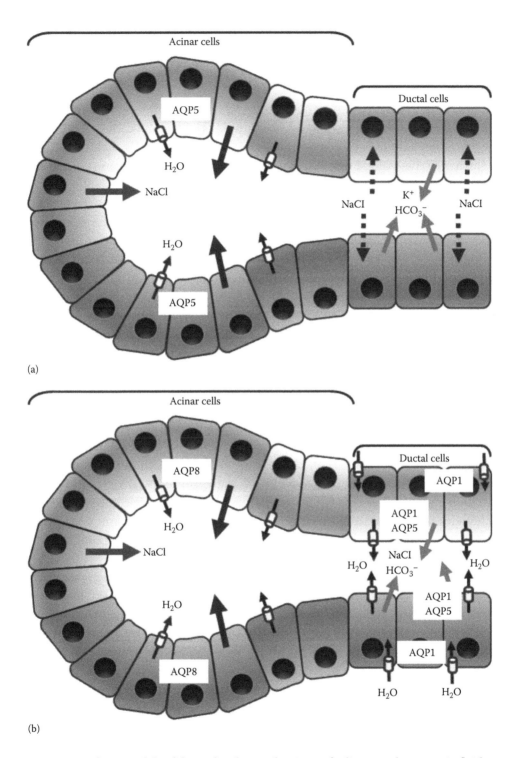

FIGURE 9.3 Working models of the molecular mechanisms of salivary and pancreatic fluid secretions. (a) Salivary fluid secretion. (b) Pancreatic fluid secretion.

NaCl, the subsequent generation of a transepithelial osmotic gradient and large water flow via apical AQP5 and possible paracellular pathways. During the second step, the composition of the primary secreted fluid is modified by the ductal cells (relatively impermeable to water) due to the reabsorption of most of the Na^+ and Cl^- and the secretion of HCO_3^- and K^+.

AQP knockout mice revealed the role of AQP5 in saliva secretion, while ruling out that of AQP1, AQP4 or AQP8 (for review, see Delporte 2014). The implication of other AQPs in saliva secretion has not been demonstrated. Besides, as ductal cells are impermeable to water, further studies will be required to assess the functional role of the possible ductal AQP5 expression.

In addition to transcellular water flow, paracellular water flow might contribute to saliva secretion (Murakami et al. 2006). A first model called the osmosensor feedback model suggests that an osmosensor, likely AQP5 in SG, controls the tonicity of the transported fluid by mixing transcellular and paracellular flows (Hill and Shachar-Hill 2006). However, a second model based on transcellular-only osmotic mechanism predicts the results of *Aqp5* knockout studies (Maclaren et al. 2013). Finally, a multiscale modelling of saliva secretion was constructed based on, among several assumptions, osmotically driven water flow across acinar cells (Sneyd et al. 2014). The precise contributions of both transcellular and paracellular water pathways to saliva secretion remain unclear.

9.4.3 Implication in Xerostomic Conditions

Saliva decline due to senescence, possibly resulting from decreased AQP5 translocation, could be reverted by cevimeline or DNA demethylation agent (for review, see Delporte 2014). In diabetes, reduced saliva flow is unlikely to result from altered AQP5 localization, expression or trafficking (Wang et al. 2011; Soyfoo et al. 2012a). In head and neck cancer radiation therapy, loss of salivation could result from reduced AQP5 expression, AQP5 translocation and/or acinar cells (Delporte et al. 1997; Asari et al. 2009). In irradiated SG from animals and humans, *AQP1* gene delivery led to *AQP1* expression in epithelial cells (Delporte et al. 1998; He et al. 1998), restoration of saliva secretion (Delporte et al. 1997; Baum et al. 2006, 2012).

Sjögren's syndrome (SS), an autoimmune disease mainly affecting salivary and lachrymal glands, displays a multifactorial pathogenesis possibly implicating AQPs (Delporte et al. 2011). Decreased AQP1 expression in LSG biopsies from SS patients is unlikely to participate to saliva secretion loss (for review, see Delporte 2013, 2014). In both SS patients and animal models for SS, the abnormal AQP5 localization (BM versus AM) (Steinfeld et al. 2001; Soyfoo et al. 2007a,b, 2012b) appears to be concomitant to the presence of inflammatory infiltrates, cytokine secretion and acinar destruction (Soyfoo et al. 2012b; Yamamura et al. 2012), and/or antoantibodies against muscarinic M3 receptors inhibiting AQP5 trafficking (Lee et al. 2013). Altered AQP5 localization could therefore participate to the pathogenesis of SS, even though it could not directly account for saliva impairment (Soyfoo et al. 2007a). DNA demethylation agents and viral vectors coding for AQPs could represent promising therapies for the treatment of SS.

9.5 EXOCRINE PANCREAS AQUAPORINS

9.5.1 Expression and Localization

The exocrine pancreas AQPs are summarized in Table 9.2 (for review, see Burghardt et al. 2003, 2006; Isokpehi et al. 2009; Delporte 2014).

9.5.2 Role in Pancreatic Secretion

The mechanisms leading to pancreatic juice secretion involve two steps; during the first one, acinar cells secrete small amount of primary isotonic fluid that reaches the ductal lumen, and water flows via AQP8, while during the second one, ductal cells secrete Na^+, Cl^- and HCO_3^- as well as most of the water through AQP1 and AQP5 (Figure 9.3b; for details, see Lee et al. 2012; Delporte 2014). *Aqp1*, *Aqp5*, *Aqp8* and *Aqp12* knockout mice display normal pancreatic function (Burghardt et al. 2003; Yang et al. 2005; Ohta et al. 2009). The possible combined role of AQPs to fluid secretion remains to be addressed.

9.5.3 Implications in Pancreatic Exocrine Diseases

Further studies are required to shed light on the involvement of AQP1 and AQP12 in exocrine pancreatic insufficiency and on the regulation of ductal AQP1 expression by the CFTR (for review, see Delporte 2014). While *Aqp12* knockout mice display increased susceptibility to caerulein-induced acute pancreatitis and larger acinar exocytotic vesicles in pancreatic acini (Ohta et al. 2009), additional studies are required to better understand its implication in pancreatitis.

9.6 ENDOCRINE PANCREAS AQUAPORINS

Table 9.2 summarizes the expression of AQPs that has only been documented in non-human β-cells (Matsumura et al. 2007; Best et al. 2009; Louchami et al. 2012).

Elevated plasma glucose level triggers subsequent molecular events leading to the secretion of insulin: glucose entry via the glucose transporter type 2 (GLUT2), glucose metabolism leading to ATP synthesis, inhibition of ATP-sensitive K^+ channels, cell membrane depolarization, opening of voltage-dependent Ca^{2+} channels and intracellular Ca^{2+} concentration increase triggering exocytosis of insulin-containing granules (Figure 9.4). Moreover, glucose metabolic amplifying pathways also stimulate the Ca^{2+}-induced exocytosis of insulin-containing granules in response to non-metabolized secretagogues (Henquin 2011). In response to glucose, the increased β-cell volume is likely to participate to insulin secretion as β-cell exposure to hypotonic stress leads to subsequent cell swelling leading to the activation of volume-regulated anion channel, cell depolarization, activation of voltage-sensitive Ca^{2+} channels, calcium entry and insulin secretion (Drews et al. 2010) (Figure 9.4).

AQP7 knockout mice display different phenotypes in terms of insulinemia and glycemia (for review, see Virreira et al. 2011; Delporte 2014). Glycerol entry in rat β-cells induces cell swelling leading to insulin release (Best et al. 2009), while glycerol metabolism accounts for the activation of β-cells (Best et al. 2009; Delporte et al. 2009). AQP7 is likely to play a dual role in the regulation of insulin release by allowing the entry or exit of glycerol, and

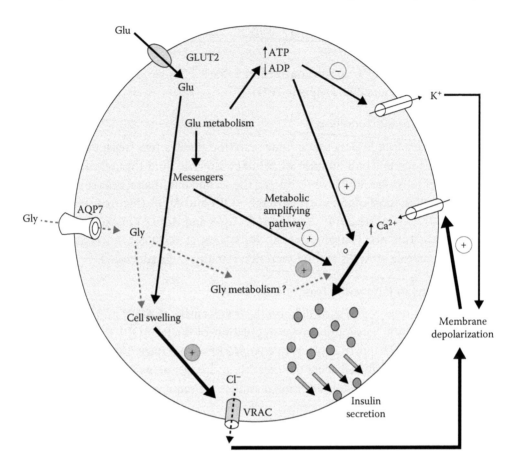

FIGURE 9.4 Suggested involvement of AQP7 in insulin secretion. The mechanisms of glucose- (plain narrow black arrows) and glycerol- (broken grey arrows) induced insulin secretion are summarized. Thick black arrows indicate common mechanisms between glucose- and glycerol-induced insulin secretion. AQP7, aquaporin 7; Glu, glucose; GLUT2, glucose transporter 2; Gly, glycerol; VRAC, volume-regulated anion channel and cell depolarization.

by acting, directly or indirectly, at a distal downstream site in the insulin exocytosis pathway (Louchami et al. 2012). AQP7 inhibitors could unravel the role of AQP7 in the insulin secretion process (Katano et al. 2014; Madeira et al. 2014).

Further work is needed to understand the regulation of AQP7 in pancreatic β-cells by shedding light on its potential contribution to the development of diabetes.

9.7 CONCLUSIONS AND FUTURE PERSPECTIVES

Investigation on hepatobiliary, salivary and pancreas AQPs is highly instructive and provides important insights into complex biological functions such as exocrine secretion of digestive fluids, energy homeostasis, ammonia detoxification, ureagenesis and mitochondrial ROS release, without leaving out a critically important function such as the control of insulin secretion. Nevertheless, deeper investigation is warranted to fully elucidate aspects regarding the mechanisms with which such AQPs are regulated in their expression and function, especially when trying to translate the information acquired with animal and cellular models to humans. Given their pleiotropic significance, additional roles for such membrane channels are sure to emerge in the coming years. The pathophysiological involvement of AQPs in the onset of human diseases is a growing new area of investigation, with important key diagnostic and therapeutic implications.

ACKNOWLEDGEMENTS

Financial support is gratefully acknowledged by grants from Regione Puglia (Rete di Laboratori Pubblici di Ricerca WAFITECH - cod. 09) to G.C., grants PIP 0244 from CONICET and PICT 1217 from Agencia Nacional de Promoción Científica y Tecnológica to R.A.M., and grants 3.4604.05 and 3.4502.09 from Fund for Medical Scientific Research (Belgium) to C.D.

REFERENCES

Agre, P. 2004. Aquaporin water channels (Nobel Lecture). *Angew Chem Int Ed* 43:4278–4290.

Antunes, F. and E. Cadenas. 2000. Estimation of H_2O_2 gradients across biomembranes. *FEBS Lett* 475:121–126.

Asari, T., Maruyama, K. and H. Kusama. 2009. Salivation triggered by pilocarpine involves aquaporin-5 in normal rats but not in irradiated rats. *Clin Exp Pharmacol Physiol* 36:531–538.

Aure, M.H., Ruus, A.K. and H.K. Galtung. 2014. Aquaporins in the adult mouse submandibular and sublingal salivary glands. *J Mol Hist* 45:69–80.

Baum, B.J., Alevizos, I., Zheng, C. et al. 2012. Early responses to adenoviral-mediated transfer of the aquaporin-1 cDNA for radiation-induced salivary hypofunction. *Proc Natl Acad Sci USA* 109:19403–19407.

Baum, B.J., Zheng, C., Cotrim, A.P. et al. 2006. Transfer of the AQP1 cDNA for the correction of radiation-induced salivary hypofunction. *Biochim Biophys Acta* 1758:1071–1077.

Best, L., Brown, P.D., Yates, A.P. et al. 2009. Contrasting effects of glycerol and urea transport on rat pancreatic beta-cell function. *Cell Physiol Biochem* 23:255–264.

Bienert, G.P. and F. Chaumont. 2014. Aquaporin-facilitated transmembrane diffusion of hydrogen peroxide. *Biochim Biophys Acta* 1840:1596–1604.

Bienert, G.P., Moller, A.L., Kristiansen, K.A. et al. 2007. Specific aquaporins facilitate the diffusion of hydrogen peroxide across membranes. *J Biol Chem* 282:1183–1192.

Bienert, G.P., Schjoerring, J.K. and T.P. Jahn. 2006. Membrane transport of hydrogen peroxide. *Biochim Biophys Acta* 1758:994–1003.

Boyer, J.L. 2013. Bile formation and secretion. *Compr Physiol* 3:1035–1078.

Burghardt, B., Elkjaer, M.L., Kwon, T.H. et al. 2003. Distribution of aquaporin water channels AQP1 and AQP5 in the ductal system of the human pancreas. *Gut* 52:1008–1016.

Burghardt, B., Nielsen, S. and M.C. Steward. 2006. The role of aquaporin water channels in fluid secretion by the exocrine pancreas. *J Membr Biol* 210:143–153.

Cai, C., Wang, C., Ji, W. et al. 2013. Knockdown of hepatic aquaglyceroporin-9 alleviates high fat diet-induced non-alcoholic fatty liver disease in rats. *Int Immunopharmacol* 15:550–556.

Calamita, G., Ferri, D., Bazzini, C. et al. 2005a. Expression and subcellular localization of the AQP8 and AQP1 water channels in the mouse gallbladder epithelium. *Biol Cell* 97:415–423.

Calamita, G., Ferri, D., Gena, P. et al. 2005b. The inner mitochondrial membrane has aquaporin-8 water channels and is highly permeable to water. *J Biol Chem* 280:17149–17153.

Calamita, G., Ferri, D., Gena, P. et al. 2008. Altered expression and distribution of aquaporin-9 in the liver of rat with obstructive extrahepatic cholestasis. *Am J Physiol* 295:G682–G690.

Calamita, G., Gena, P., Ferri, D. et al. 2012. Biophysical assessment of aquaporin-9 as principal facilitative pathway in mouse liver import of glucogenetic glycerol. *Biol Cell* 104:342–351.

Calamita, G., Gena, P., Meleleo, D., Ferri, D. and M. Svelto. 2006. Water permeability of rat liver mitochondria: A biophysical study. *Biochim Biophys Acta* 1758:1018–1024.

Calamita, G., Mazzone, A., Bizzoca, A. et al. 2001. Expression and immunolocalization of the aquaporin-8 water channel in rat gastrointestinal tract. *Eur J Cell Biol* 80:711–719.

Calamita, G., Moreno, M., Ferri, D. et al. 2007. Triiodothyronine modulates the expression of aquaporin-8 in rat liver mitochondria. *J Endocrinol* 192:111–120.

Carbrey, J.M., Gorelick-Feldman, D.A., Kozono, D., Praetorius J., Nielsen, S. and P. Agre. 2003. Aquaglyceroporin AQP9: Solute permeation and metabolic control of expression in liver. *Proc Natl Acad Sci USA* 100:2945–2950.

Carbrey, J.M., Song, L., Zhou, Y. et al. 2009. Reduced arsenic clearance and increased toxicity in aquaglyceroporin-9-null mice. *Proc Natl Acad Sci USA* 106:15956–15960.

Carreras, F.I., Gradilone, S.A., Mazzone, A. et al. 2003. Rat hepatocyte aquaporin-8 water channels are down-regulated in extrahepatic cholestasis. *Hepatology* 37:1026–1033.

Carreras, F.I., Lehmann, G.L., Ferri, D., Tioni, M.F., Calamita, G. and R.A. Marinelli. 2007. Defective hepatocyte aquaporin-8 expression and reduced canalicular membrane water permeability in estrogen-induced cholestasis. *Am J Physiol* 292:G905–G912.

Chalasani, N., Younossi, Z., Lavine, J.E. et al. 2012. The diagnosis and management of non-alcoholic fatty liver disease: Practice guideline by the American Gastroenterological Association, American Association for the Study of Liver Diseases, and American College of Gastroenterology. *Gastroenterology* 142:1592–1609.

D'Abbicco, M., Del Buono, N., Gena, P., Berardi, M., Calamita, G. and L. Lopez. 2015. A model for the hepatic glucose metabolism based on Hill and step functions. *J Comput Appl Math*.

Delporte, C. 2013. Role of aquaporins in saliva secretion. *OA Biochem* 1:14.

Delporte, C. 2014. Aquaporins in salivary glands and pancreas. *Biochim Biophys Acta* 1840:1524–1532.

Delporte, C., Hoque, A.T., Kulakusky, J.A. et al. 1998. Relationship between adenovirus-mediated aquaporin 1 expression and fluid movement across epithelial cells. *Biochem Biophys Res Commun* 246:584–588.

Delporte, C., O'Connell, B.C., He, X. et al. 1997. Increased fluid secretion after adenoviral-mediated transfer of the aquaporin-1 cDNA to irradiated rat salivary glands. *Proc Natl Acad Sci USA* 94:3268–3273.

Delporte, C., Perret, J. and M.S. Soyfoo. 2011. Primary Sjögren's syndrome: Current pathophysiological, diagnostic and therapeutic advances, in: *Autoimmune Disorders*, ed. F.P. Huang, pp. 41–66. Intech Publishers Inc., Rijeka, Croatia.

Delporte, C. and S. Steinfeld. 2006. Distribution and roles of aquaporins in salivary glands. *Biochim Biophys Acta* 1758:1061–1070.

Delporte, C., Virreira, M., Crutzen, R. et al. 2009. Functional role of aquaglyceroporin 7 expression in the pancreatic beta-cell line BRIN-BD11. *J Cell Physiol* 221:424–429.

Drews, G., Krippeit-Drews P. and M. Düfer. 2010. Electrophysiology of islet cells. *Adv Exp Med Biol* 654:115–163.

Elkjaer, M.L., Nejsum, L.N., Gresz, V. et al. 2001. Immunolocalization of aquaporin-8 in rat kidney, gastrointestinal tract, testis, and airways. *Am J Physiol* 281:F1047–F1057.

Elkjaer, M.L., Vajda, Z., Nejsum, L.N. et al. 2000. Immunolocalization of AQP9 in liver, epididymis, testis, spleen, and brain. *Biochem Biophys Res Commun* 276:1118–1128.

Ferri, D., Mazzone, A., Liquori, G.E., Cassano, G., Svelto, M. and G. Calamita. 2003. Ontogeny, distribution, and possible functional implications of an unusual aquaporin, AQP8, in mouse liver. *Hepatology* 38:947–957.

García, F., Kierbel, A., Larocca, M.C. et al. 2001. The water channel aquaporin-8 is mainly intracellular in rat hepatocytes, and its plasma membrane insertion is stimulated by cyclic AMP. *J Biol Chem* 276:12147–12152.

Gena, P., Fanelli, E., Brenner, C., Svelto, M. and G. Calamita. 2009. News and views on mitochondrial water transport. *Front Biosci* 1:352–361.

Gena, P., Mastrodonato, M., Portincasa, P. et al. 2013. Liver glycerol permeability and aquaporin-9 are dysregulated in a murine model of non-alcoholic fatty liver disease. *PLoS One* 8:e78139.

Goldblatt, M.I., Swartz-Basile, D.A., Svatek, C.L., Nakeeb, A. and H.A. Pitt. 2002. Decreased gallbladder response in leptin-deficient obese mice. *J Gastrointest Surg* 6:438–442.

Gong, A.Y., Masyuk, A.I., Splinter, P.L., Huebert, R.C., Tietz, P.S. and N.F. LaRusso. 2002. Channel-mediated water movement across enclosed or perfused mouse intrahepatic bile duct units. *Am J Physiol* 283:C338–C346.

Gradilone, S.A., Carreras, F.I., Lehmann, G.L. and R.A. Marinelli. 2005. Phosphoinositide 3-kinase is involved in the glucagon-induced translocation of aquaporin-8 to hepatocyte plasma membrane. *Biol Cell* 97:831–836.

Gradilone, S.A., García, F., Huebert, R.C. et al. 2003. Glucagon induces the plasma membrane insertion of functional aquaporin-8 water channels in isolated rat hepatocytes. *Hepatology* 37:1435–1441.

He, X., Kuijpers, G.A., Goping, G. et al. 1998. A polarized salivary cell monolayer useful for studying transepithelial fluid movement in vitro. *Pflugers Arch* 435:375–381.

Henquin, J.C. 2011. The dual control of insulin secretion by glucose involves triggering and amplifying pathways in β-cells. *Diabetes Res Clin Pract* 93(Suppl. 1):S27–S31.

Hill, A.E. and B. Shachar-Hill. 2006. A new approach to epithelial isotonic fluid transport: An osmosensor feedback model. *J Membr Biol* 210:77–90.

Holm, L.M., Jahn, T.P., Moller, A.L. et al. 2005. NH_3 and NH_4 permeability in aquaporin-expressing Xenopus oocytes. *Pflugers Arch* 450:415–428.

Huebert, R.C., Splinter, P.L., García F., Marinelli, R.A. and N.F. LaRusso. 2002. Expression and localization of aquaporin water channels in rat hepatocytes. Evidence for a role in canalicular bile secretion. *J Biol Chem* 277:22710–22717.

Hung, K.-C., Hsieh, P.-M., Hsu, C.-Y. et al. 2012. Expression of aquaporins in rat liver regeneration. *Scand J Gastroenterol* 47:676–685.

Ishibashi, K., Tanaka, Y. and Y. Morishita. 2014. The role of mammalian superaquaporins inside the cell. *Biochim Biophys Acta* 1840:1507–1512.

Isokpehi, R.D., Rajnarayanan, R.V., Jeffries, C.D., Oyeleye, T.O. and H.H. Cohly. 2009. Integrative sequence and tissue expression profiling of chicken and mammalian aquaporins. *BMC Genomics* 10(Suppl. 2):S7.

Jahn, T.P., Møller, A.L.B., Zeuthen, T. et al. 2004. Aquaporin homologues in plants and mammals transport ammonia. *FEBS Lett* 574:31–36.

Jelen, S., Gena, P., Lebeck, J. et al. 2012. Aquaporin-9 and urea transporter-A gene deletions affect urea transmembrane passage in murine hepatocytes. *Am J Physiol* 303:G1279–G1287.

Jelen, S., Wacker, S., Aponte-Santamaría, C. et al. 2011. Aquaporin-9 protein is the primary route of hepatocyte glycerol uptake for glycerol gluconeogenesis in mice. *J Biol Chem* 286:44319–44325.

Katano, T., Itoh, Y., Yasujima, T., Inoue, K. and H. Yuasa. 2014. Competitive inhibition of AQP7-mediated glycerol transport by glycerol derivatives. *Drug Metab Pharmacokinet* 29:348–351.

Ko, S.B.H., Yamamoto, A., Azuma, S. et al. 2011. Effects of CFTR gene silencing by siRNA or the luminal application of a CFTR activator on fluid secretion from guinea-pig pancreatic duct cells. *Biochem Biophys Acta* 410:904–909.

Kuriyama, H., Shimomura, I., Kishida, K. et al. 2002. Coordinated regulation of fat-specific and liver-specific glycerol channels, aquaporin adipose and aquaporin 9. *Diabetes* 51:2915–2921.

Larocca, M.C., Soria, L.R., Espelt, M.V., Lehmann, G.L. and R.A. Marinelli. 2009. Knockdown of hepatocyte aquaporin-8 by RNA interference induces defective bile canalicular water transport. *Am J Physiol Gastrointest Liver Physiol* 296:G93–G100.

Larsen, H.S., Aure, M.H., Peters, S.B., Larsen, M., Messelt, E.B. and H.K. Galtung. 2011. Localization of AQP5 during development of the mouse submandibular salivary gland. *J Mol Histol* 42:71–81.

Lebeck, J. 2014. Metabolic impact of the glycerol channels AQP7 and AQP9 in adipose tissue and liver. *J Mol Endocrinol* 52:R165–R178.

Lebeck, J., Cheema, M.U., Skowronski, M.T., Nielsen, S. and J. Praetorius. 2015. Hepatic AQP9 expression in male rats is reduced in response to PPARα agonist treatment. *Am J Physiol Gastrointest Liver Physiol* 308:G198–G205.

Lebeck, J., Gena, P., O'Neill, H. et al. 2012. Estrogen prevents increased hepatic aquaporin-9 expression and glycerol uptake during starvation. *Am J Physiol Gastrointest Liver Physiol* 302:G365–G374.

Lee, B.H., Gauna, A.E., Perez, G. et al. 2013. Autoantibodies against muscarinic type 3 receptor in Sjögren's syndrome inhibit aquaporin 5 trafficking. *PLoS One* 8:e53113.

Lee, J. and J.L. Boyer. 2000. Molecular alterations in hepatocyte transport mechanisms in acquired cholestatic liver disorders. *Semin Liver Dis* 20:373–384.

Lee, M.G., Ohana, E., Park, H.W., Yang, D. and S. Muallem. 2012. Molecular mechanism of pancreatic and salivary gland fluid and HCO_3^- secretion. *Physiol Rev* 92:39–74.

Lehmann, G.L., Carreras, F.I., Soria, L.R., Gradilone, S.A. and R.A. Marinelli. 2008. LPS induces the TNF-alpha-mediated downregulation of rat liver aquaporin-8: Role in sepsis-associated cholestasis. *Am J Physiol Gastrointest Liver Physiol* 294:G567–G575.

Lehmann, G.L. and R.A. Marinelli. 2009. Peritoneal sepsis downregulates liver expression of Aquaporin-8: A water channel involved in bile secretion. *Liver Int* 29:317–318.

Li, L., Zhang, H., Ma, T. and A.S. Verkman. 2009. Very high aquaporin-1 facilitated water permeability in mouse gallbladder. *Am J Physiol* 296:G816–G822.

Liu, K., Nagase, H., Huang, C.G., Calamita, G. and P. Agre. 2006. Purification and functional characterization of aquaporin-8. *Biol Cell* 98:153–161.

Louchami, K., Best, L., Brown, P. et al. 2012. A new role for aquaporin 7 in insulin secretion. *Cell Physiol Biochem* 29:65–74.

Maclaren, O.J., Sneyd, J. and E.J. Crampin. 2013. What do aquaporin knockout studies tell us about fluid transport in epithelia? *J Membr Biol* 246:297–305.

Madeira, A., de Almeida, A., de Graaf, C. et al. 2014. A gold coordination compound as a chemical probe to unravel aquaporin-7 function. *Chembiochem* 15:1487–1494.

Marchissio, M.J., Francés, D.E.A., Carnovale, C.E. and R.A. Marinelli. 2012. Mitochondrial aquaporin-8 knockdown in human hepatoma HepG2 cells causes ROS-induced mitochondrial depolarization and loss of viability. *Toxicol Appl Pharmacol* 264:246–254.

Marchissio, M.J., Francés, D.E.A., Carnovale, C.E. and R.A. Marinelli. 2014. Evidence for necrosis, but not apoptosis, in human hepatoma cells with knockdown of mitochondrial aquaporin-8. *Apoptosis* 19:851–859.

Marinelli, R.A., Lehmann, G.L., Soria, L.R. and M.J. Marchissio. 2011. Hepatocyte aquaporins in bile formation and cholestasis. *Front Biosci (Landmark Ed)* 17:2642–2652.

Marinelli, R.A., Pham, L.D., Agre, P. and N.F. LaRusso. 1997. Secretin promotes osmotic water transport in rat cholangiocytes by increasing aquaporin-1 water channels in plasma membrane. Evidence for a secretin-induced vesicular translocation of aquaporin-1. *J Biol Chem* 272:12984–12988.

Marinelli, R.A., Pham, L.D., Tietz, P.S. and N.F. LaRusso. 2000. Expression of aquaporin-4 water channels in rat cholangiocytes. *Hepatology* 31:1313–1317.

Marinelli, R.A., Tietz, P.S., Caride, A.J., Huang, B.Q. and N.F. LaRusso. 2003. Water transporting properties of hepatocyte basolateral and canalicular plasma membrane domains. *J Biol Chem* 278:43157–43162.

Marinelli, R.A., Tietz, P.S., Pham, L.D., Rueckert, L., Agre, P. and N.F. LaRusso. 1999. Secretin induces the apical insertion of aquaporin-1 water channels in rat cholangiocytes. *Am J Physiol* 276:G280–G286.

Marrone, J., Lehmann, G.L., Soria, L.R., Pellegrino, J.M., Molinas, S. and R.A. Marinelli. 2014. Adenoviral transfer of human aquaporin-1 gene to rat liver improves bile flow in. *Gene Ther* 21:1058–1064.

Masyuk, A.I., Masyuk, T.V., Tietz, P.S., Splinter, P.L. and N.F. LaRusso. 2002. Intrahepatic bile ducts transport water in response to absorbed glucose. *Am J Physiol* 283:C785–C791.

Matsumura, K., Chang, B.H.J., Fujimiya, M. et al. 2007. Aquaporin 7 is a beta-cell protein and regulator of intraislet glycerol content and glycerol kinase activity, beta-cell mass, and insulin production and secretion. *Mol Cell Biol* 27:6026–6037.

Mazzone, A., Tietz, P., Jefferson, J., Pagano, R. and N.F. LaRusso. 2006. Isolation and characterization of lipid microdomains from apical and basolateral plasma membranes of rat hepatocytes. *Hepatology* 43:287–296.

Méndez-Giménez, L., Rodríguez, A., Balaguer, I. and G., Frühbeck. 2014. Aquaglyceroporins and caveolins in energy and metabolic homeostasis. *Mol Cell Endocrinol* 397:78–92.

Mennone, A., Verkman, A.S. and J.L. Boyer. 2002. Unimpaired osmotic water permeability and fluid secretion in bile duct epithelia of AQP1 null mice. *Am J Physiol Gastrointest Liver Physiol* 283:G739–G746.

Miller, E.W., Dickinson, B.C. and C.J. Chang. 2010. Aquaporin-3 mediates hydrogen peroxide uptake to regulate downstream intracellular signalling. *Proc Natl Acad Sci USA* 107: 15681–15686.

Miranda, M., Ceperuelo-Mallafre, V., Lecube, A. et al. 2009. Gene expression of paired abdominal adipose AQP7 and liver AQP9 in patients with morbid obesity: Relationship with glucose abnormalities. *Metabolism* 58:1762–1768.

Murakami, M., Murdiastuti, K., Hosoi, K. and A.E. Hill. 2006. AQP and the control of fluid transport in a salivary gland. *J Membr Biol* 210:91–103.

Nielsen, S., Smith, B.L., Christensen, E.I. and P. Agre. 1993. Distribution of the aquaporin CHIP in secretory and resorptive epithelia and capillary endothelia. *Proc Natl Acad Sci USA* 90:7275–7279.

Ohta, E., Itoh, T., Nemoto, T. et al. 2009. Pancreas-specific aquaporin 12 null mice showed increased susceptibility to caerulein-induced acute pancreatitis. *Am J Physiol Cell Physiol* 297:C1368–C1378.

Padma, S., Smeltz, A.M., Banks, P.M., Iannitti, D.A. and I.H. McKillop. 2009. Altered aquaporin 9 expression and localization in human hepatocellular carcinoma. *HPB (Oxford)* 11:66–74.

Patsouris, D., Mandard, S., Voshol, P.J. et al. 2004. PPARα governs glycerol metabolism. *J Clin Invest* 114:94–103.

Poling, H.M., Mohanty, S.K., Tiao, G.M. and S.S. Huppert. 2014. A comprehensive analysis of aquaporin and secretory related gene expression in neonate and adult cholangiocytes. *Gene Expr Patterns* 15:96–103.

Portincasa, P. and G. Calamita. 2012. Water channel proteins in bile formation and flow in health and disease: When immiscible becomes miscible. *Mol Aspects Med* 33:651–664.

Portincasa, P., Di Ciaula, A., Wang, H.H. et al. 2008. Coordinate regulation of gallbladder motor function in the gut-liver axis. *Hepatology* 47:2112–2126.

Portincasa, P., Moschetta, A. and G. Palasciano. 2006. Cholesterol gallstone disease. *Lancet* 368: 230–239.

Portois, L., Zhang, Y., Ladriere, L. et al. 2012. Perturbation of glycerol metabolism in hepatocytes from n3-PUFA-depleted rats. *Int J Mol Med* 29:1121–1126.

Redman, R.S. 1987. Development of salivary glands, in: *The Salivary System*, ed. L.M. Sreebny, pp. 1–20. CRC Press, Boca Raton, FL.

Reshef, L., Olswang, Y., Cassuto, H. et al. 2003. Glyceroneogenesis and the triglyceride/fatty acid cycle. *J Biol Chem* 278:30413–30416.

Rigoulet, M., Yoboue, E.D. and A. Devin. 2011. Mitochondrial ROS generation and its regulation: Mechanisms involved in H(2)O(2) signaling. *Antioxid Redox Signal* 14:459–468.

Rodríguez, A., Catalan, V., Gomez-Ambrosi, J. et al. 2011b. Insulin- and leptin-mediated control of aquaglyceroporins in human adipocytes and hepatocytes is mediated via the PI3K/Akt/mTOR signaling cascade. *J Clin Endocrinol Metab* 96:E586–E597.

Rodríguez, A., Catalán, V., Gómez-Ambrosi, J. and G. Frühbeck. 2011a. Aquaglyceroporins serve as metabolic gateways in adiposity and insulin resistance control. *Cell Cycle* 10:1548–1556.

Rodríguez, A., Gena, P., Méndez-Gimenez, L. et al. 2014. Reduced hepatic aquaporin-9 and glycerol permeability are related to insulin resistance in non-alcoholic fatty liver disease. *Int J Obes* 38:1213–1220.

Rodriguez, M.R., Soria, L.R., Ventimiglia, M.S. et al. 2013. Endothelin-1 and -3 induce choleresis in the rat through ETB receptors coupled to nitric oxide and vagovagal reflexes. *Clin Sci (Lond)* 125:521–532.

Rojek, A.M., Skowronski, M.T., Fuchtbauer, E.M. et al. 2007. Defective glycerol metabolism in aquaporin 9 (AQP9) knockout mice. *Proc Natl Acad Sci USA* 104:3609–3614.

Saparov, S.M., Liu, K., Agre, P. and P. Pohl. 2007. Fast and selective ammonia transport by aquaporin-8. *J Biol Chem* 282:5296–5301.

Sies, H. 2014. Role of metabolic H_2O_2 generation: Redox signaling and oxidative stress. *J Biol Chem* 289:8735–8741.

Sneyd, M.J., Cameron, C. and B. Cox. 2014. Multiscale modelling of saliva secretion. *Math Biosci* 257:69–79.

Soria, L.R., Fanelli, E., Altamura, N., Svelto, M., Marinelli, R.A. and G. Calamita. 2010. Aquaporin-8-facilitated mitochondrial ammonia transport. *Biochem Biophys Res Commun* 393:217–221.

Soria, L.R., Gradilone, S.A., Larocca, M.C. and R.A. Marinelli. 2009. Glucagon induces the gene expression of aquaporin-8 but not that of aquaporin-9 water channels in the rat hepatocyte. *Am J Physiol Regul Integr Comp Physiol* 296:R1274–R1281.

Soria, L.R., Marrone, J., Calamita, G. and R.A. Marinelli. 2013. Ammonia detoxification via ureagenesis in rat hepatocytes involves mitochondrial aquaporin-8 channels. *Hepatology* 57:2061–2071.

Soria, L.R., Marrone, J., Molinas, S.M., Lehmann, G.L., Calamita, G. and R.A. Marinelli. 2014. Lipopolysaccharide impairs hepatocyte ureagenesis from ammonia: Involvement of mitochondrial aquaporin-8. *FEBS Lett* 588:1686–1691.

Soyfoo, M.S., Bolaky, N., Depoortere, I. and C. Delporte. 2012b. Relationship between aquaporin-5 expression and saliva flow in streptozotocin-induced diabetic mice? *Oral Dis* 18:501–505.

Soyfoo, M.S., De Vriese, C., Debaix, H., Martin-Martinez, M.D., Mathieu, C., Devuyst, O. et al. 2007a. Modified aquaporin 5 expression and distribution in submandibular glands from NOD mice displaying autoimmune exocrinopathy. *Arthritis Rheum* 56:2566–2574.

Soyfoo, M.S., Konno, A., Bolaky, N., Oak, J.S., Fruman, D., Nicaise C. et al. 2012a. Link between inflammation and aquaporin-5 distribution in submandibular gland in Sjögren's syndrome? *Oral Dis* 18:568–574.

Soyfoo, M.S., Steinfeld, S. and C. Delporte. 2007b. Usefulness of mouse models to study the pathogenesis of Sjögren's syndrome. *Oral Dis* 13:366–375.

Steinfeld, S., Cogan, E., King, L.S., Agre, P., Kiss, R. and C. Delporte. 2001. Abnormal distribution of aquaporin-5 water channel protein in salivary glands from Sjögren's syndrome patients. *Lab Invest* 81:143–148.

Swartz-Basile, D.A., Lu, D., Basile, D.P. et al. 2007. Leptin regulates gallbladder genes related to absorption and secretion. *Am J Physiol* 293:G84–G90.

Tani, T., Koyama, Y., Nihei, K. et al. 2001. Immunolocalization of aquaporin-8 in rat digestive organs and testis. *Arch Histol Cytol* 64:159–168.

Tietz, P., Jefferson, J., Pagano, R. and N.F. Larusso. 2005. *J Lipid Res* 46:1426–1432.

Tietz, P., Marinelli, R.A., Chen, X.-M. et al. 2003. Agonist-induced coordinated trafficking of functionally related transport proteins for water and ions in cholangiocytes. *J Biol Chem* 278:20413–20419.

Tiniakos, D.G., Vos, M.B. and E.M. Brunt. 2010. Nonalcoholic fatty liver disease: Pathology and pathogenesis. *Annu Rev Pathol* 5:145–171.

Tsukaguchi, H., Weremowicz, S., Morton, C.C. and M.A. Hediger. 1999. Functional and molecular characterization of the human neutral solute channel aquaporin-9. *Am J Physiol* 277: F685–F696.

Ueno, Y., Alpini, G., Yahagi, K. et al. 2003. Evaluation of differential gene expression by microarray analysis in small and large cholangiocytes isolated from normal mice. *Liver Int* 23:449–459.

van Erpecum, K.J., Wang, D.Q., Moschetta, A. et al. 2006. Gallbladder histopathology during murine gallstone formation: Relation to motility and concentrating function. *J Lipid Res* 47:32–41.

Virreira, M., Perret, J. and C. Delporte. 2011. Pancreatic beta-cells: Role of glycerol and aquaglyceroporin 7. *Int J Biochem Cell Biol* 43:10–13.

Wang, D., Yuan, Z., Inoue, N., Cho, G., Shono, M. and Y. Ishikawa. 2011. Abnormal subcellular localization of AQP5 and downregulated AQP5 protein in parotid glands of streptozotocin-induced diabetic rats. *Biochim Biophys Acta* 1810:543–554.

Yamamura, Y., Motegi, K., Kani, K. et al. 2012. TNF-α inhibits aquaporin 5 expression in human salivary gland acinar cells via suppression of histone H4 acetylation. *J Cell Mol Med* 16:1766–1775.

Yang, B., Song, Y., Zhao, D. and A.S. Verkman. 2005. Phenotype analysis of aquaporin-8 null mice. *Am J Physiol* 288:C1161–C1170.

Yang, B., Zhao, D., Solenov, E. and A.S. Verkman. 2006. Evidence from knockout mice against physiologically significant aquaporin 8-facilitated ammonia transport. *Am J Physiol* 291:C417–23.

Yang, B., Zhao, D. and A.S., Verkman. 2006. Evidence against functionally significant aquaporin expression in mitochondria. *J Biol Chem* 281:16202–16206.

Yokoyama, Y., Iguchi, K., Usui, S. et al. 2011. AMP-activated protein kinase modulates the gene expression of aquaporin 9 via forkhead box a2. *Arch Biochem Biophys* 515:80–88.

CHAPTER 10

Aquaporins within the Central Nervous System

Implications for Oedema Following Traumatic CNS Injury

Anna V. Leonard and Renée J. Turner

CONTENTS

Abstract   205
10.1 Distribution of Aquaporins within the CNS   206
    10.1.1 Localization of AQP1   206
    10.1.2 Localization of AQP4   206
    10.1.3 Localization of AQP9   207
10.2 Function of Aquaporins within the CNS   207
    10.2.1 Functions of AQP1   207
    10.2.2 Functions of AQP4   207
    10.2.3 Functions of AQP9   207
10.3 Traumatic CNS Injury   207
    10.3.1 Role of AQP4 in Traumatic Brain Injury   208
10.4 Role of AQP4 in Cytotoxic Models of OEdema   209
10.5 Role of AQP4 in Vasogenic Models of OEdema   210
10.6 Role of AQP4 in SCI   211
10.7 Conclusion   212
References   212

ABSTRACT

OEDEMA AND ITS SEQUELAE remain a common and life-threatening complication of traumatic brain injury (TBI) and spinal cord injury (SCI). Despite the burden of death and disability associated with post-traumatic swelling, treatments remain limited and largely inadequate. In order to effectively treat oedema and improve patient survival and outcome, alternative approaches are required. The recent discovery of aquaporin

(AQP) proteins, in particular AQP4, at tissue–fluid interfaces within the central nervous system (CNS) may provide a novel target for therapeutic intervention. AQPs are water channels that allow the flux of fluid into and out of the CNS; hence, they have been implicated in the pathological accumulation of water following injury. Indeed, over the past decade, numerous studies have demonstrated alterations in AQP expression following injury, concurrent with changes in levels of CNS tissue swelling. This review examines the body of literature supporting a role for AQP4 in the genesis and resolution of swelling following TBI and SCI.

10.1 DISTRIBUTION OF AQUAPORINS WITHIN THE CNS

Aquaporins (AQPs) are a family of homologous, bi-directional water transporting proteins found within the membrane of many epithelium, endothelium and other tissues throughout the body (Verkman 2002). Although at total of 14 isoforms of AQPs have been identified from various mammalian tissues (Venero et al. 2004), the most commonly expressed within the brain and spinal cord are AQP1 and AQP4, classified functionally as water selective, and AQP9 which is additionally selective for glycerol and is classified as a aquaglyceroporin (Lehmann et al. 2004). Each of these AQPs has their own individual pattern of expression, linked to their unique functions.

10.1.1 Localization of AQP1

AQP1 is localized to epithelial cells of the choroid plexus. The localization and therefore function of AQP1 appear to be highly species dependent with differing polarization and localization between rodent and non-human primate species (Brown et al. 2004). Within the spinal cord, AQP1 is expressed within the dorsal horn of the gray matter with much weaker expression observed within the glia limitans of the white matter. Specifically, AQP1 is co-labelled with unmyelinated neurons and endothelial cells of the blood vessels, but not with astrocytes (Oklinski et al. 2014).

10.1.2 Localization of AQP4

The most abundant and widely studied AQP within the CNS is AQP4, which is localized to brain–fluid interfaces including cerebrospinal fluid (CSF) spaces such as the ventricles and subarachnoid space and brain parenchyma (Zador et al. 2009; Papadopoulos and Verkman 2013). Specifically, AQP4 is found within astrocytic end feet surrounding the cerebral vasculature and adjacent to the cells of the glia limitans and ventricular ependymal which form the border zones between the brain and fluid compartments (Amiry-Moghaddam et al. 2004a; Zador et al. 2009; Papadoloulos and Verkman 2013). No expression of AQP4 has been observed within neurons of the brain nor cells of the meningeal layers or oligodendrocytes. However, expression of AQP4 has been noted in microglia under some conditions. Similarly, within the spinal cord, AQP4 is widely distributed in both grey and white matter astrocytes and particularly abundant within the perivascular end feet, sub-pial region and the ependymal cells surrounding the central canal (Nesic et al. 2006; Oklinski et al. 2014). Such localization implicates AQP4 as playing a role in controlling water fluxes into and out of the spinal cord parenchyma from either blood vessels or the central canal.

10.1.3 Localization of AQP9

Two isoforms of AQP9 have been discovered a short isoform (26 kDa) localized to the inner mitochondrial membrane and a long isoform (30 kDa) localized to the cell membrane (Amiry-Moghaddam et al. 2005). Within the brain, AQP9 is localized to glial cells (Badaut et al. 2001, 2004), endothelial cells and a subset of neurons predominately of the catecholaminergic type (Badaut et al. 2004). AQP9 is also widely expressed in the spinal cord and is colocalized with astrocytes, largely within the white matter (Oshio et al. 2004).

10.2 FUNCTION OF AQUAPORINS WITHIN THE CNS

10.2.1 Functions of AQP1

Given the species differences in localization of AQP1, different roles have been described. The localization of AQP1 within epithelial cells of the choroid plexus has implicated it in the formation and circulation of CSF (Brown et al. 2004). In addition, AQP1 has been observed to localize within the dorsal horn of the spinal cord and trigeminal sensory ganglia, suggesting a possible role in nociception (Shields et al. 2007).

10.2.2 Functions of AQP4

A wide range of roles for AQP4 within the CNS have been described, including the control of water flux, neural signal transduction, promotion of astrocyte migration (Tait et al. 2008) and cell migration (Hiroaki et al. 2006). More recently, the polarization of AQP4 within close proximity of the vasculature implicates a role in the exchange of gases, including O_2, CO_2 and NO (Kimelberg and Nedergaard 2010). AQP4 has also been suggested to play a role in the modulation of serum osmoregularity (Verkman 2005). In particular, co-localization with Kir4.1 facilitates water diffusion along K^+ gradients (Brown et al. 2004; Binder et al. 2006).

The location of AQP4 at CNS–fluid interfaces implicates AQP4 in playing a role in controlling water fluxes into and out of the CNS tissues from blood vessels, CSF compartments and subarachnoid space. AQP4 therefore represents an attractive target for pharmacological intervention aimed at reducing oedema following CNS injury, and accordingly the role of AQP4 following both TBI and SCI will be discussed in detail in this chapter.

10.2.3 Functions of AQP9

AQP9 is an aquaglyceroporin that facilitates the diffusion of water solutes such as glycerol, monocarboxylate and urea within the CNS (Arciénega et al. 2010). The localization of AQP9 suggests that it is integrally involved in water homeostasis and may play a role in diseases of normal water flux such as SCI and syringomyelia (Oshio et al. 2004).

10.3 TRAUMATIC CNS INJURY

Traumatic injuries to the CNS, such as TBI and SCI, are a significant cause of death and disability worldwide. Whilst TBI and SCI represent two distinct and unique insults to the CNS, predominantly due to their anatomical differences, there are many similarities in terms of injury types and development. Both TBI and SCI are characterized by two main mechanisms of injury, known as primary and secondary injury. Primary injury results from the initial mechanical injury due to local deformation and energy transformation at the time of the trauma and is not reversible. Secondary injury encompasses an entire cascade of systemic and cellular processes initiated by the primary injury event that contribute to further destruction of the CNS tissue (Sekhon and Fehlings 2001; Finnie and Blumbergs 2002). Given the delayed nature of these events, there is a potential window to prevent or limit the progression of secondary injury with pharmacological intervention.

Of these secondary factors, it is the development of oedema that serves as a principal prognostic factor for neurological outcome. Cerebral oedema following TBI can lead to elevated intracranial pressure (ICP), reduced cerebral perfusion pressure, compression of adjacent tissue and herniation (Klatzo 1987, 1994; Engelborghs et al. 1998). As such, cerebral oedema accounts for over 50% of the morbidity and mortality associated with TBI (Marmarou 2007) and is widely recognized as a major clinical management target. Similarly, oedema following SCI is thought to contribute to raised intrathecal pressure (Yashon et al. 1973; Kwon et al. 2009; Leonard et al. 2013), reduced spinal cord barrier function (Wang et al. 1993; Kwon et al. 2009) and myelin damage (Sharma 2005), all of which can lead to increased tissue damage and reduced neurological function. Given the severe complications of oedema development following both brain and spinal cord injuries, the swelling of the brain and spinal cord is an important target in the clinical management of trauma patients. However, treatment options have remained relatively unchanged in decades and of only limited efficacy in combating post-traumatic swelling. Despite this, the exact mechanisms involved in the genesis of swelling within the CNS remain poorly understood.

Although there are a number of classifications of oedema, there are two main types which occur in CNS trauma: cytotoxic and vasogenic oedema (Figure 10.1). Cytotoxic oedema occurs in the setting of bioenergetics crisis and failure of the Na^+/K^+ ATPase pump with resultant disruption to ionic gradients, massive Na^+ and H_2O influx, and cellular swelling (Klatzo 1987). Vasogenic oedema occurs in conjunction with a disrupted blood–brain or blood–spinal cord barrier. In this setting, plasma proteins move freely into the tissue, with water following to cause an intraparenchymal accumulation of fluid, the complications of which include elevated ICP, which can be life threatening (Hacke et al. 1996).

10.3.1 Role of AQP4 in Traumatic Brain Injury

Over the last decade, it has been heavily debated as to whether AQP4 plays a role in the development or resolution of oedema following traumatic injury to the CNS. However, it is becoming increasingly apparent that the exact role that AQP4 plays in the ensuing post-traumatic oedema is highly dependent upon both the injury type and the time point following injury. AQP4 is thought to play a key role in the development of cytotoxic oedema (Manley et al. 2004), whilst assisting in the clearance of water in vasogenic brain oedema (Papadopoulos et al. 2004). Studies in AQP4-null mice demonstrate reduced brain swelling and improved neurological outcome in cytotoxic oedema models whereas brain swelling and clinical outcomes were worsened in vasogenic oedema models (Verkman et al. 2006).

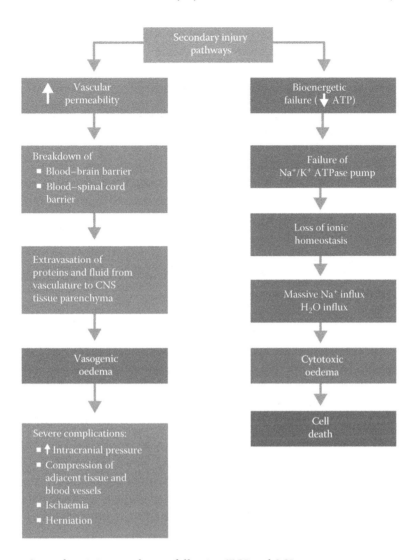

FIGURE 10.1   Secondary injury pathways following TGI and SCI.

Such studies indicate that deletion of AQP4 provides an advantageous environment following cytotoxic oedema, though it is detrimental to the clearance of vasogenic oedematous fluid. As such, it appears that AQP4 is capable of contributing to both oedema genesis and clearance, with the exact role depending on the injury model; such a premise will be further examined in relation to oedema type.

10.4 ROLE OF AQP4 IN CYTOTOXIC MODELS OF OEDEMA

In an in vitro model where cultured astrocytes were injured using a modified fluid percussion device, astrocytes demonstrated significantly upregulated AQP4 levels, maximal at 3 h post injury (Rao et al. 2011). However, modulating AQP4 function with administration of small-interfering RNA (siRNA) to the AQP4 gene 48 h prior to injury induction significantly reduced cell swelling at 3 h post injury. Indeed, in vivo cytotoxic oedema models have shown such similarity using AQP4 knockout mice. AQP4-null mice subjected to acute water intoxication, a model of cytotoxic oedema, demonstrated improved survival with reduced brain water content and swelling of peri-capillary astrocytic foot processes (Manley et al. 2000). In another study of water intoxication, perivascular AQP4 was removed via α-syntrophin deletion, resulting in delayed onset of oedema and reduced brain water content (Vajda et al. 2002; Amiry-Moghaddam et al. 2004b). Furthermore, in a model of focal cerebral ischemia, AQP4-null mice showed improved neurological outcome and decreased cerebral oedema (Manley et al. 2000). Overall, these studies indicate that AQP4 is involved in the formation of oedema in TBI models that are primarily cytotoxic in nature. However, dystrophin-null mice did not demonstrate improved survival (Vajda et al. 2002); this is in contrast to the results observed in AQP4-null mice (Manley et al. 2000) and may be attributed to morphological differences in the animal model, given that dystrophin-null mice demonstrate increased BBB permeability (Nico et al. 2004). Such permeability alterations are characteristic of vasogenic oedema, further emphasizing the protective role of AQP4 in models of vasogenic oedema.

10.5 ROLE OF AQP4 IN VASOGENIC MODELS OF OEDEMA

AQP4 is thought to assist in the clearance of excess fluid from within the brain in the setting of vasogenic oedema. Indeed, in experimental brain tumour models, AQP4-null mice demonstrated higher ICP and accelerated neurological deterioration (Papadopoulos et al. 2004). Furthermore, the expression of AQP4 within astrocytes was upregulated in oedematous brain tumors (Saadoun et al. 2002), indicating a potential protective mechanism to assist in water clearance. More recently, a study investigating the efficacy of decompressive craniectomy to treat raised ICP following fluid perfusion TBI demonstrated that decompression reduced oedema and was associated with reduced AQP4 expression, whilst the untreated group revealed significantly higher expression of AQP4 which accompanied increased brain swelling (Tomura et al. 2011). In another study investigating the effect of ethanol administration following a diffuse impact acceleration model of TBI, reduced oedema was associated with significantly reduced levels of AQP4 mRNA and protein expression (Wang et al. 2013). Although it is tempting to speculate how AQP4 may be influencing oedema in this setting, in both studies, it is difficult to determine the cause and effect and whether the increased AQP4 represents a protective compensatory mechanism to try and combat the oedema or whether in fact it was responsible for the genesis of swelling.

Following a controlled cortical impact (CCI) model of TBI in juvenile rats, oedema was found to increase by day 1 and normalize by day 3 post injury. Interestingly, AQP4 levels in the perivascular end feet were increased at the delayed time points of day 3 and 7 post injury (Fukuda et al. 2012). Such findings indicate that AQP4 may not be the driving force in the development of oedema, but rather in its resolution given that an increase in AQP4 was associated with normalized brain water content. However, subsequent investigation using the same model demonstrated that inhibition of AQP4 expression by injection of AQP4 siRNA significantly reduced oedema and improved neurological outcome post trauma (Fukuda et al. 2013). These results certainly suggest that AQP4 may in fact contribute to oedema development following CCI. In a focal model of TBI, brain water content was significantly increased at 6 h with maximal oedema observed at 24 h. Interestingly, AQP4 expression was reduced at 1 h post injury, lowest at 6 h, and then increased at 12 h post injury (Lu et al. 2013). These findings demonstrate a temporal disconnect between changes in AQP4 expression and oedema development. Indeed, such studies suggest that the exact role that AQP4 plays following trauma is not only dependent upon the type of injury but that levels of AQP4 may also change significantly depending upon the timing of assessment.

One further aspect that should be considered is the localization of AQP4 expression following injury. It is not only the levels of AQP4 expression that are important in determining the effect of fluid flow but most importantly the location of AQP4 expression. Overall, studies have shown that whilst the global expression of AQP4 following TBI is generally increased, there is a loss of polarized localization at the end-foot processes of reactive astrocytes (Ren et al. 2013). Such a depolarized state of the AQP4 channels following injury suggests that its role in either the development or resolution of oedema may be insignificant. Regardless of this, it is clear that further investigation is essential in order to truly unravel the relationship between alterations in AQP4 expression and both the genesis and clearance of oedema development following TBI.

10.6 ROLE OF AQP4 IN SCI

Similar to brain tissue, AQP4 within the spinal cord is found widely distributed in both grey and white matter astrocytes, within perivascular end feet surrounding the vasculature and ependymal cells of the central canal and the sub-pial region (Nesic et al. 2006). These locations implicate AQP4 in the control of water fluxes both into and out of the spinal cord parenchyma from either blood vessels or the central canal. Given that SCI is associated with substantial oedema, recent studies have focused on elucidating whether AQP4 plays a role in the genesis and/or resolution of traumatic spinal cord swelling. Kimura et al. (2010) found that following a contusion model of SCI, AQP4-null mice demonstrated excessive oedema development, cyst formation and worsened neurological outcome. Similarly, AQP4-null mice subjected to transection SCI demonstrated significantly higher oedema content, larger cyst volume and impaired astrocytic migration toward the lesion site (Wu et al. 2014). Such results implicate a protective role for AQP4 presumably by facilitating the removal of excess water. In contrast, forceps compression SCI in AQP4-null mice was associated with reduced oedema and improved neurological function (Saadoun et al. 2008), thereby implicating a role for AQP4 in facilitating water movement into the spinal cord tissue. Differences in experimental models may, in part, explain the observed dissimilarities in outcomes. However, upregulation of AQP4 has been associated with reduced oedema development in both contusion (Nesic et al. 2006) and clip compression models (Mao et al. 2011), further suggesting that AQP4 plays a role in water clearance following SCI. More recently, balloon compression SCI was shown to be associated with increased AQP4 levels, both within the perivascular region and in the ependymal cells of the central canal (Leonard et al. 2013). This initial increase was followed by a decrease at 2 week post-SCI. The initial increase in AQP4 immunoreactivity was associated with increased oedema development and thus may be responsible for such increased oedema. Alternatively, such an increase may be a compensatory mechanism to facilitate clearance of water from the spinal cord tissue.

Many studies have observed AQP4 overexpression in tissues surrounding fluid-filled cavities (Nesic et al. 2010) and mature syrinx cavities (Hemley et al. 2013) and common features of chronic SCI. Increased astrocytic AQP4 expression was similarly proportional to the degree of central canal expansion (Zhang et al. 2012). AQP4 has also been implicated in many other spinal cord disorders associated with water imbalances such as spinal cord ischemia (Xu et al. 2008), neuromyelitis optica (Misu et al. 2007; Hinson et al. 2010) and amyotrophic lateral sclerosis (Nicaise et al. 2009). Taken together, the exact involvement of AQP4 in oedema development or resolution following SCI is still debated, with further investigation essential to better understand such a role and to realize the potential therapeutic potential of AQP4 modulation.

10.7 CONCLUSION

Both TBI and SCI are characterised by oedema development for which AQP4 has been implicated in its either resolution or development. The literature to date has focused on studies utilising AQP4 knockout strains or simply observing changes in AQP4 expression associated with changes in oedema. These studies are unable to clearly define the cause and effect of AQP4 following TBI or SCI. As such, further studies, aimed at pharmacological modulating AQP4, are required to truly elucidate its role following CNS trauma.

REFERENCES

Amiry-Moghaddam M, Frydenlund DS and Ottersen OP. (2004a). Anchoring of aquaporin-4 in brain: Molecular mechanisms and implications for the physiology and pathophysiology of water transport. *Neuroscience* **129**, 999–1010.

Amiry-Moghaddam M, Lindland H, Zelenin S, Roberg BA, Gundersen BB, Petersen P, Rinvik E, Torgner IA and Ottersen OP. (2005). Brain mitochondria contain aquaporin water channels: Evidence for the expression of a short AQP9 isoform in the inner mitochondrial membrane. *FASEB J* **19**, 1459–1467.

Amiry-Moghaddam M, Xue R, Haug FM, Neely JD, Bhardwaj A, Agre P, Adams ME, Froehner SC, Mori S and Ottersen OP. (2004b). Alpha-syntrophin deletion removes the perivascular but not endothelial pool of aquaporin-4 at the blood-brain barrier and delays the development of brain edema in an experimental model of acute hyponatremia. *FASEB J* **18**, 542–544.

Arciénega II, Brunet JF, Bloch J and Badaut J. (2010). Cell locations for AQP1, AQP4 and 9 in the non-human primate brain. *Neuroscience* **167**, 1103–1114.

Badaut J, Hirt L, Granziera C, Bogousslavsky J, Magistretti PJ and Regli L. (2001). Astrocyte-specific expression of aquaporin-9 in mouse brain is increased after transient focal cerebral ischemia. *J Cereb Blood Flow Metab* **21**, 477–482.

Badaut J, Petit JM, Brunet JF, Magistretti PJ, Charriaut-Marlangue C and Regli L. (2004). Distribution of Aquaporin 9 in the adult rat brain: Preferential expression in catecholaminergic neurons and in glial cells. *Neuroscience* **128**, 27–38.

Binder DK, Yao X, Zador Z, Sick TJ, Verkman AS and Manley GT. (2006). Increased seizure duration and slowed potassium kinetics in mice lacking aquaporin-4 water channels. *Glia* **53**, 631–636.

Brown PD, Davies SL, Speake T and Millar ID. (2004). Molecular mechanisms of cerebrospinal fluid production. *Neuroscience* **129**, 957–970.

Engelborghs K, Verlooy J, Van Reempts J, Van Deuren B, Van de Ven M and Borgers M. (1998). Temporal changes in intracranial pressure in a modified experimental model of closed head injury. *J Neurosurg* **89**, 796–806.

Finnie JW and Blumbergs PC. (2002). Traumatic brain injury. *Vet Pathol* **39**, 679–689.

Fukuda AM, Adami A, Pop V, Bellone JA, Coats JS, Hartman RE, Ashwal S, Obenaus A and Badaut J. (2013). Posttraumatic reduction of edema with aquaporin-4 RNA interference improves acute and chronic functional recovery. *J Cereb Blood Flow Metab* **33**, 1621–1632.

Fukuda AM, Pop V, Spagnoli D, Ashwal S, Obenaus A and Badaut J. (2012). Delayed increase of astrocytic aquaporin 4 after juvenile traumatic brain injury: Possible role in edema resolution? *Neuroscience* **222**, 366–378.

Hacke W, Schwab S, Horn M, Spranger M, De Georgia M and von Kummer R. (1996). 'malignant' middle cerebral artery territory infarction: Clinical course and prognostic signs. *Archives of Neurology* **53**, 309–315.

Hemley SJ, Bilston LE, Cheng S, Chan JN and Stoodley MA. (2013). Aquaporin-4 expression in post-traumatic syringomyelia. *J Neurotrauma* **30**, 1457–1467.

Hinson SR, McKeon A and Lennon VA. (2010). Neurological autoimmunity targeting aquaporin-4. *Neuroscience* **168**, 1009–1018.

Hiroaki Y, Tani K, Kamegawa A, Gyobu N, Nishikawa K, Suzuki H, Walz T et al. (2006). Implications of the aquaporin-4 structure on array formation and cell adhesion. *J Mol Biol* **355**, 628–639.

Kimelberg HK and Nedergaard M. (2010). Functions of astrocytes and their potential as therapeutic targets. *Neurotherapeutics* **7**, 338–353.

Kimura A, Hsu M, Seldin M, Verkman AS, Scharfman HE and Binder DK. (2010). Protective role of aquaporin-4 water channels after contusion spinal cord injury. *Ann Neurol* **67**, 794–801.

Klatzo I. (1987). Pathophysiological aspects of brain edema. *Acta Neuropathol* **72**, 236–239.

Klatzo I. (1994). Evolution of brain edema concepts. *Acta Neurochir Suppl (Wien)* **60**, 3–6.

Kwon BK, Curt A, Belanger LM, Bernardo A, Chan D, Markez JA, Gorelik S et al. (2009). Intrathecal pressure monitoring and cerebrospinal fluid drainage in acute spinal cord injury: A prospective randomized trial. *J Neurosurg Spine* **10**, 181–193.

Lehmann GL, Gradilone SA and Marinelli RA. (2004). Aquaporin water channels in central nervous system. *Curr Neurovasc Res* **1**, 293–303.

Leonard AV, Thornton E and Vink R. (2013). Substance P as a mediator of neurogenic inflammation following balloon compression induced spinal cord injury. *J Neurotrauma* 30, 1812–1823.

Lu H, Lei XY, Hu H and He ZP. (2013). Relationship between AQP4 expression and structural damage to the blood-brain barrier at early stages of traumatic brain injury in rats. *Chin Med J (Engl)* **126**, 4316–4321.

Manley GT, Binder DK, Papadopoulos MC and Verkman AS. (2004). New insights into water transport and edema in the central nervous system from phenotype analysis of aquaporin-4 null mice. *Neuroscience* **129**, 983–991.

Manley GT, Fujimura M, Ma T, Noshita N, Filiz F, Bollen AW, Chan P and Verkman AS. (2000). Aquaporin-4 deletion in mice reduces brain edema after acute water intoxication and ischemic stroke. *Nat Med* **6**, 159–163.

Mao L, Wang HD, Pan H and Qiao L. (2011). Sulphoraphane enhances aquaporin-4 expression and decreases spinal cord oedema following spinal cord injury. *Brain Inj* **25**, 300–306.

Marmarou A. (2007). A review of progress in understanding the pathophysiology and treatment of brain edema. *Neurosurg Focus* **22**, E1.

Misu T, Fujihara K, Kakita A, Konno H, Nakamura M, Watanabe S, Takahashi T, Nakashima I, Takahashi H and Itoyama Y. (2007). Loss of aquaporin 4 in lesions of neuromyelitis optica: Distinction from multiple sclerosis. *Brain* **130**, 1224–1234.

Nesic O, Guest JD, Zivadinovic D, Narayana PA, Herrera JJ, Grill RJ, Mokkapati VU, Gelman BB and Lee J. (2010). Aquaporins in spinal cord injury: The Janus face of aquaporin 4. *Neuroscience* **168**, 1019–1035.

Nesic O, Lee J, Ye Z, Unabia GC, Rafati D, Hulsebosch CE and Perez-Polo JR. (2006). Acute and chronic changes in aquaporin 4 expression after spinal cord injury. *Neuroscience* **143**, 779–792.

Nicaise C, Soyfoo MS, Authelet M, De Decker R, Bataveljic D, Delporte C and Pochet R. (2009). Aquaporin-4 overexpression in rat ALS model. *Anat Rec (Hoboken)* **292**, 207–213.

Nico B, Paola Nicchia G, Frigeri A, Corsi P, Mangieri D, Ribatti D, Svelto M and Roncali L. (2004). Altered blood-brain barrier development in dystrophic MDX mice. *Neuroscience* **125**, 921–935.

Oklinski MK, Lim JS, Choi HJ, Oklinska P, Skowronski MT and Kwon TH. (2014). Immunolocalization of water channel proteins AQP1 and AQP4 in rat spinal cord. *J Histochem Cytochem* **62**, 598–611.

Oshio K, Binder DK, Yang B, Schecter S, Verkman AS and Manley GT. (2004). Expression of aquaporin water channels in mouse spinal cord. *Neuroscience* **127**, 685–693.

Papadopoulos MC, Manley GT, Krishna S and Verkman AS. (2004). Aquaporin-4 facilitates reabsorption of excess fluid in vasogenic brain edema. *FASEB J* **18**, 1291–1293.

Papadopoulos MC, Saadoun S and Verkman AS. (2008). Aquaporins and cell migration. *Pflugers Arch* **456**, 693–700.

Papadopoulos MC and Verkman AS. (2013). Aquaporin water channels in the nervous system. *Nat Rev Neurosci* **14**, 265–277.

Rao KV, Reddy PV, Curtis KM and Norenberg MD. (2011). Aquaporin-4 expression in cultured astrocytes after fluid percussion injury. *J Neurotrauma* **28**, 371–381.

Ren Z, Iliff JJ, Yang L, Yang J, Chen X, Chen MJ, Giese RN, Wang B, Shi X and Nedergaard M. (2013). 'Hit & run' model of closed-skull traumatic brain injury (TBI) reveals complex patterns of post-traumatic AQP4 dysregulation. *J Cereb Blood Flow Metab* **33**, 834–845.

Ribeiro M de C, Hirt L, Bogousslavsky J, Regli L and Badaut J. (2006). Time course of aquaporin expression after transient focal cerebral ischemia in mice. *J Neurosci Res* **83**, 1231–1240.

Saadoun S, Bell BA, Verkman AS and Papadopoulos MC. (2008). Greatly improved neurological outcome after spinal cord compression injury in AQP4-deficient mice. *Brain* **131**, 1087–1098.

Saadoun S, Papadopoulos MC, Davies DC, Krishna S and Bell BA. (2002). Aquaporin-4 expression is increased in oedematous human brain tumours. *J Neurol Neurosurg Psychiatry* **72**, 262–265.

Sekhon LH and Fehlings MG. (2001). Epidemiology, demographics, and pathophysiology of acute spinal cord injury. *Spine* **26**, S2–S12.

Sharma HS. (2005). Pathophysiology of blood-spinal cord barrier in traumatic injury and repair. *Curr Pharm Des* **11**, 1353–1389.

Shields SD, Mazario J, Skinner K and Basbaum AI. (2007). Anatomical and functional analysis of aquaporin 1, a water channel in primary afferent neurons. *Pain* **131**, 8–20.

Tait MJ, Saadoun S, Bell BA and Papadopoulos MC. (2008). Water movements in the brain: Role of aquaporins. *Trends Neurosci* **31**, 37–43.

Tomura S, Nawashiro H, Otani N, Uozumi Y, Toyooka T, Ohsumi A and Shima K. (2011). Effect of decompressive craniectomy on aquaporin-4 expression after lateral fluid percussion injury in rats. *J Neurotrauma* **28**, 237–243.

Vajda Z, Pedersen M, Fuchtbauer EM, Wertz K, Stodkilde-Jorgensen H, Sulyok E, Doczi T et al. (2002). Delayed onset of brain edema and mislocalization of aquaporin-4 in dystrophin-null transgenic mice. *Proc Natl Acad Sci USA* **99**, 13131–13136.

Venero JL, Machado A and Cano J. (2004). Importance of aquaporins in the physiopathology of brain edema. *Curr Pharm Des* **10**, 2153–2161.

Verkman AS. (2002). Aquaporin water channels and endothelial cell function. *J Anat* **200**, 617–627.

Verkman AS. (2005). More than just water channels: Unexpected cellular roles of aquaporins. *J Cell Sci* **118**, 3225–3232.

Verkman AS, Binder DK, Bloch O, Auguste K and Papadopoulos MC. (2006). Three distinct roles of aquaporin-4 in brain function revealed by knockout mice. *Biochim Biophys Acta* **1758**, 1085–1093.

Wang R, Ehara K and Tamaki N. (1993). Spinal cord edema following freezing injury in the rat: Relationship between tissue water content and spinal cord blood flow. *Surg Neurol* **39**, 348–354.

Wang T, Chou DY, Ding JY, Fredrickson V, Peng C, Schafer S, Guthikonda M, Kreipke C, Rafols JA and Ding Y. (2013). Reduction of brain edema and expression of aquaporins with acute ethanol treatment after traumatic brain injury. *J Neurosurg* **118**, 390–396.

Wu Q, Zhang YJ, Gao JY, Li XM, Kong H, Zhang YP, Xiao M, Shields CB and Hu G. (2014). Aquaporin-4 mitigates retrograde degeneration of rubrospinal neurons by facilitating edema clearance and glial scar formation after spinal cord injury in mice. *Mol Neurobiol* **49**, 1327–1337.

Xu WB, Gu YT, Wang YF, Lu XH, Jia LS and Lv G. (2008). Bradykinin preconditioning modulates aquaporin-4 expression after spinal cord ischemic injury in rats. *Brain Res* **1246**, 11–18.

Yashon D, Bingham WG, Jr., Faddoul EM and Hunt WE. (1973). Edema of the spinal cord following experimental impact trauma. *J Neurosurg* **38**, 693–697.

Zador Z, Stiver S, Wang V and Manley GT. (2009). Role of aquaporin-4 in cerebral edema and stroke. *Handb Exp Pharmacol* **190**, 159–170.

Zhang Y, Zhang YP, Shields LB, Zheng Y, Xu XM, Whittemore SR and Shields CB. (2012). Cervical central canal occlusion induces noncommunicating syringomyelia. *Neurosurgery* **71**, 126–137.

CHAPTER 11

Aquaporins in Carcinogenesis

Water and Glycerol Channels as New Potential Drug Targets

Chulso Moon and David Moon

CONTENTS

Abstract   217
11.1 Introduction   217
11.2 Role of Each Aquaporin in Human Carcinogenesis   218
    11.2.1 Aquaporin 1   218
    11.2.2 Aquaporin 3   220
    11.2.3 Aquaporin 4   222
    11.2.4 Aquaporin 5   224
    11.2.5 Other Aquaporins   225
11.3 Aquaporins as Therapeutic Targets and Prognostic Markers   225
References   228

ABSTRACT

THE AQUAPORINS (AQPs) ARE a family of transmembrane water channel proteins and play a crucial role in transcellular and transepithelial water movement. Recent evidence indicates that AQPs may be involved in cell proliferation, migration and angiogenesis, each of which plays an important role in human carcinogenesis. This chapter summarizes recent data concerning the involvement of AQPs in tumour growth, angiogenesis and metastasis. Furthermore, we briefly discuss various potential therapeutic approaches by antagonizing their biological activity.

11.1 INTRODUCTION

The aquaporins (AQPs) represent a family of transmembrane water channel proteins widely distributed in various tissues throughout the body. Mostly, they play a major role in both transcellular and transepithelial water movement [1]. AQPs are present in many epithelial, endothelial and other tissues with at least 13 AQPs being described in mammals. There are two groups of AQP depending on their water transport and other transporter capabilities.

For example, AQP1, AQP2, AQP4, AQP5 and AQP8 are primarily water selective, whereas AQP3, AQP7, AQP9 and AQP10 (called 'aquaglyceroporins') also transport glycerol and other small solutes [2]. While a majority of AQPs are localized in the plasma membrane, some isoforms are present in the cytoplasmic compartments, particularly in endoplasmic reticulum and their translocation to the plasma membrane is crucial in the regulation of water transfer [3,4]. Besides their role in transport fluid and regulating the osmotic balance, AQPs were shown to be involved in cell signalling pathways and cellular migration [2–6]. Tumour cell growth largely depends on signalling from extracellular space to the intracellular space, as seen in EGF to epidermal growth factor receptor (EGFR)-mediated signalling pathways [7]. Likewise, cell migration is involved in tumour growth and metastatic process [8]. Fundamentally, endothelial cell migration plays a key role in angiogenesis in the development of normal organs and growth of tumour cells. In the last 10 years, with the initial report from our group for the role of AQP during colorectal development [9], several tumour cell types were shown to express AQPs in vivo in humans and rodents. Furthermore, prognostic implications have been established based on AQP expression in lung and brain tumours and human glioblastoma [3,4,10]. Here, we review various report regarding each expression of AQPs in human and rodent models and discuss about their potential role in human carcinogenesis. Then, we discuss about the present data and future expectations of AQPs based on onco-therapeutics and use of AQP expression as a valuable prognostic markers.

11.2 ROLE OF EACH AQUAPORIN IN HUMAN CARCINOGENESIS

11.2.1 Aquaporin 1

Expression of AQP1 has been initially reported during colorectal cancer development. In this study, by reverse transcriptase polymerase chain reaction analysis, expression of AQP1, AQP3 and AQP5 was examined in seven colon and colorectal cancer cell lines. Western blot analysis confirmed their expression in four of these cell lines. In situ hybridization demonstrated that during colorectal carcinogenesis, the expression of AQP1 and AQP5 was induced in early-stage disease (early dysplasia) and maintained through the late stages of colon cancer development (Figure 11.1). Expression of AQP1 and AQP5 was maintained even in metastatic lesions in the liver. These findings demonstrate that the expression of several AQPs is found in tumour cells and is associated with an early stage of colorectal cancer development. These initial observations suggest that multiple AQP expression may be advantageous to human carcinogenesis. Since these initial observations, various reports have suggested expression of AQP1 in mammary carcinomas, brain tumours, hemangioblastomas, multiple myeloma and breast cancer [3,6–11]. Further mechanistic studies by our groups have demonstrated that AQP1 is overexpressed in lung cancer cells and stimulates NHI-3T3 cell proliferation and anchorage-independent growth [12]. Additionally, the mechanisms behind this observation seem to be at least depending on 'significant activation of ERK2'.

An initial report by Shanahan et al. has suggested that AQP1 can play an important role in the water permeability of smooth muscle cells from vascular wall. More detailed studies have subsequently followed [13]. AQP1 is shown to be expressed in endothelial cells of non-fenestrated capillaries [14] and human arteries [13]. Moreover, the involvement of AQP1 in

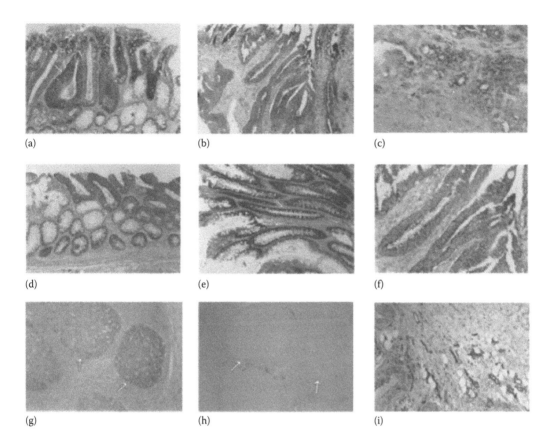

FIGURE 11.1 **(See colour insert.)** Induced expression of aquaporin (AQP)1 and AQP5 during colorectal carcinogenesis. In situ hybridization with AQP1 and AQP5 probes in the multistep carcinogenesis model of colon cancer development. The AQP1 antisense riboprobe clearly stains not only colonic adenoma (a, arrow) but also the primary colon cancer (b, arrow) and metastatic lesions in the liver (c, arrow). Likewise, the AQP5 antisense riboprobe clearly stains early adenoma with moderately dysplastic cells (d, arrow), late adenoma with severe dysplastic cells (e, arrow), and adenocarcinoma (f, arrow). There is almost no staining in the surrounding normal colonic mucosa when probed with AQP1 and AQP5 antisense riboprobes (a and d, star). As positive controls, the germinal centers of the tonsils were stained with antisense riboprobes of AQP1 (data not shown) and AQP5 (g). Sense riboprobes of AQP1 (data not shown) and AQP5 (h) were used as negative controls. As an internal control, vascular endothelium was stained with the AQP1 antisense riboprobe (i). (Adapted from Moon, C. et al., *Oncogene*, 22, 6699, 2003.)

transendothelial water transport of non-fenestrated endothelium has been shown in the descending vasa recta [14–16] and in peritoneal capillaries [17], suggesting the role of peritoneal metastasis seen most of terminal cancers. AQP1 is strongly expressed in proliferating tumour microvessels in human [3,8] and rat [6] and in the chick embryo chorioallantoic membrane microvessels [18]. Subsequent pioneering study done by the Verkman's group has established an unambiguous role of AQP1 during angiogenesis. In this study, based on wild-type and AQP1-null mice subcutaneously implanted with B16F10 melanoma cells, his group initially demonstrated that tumour growth was reduced in AQP1-null mice due to impaired angiogenesis [19]. Further histological examination has proved that in tumours of AQP1-null mice, there was a much reduced density of microvessels and the presence of islands of viable tumour cells surrounded by necrotic tissue. Finally, the same group has demonstrated that tumour cell migration and metastatic potential greatly increased in two mouse tumour cell lines with AQP1 expression as compared to the controls [19,20]. Subsequent functional studies have followed. In the commonly used chicken models, AQP1-specific dsRNA oligonucleotides (siRNA) caused a significant reduction in the growth of new blood vessels in the cell adhesion molecule (CAM) [21]. While Hoque et al. have initially reported expression of AQP1 in resected lung cancer samples, Yang et al. [22,23] demonstrated a positive correlation between expression of AQP1 and AQP5 and intratumoral microvessel density in epithelial ovarian tumours and between expression of AQP1 and AQP5 and ascites formation. Pan et al. [24] then have demonstrated a positive correlation between AQP1 and intratumoral microvascular density. Furthermore, a correlation of vascular endothelial growth factor (VEGF) expression with AQP1expression has been suggested in tumour progression of endometrial adenocarcinoma. This report provided a key linkage of AQP1 expression and VEGF, which is an essential component of angiogenesis. Kaneko et al. [25] further provided more detailed functional evidences based on retinal vascular endothelial cells cultured under hypoxic conditions. In this report, the levels of AQP1 mRNA and protein expression were shown to be increased under hypoxia, and inhibition of VEGF signalling did not affect AQP1 expression. Moreover, they have demonstrated that reduced expression of AQP1 and inhibition of VEGF signalling both significantly inhibited vascular tube formation. More recently, Jiang et al. [26] investigated the expression of AQP1 in human HT20 colon cancer and characterized its function in cell migration by using adenovirus-mediated AQP1 transfer. In this report, adenovirus-mediated high expression of AQP1 increased relative plasma membrane water permeability and migration rate in both wound healing and invasive transwell migration assay.

11.2.2 Aquaporin 3

AQP3 is strongly expressed at the plasma membranes of basal epidermal cells in the skin. It was known that human skin squamous cell carcinoma strongly overexpresses AQP3 [27]. A novel role for AQP3 in skin tumorigenesis was discovered using mice with targeted AQP3 gene disruption by Verkman et al. [27,28]. In this report, AQP3-null mice were shown to be remarkably resistant to the development of skin tumours following exposure to a tumour initiator and phorbol ester promoter. Though tumour initiator challenge produced comparable apoptotic responses in wild-type and AQP3-null mice, promoter-induced cell proliferation was greatly impaired in the AQP3-null epidermis. Reductions of epidermal cell glycerol, its metabolite glycerol-3-phosphate, and ATP were observed in AQP3 deficiency without impairment of mitochondrial function. Glycerol supplementation corrected the reduced proliferation and ATP content in AQP3 deficiency, with cellular glycerol, ATP and proliferative ability being closely correlated. This report provides evidence of the role of AQP in the energy production. Specifically, involvement of AQP3-facilitated glycerol transport in epidermal cell proliferation and tumorigenesis is explained by a novel mechanism implicating cellular glycerol as a key determinant of cellular ATP energy. In Figure 11.2,

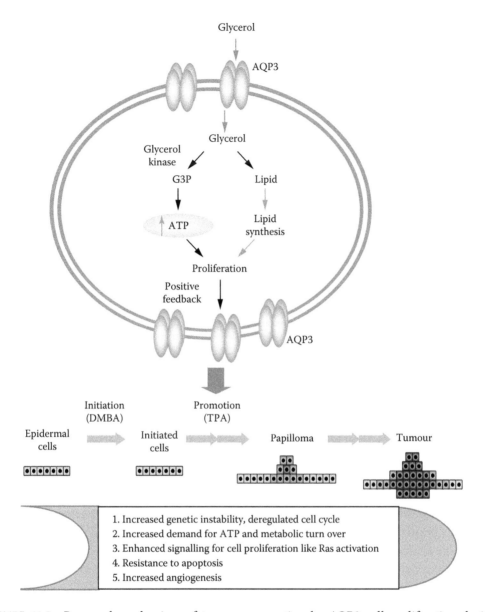

FIGURE 11.2   Proposed mechanism of tumour promotion by AQP3 cell proliferation during skin tumorigenesis. AQP3-mediated glycerol transport–dependent ATP production is proposed to play an important role in the transformation of normal skin epithelial cells and growth of skin cancers. *Abbreviations:* G3P, glycerol 3-phosphate; DMBA, 1,3-dimethylbutylamine; TPA, 12-*O*-tetradecanoylphorbol-13-acetate. (Modified from Hara-Chikuma, M. and Verkman, A.S., *Mol. Cell Biol.,* 98, 326, 2008.)

AQP3-mediated glycerol passage transport–dependent ATP production is proposed to play an important role in the transformation of normal skin epithelial cells and growth of skin cancers. In the case of colon cancer, an interesting study was recently reported. In the report by Li et al., human epidermal growth factor (hEGF) increased the expression of AQP3 and, subsequently, the migration ability of human colorectal carcinoma cells

HCT116 in a dose- and time-dependent manner [29]. The expression intensity of AQP3 was associated with the differentiation, lymph node and distant metastasis of colorectal carcinoma patients. In the case of lung cancer, Liu et al. [30] recently demonstrated that AQP3 is not only expressed in the normal respiratory tract and plays an important role in the maintenance of water homeostasis but, like AQP1, also may play an important role in lung carcinomatosis. In prostate cancer, it seems that inhibition of AQP3 increases the sensitivity of prostate cancer cells to cryotherapy [31], although more extensive samples need to be examined. Finally, a large-scale expression study was reported by Niu et al. [32, Histopathology]. In this study, expression of AQP3 was investigated among 798 neoplastic tissues using immunohistochemistry with anti-AQP3 antibody. Overall, a high positive frequency of AQP3 immunoreactivity in pituitary adenomas; salivary gland tumours; thymic tumours; adenocarcinoma of the lung and prostate; squamous cell carcinomas of the skin, oesophagus, and uterine cervix; apocrine carcinoma of the breast; germinal cell tumours of the ovary and testis; and urothelial carcinoma of the bladder. Of note, none of the sarcomas or central nervous system tumours showed AQP3 immunoreactivity. Most tumours with a high frequency of AQP3 positivity had corresponding or surrounding normal cells that also expressed AQP3.

11.2.3 Aquaporin 4

The study of AQP4 in human pathophysiology was initially based on its role in pure water homeostasis in the brain. AQP4 was found to be expressed predominantly in the brain astrocytes, their foot processes near blood vessels. From the study of AQP4 KO mice, two common brain pathologies, pseudotumor cerebri and cerebral oedema [33], were hypothesized to be involved by AQP4. For example, AQP4-null mice were shown to be resistant to water intoxication, meningitis and brain ischemia [34]. Furthermore, AQP4-null mice have a significantly greater increase in brain water content and intracranial pressure than the wild-type mice, suggesting that brain water elimination is defective after AQP4 deletion [35,36].

As many brain pathophysiology including brain tumour involves rapid water movement and a series of process leading to angiogenesis, a role of AQP4 in vasogenic oedema were examined. Basically, AQP4 can control bidirectional water flux and therefore is responsible for the formation of cellular brain oedema. However, expression of AQP4 counteracts vasogenic oedema [3,8]. Expression of AQP4 was shown to play a protective role by increasing brain water clearance. In contrast, in cytotoxic oedema, it is the main contributor to astrocytic cell swelling [34,35,37]. In fact, water-intoxicated AQP4-null mice showed a significant reduction in astrocytic foot process swelling and a decrease in brain water content [38]. Presently, it is believed that this dual function of AQP4 may play an important role in any processes involving brain swelling. Melanoma has a high propensity to invade the brain and often causes a significant peritumoral oedema. In fact, in mice model, melanoma cells implanted into the striatum of wild-type and AQP4-null mice produce oedema surrounding tumour mass and comparable-sized tumours in both groups after a week. However, the AQP4-null mice have a higher intracerebral pressure and water content [34], which happens in various terminal cancer patients with brain metastasis and patients who undergoes radiation therapy.

The expression of AQP4 was shown to be upregulated in certain form of astrocytoma and metastatic tumours [3,8]. For example, AQP4 expression was shown to be increased in glioblastoma multiforme (GBM). Moreover, often times, this is accompanied by AQP4 redistribution in glioma cells [4,5,39]. In the initial report by Warth et al. based on 189 WHO grade I–IV gliomas [5], a significant expression of AQP4 was demonstrated with its prognostic significances. In gliomas, again, a redistributed expression of AQP4 was observed in comparison with normal central nervous system tissue. In fact, the highest membranous staining levels were seen in pilocytic astrocytoma WHO grade I, which carries better prognosis and grade IV glioblastomas, which carries worst prognosis. Nico et al. [40] have examined the relation of AQP4 expression and its angiogenesis potential in resected GBM samples. In this study, AQP4 expression and VEGF-VEGFR-2 expression is shown to be closely correlated to each other. Moreover, this report has demonstrated that reduced expression of AQP4 parallels with VEGF-VEGFR-2 expression. Of note, in the peripheral areas of relapsed tumours, AQP4 expression assumed normal findings of perivascular expression pattern. While the interpretation of this important observation needs to be examined in larger samples, in GBM, chemotherapy and radiotherapy induce a downregulation of AQP4 expression and can restore its perivascular rearrangement. While the clinical meaning of this observation remains unclear, it was suggested that a polarized rearrangement of AQP4 expression in peritumoral area specimens after combined chemotherapy and radiotherapy can lead into the normalization of tumour blood vessels.

Tumour implantation experiments into AQP4-null mice have further solidified the role of AQP4 in the development of glioma. As expected, various studies demonstrated that AQP4 KO mice have an increased intracranial pressure than wild-type controls [33,35].

McCoy et al. [41] had recently provided an initial mechanistic insight. In this report, using D54MG glioma cells stably transfected with either AQP1 or AQP4, it was clearly demonstrated that protein kinase C (PKC) activity regulates water permeability and that this phenomenon can be modulated by phosphorylation of AQP4. For example, biochemical activation of PKC with two known PKC activators, either phorbol 12-myristate 13-acetate or thrombin, enhanced AQP4 phosphorylation, which further reduced water permeability and significantly decreased tumour cell invasion. Likewise, inhibition of PKC activity with chelerythrine reduced AQP4 phosphorylation and enhanced water permeability and tumour cell invasion. Recently, therapeutic effects of Temodar can be mediated by reduced activation of AQP4. While PKC-mediated, AQP4-dependent oncogenic activity was not considered in terms of Temodar (which is the most commonly used chemotherapeutic agent) response, we surmise that at least some of therapeutic effects of Temodar may be mediated by reduced phosphorylation of 'AQP4', as mediated by PKC.

In terms of meningioma, only one study has reported based on small samples [42]. However, being meningioma as the most common adult brain tumour, it is worthwhile to discuss in some details. Ng et al. [42] have studied 17 human meningioma specimens for the expression of AQP4 and studied peritumoral oedema based on brain MRI. Basically, overexpression of AQP4 was associated with significant peritumoral oedema and immunohistochemistry showed upregulation of AQP4 throughout the specimens. This study further concludes that increased expression of AQP4 is associated with peritumoral oedema in meningioma, which can be a challenging problem in the management of certain forms of meningioma. The inhibition of AQP4 water channels therefore can be a potential therapeutic option to reduce the adverse effects of peritumoral oedema associated with meningioma.

11.2.4 Aquaporin 5

Increased expression of AQP5 was initially reported in colon and pancreatic cancer [7,43,44]. As described earlier, in situ hybridization demonstrated that during colorectal carcinogenesis, the expression of AQP1 and AQP5 was induced in early-stage disease (early dysplasia) and maintained through the late stages of colon cancer development [7,44]. These observations lead us to study molecular mechanisms behind 'AQP5-induced oncogenesis'. In the report by Woo et al., the overexpression of AQP5 in NIH3T3 cells demonstrated a significant activation of Ras pathways and AQP5-mediated activation of Ras was shown to be mediated by phosphorylation of the protein kinase A (PKA) consensus site of AQP5 [4,45]. We believe that this is the first evidence demonstrating an association between AQP5 and Ras signal transduction pathway, which may be the basis of the oncogenic properties in AQP-overexpressing cells.

In fact, expression of AQP5 also seems to be regulated by PKA-mediated phosphorylation of AQP5 [45]. Another study by Woo et al. has demonstrated that ectopic expression of human AQP5 (hAQP5) in BEAS-2 cells induces many phenotypic changes characteristic of transformation both in vitro and in vivo. Furthermore, the cell proliferative ability of AQP5 appears to be dependent upon the phosphorylation of a cyclic adenosine monophosphate (cAMP)-PKA consensus site located in a cytoplasmic loop of AQP5 [46]. Importantly, phosphorylation of the PKA consensus site was found to be phosphorylated preferentially in tumours. Both of these findings indicate that hAQP5 plays an important role in human carcinogenesis through 'PKA-dependent phosphorylation of AQP5', which can activate Ras signalling pathways. Likewise, Kang et al. did more focused study in colorectal carcinogenesis [44]. As is seen in NIH3T3 cell and BEAS-2 cell, overexpression of wild-type hAQP5 increased proliferation and phosphorylation of extracellular signal-regulated kinase-1/2 in HCT116 colon cancer cells. Of note, these phenomena in hAQP5 mutants (N185D and S156A) were diminished, indicating that both membrane association and serine/threonine phosphorylation of AQP5 are required for proper function. Interestingly, while overexpression of AQP1 and AQP3 showed no differences in extracellular signal-regulated kinase-1/2 phosphorylation, suggesting that AQP5, unlike AQP1 or AQP3, can modulate signal transduction. Importantly, hAQP5-overexpressing cells showed an increase in retinoblastoma protein phosphorylation through the formation of a nuclear complex with cyclin D1 and CDK4. These data had provided a unique molecular mechanism for colon cancer development through the interaction of hAQP5 with the Ras/extracellular signal-regulated kinase/ retinoblastoma protein signalling pathway, extending AQP5 expression on the regulation of cell cycles. In the case of lung cancer, Chae et al. [46] have reported that, among more than 400 resected non-small cell lung cancer samples, various degree of AQP5 expression has been observed with significant prognostic implications. In vitro invasion assay using BEAS-2B and NIH3T3 cells stably transfected with various AQP5 expression constructs (wild type and two mutants, N185D or S156A) demonstrated that wild-type AQP5 expression can induce cell invasions. In fact only wild-type AQP5 caused a spindle-like and fibroblastic morphologic change in addition to losses of cell–cell contacts and cell polarity [46]. Of note, most recently, the relationship between expressions of AQP1, AQP3 and AQP5 in various resected lung cancer samples (a total of 170 samples) have been investigated [47]. In this study, AQP1, AQP3 and AQP5 were expressed in tumour cells in 71%, 40% and 56% of lung cancers, respectively. Similar to Chae et al.'s report, AQP expressions were frequent in adenocarcinomas, whereas AQP1 and AQP5 were negative in squamous cell carcinomas. Furthermore, this report has demonstrated that expression patterns of AQP1, AQP3 and AQP5 in lung cancer cells are mostly associated with cellular differentiation. Additionally, the expression of AQP1 and AQP5 was shown to be upregulated in invading lung cancer cells and the overexpression of AQP1 with loss of subcellular polarization was suggested to be involved in their invasive and metastatic potential.

11.2.5 Other Aquaporins

While AQP1, AQP4 and AQP5 have studied rather extensively, the role of other AQPs in the human carcinogenesis has been scarcely investigated. A decreased expression of AQP8 and AQP9 leads to an increased resistance to apoptosis in hepatocellular carcinoma (HCC) [48]. Likewise, in human glioblastoma, most glioma cells throughout the tumour revealed strong anti-AQP9 immunoreactivity across the whole surface of the cell. It is assumed that an upregulation of AQP9 can counteract the glioma-associated lactic acidosis by clearance of glycerol and lactate from the extracellular space.

11.3 AQUAPORINS AS THERAPEUTIC TARGETS AND PROGNOSTIC MARKERS

It has been known for a long time that cell viability and motility are critical for cancer progression. Therefore, a balance of water and monovalent metal cations plays a pivotal role in the dynamics of focal contacts and cytoskeletal rearrangements at the cell's leading edge. Furthermore, fundamental components of cell survival require the optimal concentration of water and solutes. Most of the data so far suggest that AQPs play a crucial role during this process. In the last 10 years, the inhibition of AQP expression and/or AQP-mediated water influx, therefore, has been initially speculated as a way of treating cancer. Various chemicals including acetazolamide, topiramate, cyclophosphamide, thiopental, phenobarbital and propofol have been tested in this setting and have shown some therapeutic efficacy by modulating AQP activities [49]. While different mechanisms are likely to be involved, these chemicals seem to exercise their effects also by downregulation of water transport [50].

Acetazolamide as a sulfonamidic carbonic anhydrase inhibitor is mainly used for edematous diseases such as glaucoma, mountain sickness and congestive heart failure–induced or drug-induced oedema. Xue-jun Li's group previously indicated that acetazolamide inhibited gene expression of AQP1 in rat kidney [51]. In fact, preliminary study has shown that both acetazolamide and cyclophosphamide can suppress tumour metastasis and related protein expression in mice bearing Lewis lung carcinoma. Based on these initial observations, the same group conducted an elegant study to test acetazolamide as a potential 'AQP-modifying anticancer agent'. In this report by Xiang et al. [52], AQP1 in mice tumour was dye labelled in capillaries, post-capillary venules and endothelial cells, and, after treatment with acetazolamide, the number of capillaries and post-capillary venules was significantly decreased in tumour tissue. Furthermore, acetazolamide showed significant inhibitory effect on angiogenesis in CAM and endothelial cell proliferation. Inhibition of AQP1 expression with another sulfonamide-type compound, topiramate, attenuated water influx at the leading edge, thereby affecting membrane protrusion, cell migration and metastasis [53]. Interestingly, both acetazolamide and topiramate can suppress metastasis in patients with Lewis lung carcinoma by inhibiting AQP1 expression in tumoral endothelial cells [52,53]. In the case of AQP3, four recent studies draw our attention. First, matrix metalloproteinase (MMP) expression, which are associated with tumour progression including invasion, migration, angiogenesis and metastasis, was shown to be modulated by AQP3 over-expression or knockdown systems [54]. By using this system, Xu et al. have demonstrated that AQP3 might positively regulate MMP expression through PI3K/AKT signal pathway in human gastric carcinoma SGC7901 cells [54]. As MMP is a crucial component of cancer metastasis, future AQP3 antagonists might inhibit cancer cells with robust AQP3 expression for their ability for metastasis. Second, Trigueros-Motos et al. [55] have proposed that the AQP3 is required for cytotoxic activity of 5′-DFUR and gemcitabine in the breast cancer cell line MCF7 and the colon adenocarcinoma cell line HT29 and is implicated in cell volume increase and cell cycle arrest. This finding is interesting as modulation of AQP3 expression might provide a unique opportunity for improved chemosensitivity. Third, Serna et al. reported that the gold(III) complex Auphen is a very selective and potent inhibitor of AQP3's glycerol permeability (Pgly) and that Auphen can cause inhibitory effect on proliferation of various cells expressing AQP3 [56]. This study is exciting, as it has demonstrated that a targeted therapeutic effect on carcinomas with large AQP3 expression may be possible by compromising AQP3 function. Fourth, Xia et al. [57] have demonstrated that using a subcutaneous xenograft mouse model of non small cell lung cancer (NSCLC), AQP3 knockdown by 'short-hairpin RNA targeting AQP3' (AQP3-sh RNA) can inhibit tumour growth and prolonged survival of mice with tumours. Of note, AQP3 knockdown reduced cellular glycerol content and suppressed mitochondrial ATP formation. Thus AQP3–shRNA was suggested to be one of the novel therapeutic strategies for NSCLC. In the last 10 years, the development of novel therapeutics has largely focused on the inhibition of tyrosine kinase pathways including EGFR and BCR-ABL pathways [58]. Importantly, we have demonstrated AQP5 can be a novel therapeutic target by blocking downstream of certain tyrosine kinase pathways. For example, Chae et al. have reported that in a human SH3-domains protein array, cellular extracts from BEAS-2B with AQP5 showed a robust binding activity to SH3-domains of the c-Src, Lyn, and Grap2 C-terminal [46]. This study is important as it not only provided evidence of a crucial role of AQP5 in the metastasis of lung cancer but also provided mechanistic insight for 'AQP5-induced epithelial to mesenchymal transformations'. Likewise, Chae et al. [59] have demonstrated that AQP5 is not only overexpressed in some of the chronic myelogenous leukemia (CML) cells but also plays an important role in promoting leukemic cell proliferation and inhibiting apoptosis. Furthermore, AQP5 expression levels increased with the emergence of imatinib mesylate treatment resistance. These observations carry important meanings for the role of AQP5 in the development of CML but also provide AQP5 as a unique therapeutic target not based on BCR-ABL pathway, as there are urgent needs for CML patients with imatinib mesylate resistances [60].

While various strategies can be employed in the modulation of AQP5 channel activity or their downstream pathways, it is becoming increasingly clear that AQP family proteins will be a novel therapeutic target as we have discussed earlier. As the trend of novel therapeutics largely depends on novel targets, we expect an early stage of clinical development for AQP targeting agent in the very near future. Likewise, while the different putative roles of various AQPs and aquaglyceroporins have been investigated in the development of various human cancer, therapeutic strategies employed in each AQP can be diverse. Thus, on one hand, the role of water transport itself can be exploited in most conventional AQPs. In the case of ATP production from glycerol, intervention of downstream pathways of AQP3 can be exploited (Figure 11.2). In the case of AQP5, various kinase pathways activated by AQP5 can be suppressed by blocking downstream of AQP5. In this respect, we suggest the advantages of targeting AQP5 and possibly targeting aquaglyceroporins such as AQP3.

Since the report by Moon et al., which has suggested the role of AQPs in the progression and metastasis of colon cancer [7], various reports did suggest the role of AQPs as prognostic markers, some of which are mentioned previously in each sections of AQPs. For example, Sekine et al. [61] found that the survival of biliary tract carcinoma patients with high AQP5 expression was longer compared to that of patients with low AQP5 expression and that there was correlation between AQP5 expression and tumour size. Zhang et al. [62] showed that the high AQP5 protein expression in intestinal type of adenocarcinoma was significantly associated with lymph node metastasis and lymphovascular invasion in patients. Watanabe et al. [63] also found that upregulation of AQP5 might be involved in differentiation of human gastric cancer cells. Yang et al. [64] have demonstrated that AQP5 expression in ovarian malignant and borderline tumours was significantly higher than that of benign ovarian tumours and normal ovarian tissue and that the increased AQP5 protein level was associated with lymph node metastasis and ascites. Li et al. [65] reported that AQP3 overexpression could facilitate colorectal carcinoma cell migration and that AQP3 may be considered a potential indicator and therapeutic target for colon tumour metastasis and prognosis. Otto et al. [66] indicated that loss of AQP3 protein expression in pT1 bladder cancer may play a key role in disease progression and is associated with worse progression-free survival. Most recently, Guo et al. studied the expression of both AQP3 and AQP5 in the resected samples from 170 cases of HCC patients [67] and demonstrated that the overexpression of both AQP3 and AQP5 was associated with advanced tumour stage, positive distant metastasis and unfavourable prognosis. Overall, these results indicated that AQP3 and AQP5 are involved in the development of HCC and that the two proteins function as tumour promoter or tumour suppressor in different tumour types. Fifteen years ago, Hanahan and Weinberg reported six functional capabilities of cancers, described as hallmarks of cancer [68]. These six hallmarks were self-sufficiency in growth signals, insensitivity to anti-growth signals, evasion of apoptosis, sustained angiogenesis, tissue invasion and metastasis and unlimited replicative potential. Since the initial description, the models have been revised to provide a better understanding of the molecular basis for these hallmarks. An additional concept was also introduced that described how mutations leading to the hallmarks did not have to be acquired in any specific order, while a detailed molecular pathway has been outlined. It is interesting to note that in the original six hallmarks of cancer model, genomic instability was considered separately in that it is not a functional capability of cancer per se, but a property that enables the acquisition of the hallmarks. In other words, genomic instability is a property driving tumorigenesis. In the case of AQPs, there are multiple published data, as extensively discussed previously, which suggest that AQPs do play an important role in the described six hallmarks, rendering normal cells to be tumour cells and changing behaviours of tumour cells. Moreover, we propose that AQP expression can change the behaviours of normal cells, at least by promoting genetic instability which can lead into unlimited growth. Further mechanistic understanding about the role of AQP in these six hallmarks and genetic instabilities of cancer cells will likely to facilitate clinical application of AQPs in the management of various tumours. For example, in the near future, we expect to see the 'clinical validation study' for the role of novel AQP antagonist in the management of various human cancers [58,60,68].

REFERENCES

1. L.S. King, P. Agre, Pathophysiology of the aquaporin water channels, *Annu. Rev. Physiol.* 58 (1996) 619–648.
2. A.S. Verkman, More than just water channels: Unexpected cellular roles of aquaporins, *J. Cell Sci.* 118 (2005) 3225–3232.
3. S. Saadoun, M.C. Papadopoulos, D.C. Davies, S. Krishna, B.A. Bell, Aquaporin-4 expression is increased in oedematous human brain tumours, *J. Neurol. Neurosurg. Pscychiatry* 72 (2002) 262–265.
4. J. Woo, J. Lee, S.L. Jang, D. Sidransky, C. Moon, The effect of aquaporin 5 overexpression on the Ras signaling pathway. *Biochem. Biophys. Res. Commun.* 367 (2008) 291–298.
5. A. Warth et al., Expression pattern of the water channel aquaporin-4 in human gliomas is associated with blood-brain barrier disturbance but not with patient survival, *J. Neurosci. Res.* 85 (2007) 1336–1345.
6. M. Endo, R.K. Jain, B. Witwer, D. Brown, Water channel (aquaporin 1) expression and distribution in mammary carcinomas and glioblastomas, *Microvasc. Res.* 58 (1999) 89–98.
7. C. Moon, J.C. Soria, S.J. Jang, J. Lee, M.O. Hoque, M. Sibonu, B. Trink, Y.S. Chang, D. Sidransky, L. Mao, Involvement of aquaporins in colorectal carcinogenesis, *Oncogene* 22 (2003) 6699–8703.
8. S. Saadoun, M.C. Papadopoulos, D.C. Davies, B.A. Bell, S. Krishna, Increased aquaporin 1 water channel expression in human brain tumours, *Br. J. Cancer* 87 (2002) 621–623.
9. Y. Chen, O. Tachibana, M. Oda, R. Xu, J. Hamada, J. Yamashita, N. Hashimoto, J.A. Takahashi, Increased expression of aquaporin 1 in human hemangioblastomas and its correlation with cyst formation, *J. Neurooncol.* 80 (2006) 219–225.
10. A. Vacca, A. Frigeri, D. Ribatti, G.P. Nicchia, B. Nico, R. Ria, M. Svelto, F. Dammacco, Microvessel overexpression of aquaporin-1 parallels bone marrow angiogenesis in patients with active multiple myeloma, *Br. J. Haematol.* 113 (2001) 415–421.

11. F. Otterbach, R. Calliers, R. Kimmig, K.W. Schmid, A. Bankfalvi, Aquaporin-1 expression in invasive breast carcinoma, *Pathologie* 29(Suppl. 2) (2008) 357–362.
12. M.O. Hoque et al., Aquaporin 1 is overexpressed in lung cancer and stimulates NIH-3T3 cell proliferation and anchorage-independent growth, *Am. J. Pathol.* 168 (2001) 1345–1353.
13. C.M. Shanahan, D.L. Commolly, K.L. Tyson, N.R.B. Carry, J.K. Osborn, P. Agree, P.L. Weissberg, Aquaporin-1 is expressed by vascular smooth muscle cells and mediates rapid water transport across vascular cell membranes, *J. Vasc. Res.* 36 (1999) 353–362.
14. S. Nilesen, B.L. Smith, E.I. Christensen, M.A. Knepper, P. Agree, CHIP 28 water channels are localized in constitutively water-permeable segments of the nephron, *J. Cell Biol.* 120 (1993) 371–383.
15. T.L. Pallone, B.K. Kishore, S. Nielsen, P. Agree, M.A. Knepper, Evidence that aquaporin-1 mediates NaCl-induced water flux across descending vasa recta, *Am. J. Physiol.* 36 (1994) R260–R267.
16. T.L. Pallone, A. Edwards, E.P. Silldorrff, A.S. Vermann, Requirement of aquaporin-1 for NaCl-driven water transport across descending vasa recta, *J. Clin. Invest.* 105 (2000) 215–222.
17. O. Devust, S. Nilesen, J.P. Cosyns, B.L. Smith, P. Agree, J.P. Squifflet, D. Pouhtier, E. Goffin, Aquaporin-1 and endothelial nitric oxide synthase expression in capillary endothelial of human peritoneum, *Am. J. Physiol.* 275 (1998) H234–H242.
18. D. Ribatti, A. Frigeri, B. Nico, G.P. Nicchia, M. De Giorgis, L. Roncali, M. Svelto, Aquaporin-1 expression in the chick embryo chorioallantoic membrane, *Anat. Rec.* 268 (2002) 85–89.
19. S. Saadoun, M.C. Papadoulos, M. Hara-Chikuma, S. Verkman, Impairment of angiogenesis and cell migration by targeted aquaporin-1 gene disruption, *Nature* 434 (2005) 786–792.
20. J. Hu, S. Verkman, Increased migration and metastatic potential of tumour cells expressing aquaporin water channels, *FASEB J.* 20 (2006) 1892–1894.
21. G.M. Camerino, G.P. Nicchia, M.M. Dinardo, D. Ribatti, M. Svelto, A. Frigeri, In vivo silencing of aquaporin-1 by RNA interference inhibits angiogenesis in the chick embryo chorioallantoic membrane assay, *Cell Mol. Biol.* 52 (2006) 51–56.
22. J.H. Yang, Y.F. Shi, X.D. Chen, W.J. Qi, The influence of aquaporin-1 and microvessel density on ovarian carcinogenesis and ascites formation, *Int. J. Gynecol. Cancer* 16 (Suppl. 1) (2006) 400–405.
23. J.H. Yang, Y.F. Shi, Q. Cheng, L. Deng, Expression and localization of aquaporin-5 in the epithelial ovarian tumours, *Gynecol. Oncol.* 100 (2006) 294–299.
24. H. Pan, C.C. Sun, C.Y. Zhou, H.F. Huang, Expression of aquaporin-1 in normal, hyperplastic and carcinomatous endometria, *Int. J. Gynaecol. Obstet.* 101 (2008) 239–244.
25. K. Kaneko, K. Yagui, A. Tanaka, K. Yoshihara, K. Ishikawa, K. Takahashi, H. Bujo, K. Sakurai, Y. Saito, Aquaporin 1 is required for hypoxia-inducible angiogenesis in human retinal vascular endothelial cells, *Microvasc. Res.* 75 (2008) 297–301.
26. Y. Jiang, Aquaporin-1 activity of plasma membrane affects HT20 colon cancer migration, *IUBMB Life* 61 (2009) 1001–1009.
27. M. Hara-Chikuma, A.S. Verkman, Prevention of skin tumorigenesis and impairment of epidermal cell proliferation by targeted aquaporin 3 gene disruption, *Mol. Cell Biol.* 98 (2008) 326–332.
28. M. Hara-Chikuma, A.S. Verkman, Roles of aquaporin-3 in the epidermis, *J. Invest. Dermatol.* 128 (2008) 2145–2151.
29. A. Li, D. Lu, Y. Zhang, J. Li, Y. Fang, F. Li, J. Sun, Critical role of aquaporin-3 in epidermal growth factor-induced migration of colorectal carcinoma cells and its clinical significance, *Oncol. Rep.* 29.2 (2013) 535–540.
30. Y.L. Liu et al., Expression of aquaporin-3 (AQP3) in normal and neoplastic lung tissues, *Hum. Pathol.* 38 (2007) 171–178.
31. M. Ismail, S. Bokaee, R. Morgan, J. Davies, K.J. Harrington, H. Pandaia, Inhibition of the aquaporin 3 water channel increases the sensitivity of prostate cancer cells to cryotherapy, *Br. J. Cancer* 100 (2009) 1889–1895.

32. D. Niu, T. Kondo, T. Nakazawa, T. Yamane, K. Mochizuki, T. Kawasaki, T. Matsuzaki, K. Takata, R. Katoh, Expression of aquaporin3 in human neoplastic tissues. *Histopathology* 61 (2012) 543–551.

33. J. Badaut, F. Lasbennes, P.J. Magistretti, L. Regli, Aquaporins in brain: Distribution, physiology and pathophysiology, *J. Ceram. Blood Flow Metab.* 22 (2002) 367–378.

34. G.T. Manley, D.K. Binder, M.C. Papafopoulos, A.S. Verkman, New insights into water transport and oedema in the central nervous system from phenotype analysis of aquaporin-4 null mice, *Neuroscience* 129 (2004) 983–991.

35. M.C. Papadopoulos, G.T. Manley, S. Krishna, A.S. Verkman, Aquaporin-4 facilitates reabsorption of excess fluid in vasogenic brain oedema, *FASEB J.* 18 (2004) 1291–1293.

36. M.C. Papadopoulos, A.S. Verkman, Aquaporin-4 and brain oedema, *Pediatr. Nephrol.* 22 (2007) 778–784.

37. M.C. Papadopoulos, S. Saadoun, D.K. Binder, G.T. Manley, S. Krishna, A.S. Verkman, Molecular mechanism of brain tumour oedema, *Neuroscience* 129 (2004) 1011–1020.

38. G.T. Manley, M. Fujimura, T. Ma, N. Noshita, F. Filiz, A.W. Bollen, P. Cahn, A.S. Verkman, Aquaporin-4 deletion in mice reduces brain oedema after acute water intoxication and ischemic stroke, *Nat. Med.* 6 (2000) 159–163.

39. A. Warth, M. Mittelbronn, H. Wolburg, Redistribution of the water channel protein aquaporin-4 and the K^+ channel protein Kir 4.1 differs in low- and high-grade human brain tumours, *Acta Neuropathol.* 109 (2005) 418–426.

40. B. Nico, D. Mangieri, R. Tamma, V. Longo, T. Annese, E. Crivellato, B. Pollo, E. Maderna, D. Ribatti, A. Salmaggi, Aquaporin-4 contributes to the resolution of peritumoral brain oedema in human glioblastoma multiforme after combined chemotherapy and radiotherapy, *Eur. J. Cancer* 45 (2009) 3315–3325.

41. E.S. McCoy, B.R. Haas, H. Sontheimer, Water permeability through aquaporin-4 is regulated by protein kinase C and becomes rate limiting for glioma invasion, *Neuroscience* 68(4) (July 2010) 971–981.

42. W.H. Ng, J.W. Hy, W.L. Tan, D. Liev, T. Lim, B.T. Ang, V. Ng, Aquaporin-4 expression is increased in edematous meningiomas, *J. Clin. Neurosci.* 16 (2009) 441–443.

43. B. Burghardt, M.L. Elkaer, T.H. Kwon, G.Z. Racz, G. Varga, M.C. Steward, S. Nielsen, Distribution of aquaporin water channels AQP1 and AQP5 in the ductal system of the human pancreas, *Gut* 52 (2003) 1008–1016.

44. S.K. Kang, Y.K. Chae, J. Woo, M.S. Kim, J.C. Park, J. Lee, J.C. Soria, S.J. Jang, D. Sidransky, C. Moon, Role of human aquaporin-5 in colorectal carcinogenesis, *Am. J. Pathol.* 173 (2008) 518–525.

45. J. Woo et al., Overexpression of AQP5, a putative oncogene, promotes cell growth and transformation, *Cancer Lett.* 264 (2008) 54–62.

46. Y.K. Chae et al., Expression of aquaporin 5 (AQP5) promotes tumour invasion in human non small cell lung cancer, *PLoS One* 3 (2008) e2162.

47. M. Yuichiro et al. Relationship of aquaporin 1, 3, and 5 expression in lung cancer cells to cellular differentiation, invasive growth, and metastasis potential, *Hum. Pathol.* 42(5) (2011), 669–678.

48. E.M. Jablonski, A.M. Mattocks, E. Sokolov, L.G. Koniaris, F.M. Hughes Jr., N. Fausto, R.H. Pierce, I.H. McKillop, Decreased aquaporin expression leads to increased resistance to apoptosis in hepatocellular carcinoma, *Cancer Lett.* 250 (2007) 36–46.

49. E. Monzani, A.A. Shtil, C.A. La Porta, The water channels, new druggable targets to combat cancer cell survival, invasiveness and metastasis, *Curr. Drug Targets* 8 (2007) 1132–1137.

50. S.F. Pedersen, E.K. Hoffmann, J.W. Millis, The cytoskeleton and cell volume regulation, *Comp. Biochem. Physiol.* 130 (2001) 385–399.

51. B. Ma, Y. Xiang, T. Li, H.M. Yu, X.J. Li, Inhibitory effect of topiramate on Lewis lung carcinoma metastasis and its relation with AQP1 water channel, *Acta Pharmacol. Sin.* 25 (2004) 54–60.

52. Y. Xiang, B. Ma, T. Li, Acetazolamide suppresses tumour metastasis and related protein expression in mice bearing Lewis lung carcinoma, *Acta Pharmacol. Sin.* 23 (2002) 751–754.
53. B. Ma, Y. Xinag, T. Li, Inhibitory effect of topiramate on Lewis lung carcinoma metastasis and its relation with AQP1 water channel, *Acta Pharmacol. Sin.* 25 (2004) 54–60.
54. H. Xu, Y. Xu, W. Zhang, L. Shen, L. Yang, Z. Xu, Aquaporin-3 positively regulates matrix metalloproteinases via PI3K/AKT signal pathway in human gastric carcinoma SGC7901 cells, *J. Exp. Clin. Cancer Res.* 30 (2011) 86.
55. L. Trigueros-Motos, S. Pérez-Torras, F. Javier Casado, M. Molina-Arcas, M. Pastor-Angladal, Aquaporin 3 participation in the cytotoxic response to nucleoside-derived drugs, *BMC Cancer* 12 (2012) 434.
56. A. Serna, A. Galán-Cobo, C. Rodrigues, I. Sánchez-Gomar, J.J. Toledo-Aral, T.F. Moura, A. Casini, G. Soveral, M. Echevarría, Functional inhibition of aquaporin-3 with a gold-based compound induces blockage of cell proliferation, *J. Cell Physiol.* 229(11) (November 2014) 1787–1801.
57. H. Xia, Y.-F. Ma, C.-H. Yu, Y.-J. Li, J. Tang, J.-B. Li, Y.-N. Zhao, Y. Liu, Aquaporin 3 knockdown suppresses tumour growth and angiogenesis in experimental non-small cell lung cancer, *Exp. Physiol.* 99 (2014) 974–984.
58. C. Moon, J. Lee, Targeting epidermal growth factor receptor in head and neck cancer: Lessons learned from cetuximab, *Exp. Biol. Med. (Maywood)* 235 (2010) 907–920.
59. Y.K. Chae et al., Human AQP5 plays a role in the progression of chronic myelogenous leukemia (CML), *PLoS One* 3 (2008) e2594.
60. H. Herrmann, I. Sadovnik, S. Cerny-Reiterer, et al. Dipeptidylpeptidase IV (CD26) defines leukemic stem cells (LSC) in chronic myeloid leukemia. *Blood.* 123(25) (2014) 3951–3962.
61. S. Sekine et al., Prognostic significance of aquaporins in human biliary tract carcinoma, *Oncol. Rep.* 27 (2012) 1741–1747.
62. Z.-Q. Zhang, Z.-X. Zhu, C.-X. Bai, Z.-H. Chen, Aquaporin 5 expression increases mucin production in lung adenocarcinoma, *Oncol. Rep.* 25(6) (2011) 1645–1650.
63. T. Watanabe et al., Involvement of aquaporin-5 in differentiation of human gastric cancer cells, *J. Physiol. Sci.*, 59(2) (2009) 113–122.
64. J.-H. Yang, Y.-F. Shi, Q. Cheng, L. Deng, Expression and localization of aquaporin-5 in the epithelial ovarian tumours, *Gynecol. Oncol.* 100(2) (2006), 294–299.
65. J. Li, H. Tang, X. Hu, M. Chen, and H. Xie. Aquaporin-3 gene and protein expression in sun-protected human skin decreases with skin ageing. The Australasian journal of dermatology. *Australas J Dermatol.* 51(2) (May 2010) 106–112.
66. W. Otto et al., Loss of aquaporin 3 protein expression constitutes an independent prognostic factor for progression-free survival: An immunohistochemical study on stage pT1 urothelial bladder cancer, *BMC Cancer,* 12 (2012) article 459.
67. X. Guo, T. Sun, M. Yang, Z. Li, Z. Li, Y. Gao, Prognostic value of combined aquaporin 3 and aquaporin 5 overexpression in hepatocellular carcinoma, *Biomed. Res. Int.* (2013) 206525.
68. D. Hanahan, R.A. Weinberg, The hallmarks of cancer, *Cell* 100(1) (2000) 57–70.

CHAPTER 12

Attacking Aquaporin Water and Solute Channels of Human-Pathogenic Parasites

New Routes for Treatment?

Julia von Bülow and Eric Beitz

CONTENTS

Abstract	233
12.1 Features of the Parasite AQP Protein Structures	234
12.2 Confirmed and Putative Roles of AQPs in the Physiology of Parasites	237
12.2.1 Malaria Parasites, *Plasmodium* spp. AQPs, PfAQP	237
12.2.2 Toxoplasmosis parasites, *Toxoplasma gondii* AQP, TgAQP	238
12.2.3 Sleeping Sickness Parasites, *Trypanosoma brucei* AQPs	238
12.2.4 Chagas Disease Parasites, *Trypanosoma cruzi* AQPs, TcAQPs	239
12.2.5 Leishmaniasis Parasites, *Leishmania* spp. AQPs, LmAQP1	239
12.3 Using AQPs to Shuttle Cytotoxic Compounds into Parasites	239
12.4 Challenges and Potential of AQP Inhibitors for Anti-Infectious Therapy	241
References	243

ABSTRACT

Human-infecting parasite populations causing fatal diseases, such as sleeping sickness or malaria, undergo cycles of retreat and spreading. Newly gained drug resistance is one key initiator for peaks in disease incidents. Timely and costly efforts required for the development of anti-parasitic drugs are at odds with their expected time frame of usability. Looking at the sites of action of the current anti-parasitic drugs reveals a common theme: virtually all act on intracellular, mainly metabolic processes. As a consequence, this opens up multiple options for the parasite to defend itself against the treatment, that is by chemical modification of the compound, trapping it in intracellular compartments or expulsion from the cell. An outside attack to the parasites' periphery would increase the chance of sustained usability of respective drugs. Transporters and channels that govern the uptake of vital nutrients and release of metabolic waste represent promising candidates as novel drug targets against parasitic diseases. In this regard, aquaporin (AQP) channels for water and small, uncharged solutes, such as glycerol, urea, or ammonia, are being studied. Biochemical and physiological data hint at several crucial functions of parasite AQPs: (1) alleviation of osmotic stress, for example during passage of the salt-laden kidneys or during transmission between the insect and the human host; (2) uptake of glycerol from the host blood serum as a precursor for glycerolipid synthesis enabling rapid parasite growth by extension of the lipid plasma membrane; (3) release of nitrogen waste, that is urea and ammonia, and of aldehyde metabolites, that is methylglyoxal, preventing self-intoxication of the parasite; (4) new data that suggest that cell motility of amoeba may depend on AQP water permeability and (5) drug resistance mechanisms which have been directly linked to AQP in *Leishmania* and *Trypanosoma*. This chapter gives an overview on the research field and provides an outlook at potential therapeutic exploitation of parasite AQPs.

12.1 FEATURES OF THE PARASITE AQP PROTEIN STRUCTURES

As the general protein fold of the aquaporin (AQP) family has been discussed in depth in other chapters of this book, the description here will focus on structural commonalities and peculiarities of the AQP isoforms expressed by human-pathogenic parasites. It is quite striking that all characterized plasma membrane–residing parasite AQPs exhibit biochemical properties of the aquaglyceroporin subfamily and a broad permeability profile (reviewed in Song et al. 2014). This includes, besides water, relevant metabolites, such as glycerol, urea, ammonia and various carbonyl compounds (Figure 12.1). AQPs that are localized in intracellular structures, so-called contractile vacuoles, as part of the osmoregulation system, among others, of *Trypanosoma cruzi* (Montalvetti et al. 2004), non-pathogenic *Dictyostelium discoideum* (von Bülow et al. 2012) and *Amoeba proteus* (Nishihara et al. 2008) appear to be water specific. It seems fair to conclude that establishing the permeability profile of parasite AQPs provides insight into physiological functions. This topic will be addressed right after delineating the structural aspects of parasite AQPs.

The best characterized parasite AQP is the aquaglyceroporin from the malaria parasite *Plasmodium falciparum* (PfAQP; Hansen et al. 2002), so far representing the only AQP in the field whose crystal structure has been elucidated (Newby et al. 2008). Despite high structural similarity with the prototypical aquaglyceroporin from *Escherichia coli* (GlpF; Fu et al. 2000), PfAQP is bi-functional and exhibits high rates of glycerol and water conductance (Hansen et al. 2002, Beitz et al. 2004, 2009, Newby et al. 2008), which prompted a series of in vivo, in vitro and in silico experiments.

The composition of the pore-lining residues and the critical AQP filter regions, that is the aromatic/arginine selectivity filter (ar/R; Beitz et al. 2006) and the Asp-Pro-Ala (NPA; Wu et al. 2009) region, are almost identical in PfAQP and GlpF (Hansen et al. 2002). The ar/R constrictions are equally composed of Arg, Trp and Phe; the NPA motifs of PfAQP deviate only in the functionally irrelevant, positions with Asn-Leu-Ala and Asn-Pro-Ser (Hansen et al. 2002). The initially surprising differences in GlpF and PfAQP water permeability were resolved by determining the contribution of a structure that has not been linked to

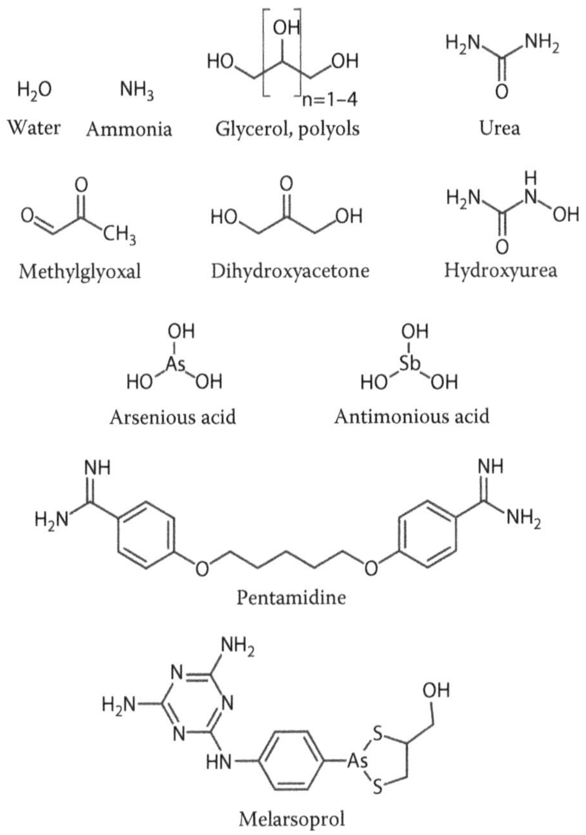

FIGURE 12.1 Structures of compounds that have been shown to directly pass aquaglyceroporins and of drugs that are related to parasite resistance when they lack certain aquaporins.

permeability properties before, that is the extracellular connecting loop C (Beitz et al. 2004). A triad of loop C residues (Trp124, Glu125, Thr126 in PfAQP vs. Phe135, Ser136, Thr137 in GlpF) is located close to the ar/R region (Figure 12.2). Mutation in PfAQP of the central Glu125 to Ser, that is the corresponding residue of GlpF, hardly affected glycerol permeability yet reduced water permeability almost to background (Beitz et al. 2004). Simultaneously, the activation energy increased by about 3 kcal mol^{-1} hinting at a more avid binding of passing water molecules via hydrogen bonds. The PfAQP crystal structure later revealed that the hydrogen bond network around the pore arginine differs from that of *E. coli* GlpF (Figure 12.2). In PfAQP, all hydrogen bond donor sites of Arg196 are saturated by neighbouring residues with a major contribution of Trp134 of loop C and glycerol. In GlpF, however, the backbone carbonyl of Phe135 of loop C is shifted to the left (arrow head in Figure 12.2) allowing for the formation of only one hydrogen bond to Arg206 and leaving it with a free hydrogen (frame in Figure 12.2). It is thought that this extra valence enhances binding of water molecules to the ar/R filter resulting in low permeability rates (Newby et al. 2008).

Further, the structural differences in the loop C of PfAQP and GlpF directly affect the shape and affinity for polyol solutes of the respective vestibules that funnel permeants to

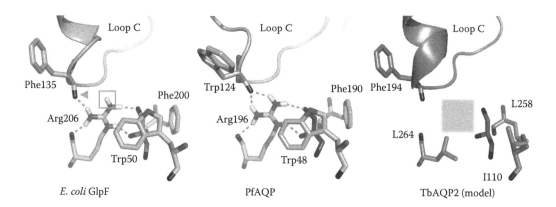

FIGURE 12.2 **(See colour insert.)** Details of the protein structures of the aquaglyceroporins from *Escherichia coli*, GlpF (protein data bank PDB #1FX8), *Plasmodium falciparum*, PfAQP (PDB #3C02) and *Trypanosoma brucei*, TbAQP2 (modelled on the basis of GlpF). Shown are the aromatic/arginine selectivity filter regions, typically comprising an arginine residue in an aromatic environment, holding a glycerol molecule, and the connecting loop C, which makes contact to the pore arginine via hydrogen bonds (dashed lines). Full saturation of the arginine hydrogen donor sites as in PfAQP or absence of such sites as in TbAQP2 (square) is thought to enhance water permeability.

the ar/R constriction (Song et al. 2012). Contrary to GlpF, water permeability of PfAQP strongly depends on the nature of the external osmolyte with salt and saccharose yielding high PfAQP water permeability (Hansen et al. 2002, Beitz et al. 2004, Newby et al. 2008, Song et al. 2012), whereas highly concentrated sorbitol inhibits PfAQP water permeability due to binding and clogging the permeation path in the vestibule area (Hedfalk et al. 2008, Song et al. 2012). This effect is noteworthy with regard to evaluating the possibility of developing PfAQP inhibitors as potential drugs.

With regard to therapeutic exploitation, parasite proteins must be seen in the context of homologous human proteins. The aim is certainly to generate potent drug molecules that bind the parasite protein but leave the human version untouched in order to minimize unwanted side effects. The human aquaglyceroporins AQP3, AQP7, AQP9 and AQP10 fulfill physiological functions, among others, in skin moistening, fat metabolism and transport in the liver and the blood–brain barrier; for a recent and comprehensive review, see Verkman et al. (2014). AQP3 is further localized in the erythrocyte membrane where it enables the malaria parasite in conjunction with PfAQP to take up glycerol from the host blood. Unfortunately, protein structures of AQP3, AQP7, AQP9 and AQP10 are not available. Sequence wise, the human aquaglyceroporins are very close to *E. coli* GlpF; actually they are even closer to GlpF than PfAQP because they carry a connecting loop C that resembles that from GlpF in terms of length and amino acid composition (Beitz et al. 2004). The peculiar loop C in PfAQP may help in the design of specific inhibitors. For drug selectivity, even single amino acids can make a difference. AQP3, for instance, carries a cysteine residue, Cys40, just above the ar/R region, which can act as an anchor for organo-gold (III) complexes (Martins et al. 2013). This cysteine is absent in PfAQP (Hansen et al. 2002).

Biophysical permeability assays are typically carried out in an as simple setup as possible meaning that a single type of AQP is tested in a defined buffer containing a single permeant. In vivo, however, AQPs are exposed to mixtures of solutes. Simulating such a situation indeed showed unexpected effects when glycerol and urea permeability was assayed simultaneously. Under isotonic conditions, any ratio of glycerol and urea passed PfAQP at equal rates. Yet, in countercurrent with water generated by a hypertonic buffer, glycerol was clearly preferred over urea (Song et al. 2012). The preference of PfAQP for glycerol was also seen in molecular dynamics simulations and energy calculations (Aponte-Santamaria et al. 2010). Hence, the data are suggestive of PfAQP being optimized for glycerol facilitation in vivo. In support of this, deletion of the PfAQP-homologuous gene in the rodent malaria parasite *Plasmodium berghei* produced a strain with impaired glycerol import. Its growth rate and virulence was reduced by half probably due to lack of access to glycerol from the host blood and consequently diminished capability of glycerolipid synthesis (Promeneur et al. 2007).

A mutational analysis of the aquaglyceroporin (LmAQP1) from the causative agent of leishmaniasis, *Leishmania major*, underscored the role of loop C in AQP selectivity (Mukhopadhyay et al. 2011). As most other aquaglyceroporins, LmAQP1 conducts trivalent arsenious acid, $As(OH)_3$ and antimonious acid, $Sb(OH)_3$, due to some similarity to glycerol (Figure 12.1). The triad of LmAQP1 loop C residues that are located close to the ar/R constriction comprises Phe162, Ala163 and Thr164. Replacement of Ala163 by a larger glutamate (to mimic the situation of PfAQP), aspartate or glutamine abolished glycerol permeability, whereas with threonine or serine passage of glycerol was unaffected. Yet conductance for arsenious acid and antimonious acid were reduced. Permeability rates of the latter two compounds inversely corresponded to the volume of the residue side chains (Ala163Thr < Ala163Ser) and permeant ($Sb(OH)_3$ < $As(OH)_3$) indicating a selectivity mechanism by size.

Structural or mutational data of other parasite AQPs are not yet available. Sequence comparisons show that AQPs from *Toxoplasma gondii* (TgAQP), *Trypanosoma brucei* (TbAQP2), *T. cruzi* (TcAQPα-δ) and *L. major* (LmAQPα-δ) lack the arginine in the ar/R selectivity filter but carry a neutral, aliphatic valine or leucine instead (Figure 12.2; Beitz 2005). These AQPs fully lack hydrogen bond donor sites in the ar/R region (square in Figure 12.2). In line with the positive effect of the absence of a single valence in the hydrogen bond network around the pore arginine (Beitz et al. 2004), TgAQP and TbAQP2 exhibit considerable water permeability besides a capability of conducting solutes, such as glycerol and urea (Pavlovic-Djuranovic et al. 2003, Uzcategui et al. 2004). TcAQP has been reported to be water specific, yet permeability was rather low in the employed expression system (Montalvetti et al. 2004). Permeability and selectivity profiles of TcAQPβ-δ and LmAQPα-δ are not reported.

12.2 CONFIRMED AND PUTATIVE ROLES OF AQPs IN THE PHYSIOLOGY OF PARASITES

12.2.1 Malaria Parasites, *Plasmodium* spp. AQPs, PfAQP

The genomes of the various *Plasmodium* malaria parasite species encode a single aquaglyceroporin, that is PfAQP in *P. falciparum* (Gardner et al. 2002, Hansen et al. 2002). PfAQP is constitutively expressed during all stages of the parasite's asexual developmental cycle in the human host as well as in the sexual gametocyte form and is localized to the plasma membrane (Hansen et al. 2002). PfAQP is bi-functional facilitating water and physiological solutes, such as sugar, alcohols, urea, methylglyoxal and ammonia (Figure 12.1) putting PfAQP into various metabolic cellular processes (Hansen et al. 2002, Pavlovic-Djuranovic et al. 2006, Zeuthen et al. 2006). A major function of PfAQP is most likely in the uptake of glycerol from the host blood as a precursor for glycerolipid synthesis for cell membrane biogenesis, which enables plasmodia to proliferate rapidly (Promeneur et al. 2007). Rapid growth is accompanied by high rates of glucose consumption for energy generation, albeit rather inefficiently, because plasmodia produce ATP anaerobically from glycolysis alone yielding only 2 mol of ATP from 1 mol of glucose (Beitz 2007). A non-enzymatic, toxic side product of glycolysis is the chemically highly reactive methylglyoxal (Figure 12.1). PfAQP exhibits high permeability for methylglyoxal and may thus serve as an exit pathway maintaining health of the parasite cell (Pavlovic-Djuranovic et al. 2006). Other permeants of PfAQP originate from nitrogen metabolism, that is urea from arginine degradation via ornithine (Hansen et al. 2002) and ammonia from the conversion of amino acids to α-ketocarbonic acids (Zeuthen et al. 2006), and release via PfAQP may further positively affect parasite functionality. During kidney passages and transmission between the human host and the mosquito vector, the parasite is exposed to osmotic stress, which might be alleviated by rapid volume adaptation via water flux through PfAQP.

12.2.2 Toxoplasmosis parasites, *Toxoplasma gondii* AQP, TgAQP

Being an apicomplexan parasite as well, *T. gondii* is closely related to *Plasmodium* and equally expresses a single AQP constitutively throughout development (Pavlovic-Djuranovic et al. 2003). Despite the differences in the layout of the ar/R selectivity filter pointed out earlier, TgAQP exhibits similar water and solute permeability properties as PfAQP calling for related cellular functions in osmotic protection and metabolism (Pavlovic-Djuranovic et al. 2003, Zeuthen et al. 2006). However, sequence comparison further shows a surprisingly high similarity of TgAQP with intracellular plant AQPs of the tonoplast intrinsic protein family (Pavlovic-Djuranovic et al. 2003). Apicomplexan parasites are characterized by carrying a special organelle, the apicoplast. It is thought to be derived from the endosymbiotic ingestion of a green alga providing a source for plantlike proteins in these parasites. Recently, similarity even on the cell morphological level has been described, that is a plantlike vacuole in *T. gondii* to which TgAQP is localized (Miranda et al. 2010).

12.2.3 Sleeping Sickness Parasites, *Trypanosoma brucei* AQPs

Different from *Plasmodium* and *Toxoplasma*, three aquaglyceroporins (TbAQP1, TbAQP2 and TbAQP3) are expressed in *T. brucei*, that is the parasite that causes human African trypanosomiasis or sleeping sickness (Uzcategui et al. 2004). The need for more than one AQP of this parasite may be derived from the extracellular lifestyle and its direct exposure to the environment. This complexity may have driven a higher degree of differentiation in morphology and functionality of the parasite–host interface. TbAQP1 is localized in the membrane of the propelling flagellum (Bassarak et al. 2011), and TbAQP2 is restricted to the base of the flagellum, that is the flagellar pocket (Baker et al. 2012), whereas TbAQP3 is distributed around the plasma membrane (Bassarak et al. 2011). *T. brucei* has developed a unique anaerobic pathway of glucose metabolism making use of a parasite-specific organelle, the glycosome, where glucose is converted into equimolar amounts of pyruvate and glycerol phosphate. Under catalysis by glycerol kinase, the phosphate moiety can be transferred to ADP in order to yield ATP given that glycerol is readily removed from the equilibrium by facilitated diffusion across the membrane (Bakker et al. 2000, Steinborn et al. 2000). Here, the glycerol-facilitating TbAQPs may be directly linked to anaerobic energy generation (Uzcategui et al. 2004).

12.2.4 Chagas Disease Parasites, *Trypanosoma cruzi* AQPs, TcAQPs

Trypanosoma cruzi is an extracellular parasite causing Chagas disease in South America. There are four open reading frames in the genome encoding AQP-like proteins, TcAQP and TcAQP β-δ. Only TcAQP has been heterologously expressed and characterized and exhibited low water permeability. Glycerol was not found to pass TcAQP. Immunocytochemistry localized TcAQP at intracellular, so-called acidocalcisomes and the contractile vacuole complex of *T. cruzi* (Montalvetti et al. 2004). Acidocalcisomes harbour a calcium and phosphate-rich milieu probably for storage, as well as acidic pH. It is assumed that the organelles are responsible for cellular pH and osmoregulation (Rohloff et al. 2004, Docampo et al. 2005). Hence, a water-specific AQP would fit well into such a scenario. Cell volume controlling contractile vacuoles are present in other protozoa as well, such as *A. proteus* or *D. discoideum*. Both express at least one water-specific AQP, which is localized in the vacuoles (Nishihara et al. 2008, von Bülow et al. 2012). There is evidence that in the single-celled, amoeboidal developmental stage of *D. discoideum*, AQP-facilitated water permeability is involved in cell motility allowing for water influx at the sites of the plasma membrane where membrane protrusions form during the initiation of lamellipodia (von Bülow et al. 2012, 2015).

12.2.5 Leishmaniasis Parasites, *Leishmania* spp. AQPs, LmAQP1

Leishmaniasis is caused by a multitude of related *Leishmania* species. AQPs from the species *L. major* (Figarella et al. 2007) and *L. donovani* (Biyani et al. 2011) have been studied. Although there are five open reading frames coding for AQP-like proteins, for example LmAQP1 and LmAQPα-δ in *L. major* (Beitz 2007), only LmAQP1 has been shown to be functional as an aquaglyceroporin (Figarella et al. 2007). In the promastigote insect form, LmAQP1 was found at the flagellum, whereas in the amastigote human form, LmAQP1 was shown to reside in the flagellar pocket and the contractile vacuoles (Figarella et al. 2007). Its localization and permeability profile (water, glycerol, urea, methylglyoxal) are compatible with functions in osmoprotection, for example during transmission, and metabolism (Figarella et al. 2007, Mandal et al. 2012).

12.3 USING AQPs TO SHUTTLE CYTOTOXIC COMPOUNDS INTO PARASITES

As pointed out above, virtually all AQPs of parasite plasma membranes are of the aquaglyceroporin type exhibiting permeability for relatively small, uncharged solutes (Figure 12.1; Wu and Beitz 2007). Polyols up to five, in some cases six, carbons in length can pass aquaglyceroporins. It will be challenging to identify specific and potent drug molecules in that structural category. Three small organic compounds with antiparasitic properties due to a more general cytotoxicity and antineoplastic activity have been shown to pass the aquaglyceroporins from *P. falciparum*, *T. gondii* and *T. brucei* (Figure 12.1), that is hydroxyurea (Pavlovic-Djuranovic et al. 2003), dihydroxyacetone (Pavlovic-Djuranovic et al. 2006, Uzcategui et al. 2007) and methylglyoxal (Pavlovic-Djuranovic et al. 2006). Hydroxyurea is used in anticancer therapy due to its toxicity to replicating eukaryotic cells and was also shown to inhibit proliferation of *T. gondii* (Kasper and Pfefferkorn 1982), *Leishmania mexicana* (Martinez-Rojano et al. 2008) and *P. falciparum* (Pino et al. 2006). The highly reactive dicarbonyl methylglyoxal as a non-enzymatic glycolytic side product is made responsible for cellular senescence by generation of advanced glycosylation end products (Desai et al. 2010) and has mutagenic potency (Sousa Silva et al. 2012). Already micromolar concentrations in an in vitro culture setup inhibited growth of *P. falciparum* (Pavlovic-Djuranovic et al. 2006), *T. brucei*, *T. cruzi* and *L. major* (Greig et al. 2009). Its general toxicity certainly prevents the compound from a therapeutic use. The less reactive monocarbonyl dihydroxyacetone is not toxic to most cells; it can even be used as a nutrient after phosphorylation by dihydroxyacetone kinase and entering the glycolysis pathway at the three-carbon stage. It turned out that some parasites lack dihydroxyacetone kinase, which prolongs the residence time of the compound in the cell resulting in chemical DNA or protein modification and eventually fatal cell damage. An antiproliferative effect of dihydroxyacetone, albeit in the millimolar range, has been shown for *P. falciparum* (Pavlovic-Djuranovic et al. 2006) and *T. brucei* (Uzcategui et al. 2007).

Due to the lack of alternatives, the antimonials sodium stibogluconate (Pentostam) and meglumine antimoniate (Glucantime) are still the first-line treatment of leishmaniasis (Gourbal et al. 2004). The drug compounds contain pentavalent antimony, which is reduced to trivalent antimonious acid (Figure 12.1) in phagocytic cells of the immune system where *Leishmania* reside after an infection. $Sb(OH)_3$ readily diffuses into the parasite via LmAQP1 (Gourbal et al. 2004, Marquis et al. 2005, Bhattacharjee et al. 2009). This way, a certain degree of selectivity is achieved protecting the cells of the human host. A similar mechanism is used in the treatment of a rare type of leukemia, acute promyelocytic leukemia, using arsenic trioxide, which converts into arsenious acid in aqueous solution (Figure 12.1). Here, human AQP9 is responsible for the accumulation of arsenic in affected leukocytes initiating apoptosis (Kwong and Todd 1997, Iriyama et al. 2013).

The treatment of human African trypanosomiasis also depends on old drugs from the 1940s, such as pentamidine and melarsoprol, the latter being an arsenical (Figure 12.1). It came rather unexpected that the uptake of these compounds is connected to the presence of TbAQP2 in the parasite plasma membrane (Alsford et al. 2012). Downregulation, knockout or mutation of TbAQP2 leads to pentamidine and melarsoprol resistance (Alsford et al. 2012, Baker et al. 2012) suggesting that TbAQP2 represents an uptake pathway for the compounds. TbAQP2 is most likely not the drug target itself because the knockout is perfectly viable, at least under in vitro culture conditions. Further, melarsoprol and pentamidine cross the parasite membrane in their intact form (Fairlamb et al. 1989, Damper and Patton 1976). Intracellularly, melarsoprol has been shown to bind and inactivate the glutathione-analogue trypanothione affecting the parasite's defense system against oxidative stress (Fairlamb and Cerami 1992), whereas pentamidine is thought to interfere mainly with the kinetoplast DNA (Shapiro and Englund 1990), while other mechanisms of action are discussed as well. A direct passage of melarsoprol or pentamidine through TbAQP2 has not been shown yet. It is also discussed that the TbAQP2 effect on pentamidine uptake may be indirect via a so far unidentified high-affinity pentamidine transporter (Baker et al. 2013). The trypanosomal P2 aminopurine transporter and a non-identified low-affinity pentamidine transporter represent established uptake pathways (Teka et al. 2011). Elucidation of the exact mode of pentamidine transport into trypanosomes will probably provide knowledge usable in the design of modern drugs that make use of the shuttle system.

12.4 CHALLENGES AND POTENTIAL OF AQP INHIBITORS FOR ANTI-INFECTIOUS THERAPY

The potential of addressing AQPs as therapeutic targets in human diseases (Beitz and Schultz 1999) and for chemotherapy of infections by parasites (Beitz 2005) has been pointed out soon after their discovery. Yet a breakthrough in the design of potent, specific and druggable AQP inhibitors is still to come. Apparently, there are peculiar difficulties linked to AQP proteins, which hamper the design of adequate small-molecule inhibitors.

One challenge lies in the narrow channel structure of AQPs (Figure 12.3). Aquaglyceroporins allow for the entry of linear carbon hydroxyls, whereas water-specific AQPs are too narrow to accommodate any organic molecule (Wu and Beitz 2007, Seeliger et al. 2013). A major site of interaction must therefore lie within the extracellular vestibule (Mukhopadhyay and Beitz 2010). Alternatively, an interaction could occur via the intracellular channel entry, which is wider even in the case of water-specific AQPs and would allow organic compounds to travel up even to the central NPA region. This approach, however, would counteract the advantages, especially in the parasite scenario, of an attack from the outside of the cell. The outer AQP vestibule is typically more shallow than pocket like. Hence, the shielding of the interaction sites between inhibitor and protein from competing solvent molecules is hard to achieve, yet required for high-affinity binding. From what is known from the available crystal structures, the outer vestibules of aquaglyceroporins appear deeper than those of water-specific AQPs enhancing the chance of success in finding inhibitors. A second challenge lies in the hidden, intracellular lifestyle of many parasites, for example of *T. gondii* or apicomplexan parasites of the *Plasmodium* species. This poses special requirements on pharmacokinetics, because the compounds need to cross several membranes to reach their target (Cossum 1988).

Finally, it is questionable whether the inhibition of a parasite AQP alone suffices to cure the respective infectious disease, because knockout or knockdown experiments have yielded mild or no phenotypes. For instance, the rodent malaria parasite *P. berghei* lacking its single aquaglyceroporin due to a gene deletion is highly deficient in glycerol transport but is viable. Its proliferation rate reduced by half and infected mice survive twice as long as their mates infected with the wild-type parasites (Promeneur et al. 2007). Knockdown of AQP1 or AQP3 in *T. brucei* did not affect survival under hypo-osmotic stress conditions

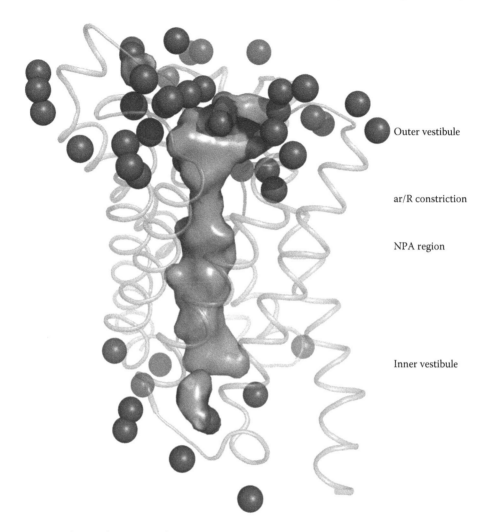

FIGURE 12.3 **(See colour insert.)** Vestibules and transduction channel of PfAQP (PDB #3C02). The design of aquaporin inhibitors with high affinity and selectivity is challenged by spatially restricted binding sites. Solvent water molecules are shown as spheres.

(Bassarak et al. 2011) and an AQP2/AQP3 double-deletion strain grew normally under standard culture conditions (Alsford et al. 2012). Similarly, *L. major* parasites lacking LmAQP1 showed unaltered viability in vitro (Gourbal et al. 2004).

A reduced proliferation rate of a parasite, however, indicates that the loss of AQP functionality affects the metabolic turnover either by reduced uptake of precursors, such as glycerol, or release of waste, such as ammonia or methylglyoxal. Concomitantly, the level of metabolism-derived cellular stress (oxidative stress, activation of alternate biosynthesis pathways and detoxification programmes) can be expected to be risen. AQP inhibitors may, thus, be useful in combination with therapy regimes, which affect related functions. Slower growth of parasites cells may further provide that amount of extra time required by the host's immune system to fight the infection eventually (Schmid-Hempel 2008).

REFERENCES

Alsford, S., Eckert, S., Baker, N., Glover, L., Sanchez-Flores, A., Leung, K.F., Turner, D.J., Field, M.C., Berriman, M., Horn, D. 2012. High-throughput decoding of antitrypanosomal drug efficacy and resistance. *Nature* 482:232–236.

Aponte-Santamaria, C., Hub, J.S., de Groot, B.L. 2010. Dynamics and energetics of solute permeation through the *Plasmodium falciparum* aquaglyceroporin. *Phys. Chem. Chem. Phys.* 12:10246–10254.

Baker, N., de Koning, H.P., Maser, P., Horn, D. 2013. Drug resistance in African trypanosomiasis: The melarsoprol and pentamidine story. *Trends Parasitol.* 29:110–118.

Baker, N., Glover, L., Munday, J.C., Aguinaga Andres, D., Barrett, M.P., de Koning, H.P., Horn, D. 2012. Aquaglyceroporin 2 controls susceptibility to melarsoprol and pentamidine in African trypanosomes. *Proc. Nat. Acad. Sci. USA* 109:10996–11001.

Bakker, B.M., Westerhoff, H.V., Opperdoes, F.R., Michels, P.A. 2000. Metabolic control analysis of glycolysis in trypanosomes as an approach to improve selectivity and effectiveness of drugs. *Mol. Biochem. Parasitol.* 106:1–10.

Bassarak, B., Uzcategui, N.L., Schonfeld, C., Duszenko, M. 2011. Functional characterization of three aquaglyceroporins from *Trypanosoma brucei* in osmoregulation and glycerol transport. *Cell. Physiol. Biochem.* 27:411–420.

Beitz, E. 2005. Aquaporins from pathogenic protozoan parasites: Structure, function and potential for chemotherapy. *Biol. Cell* 97:373–383.

Beitz, E. 2007. Jammed traffic impedes parasite growth. *Proc. Nat. Acad. Sci. USA* 104:13855–13856.

Beitz, E., Becker, D., von Bülow, J., Conrad, C., Fricke, N., Geadkaew, A., Krenc, D., Song, J., Wree, D., Wu, B. 2009. In vitro analysis and modification of aquaporin pore selectivity. *Handb. Exp. Pharmacol.* 190:77–92.

Beitz, E., Pavlovic-Djuranovic, S., Yasui, M., Agre, P., Schultz, J.E. 2004. Molecular dissection of water and glycerol permeability of the aquaglyceroporin from *Plasmodium falciparum* by mutational analysis. *Proc. Nat. Acad. Sci. USA* 101:1153–1158.

Beitz, E., Schultz, J.E. 1999. The mammalian aquaporin water channel family: A promising new drug target. *Curr. Med. Chem.* 6:457–467.

Beitz, E., Wu, B., Holm, L.M., Schultz, J.E., Zeuthen, T. 2006. Point mutations in the aromatic/arginine region in aquaporin 1 allow passage of urea, glycerol, ammonia, and protons. *Proc. Nat. Acad. Sci. USA* 103:269–274.

Bhattacharjee, H., Rosen, B.P., Mukhopadhyay, R. 2009. Aquaglyceroporins and metalloid transport: Implications in human diseases. *Handb. Exp. Pharmacol.* 190:309–325.

Biyani, N., Mandal, S., Seth, C., Saint, M., Natarajan, K., Ghosh, I., Madhubala, R. 2011. Characterization of *Leishmania donovani* aquaporins shows presence of subcellular aquaporins similar to tonoplast intrinsic proteins of plants. *PloS One* 6:e24820.

Cossum, P.A. 1988. Role of the red blood cell in drug metabolism. *Biopharm. Drug Dispos.* 9:321–326.

Damper, D., Patton, C.L. 1976. Pentamidine transport and sensitivity in brucei-group trypanosomes. *J. Protozool.* 23:349–356.

Desai, K.M., Chang, T., Wang, H., Banigesh, A., Dhar, A., Liu, J., Untereiner, A., Wu, L. 2010. Oxidative stress and aging: Is methylglyoxal the hidden enemy? *Can. J. Physiol. Pharmacol.* 88:273–284.

Docampo, R., de Souza, W., Miranda, K., Rohloff, P., Moreno, S.N. 2005. Acidocalcisomes—Conserved from bacteria to man. *Nat. Rev. Microbiol.* 3:251–261.

Fairlamb, A.H., Cerami, A. 1992. Metabolism and functions of trypanothione in the Kinetoplastida. *Ann. Rev. Microbiol.* 46:695–729.

Fairlamb, A.H., Henderson, G.B., Cerami, A. 1989. Trypanothione is the primary target for arsenical drugs against African trypanosomes. *Proc. Nat. Acad. Sci. USA* 86:2607–2611.

Figarella, K., Uzcategui, N.L., Zhou, Y., LeFurgey, A., Ouellette, M., Bhattacharjee, H., Mukhopadhyay, R. 2007. Biochemical characterization of *Leishmania major* aquaglyceroporin LmAQP1: Possible role in volume regulation and osmotaxis. *Mol. Microbiol.* 65:1006–1017.

Fu, D., Libson, A., Miercke, L.J., Weitzman, C., Nollert, P., Krucinski, J., Stroud, R.M. 2000. Structure of a glycerol-conducting channel and the basis for its selectivity. *Science* 290:481–486.

Gardner, M.J., Hall, N., Fung, E., White, O., Berriman, M., Hyman, R.W., Carlton, J.M. et al. 2002. Genome sequence of the human malaria parasite *Plasmodium falciparum*. *Nature* 419:498–511.

Gourbal, B., Sonuc, N., Bhattacharjee, H., Legare, D., Sundar, S., Ouellette, M., Rosen, B.P., Mukhopadhyay, R. 2004. Drug uptake and modulation of drug resistance in *Leishmania* by an aquaglyceroporin. *J. Biol. Chem.* 279:31010–31017.

Greig, N., Wyllie, S., Patterson, S., Fairlamb, A.H. 2009 A comparative study of methylglyoxal metabolism in trypanosomatids. *FEBS J.* 276:376–386.

Hansen, M., Kun, J.F., Schultz, J.E., Beitz, E. 2002. A single, bi-functional aquaglyceroporin in blood-stage *Plasmodium falciparum* malaria parasites. *J. Biol. Chem.* 277:4874–4882.

Hedfalk, K., Pettersson, N., Oberg, F., Hohmann, S., Gordon, E. 2008. Production, characterization and crystallization of the *Plasmodium falciparum* aquaporin. *Protein Expr. Purif.* 59:69–78.

Iriyama, N., Yuan, B., Yoshino, Y., Hatta, Y., Horikoshi, A., Aizawa, S., Takeuchi, J., Toyoda, H. 2013. Aquaporin 9, a promising predictor for the cytocidal effects of arsenic trioxide in acute promyelocytic leukemia cell lines and primary blasts. *Oncol. Rep.* 29:2362–2368.

Kasper, L.H., Pfefferkorn, E.R. 1982. Hydroxyurea inhibition of growth and DNA synthesis in *Toxoplasma gondii*: Characterization of a resistant mutant. *Mol. Biochem. Parasitol.* 6:141–150.

Kwong, Y.L., Todd, D. 1997. Delicious poison: Arsenic trioxide for the treatment of leukemia. *Blood* 89:3487–3488.

Mandal, G., Sharma, M., Kruse, M., Sander-Juelch, C., Munro, L.A., Wang, Y., Vilg, J.V. et al. 2012. Modulation of *Leishmania major* aquaglyceroporin activity by a mitogen-activated protein kinase. *Mol. Microbiol.* 85:1204–1218.

Marquis, N., Gourbal, B., Rosen, B.P., Mukhopadhyay, R., Ouellette, M. 2005. Modulation in aquaglyceroporin AQP1 gene transcript levels in drug-resistant *Leishmania*. *Mol. Microbiol.* 57:1690–1699.

Martinez-Rojano, H., Mancilla-Ramirez, J., Quinonez-Diaz, L., Galindo-Sevilla, N. 2008. Activity of hydroxyurea against *Leishmania mexicana*. *Antimicrob. Agents Chemother.* 52:3642–3647.

Martins, A.P., Ciancetta, A., de Almeida, A., Marrone, A., Re, N., Soveral, G., Casini A. 2013. Aquaporin inhibition by gold(III) compounds: New insights. *ChemMedChem* 8:1086–1092.

Miranda, K., Pace, D.A., Cintron, R., Rodrigues, J.C., Fang, J., Smith, A., Rohloff, P. et al. 2010. Characterization of a novel organelle in *Toxoplasma gondii* with similar composition and function to the plant vacuole. *Mol. Microbiol.* 76:1358–1375.

Montalvetti, A., Rohloff, P., Docampo, R. 2004. A functional aquaporin co-localizes with the vacuolar proton pyrophosphatase to acidocalcisomes and the contractile vacuole complex of *Trypanosoma cruzi*. *J. Biol. Chem.* 279:38673–38682.

Mukhopadhyay, R., Beitz, E. 2010. Metalloid transport by aquaglyceroporins: Consequences in the treatment of human diseases. *Adv. Exp. Med. Biol.* 679:57–69.

Mukhopadhyay, R., Mandal, G., Atluri, V.S.R., Figarella, K., Uzcategui, N.L., Zhou, Y., Beitz, E., Ajees, A.A., Bhattacharjee, H. 2011. The role of alanine 163 in solute permeability of *Leishmania major* aquaglyceroporin LmAQP1. *Mol. Biochem. Parasitol.* 175:83–90.

Newby, Z.E., O'Connell, 3rd., J., Robles-Colmenares, Y., Khademi, S., Miercke, L.J., Stroud, R.M. 2008. Crystal structure of the aquaglyceroporin PfAQP from the malarial parasite *Plasmodium falciparum*. *Nat. Struct. Mol. Biol.* 15:619–625.

Nishihara, E., Yokota, E., Tazaki, A., Orii, H., Katsuhara, M., Kataoka, K., Igarashi, H., Moriyama, Y., Shimmen, T., Sonobe, S. 2008. Presence of aquaporin and V-ATPase on the contractile vacuole of *Amoeba proteus*. *Biol. Cell* 100:179–188.

Pavlovic-Djuranovic, S., Kun, J.F., Schultz, J.E., Beitz, E. 2006. Dihydroxyacetone and methylglyoxal as permeants of the *Plasmodium* aquaglyceroporin inhibit parasite proliferation. *Biochim. Biophys. Acta* 1758:1012–1017.

Pavlovic-Djuranovic, S., Schultz, J.E., Beitz, E. 2003. A single aquaporin gene encodes a water/glycerol/urea facilitator in *Toxoplasma gondii* with similarity to plant tonoplast intrinsic proteins. *FEBS Lett* 555:500–504.

Pino, P., Taoufiq, Z., Brun, M., Tefit, M., Franetich, J.F., Ciceron, L., Krishnamoorthy, R., Mazier, D. 2006. Effects of hydroxyurea on malaria, parasite growth and adhesion in experimental models. *Parasite Immunol.* 28:675–680.

Promeneur, D., Liu, Y., Maciel, J., Agre, P., King, L.S., Kumar, N. 2007. Aquaglyceroporin PbAQP during intraerythrocytic development of the malaria parasite *Plasmodium berghei*. *Proc. Nat. Acad. Sci. USA* 104:2211–2216.

Rohloff, P., Montalvetti, A., Docampo, R. 2004. Acidocalcisomes and the contractile vacuole complex are involved in osmoregulation in *Trypanosoma cruzi*. *J. Biol. Chem.* 279:52270–52281.

Schmid-Hempel, P. 2008. Parasite immune evasion: A momentous molecular war. *Trends Ecol. Evol.* 23:318–326.

Seeliger, D., Zapater, C., Krenc, D., Haddoub, R., Flitsch, S., Beitz, E., Cerda, J., de Groot, B.L. 2013. Discovery of novel human aquaporin-1 blockers. *ACS Chem. Biol.* 8:249–256.

Shapiro, T.A., Englund, P.T. 1990. Selective cleavage of kinetoplast DNA minicircles promoted by antitrypanosomal drugs. *Proc. Nat. Acad. Sci. USA* 87:950–954.

Song, J., Almasalmeh, A., Krenc, D., Beitz, E. 2012. Molar concentrations of sorbitol and polyethylene glycol inhibit the *Plasmodium* aquaglyceroporin but not that of *E. coli*: Involvement of the channel vestibules. *Biochim. Biophys. Acta* 1818:1218–1224.

Song, J., Mak, E., Wu, B., Beitz, E. 2014. Parasite aquaporins: Current developments in drug facilitation and resistance. *Biochim. Biophys. Acta* 1840:1566–1573.

Sousa Silva, M., Ferreira, A.E., Gomes, R., Tomas, A.M., Ponces Freire, A., Cordeiro, C. 2012. The glyoxalase pathway in protozoan parasites. *Int. J. Med. Microbiol.* 302:225–259.

Steinborn, K., Szallies, A., Mecke, D., Duszenko, M. 2000. Cloning, heterologous expression and kinetic analysis of glycerol kinase (TbGLK1) from *Trypanosoma brucei*. *Biol. Chem.* 381:1071–1077.

Teka, I.A., Kazibwe, A.J., El-Sabbagh, N., Al-Salabi, M.I., Ward, C.P., Eze, A.A., Munday, J.C. et al. 2011. The diamidine diminazene aceturate is a substrate for the high-affinity pentamidine transporter: Implications for the development of high resistance levels in trypanosomes. *Mol. Pharmacol.* 80:110–116.

Uzcategui, N.L., Carmona-Gutierrez, D., Denninger, V., Schoenfeld, C., Lang, F., Figarella, K., Duszenko, M. 2007. Antiproliferative effect of dihydroxyacetone on *Trypanosoma brucei* bloodstream forms: Cell cycle progression, subcellular alterations, and cell death. *Antimicrob. Agents Chemother.* 51:3960–3968.

Uzcategui, N.L., Szallies, A., Pavlovic-Djuranovic, S., Palmada, M., Figarella, K., Boehmer, C., Lang, F., Beitz, E., Duszenko, M. 2004. Cloning, heterologous expression, and characterization of three aquaglyceroporins from *Trypanosoma brucei*. *J. Biol. Chem.* 279:42669–42676.

Verkman, A.S., Anderson, M.O., Papadopoulos, M.C. 2014. Aquaporins: Important but elusive drug targets. *Nat. Rev. Drug Discov.* 13:259–277.

von Bülow, J., Golldack, A., Albers, T., Beitz, E. 2015. The amoeboidal *Dictyostelium* aquaporin AqpB is gated via Tyr216 and aqpB gene deletion affects random cell motility. *Biol. Cell.* 107:78–88.

von Bülow, J., Müller-Lucks, A., Kai, L., Bernhard, F., Beitz, E. 2012. Functional characterization of a novel aquaporin from *Dictyostelium discoideum* amoebae implies a unique gating mechanism. *J. Biol. Chem.* 287:7487–7494.

Wu, B., Beitz, E. 2007. Aquaporins with selectivity for unconventional permeants. *Cell. Mol. Life Sci.* 64:2413–2421.

Wu, B., Steinbronn, C., Alsterfjord, M., Zeuthen, T., Beitz, E. 2009. Concerted action of two cation filters in the aquaporin water channel. *EMBO J.* 28:2188–2194.

Zeuthen, T., Wu, B., Pavlovic-Djuranovic, S., Holm, L.M., Uzcategui, N.L., Duszenko, M., Kun, J.F., Schultz, J.E., Beitz, E. 2006. Ammonia permeability of the aquaglyceroporins from *Plasmodium falciparum*, *Toxoplasma gondii* and *Trypanosoma brucei*. *Mol. Microbiol.* 61:1598–1608.

III

Aquaporins as Drug Targets

CHAPTER 13

Aquaporins

Chemical Inhibition by Small Molecules

Vincent J. Huber, Sören Wacker and Michael Rützler

CONTENTS

Abstract	249
13.1 Introduction	250
13.2 AQPs as Therapeutic Targets	251
13.2.1 Oncology	251
13.2.2 Kidney Diseases	252
13.2.3 Glaucoma	253
13.2.4 Osmotic Disequilibrium Pathologies of the Brain	254
13.2.5 AQPs in Diabetes and Obesity	255
13.2.6 Disuse Osteoporosis	256
13.2.7 AQP Inhibitors as Immunosuppressants	256
13.2.8 Medical Imaging	257
13.3 Methods for AQP Inhibitor Identification	257
13.3.1 *Xenopus laevis* Oocyte Assay	257
13.3.2 Proteoliposome Assays	258
13.3.3 Calcein Quenching Assays	258
13.3.4 End Point Hypotonic Shock Assays	259
13.4 In Silico Methods for AQP Inhibitor Optimization (and Potentially *Ab Initio* Identification)	259
13.4.1 Molecular Docking	259
13.4.2 Molecular Dynamics Simulations	262
13.5 Conclusions	263
References	263

ABSTRACT

THE HUMAN GENOME ENCODES 13 aquaporin isoforms with characteristic substrate specificity that are expressed at specific locations throughout the body. Of these isoforms, AQPs 1–4 serve important functions in renal water reabsorption. Consequently, specific AQP inhibitors have been proposed as 'aquaretics', a new class of drugs suitable to induce diuresis without concomitant salt wasting. Furthermore, animal experiments suggested that AQP4 inhibitors could be useful to treat some forms of brain edema. Other proposed applications for AQP inhibitors involve amongst others treatment of diabetes, inflammatory skin diseases and cancer. However, few of these putative applications have been critically evaluated against current forms of therapy. Furthermore, development of AQP inhibitors remains difficult and despite numerous efforts during at least the last 15 years very few AQP inhibitors have been described. Moreover, none of the hitherto described substances have been developed to a level where meaningful verification of proposed AQP drug targets in preclinical or clinical settings was possible. Nonetheless, encouraging progress towards development of such substances has been made during recent years. Novel cell-based assays facilitate high throughput screening of chemical compound libraries for hit discovery. AQP 3D structures have been solved for 10 isoforms, which can support rapidly evolving computational hit discovery methods, as well as hit to lead programs. In this chapter, we will provide a critical review of current evidence supporting relevance of AQPs as drug targets, describe current methods for AQP inhibitor discovery and will try to highlight challenges that remain before successful AQP inhibitor development.

13.1 INTRODUCTION

The aquaporin (AQP) family of membrane intrinsic proteins is a large superfamily of water and neutral solute transporters existing in virtually all organisms (Heymann and Engel 1999; Agre 2004; Zardoya 2005; Kruse et al. 2006). Thirteen AQP isoforms have been identified in mammals (Ishibashi et al. 2009) and represent water-specific and water and neutral solute transporters. All have been confirmed as being bidirectional in terms of their substrate trafficking, normally in response to a gradient across the plasma membrane. The water-specific AQPs are AQP0, 1, 2, 4 and 5, while AQP3, 7, 8, 9 and 10 are recognized as also being permeable to glycerol, urea and other neutral solutes (Hara-Chikuma and Verkman 2006; Rojek et al. 2008). The selectivity profiles of the most recently identified mammalian AQPs, AQP11 (Yakata et al. 2007) and AQP12 (Itoh et al. 2005), have not yet been fully identified, even though a recent study suggests glycerol permeability for AQP11 (Madeira et al. 2014). AQP proteins are distributed throughout the human body and are involved in a variety of physiological processes. They have also been implicated in disease processes and consequently have been suggested as targets for drug development. This area has received a considerable amount of interest and has garnered a number of significant reviews (Verkman 2001; Agre et al. 2002; Castle 2005; Wang et al. 2006; Huber et al. 2012; Akdemir et al. 2014; Verkman et al. 2014).

Despite the intense interest in using AQPs as therapeutic drug targets, few small molecule organic modulators have been reported, and an even smaller number of compounds have been reported as being active in vivo. Based on this limited number of compounds, one might conclude that AQPs are difficult targets for small molecules (Verkman et al.). However, recent reports give some hope for success in this area. Therefore, the goal of this chapter is to highlight specific indications where small molecule modulators of AQPs can be expected to be therapeutically useful. Furthermore, methods for identifying AQP modulators will be discussed, and examples for promising ligands will be given.

13.2 AQPs AS THERAPEUTIC TARGETS

13.2.1 Oncology

The WHO estimates that 14 million people were diagnosed with cancer in 2012 and 8.2 million people died as a result of it during that year. Moreover, the increase in cancer incidence is expected to outpace population growth over the next few decades. Therefore, identification of new treatments for cancer and tumours is one of the primary driving forces in pharmaceutical research. Changes to AQP expression and distribution in relation to tumorigenesis were observed for a number of isoforms and tumour types (Clapp and Martinez de la Escalera 2006; Monzani et al. 2007; Verkman et al. 2008; Nico and Ribatti 2010; Papadopoulos and Saadoun 2014). And while specific roles of these AQPs in tumour growth are only now coming into clearer focus, inhibitors of those AQPs remain of significant interest for oncotherapy (Papadopoulos and Saadoun 2014).

The role of AQP1 in carcinogenesis has been well documented: Increased AQP1 expression was found in the microvessels of brain (Endo et al. 1999; Saadoun et al. 2002), breast (Endo et al. 1999) and ovarian cancers (Yang et al. 2006c), as well as in myeloma (Vacca et al. 2001). Furthermore, ectopic expression of AQP1 increased the number of lung metastases in a mouse tumour-cell homograft model (Hu and Verkman 2006). Targeted disruption of the AQP1 encoding gene was found to dramatically impair tumour growth in subcutaneous and intracranial tumour implant models via reduced cell migration and angiogenesis (Saadoun et al. 2005). Similarly, decreased migration capacity was found in human endothelial and melanoma cells following AQP1 gene silencing (Monzani et al. 2009). Therefore, AQP1 is now generally regarded as a promoter of tumour angiogenesis (Clapp and Martinez de la Escalera 2006; Nico and Ribatti 2010). Accordingly, AQP1 inhibitors remain of considerable interest for oncotherapy, and compounds capable of testing this theory are greatly anticipated (Saadoun et al. 2005; Verkman 2006; Verkman et al. 2014).

Peri-tumoral and tumoral over-expression has been observed for AQP4 in oedematous astrocytomas, gliomas and glioblastoma multiforme (GBM) (Saadoun 2002; Sawada et al. 2007). While altered peri-tumoral AQP4 expression appears related to cerebral edema, recent reports indicate that the increased tumoral AQP4 expression in GBM may be partly responsible for its aggressive invasion into surrounding tissue (Ding et al. 2011). Those authors found that siRNA-mediated knockdown of AQP4 expression impaired GBM cell migration, which suggests that AQP4 inhibitors could be used to achieve a similar effect. Additionally, glioblastoma cell apoptosis was induced by AQP4 knockdown; although, the specific mechanism by which that occurs remains unknown (Ding et al. 2013). On the basis of these reports, AQP4 appears to be a promising target for treating malignant gliomas and GBM, and CNS-penetrant AQP4 inhibitors are greatly anticipated to test this hypothesis.

Elevated AQP5 expression has been found in several tumours, including pulmonary adenocarcinomas (Zhang et al. 2010), pancreatic and colon cancer (Kang et al. 2008; Woo et al. 2008; Wang et al. 2012), myelogenous leukemia (Chae et al. 2008a; Zhang et al. 2010; Machida et al. 2011), ovarian cancer (Yang et al. 2012) and early breast cancer (Lee et al. 2014). AQP5 has been linked with the activation of epidermal growth factor receptor/extracellular signal-related kinase/p38 mitogen-activated protein kinase signalling pathway in tumour cells (Chae et al. 2008a; Zhang et al. 2010). And abnormal expression of AQP5 in tumours appears to be predictive of differentiation and metastasis (Wang et al. 2012). The specific mechanism by which AQP5 is involved in tumour growth, metastasis and drug resistance remains poorly understood. However, it was recently shown that AQP5 silencing was able to reduce p38 MAPK signalling and reduced drug resistance in colon cancer cells (Shi et al. 2014). Unfortunately, no inhibitors have been reported that are suitable to further establish the value of AQP5 as a therapeutic target in cancer models.

13.2.2 Kidney Diseases

Kidney function is largely made possible by the expression of seven AQP isoforms, each of which have a specific distribution in individual nephron segments and consequently distinct roles in reabsorption of primary urine (Verkman 2001; Nielsen et al. 2002; King et al. 2004). AQPs 1–4 are involved in urinary concentration by facilitating near iso-osmolar water uptake across renal tubular cell membranes. Accordingly, AQP loss of function in humans and knockout mice causes increased production of dilute urine. It has therefore been proposed that AQP inhibitors could be suitable to treat conditions where enhanced water excretion is desirable. Such aquaretics (Stassen et al. 1985) should be safer than conventional diuretics when reduction of body water volume must be achieved, while salt should be retained, for example in hyponatremia.

AQP1 in the renal proximal tubule and the descending limb of Henle is involved in constitutive water reabsorption that accounts for approximately 90% of the urinary concentrating process (Wesche et al. 2012). However, individuals of the Colton-null phenotype that are naturally deficient in AQP1 display few urine-concentrating defects (Preston et al. 1994; Ma et al. 1998). Moreover, similarly mild urinary concentration defects were observed in AQP1 knockout mice (Ma et al. 1998). This is likely due to a high capacity for regulated water reabsorption in renal collecting ducts involving AQP2, 3 and 4. Together, these observations suggest that AQP1 is not a well-suited target for the development of aquaretics.

AQP2 is primarily expressed in the renal connecting tubule and collecting duct principal cells (Nielsen et al. 1993) where it is crucial for regulated urine reabsorption. AQP2 localization in apical cell membranes is controlled by the antidiuretic hormone (ADH) vasopressin and determines water permeability of collecting duct principle cells. In mice, loss of AQP2 function causes severe urine-concentrating defects (Yang et al. 2006b). Similarly, AQP2 mutations in humans comprise approximately 10% of patients suffering from inherited nephrogenic diabetes insipidus (NDI). The majority of these cases are caused by mutations in vasopressin type 2 receptor (V2R). NDI is characterized by diuresis without electrolyte loss suggesting AQP2 and V2R as targets for the development of aquaretics. Currently, vasopressin-receptor antagonists have been developed to treat conditions where elevated water excretion could alleviate hyponatremia in pathologies such as congestive heart failure, hepatic cirrhosis, syndrome of inappropriate secretion of ADH and polycystic kidney disease (Greenberg and Verbalis 2006; Verbalis 2006; Ali et al. 2007; Wang et al. 2010; Kortenoeven and Fenton 2014). The vaptan family of V_{1A} and V_2 receptor antagonists has been introduced to treat these conditions and includes conivaptan, lixivaptan, satavaptan and tolvaptan (Ali et al. 2007; Miyazaki et al. 2007; Ku et al. 2009; Ferguson-Myrthil 2010), several of which are now approved for clinical use or undergoing clinical trials. However, studies on the long-term clinical outcome of vaptan treatment for congestive heart failure showed little improvement in quality of life scores, long-term mortality or rehospitalization rates, partly attributed to increased thirst and vasopressin secretion, as well as other potential treatment adaptations (Robertson 2011). Since AQP2 membrane localization depends on vasopressin signalling, AQP2 inhibitors may present similar benefits and drawbacks as vaptan therapies. No AQP2 inhibitors are available at present to test these hypotheses.

While AQP2 localization determines connecting tubule and collecting duct cell water permeability on the apical side, AQPs 3 and 4 are expressed on the basolateral and basal side of these cells, respectively. AQP4 gene deletion causes very mild to no diuresis in mice, while loss of AQP3 function causes dramatic polyuria, which was not affected by a synthetic V2R agonist (Ma et al. 2000). Consequently, AQP3 inhibitors could be useful as aquaretics that are less sensitive to adaptive changes in vasopressin signalling. However, a splicing mutation that deletes exon 5 from AQP3 does not cause urinary concentrating defects in humans (Roudier et al. 2002), suggesting human-specific expression of compensating AQP isoforms or other differences in the urinary concentrating processes between humans and mice. Therefore, AQP3 inhibitors are not likely useful for treating human water imbalance–related disorders. Human gene mutations disrupting AQP4 function have not been reported (Kortenoeven and Fenton 2014).

Additionally, there is some question about the medical need for new aquaretics, given the availability of low-cost alternatives. Oral urea can potently enhance water excretion by elevating renal tubular osmolality. The safety and efficacy of oral urea in long-term treatment of chronic hyponatremia has also been demonstrated (Soupart et al. 2012). The preferred acute treatment for hyponatremia is sodium chloride (Overgaard-Steensen 2011). Besides AQPs, urea channels have been considered as drug targets in water retention disorders and robust urea channel inhibitors have been discovered (Levin et al. 2007; Esteva-Font et al. 2013). Accordingly, the future of NDI therapeutics may belong to 'urearectics'.

Inherently, patients suffering from NDI respond poorly, if at all, to treatment with vasopressin or vasopressin-receptor agonists (Wesche et al. 2012). Drugs that are able to bypass the vasopressin pathway to elevate AQP2 localization to the apical membrane may be therapeutically useful when NDI is caused by V2 receptor mutations. Several approaches have been presented recently: In one study, prostaglandin receptor EP2 agonists were exploited to stimulate ADH-independent elevation of intracellular cyclic adenosine monophosphate that targets AQP2 to the plasma membrane (Olesen et al. 2011). Also, statins have been noted to promote AQP2 membrane accumulation (Procino et al. 2010). Furthermore, utilizing high-throughput screening of small molecule libraries, several compounds, for example AG-490, have been identified that enhance vasopressin-receptor independent membrane localization of AQP2 in vitro and in vivo (Nomura et al. 2014).

13.2.3 Glaucoma

AQP1 and 4 are expressed in the iris and ciliary epithelium of the human eye (Frigeri et al. 1995) and are thought to play roles in the regulation and production of aqueous humour, which is in turn related to intraocular pressure (IOP). Rodents deficient in AQP1 or both,

AQP1 and AQP4, were found to have reduced IOP compared to wild-type animals due to a reduction in the rate of aqueous humour secretion from the ciliary epithelium (Zhang et al. 2002). This suggests that small molecule AQP1 inhibitors would be useful for the treatment of glaucoma.

13.2.4 Osmotic Disequilibrium Pathologies of the Brain

Intracerebral osmotic balance is carefully maintained by the function of AQPs expressed in the CNS. AQP4 is widely expressed within the brain, in astroglial cells lining the ependymal, perivascular and sub-pial surfaces in contact with the CSF and blood–brain barrier, as well as in glial cells forming the edge of the cerebral cortex and brainstem, in vasopressin secretory neurons in the supraoptic and periventricular nuclei of the hypothalamus and in Purkinje cells of the cerebellum (Mobasheri et al. 2007; Tait et al. 2008). AQP1 is primarily expressed in the choroid plexus epithelium (Nielsen et al. 1993). AQP9 expression has been confirmed in catecholaminergic neurons (Mylonakou et al. 2009). A further three AQPs (AQP3, 5, 8) were identified in the CNS by reverse transcription polymerase chain reaction, but their distribution and function remain poorly understood (Badaut et al. 2007).

Thus, much of the discussion surrounding AQPs as drug targets within the CNS has focused on AQP4 and specifically on disorders that disturb brain water balance (Kimelberg 2004). The main clinical indications for AQP4 inhibitors were considered to be acute cytotoxic edema, in particular that occurring as a result of ischemic injury, as well as severe, acute hyponatremia. Studies conducted using AQP4 null mice, or mice lacking dystrophin, which localizes AQP4 along the perivascular and subpial end feet of astroglia, found reduced intracellular swelling (cytotoxic edema) following water intoxication (Manley et al. 2000; Vajda et al. 2002). Furthermore, AQP4 gene deletion reduced cytotoxic edema in the models of ischemic stroke (Manley et al. 2000; Katada et al. 2013). These experiments strongly argue that AQP4 function is partly responsible for brain edema formation, and, therefore, AQP4 inhibitors may be able to reduce cerebral edema.

A recent report described the effect of the AQP4 inhibitor TGN-020 on brain swelling in rodents following an ischemic insult (Igarashi et al. 2011). That study demonstrated that administration of an AQP4 inhibitor prior to the ischemic insult led to decreased brain swelling and hemispherical lesion volume. However, the inhibitor was administered prior to the ischemic insult, which is reverse of the order necessitated in a clinical context. Therefore, it will be important to clarify if a window of opportunity exists following the ischemic event where administration of an AQP4 inhibitor has therapeutic value. While this area remains promising, a greater number of inhibitors and model studies are necessary before the therapeutic value of AQP4 inhibitors in cerebral ischemia can be assessed.

AQP4 has also received attention as possible drug target for treating the effects of traumatic brain injury (TBI). Depending on severity, the resulting brain injuries present a broad spectrum of clinical outcomes. At present, there are few effective treatment options for TBI, and this is a significant unmet clinical need. Alterations in CNS levels and distribution of AQP4 were identified following non-penetrating brain insults in animal models of TBI, although the specific involvement of AQP4 in the progression of post-contusion brain injuries and neurological damage was not clearly identified

(Ke et al. 2001; Kiening et al. 2002; Fukuda et al. 2012; Ren et al. 2013). Indeed, inconsistencies in the reported AQP4 dysregulation were reported in a number of cases, which may suggest still incompletely identified dependencies on impact severity, time and location with respect to the impact focus. Reduction in AQP4 expression was also found to correlate with reduced edema, smaller neurological deficits or both following cortical impact brain injuries (Guo et al. 2006; Shenaq et al. 2012; Fukuda et al. 2013; Wang et al. 2013). The relationship between AQP4 attenuation and decreased injury severity remains unclear; nonetheless, this does suggest that AQP4 inhibition may have some therapeutic benefit to ameliorate the severity of TBI. However, increased accumulation of tau neurofibrillary tangles was also found in a closed-skull, hit-and-run brain trauma model, which was exacerbated in AQP4 knockout rodents (Iliff et al. 2014). Therefore, while this remains a potentially promising area for drug discovery, careful consideration of the potential therapeutic window for AQP4 inhibitors in treating TBI is necessary once suitable compounds are available for testing. Likewise, the potential for AQP4 inhibitors to increase localized neurofibrillary tangles must also be thoroughly evaluated, preferably during early-preclinical studies.

Epilepsy has been proposed as another potential therapeutic area for AQP4 inhibitors (Binder et al. 2012). However, AQP4-disrupted mice demonstrated an increased resistance to seizure initiation compared to wild-type models (Binder et al. 2004, 2006). The greater seizure duration found for AQP4-disrupted mice (Binder et al. 2006) and reduced seizure threshold following TBI (Lu et al. 2011) argue against an AQP4 inhibition as a means of treatment.

13.2.5 AQPs in Diabetes and Obesity

Biological membranes in general are only moderately permeable to glycerol. However, the human genome encodes four AQP isoforms that facilitate high glycerol permeability (Ishibashi et al. 1997, 2002; Kuriyama et al. 1997; Tsukaguchi et al. 1998). Immunolabeling studies indicated the expression of AQP3, 7, 9 and 10 in human adipocytes (Rodríguez et al. 2011; Laforenza et al. 2013) and hepatocytes (Rodríguez et al. 2011, 2013). Expression as well as glycerol permeability of the atypical AQP11 has also been described in human adipocytes (Madeira et al. 2014). The significance of these expression patterns requires further evaluation.

Studies in mice provide conclusive evidence of AQP7 function in adipose and AQP9 in hepatic tissue. AQP7 facilitates glycerol release from adipocytes during lipolysis, while AQP9 is crucial for glycerol entry into gluconeogenesis (Maeda et al. 2004; Jelen et al. 2011). AQP7 knockout mice display a phenotypic predisposition toward adipocyte hypertrophy and obesity (Maeda et al. 2004; Hara-Chikuma et al. 2005; Hibuse et al. 2005); therefore, inhibitors of this channel appear unsuitable as therapeutics for treating metabolic diseases. This conclusion is supported by a human genome-wide association study that identified a linkage between obesity in females and a SNP causing reduced AQP7 expression (Prudente et al. 2007). Furthermore, based on these observations, weight gain may be a side effect of AQP7 inhibitors that will need to be considered in potential applications requiring long-term systemic treatment.

AQP9 is essential for gluconeogenesis from glycerol, at least in mouse hepatocytes (Jelen et al. 2011). Moreover, AQP9 gene deletion resulted in relative normalization of postprandial blood glucose levels in a mouse model of type 2 diabetes. The potential benefit of AQP9 inhibitors in type 2 diabetes therapy has therefore been suggested (Rojek et al. 2007). It is, however, currently unknown if AQP9 is similarly essential for glycerol uptake into human hepatocytes, as in murine hepatocytes. If the described legion expression of additional glycerol-permeable AQP isoforms in human hepatocytes can be confirmed, AQP9 may be dispensable for glycerol gluconeogenesis in humans. Furthermore, the origin of the normalized blood glucose phenotype in AQP9 knockout, type 2 diabetes mice is poorly understood. While the phenotype may be explained by reduced hepatic glycerol gluconeogenesis, this interpretation may be too simplistic. Adaptations in metabolic or hormonal regulation of glucose metabolism that might contribute to the phenotype and the role of renal glycerol gluconeogenesis are not well defined. In order to determine the use of AQP9 inhibitors in type 2 diabetes therapy, further studies in a variety of animal models are required.

Another aspect that requires attention is evidence from murine and human studies suggesting that AQP9 expression is downregulated in type 2 diabetes (Catalan et al. 2008; Gena et al. 2013). It has been hypothesized that this downregulation is a protective mechanism against type 2 diabetes and obesity. In agreement with these considerations, siRNA-mediated reduction of AQP9 expression partly protected rats from non-alcoholic hepatic steatosis (Cai et al. 2013). If these observations can be confirmed, AQP9 inhibitors may be useful for reducing hepatic steatosis.

13.2.6 Disuse Osteoporosis

Osteoporosis is a disease of low bone density and altered bone architecture that carries an increased risk of fractures. An important criterion in the diagnosis of human osteoporosis is a bone mineral density below 2.5 standard deviations of the control population (Kanis et al. 1997). The disease is primarily associated with aging and decreased gonadal function, especially decreased estrogen levels; however, osteoporosis can also be caused by disuse of the musculoskeletal system, for example during prolonged confinement to bed, paralysis and space flight. While effective treatment options exist for primary osteoporosis, due to separate etiologies and pathophysiologies, no effective treatments are currently available for disuse osteoporosis (Lau and Guo 2011). AQP9 knockout mice were found to have partially ameliorated bone mineral density loss by one standard deviation compared to wild-type controls in a disuse osteoporosis model (Bu et al. 2012). In contrast, AQP9 gene deletion was found to have no effect on bone density loss in ageing mice (Liu et al. 2009). Together, these observations suggest that AQP9 inhibitors may find possible applications in delaying bone loss during prolonged confinement to bed, temporary paralysis or during space flight (Lau and Guo 2011).

13.2.7 AQP Inhibitors as Immunosuppressants

Expression of AQPs has been described in cells of the immune system, including AQP3 in T-cells (Hara-Chikuma et al. 2012b), AQP7 in skin dendritic cells (Hara-Chikuma et al. 2012a) and AQP9 in neutrophils (Jelen et al. 2013). These isoforms may promote cell protrusion formation, cell migration and endocytosis (Hara-Chikuma et al. 2012a,b; Karlsson et al. 2013) and are important in leukocyte function at multiple levels. While no general

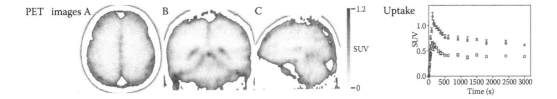

FIGURE 13.1 (left) [^{11}C]TGN-020 human PET Images (axial (A), coronal (B), sagittal (C)) constructed from frames acquired 15-60 min post injection. Standardized uptake value (SUV) scale 0-1.2, with the skull uptake set to white for visual ease. (right) Time course uptake values for [^{11}F]TGN-020 in the cortex (triangle) and choroid plexus (square). Error bars represent SEM for n = 5 volunteers. (Data courtesy of Dr. Y. Suzuki, adapted from Suzuki, Y., et al., *J Neuroimaging* 23(2), 219, 2013.)

immune deficiencies have been described in any AQP knockout mice, reduced immune responses in models of allergic contact dermatitis were found in AQP3 and AQP7 knockout mice (Hara-Chikuma et al. 2012a,b). Use of AQP3 and AQP7 inhibitors as an alternative to corticosteroid treatment of severe allergic contact dermatitis is therefore conceivable.

13.2.8 Medical Imaging

In addition to their potential as therapeutics, the use of AQP ligands as imaging agents can bring significant benefit to the treatment of patients. Changes in the expression and distribution of several AQPs are associated with disease progression that can potentially be monitored as biomarkers of those disorders. AQP-selective ligands can be used as tracers or contrast agents in magnetic resonance imaging or positron emission tomography (PET), which may be useful for diagnosing, assessing clinical outcome or monitoring treatment. The first practical example of this is the ligand [^{11}C]TGN-020, which was developed for in vivo visualization of AQP4 distribution using PET imaging (Figure 13.1). Proof-of-principle studies involving rodent models and human volunteers have been reported (Nakamura et al. 2011; Suzuki et al. 2013).

13.3 METHODS FOR AQP INHIBITOR IDENTIFICATION

13.3.1 *Xenopus laevis* Oocyte Assay

Much like other membrane surface channel proteins, AQPs can be expressed in *Xenopus laevis* oocytes (Preston et al. 1992) and used to study substrate permeation. A simple assay can be constructed by determining the swelling of individual oocytes. Ligand-mediated changes to flux can be identified by comparing sham (water injected), blank (AQP expressing with no test compound) and test (AQP expressing with test compound) oocytes. Similar experiments can be designed to measure ligand washout. While the method is very basic and can be done in virtually any laboratory, it requires a large effort, has a low screening throughput and has proven difficult to compare with other assay methodologies (Beitz et al. 2009; Verkman et al. 2014).

This assay has, nevertheless, been used in several studies to identify modulators of AQP1 and AQP4: TGN-020, described earlier, was initially identified using this type of assay

(Huber et al. 2009). The non-selective AQP1 and AQP4 inhibitor AqB013 was identified from an oocyte-screening study and was reported to be cell penetrant and selective toward AQP channels over competitor proteins, such as NKCC1 (Migliati et al. 2009). A further set of compounds was identified in a hypotonic oocyte swelling assay as AQP1 inhibitors and was found to have IC_{50} values of 8.1 ± 0.8, 17.0 ± 0.5 and 17.5 ± 0.5 µM, respectively, for three of the compounds described therein (Seeliger et al. 2013).

13.3.2 Proteoliposome Assays

Volume changes in response to osmotic shifts have also been used to study AQP function in reconstituted proteoliposomes (Zeidel et al. 1992). In concept, assays using proteoliposomes are similar to the oocyte assays described earlier. However, proteoliposomes have a volume-to-surface area ratio that is three or more orders of magnitude smaller than in oocytes. Consequently, post-osmotic shock re-equilibrium happens in less than a second, which necessitates stopped flow techniques for suitable assay control. This method has been used to successfully confirm AQP4 inhibition by acetazolamide (Tanimura et al. 2009); however, it remains unclear if proteoliposome-based assays are generally useful for identifying small molecule AQP4 inhibitors (Yang et al. 2006a, 2008).

13.3.3 Calcein Quenching Assays

Fluorescence self-quenching has long been used to study volume changes and water permeability in cell-free systems (Oku et al. 1982). This property of the calcein fluorophore was exploited by Hamann and colleagues, who used fluorescence quenching as a means of measuring fast volume changes in cultured retinal pigment cells (Hamann et al. 2002), although it was subsequently shown that calcein fluorescence intensity in living cells is instead modulated by cytosolic protein concentration (Solenov et al. 2004). Nevertheless, based on these observations, time course measurements of calcein fluorescence intensity were utilized to study AQP-mediated water and solute permeability in plate readers following an osmotic challenge (Fenton et al. 2010).

Using this method, 3575 diverse compounds were screened (Mola et al. 2009), and a small molecule inhibitor, NSC670229, was identified as a non-selective AQP1 and AQP4 inhibitor, which may be a suitable starting point for optimization studies. Similarly, a screen of 1920 structurally diverse, drug-like small molecules yielded a high number (~50) inhibitors of mouse AQP9 (Jelen et al. 2011). Most of these substances were very weak hits that merely doubled cell shrinking times (i.e. 50% inhibition) at 100 µM. However, one compound, RF03176, was found to have an IC_{50} near 1 µM. Analogue screening of another identified chemical scaffold yielded HTS13286, an inhibitor of mouse and human AQP9 water permeability in recombinant Chinese hamster ovary (CHO) cells with an IC_{50} in the range of 100–500 nM. Furthermore, both substances reduced cell water permeability to a level similar of CHO control cells without recombinant AQP expression. Both substances were highly selective for mouse AQP9, similar to phloretin, but were significantly more potent and effective than phloretin. HTS13286 also suppressed glycerol gluconeogenesis, but not gluconeogenesis from other substrates, in primary hepatocytes and perfused liver preparations. Unfortunately, these substances lacked sufficient solubility to facilitate in vivo studies in animal models. The AQP9 inhibitors identified in this study, including the moderate inhibitors, yielded suitable starting points for hit identification of human AQP9 inhibitors, as well as hit expansion by a novel in silico approach (Wacker et al. 2013).

13.3.4 End Point Hypotonic Shock Assays

Outside of the peer-reviewed literature, two methods with similar concept have been reported in patent applications. These methods were used to identify AQP1 and non-selective AQP2/4 inhibitors by high-throughput screening of small molecule libraries (Verkman 2006; Pelletier et al. 2013). Thereby, cells exposed to hypotonic shock undergo cell lysis. AQP inhibitors are identified based on reduction of cell lysis.

The clearest example of successful end point hypotonic shock screening was recently used to identify non-selective AQP2/4 inhibitors (Pelletier et al. 2013). Phenylbenzamide compounds were identified as AQP inhibitors based on delayed cell swelling and a reduction in the number of lysed cells at the end point time. Interestingly, we note chemical similarity between phenylbenzamides, as well as formylpyridines such as TGN-020 and AqB013 and even the phenylurea AQP9 inhibitor RF03176 (see Table 13.1). AQP2/4 inhibitor hit compounds from this screen having moderate potency (1–10 μM compound efficacy, ~50% inhibition) were reported as reducing cerebral edema in in vivo models of water intoxication and cerebral ischemia.

These types of cell lysis assays appear to facilitate much higher throughput, compared to time-resolved measurements of calcein quenching, and may help overcome the major bottleneck in developing AQP inhibitor-based therapies: the unbiased identification of true starting points for hit-to-lead drug development programmes and structure-based, in silico compound optimization.

13.4 IN SILICO METHODS FOR AQP INHIBITOR OPTIMIZATION (AND POTENTIALLY *AB INITIO* IDENTIFICATION)

Computer-based modelling techniques are a standard method in contemporary drug discovery. Two computational methods, in particular, have been used to identify or model AQP-related ligands: molecular docking and molecular dynamics (MD) simulations. These methods are powerful tools for identifying key molecular interactions and provide valuable insights for the development and optimization of AQP-modulating drugs.

13.4.1 Molecular Docking

Molecular docking (docking) has been extensively used to identify new ligands from virtual libraries (virtual screening), to dock known ligands in order to suggest a putative binding pose and to identify and validate likely binding sites. Recently, docking virtual screening techniques have been applied to identify AQP ligands. Seeliger et al. (2013) identified three high affinity inhibitors of human AQP1 using the docking software LeadIT (Rarey et al. 1996). Wacker et al. (2013) used a homology model of AQP9 for virtual screening and identified novel AQP9 blockers with IC50s in the low μM range.

In the following publications docking was used to model a known ligand and to explore possible binding sites: Tradtrantip et al. (2012) used a two-step docking approach to

TABLE 13.1 Summary of Reviewed Pathologies and AQPs as Therapeutic Targets

AQP Function/ Mechanism	Potential Therapeutic Applications	Evidence for Utility as Drug Target	Available Chemical Modulators	Example Structures	References
AQP1 facilitates angiogenesis.	Solid malignant tumours	Tumor homo and allograft models in AQP1 knockout mouse	AQP1 inhibitors with weak Hit[a] characteristics	(3-bromo-5-{(E)-[(1,5-dimethyl-3-oxo-2-phenyl-2,3-dihydro-1H-pyrazol-4-yl)imino]methyl}-4-hydroxyphenyl) (hydroxy)oxoammonium.	Saadoun et al. (2005), Verkman (2006), Seeliger et al. (2013), Verkman et al. (2014)
AQP1 and 4 contribute to ocular fluid formation.	Glaucoma	Knockout mouse model	AQP1 inhibitors as mentioned earlier, AQP4 inhibitors, for example TGN-020, AqB013, NSC164914, phenylbenzamides	TGN-020	Zhang et al. (2002), Huber et al. (2009), Migliati et al. (2009), Mola et al. (2009), Pelletier et al. (2013)
AQP2 facilitates fluid uptake from renal collecting ducts.	Nephrogenic diabetes insipidus	Human functional mutations, knockout mouse	EP2 agonists, statins, AG-490	AG-490	Procino et al. (2010), Olesen et al. (2011), Bogum et al. (2013), Nomura et al. (2014)
AQP2 and AQP3 facilitate fluid uptake from renal collecting ducts.	Congestive heart failure, hyponatremia	Human functional mutations, knockout mouse phenotypes	Phenylbenzamides (AQP2)	5-chloro-N-(3,5-dichlorophenyl)-2-hydroxybenzamide	Pelletier et al. (2013)

(Continued)

TABLE 13.1 (Continued)   Summary of Reviewed Pathologies and AQPs as Therapeutic Targets

AQP Function/Mechanism	Potential Therapeutic Applications	Evidence for Utility as Drug Target	Available Chemical Modulators	Example Structures	References
AQP3 and AQP7 contribute to immune cell migration/function.	Severe contact dermatitis	AQP3 and AQP7 knockout mice	Gold-based AQP3/7 inhibitors	Auphene	Hara-Chikuma et al. (2012a,b)
AQP4 may promote glioma cell tissue invasion.	Glioblastoma multiforme	Cell-based assays/siRNA to AQP4	AQP4 inhibitors, as mentioned earlier	AqB013	Ding et al. (2011b)
AQP4 enhances astrocyte water permeability.	Cytotoxic brain edema, traumatic brain injury	Knockout mouse/in vivo siRNA phenotypes	AQP4 inhibitors, as mentioned earlier		Manley et al. (2000), Vajda et al. (2002), Fukuda et al. (2013)
AQP5 may promote tumor progression.	Cancers	Correlation of AQP5 expression with outcome	None		Chae et al. (2008b)
AQP9 facilitates hepatocyte glycerol uptake.	Diabetes	AQP9 db/db knockout mouse model	AQP9 inhibitors, for example T6963384, RF03176	RF03176	Rojek et al. (2007), Jelen et al. (2011), Wacker et al. (2013)
AQP9 has unknown role in osteoclast function.	Reduce bone loss during temporary paralysis	Knockout mouse model	AQP9 inhibitors, as mentioned earlier		Bu et al. (2012)

a We define a Hit as a confirmed active compound, which is usually discovered in a compound screen.

suggest an extracellular binding site of compounds that inhibit autoantibodies (NMO-IgG) binding to AQP4. Martins et al. (2012) used standard docking parameters to analyze the binding to a proposed binding site of AQP3 for the gold compounds Auphen and Audien. Madeira et al. (2014) proposed a binding site of Auphen for AQP7 based on docking results, also using a homology model.

The studies show that AQP ligands can be identified even with low experimental effort based on docking results. It seems that the highly conserved 3D structures of AQPs are beneficial for homology modelling. The compounds binding poses purely based on docking, however, should be interpreted carefully and validated, for example with MD simulations or mutagenesis studies.

Docking will surely remain an essential method in structure-based drug design for the foreseeable future, since it is efficient and easily applicable. This general method will surely benefit from the integration of machine learning principles into the optimization of standard and target-specific scoring functions (Yuriev and Ramsland 2013).

13.4.2 Molecular Dynamics Simulations

MD simulations provide dynamic information about proteins, DNA and other biological molecules (Pronk et al. 2013) and play an increasingly important role in the field of computational drug development. MD simulations use molecular mechanics force fields and Newton's equations of motion to simulate the dynamics of molecules in silico. Exceedingly valuable insights have been gained by MD simulations with respect to AQP function and the mechanisms of solute permeation (de Groot and Grubmuller 2001; Roux and Schulten 2004; Hub et al. 2009).

Complete receptor–ligand complex formation was simulated in silico using free and unbiased MD simulation (Dror et al. 2011). However, such simulations are exceptionally rare since they require special hardware that is usually unavailable. Typical timescales involved in drug binding and unbinding are in the range of microseconds to seconds or even minutes (Copeland 2011), whereas typical contemporary MD simulations cover ranges of several hundreds of nanoseconds up to microseconds and in rare cases milliseconds (Lane et al. 2013). Therefore, free MD simulations are useful to refine premodelled states of receptor-ligand complexes, for example those generated from docking. And the identification of a stable ligand pose in an MD simulation should be considered a minimum criterion for accepting a predicted receptor–ligand complex structure (Seeliger et al. 2013).

The estimation of binding affinities based on MD simulations is usually computationally more efficient than MD simulations but are also technically more challenging. Techniques like *umbrella sampling* (Torrie and Valleau 1977; Hub et al. 2009) and *free energy perturbation* methods are used (Bruckner and Boresch 2011; Gapsys et al. 2015) for these studies. The free energy difference between the bound and the unbound state defines the affinity of a compound to a receptor. Since free energy calculations are, in general, computationally demanding, they are usually reserved for comparisons between closely related states where only a few atoms in the systems are changed, for example a ligand functional group or a protein side chain. Seeliger et al. (2013) used non-equilibrium free energy calculations to estimate the energetic contribution of residue K36 to AQP1 inhibitor binding using the

Crooks–Gaussian intersection protocol (Goette and Grubmuller 2009). The subsequently generated mutant hAQP1-K36A indeed showed no ligand-mediated water flux decrease in an oocytes swelling assay, which was consistent with inhibitor binding at the proposed site. Furthermore, the structural factors involved in AQP1 inhibition, as suggested by these results, are expected to be valuable for the design of future AQP1 related drugs.

Estimation of the drug efficacy, rather than affinity, of AQP inhibitors was attempted by calculating ligand-mediated changes in the osmotic permeation rate using equilibrium MD simulations (Muller et al. 2008; Hirano et al. 2010; Seeliger et al. 2013; Wacker et al. 2013). Seeliger et al. (2013) and Wacker et al. (2013) demonstrated that water flux through AQP4 and AQP9 was reduced in the presence of an inhibitor using a collective coordinate approach (Zhu et al. 2004), which further support the hypothesized binding poses.

MD simulation–based methods will surely occupy a larger role in computer-aided drug discovery. Simulations are significantly more complex than molecular docking. However, new methods for automating the simulation setup are under development, which will make more complex studies, for example free energy simulations, easier to run. (Pronk et al. 2013). Frameworks are already available for the automated parameterization of a broad range of organic compounds (Wang et al. 2004; Vanommeslaeghe and MacKerell 2012; Vanommeslaeghe et al. 2012). MD simulations can be performed with explicit solvent molecules at atomistic spatial and femtoseconds temporal resolution (Pronk et al. 2013). Such studies allow for full ligand and receptor flexibility and are able to capture the ligand binding process much more realistically compared to molecular docking. Consequently, MD simulations already are an increasingly versatile and reliable tool for computational drug design of AQP inhibitors.

13.5 CONCLUSIONS

Widespread enthusiasm and optimism remain for using AQP modulators as therapeutic agents in the treatment of human diseases. However, it is clear that an insufficient number of compounds have been identified in in vitro screens and in vivo disease model studies to fully evaluate the utility of aquaporin modulation in clinical pharmacy. Furthermore, we note that aquaporin inhibitor isoform specificity has been established in few studies. Such knowledge is an important pre-requisite in an eventual drug development program.

REFERENCES

Agre, P. 2004. Nobel Lecture. Aquaporin water channels. *Biosci Rep*, 24:127–63.

Agre, P., L.S. King, M. Yasui, W.B. Guggino, O.P. Ottersen, Y. Fujiyoshi, A. Engel and S. Nielsen. 2002. Aquaporin water channels—From atomic structure to clinical medicine. *J Physiol* 542(Part 1):3–16.

Akdemir, G., J. Ratelade, N. Asavapanumas and A.S. Verkman. 2014. Neuroprotective effect of aquaporin-4 deficiency in a mouse model of severe global cerebral ischemia produced by transient 4-vessel occlusion. *Neurosci Lett* 574:70–75.

Ali, F., M. Guglin, P. Vaitkevicius and J.K. Ghali. 2007. Therapeutic potential of vasopressin receptor antagonists. *Drugs* 67(6):847–858.

Badaut, J., J.F. Brunet and L. Regli. 2007. Aquaporins in the brain: From aqueduct to "multi-duct". *Metab Brain Dis* 22(3–4):251–263.

Beitz, E., D. Becker, J. von Bulow, C. Conrad, N. Fricke, A. Geadkaew, D. Krenc, J. Song, D. Wree and B. Wu. 2009. In vitro analysis and modification of aquaporin pore selectivity. *Handb Exp Pharmacol* 2009(190):77–92.

Binder, D.K., E.A. Nagelhus and O.P. Ottersen. 2012. Aquaporin-4 and epilepsy. *Glia* 60(8):1203–1214.

Binder, D.K., K. Oshio, T. Ma, A.S. Verkman and G.T. Manley. 2004. Increased seizure threshold in mice lacking aquaporin-4 water channels. *Neuroreport* 15(2):259–262.

Binder, D.K., X. Yao, Z. Zador, T.J. Sick, A.S. Verkman and G.T. Manley. 2006. Increased seizure duration and slowed potassium kinetics in mice lacking aquaporin-4 water channels. *Glia* 53(6):631–636.

Bogum, J., D. Faust, K. Zuhlke, J. Eichhorst, M.C. Moutty, J. Furkert, A. Eldahshan et al. 2013. Small-molecule screening identifies modulators of aquaporin-2 trafficking. *J Am Soc Nephrol* 24(5):744–748.

Bruckner, S. and S. Boresch. 2011. Efficiency of alchemical free energy simulations. II. Improvements for thermodynamic integration. *J Comput Chem* 32(7):1320–1333.

Bu, G., F. Shuang, Y. Wu, D. Ren and S. Hou. 2012. AQP9: A novel target for bone loss induced by microgravity. *Biochem Biophys Res Commun* 419(4):774–778.

Cai, C., C. Wang, W. Ji, B. Liu, Y. Kang, Z. Hu and Z. Jiang. 2013. Knockdown of hepatic aquaglyceroporin-9 alleviates high fat diet-induced non-alcoholic fatty liver disease in rats. *Int Immunopharmacol* 15(3):550–556.

Castle, N.A. 2005. Aquaporins as targets for drug discovery. *Drug Discov Today* 10(7):485–493.

Catalan, V., J. Gomez-Ambrosi, C. Pastor, F. Rotellar, C. Silva, A. Rodriguez, M.J. Gil et al. 2008. Influence of morbid obesity and insulin resistance on gene expression levels of AQP7 in visceral adipose tissue and AQP9 in liver. *Obes Surg* 18(6):695–701.

Chae, Y.K., J. Woo, M.J. Kim, S.K. Kang, M.S. Kim, J. Lee, S.K. Lee et al. 2008a. Expression of aquaporin 5 (AQP5) promotes tumor invasion in human non small cell lung cancer. *PLoS One* 3(5):e2162.

Clapp, C. and G. Martinez de la Escalera. 2006. Aquaporin-1: A novel promoter of tumor angiogenesis. *Trends Endocrinol Metab* 17(1):1–2.

Copeland, R.A. 2011. Conformational adaptation in drug–target interactions and residence time. *Future Med Chem* 3(12):1491–1501.

de Groot, B.L. and H. Grubmuller. 2001. Water permeation across biological membranes: Mechanism and dynamics of aquaporin-1 and GlpF. *Science* 294(5550):2353–2357.

Ding, T., Y. Ma, W. Li, X. Liu, G. Ying, L. Fu and F. Gu. 2011. Role of aquaporin-4 in the regulation of migration and invasion of human glioma cells. *Int J Oncol* 38(6):1521–1531.

Ding, T., Y. Zhou, K. Sun, W. Jiang, W. Li, X. Liu, C. Tian et al. 2013. Knockdown a water channel protein, aquaporin-4, induced glioblastoma cell apoptosis. *PLoS One* 8(8):e66751.

Dror, R.O., A.C. Pan, D.H. Arlow, D.W. Borhani, P. Maragakis, Y. Shan, H. Xu and D.E. Shaw. 2011. Pathway and mechanism of drug binding to G-protein-coupled receptors. *Proc Natl Acad Sci USA* 108(32):13118–13123.

Endo, M., R.K. Jain, B. Witwer and D. Brown. 1999. Water channel (aquaporin 1) expression and distribution in mammary carcinomas and glioblastomas. *Microvasc Res* 58(2):89–98.

Esteva-Font, C., P.-W. Phuan, M.O. Anderson and A.S. Verkman. 2013. A small molecule screen identifies selective inhibitors of urea transporter UT-A. *Chem Biol* 20(10):1235–1244.

Fenton, R.A., H.B. Moeller, S. Nielsen, B.L. de Groot and M. Rutzler. 2010. A plate reader-based method for cell water permeability measurement. *Am J Physiol Ren Physiol* 298(1):F224–F230.

Ferguson-Myrthil, N. 2010. Novel agents for the treatment of hyponatremia: A review of conivaptan and tolvaptan. *Cardiol Rev* 18(6):313–321.

Frigeri, A., M.A. Gropper, C.W. Turck and A.S. Verkman. 1995. Immunolocalization of the mercurial-insensitive water channel and glycerol intrinsic protein in epithelial cell plasma membranes. *Proc Nat Acad Sci* 92(10):4328–4331.

Fukuda, A.M., A. Adami, V. Pop, J.A. Bellone, J.S. Coats, R.E. Hartman, S. Ashwal, A. Obenaus and J. Badaut. 2013. Posttraumatic reduction of edema with aquaporin-4 RNA interference improves acute and chronic functional recovery. *J Cereb Blood Flow Metab* 33:1621–1632.

Fukuda, A.M., V. Pop, D. Spagnoli, S. Ashwal, A. Obenaus and J. Badaut. 2012. Delayed increase of astrocytic aquaporin 4 after juvenile traumatic brain injury: Possible role in edema resolution? *Neuroscience* 222:366–378.

Gapsys, V., S. Michielssens, J.H. Peters, B.L. de Groot and H. Leonov. 2015. Calculation of binding free energies. *Methods Mol Biol* 1215:173–209.

Gena, P., M. Mastrodonato, P. Portincasa, E. Fanelli, D. Mentino, A. Rodríguez, R.A. Marinelli et al. 2013. Liver glycerol permeability and aquaporin-9 are dysregulated in a murine model of non-alcoholic fatty liver disease. *PLoS One* 8(10):e78139.

Goette, M. and H. Grubmuller. 2009. Accuracy and convergence of free energy differences calculated from nonequilibrium switching processes. *J Comput Chem* 30(3):447–456.

Greenberg, A. and J.G. Verbalis. 2006. Vasopressin receptor antagonists. *Kidney Int* 69(12): 2124–2130.

Guo, Q., I. Sayeed, L.M. Baronne and S.W. Hoffman. 2006. Progesterone administration modulates AQP4 expression and edema after traumatic brain injury in male rats. *Experimental.* 198, 469–478.

Hamann, S., J.F. Kiilgaard, T. Litman, F.J. Alvarez-Leefmans, B.R. Winther and T. Zeuthen. 2002. Measurement of cell-volume changes by fluorescence self-quenching. *J Fluoresc* 12(2): 139–145.

Hara-Chikuma, M., S. Chikuma, Y. Sugiyama, K. Kabashima, A. Verkman, S. Inoue and Y. Miyachi. 2012b. Chemokine-dependent T cell migration requires aquaporin-3-mediated hydrogen peroxide uptake. *J Exp Med* 209(10):1743–1752.

Hara-Chikuma, M., E. Sohara, T. Rai, M. Ikawa, M. Okabe, S. Sasaki, S. Uchida and A.S. Verkman. 2005. Progressive adipocyte hypertrophy in aquaporin-7-deficient mice: Adipocyte glycerol permeability as a novel regulator of fat accumulation. *J Biol Chem* 280(16):15493–15496.

Hara-Chikuma, M., Y. Sugiyama, K. Kabashima, E. Sohara, S. Uchida, S. Sasaki, S. Inoue and Y. Miyachi. 2012a. Involvement of aquaporin-7 in the cutaneous primary immune response through modulation of antigen uptake and migration in dendritic cells. *FASEB J* 26(1): 211–218.

Hara-Chikuma, M. and A.S. Verkman. 2006. Physiological roles of glycerol-transporting aquaporins: The aquaglyceroporins. *Cell Mol Life Sci* 63(12):1386–1392.

Heymann, J.B. and Engel, A. 1999. Aquaporins: phylogeny, structure, and physiology of water channels. *News Physiol Sci*, 14:187–193.

Hibuse, T., N. Maeda, T. Funahashi, K. Yamamoto, A. Nagasawa, W. Mizunoya, K. Kishida et al. 2005. Aquaporin 7 deficiency is associated with development of obesity through activation of adipose glycerol kinase. *Proc Natl Acad Sci USA* 102(31):10993–10998.

Hirano, Y., N. Okimoto, I. Kadohira, M. Suematsu, K. Yasuoka and M. Yasui. 2010. Molecular mechanisms of how mercury inhibits water permeation through aquaporin-1: Understanding by molecular dynamics simulation. *Biophys J* 98(8):1512–1519.

Hu, J. and A.S. Verkman. 2006. Increased migration and metastatic potential of tumor cells expressing aquaporin water channels. *FASEB J* 20(11):1892–1894.

Hub, J.S., H. Grubmuller and B.L. de Groot. 2009. Dynamics and energetics of permeation through aquaporins. What do we learn from molecular dynamics simulations? *Handb Exp Pharmacol* 190:57–76.

Huber, V.J., M. Tsujita and T. Nakada. 2009. Identification of aquaporin 4 inhibitors using in vitro and in silico methods. *Bioorg Med Chem* 17(1):411–417.

Huber, V.J., M. Tsujita and T. Nakada. 2012. Aquaporins in drug discovery and pharmacotherapy. *Mol Aspects Med* 33(5–6):691–703.

Iliff, J.J., Chen, M.J., Plog, B.A., Zeppenfeld, D.M., Soltero, M., Yang, L., Singh, I., Deane, R. and Nedergaard, M. 2014. Impairment of glymphatic pathway function promotes tau pathology after traumatic brain injury. *J Neurosci*, 34:16180–93.

Igarashi, H., V.J. Huber, M. Tsujita and T. Nakada. 2011. Pretreatment with a novel aquaporin 4 inhibitor, TGN-020, significantly reduces ischemic cerebral edema. *Neurol Sci* 32(1):113–116.

Ishibashi, K., S. Hara and S. Kondo. 2009. Aquaporin water channels in mammals. *Clin Exp Nephrol* 13(2):107–117.

Ishibashi, K., M. Kuwahara, Y. Gu and Y. Kageyama. 1997. Cloning and functional expression of a new water channel abundantly expressed in the testis permeable to water, glycerol and urea. *J Biol Chem* 272:20782–20786.

Ishibashi, K., T. Morinaga, M. Kuwahara, S. Sasaki and M. Imai. 2002. Cloning and identification of a new member of water channel (AQP10) as an aquaglyceroporin. *Biochim Biophys Acta-Gene Struct Expr* 1576(3):335–340.

Itoh, T., T. Rai, M. Kuwahara, S.B. Ko, S. Uchida, S. Sasaki and K. Ishibashi. 2005. Identification of a novel aquaporin, AQP12, expressed in pancreatic acinar cells. *Biochem Biophys Res Commun* 330(3):832–838.

Jelen, S., B. Parm Ulhoi, A. Larsen, J. Frokiaer, S. Nielsen and M. Rutzler. 2013. AQP9 expression in glioblastoma multiforme tumors is limited to a small population of astrocytic cells and CD15(+)/CalB(+) leukocytes. *PLoS One* 8(9):e75764.

Jelen, S., S. Wacker, C. Aponte-Santamaria, M. Skott, A. Rojek, U. Johanson, P. Kjellbom, S. Nielsen, B.L. de Groot and M. Rützler. 2011. Aquaporin-9 protein is the primary route of hepatocyte glycerol uptake for glycerol gluconeogenesis in mice. *J Biol Chem* 286(52): 44319–44325.

Kang, S.K., Y.K. Chae, J. Woo, M.S. Kim, J.C. Park, J. Lee, J.C. Soria, S.J. Jang, D. Sidransky and C. Moon. 2008. Role of human aquaporin 5 in colorectal carcinogenesis. *Am J Pathol* 173(2): 518–525.

Kanis, J.A., P. Delmas, P. Burckhardt, C. Cooper and D. Torgerson. 1997. Guidelines for diagnosis and management of osteoporosis. The European Foundation for Osteoporosis and Bone Disease. *Osteoporos Int* 7(4):390–406.

Karlsson, T., B.C. Lagerholm, E. Vikström, V.M. Loitto and K.-E.E. Magnusson. 2013. Water fluxes through aquaporin-9 prime epithelial cells for rapid wound healing. *Biochem Biophys Res Commun* 430(3):993–998.

Katada, R., G. Akdemir, N. Asavapanumas, J. Ratelade, H. Zhang and A.S. Verkman. 2013. Greatly improved survival and neuroprotection in aquaporin-4-knockout mice following global cerebral ischemia. *FASEB J* 28, 705–714.

Ke, C., W.S. Poon, H.K. Ng, J.C. Pang and Y. Chan. 2001. Heterogeneous responses of aquaporin-4 in oedema formation in a replicated severe traumatic brain injury model in rats. *Neurosci Lett* 301(1):21–24.

Kiening, K.L., F.K. van Landeghem, S. Schreiber, U.W. Thomale, A. von Deimling, A.W. Unterberg and J.F. Stover. 2002. Decreased hemispheric aquaporin-4 is linked to evolving brain edema following controlled cortical impact injury in rats. *Neurosci Lett* 324(2):105–108.

Kimelberg, H.K. 2004. Water homeostasis in the brain: Basic concepts. *Neuroscience* 129(4): 851–860.

King, L.S., D. Kozono and P. Agre. 2004. From structure to disease: The evolving tale of aquaporin biology. *Nat Rev Mol Cell Biol* 5(9):687–698.

Kortenoeven, M.L. and R.A. Fenton. 2014. Renal aquaporins and water balance disorders. *Biochim Biophys Acta* 1840(5):1533–1549.

Kruse, E., Uehlein, N. and Kaldenhoff, R. 2006. The aquaporins. *Genome Biol*, 7:206.

Ku, E., N. Nobakht and V.M. Campese. 2009. Lixivaptan: A novel vasopressin receptor antagonist. *Expert Opin Investig Drugs* 18(5):657–662.

Kuriyama, H., S. Kawamoto, N. Ishida, I. Ohno, S. Mita, Y. Matsuzawa, K. Matsubara and K. Okubo. 1997. Molecular cloning and expression of a novel human aquaporin from adipose tissue with glycerol permeability. *Biochem Biophys Res Commun* 241(1):53–58.

Laforenza, U., M.F. Scaffino and G. Gastaldi. 2013. Aquaporin-10 represents an alternative pathway for glycerol efflux from human adipocytes. *PLoS One* 8(1):e54474.

Lane, T.J., D. Shukla, K.A. Beauchamp and V.S. Pande. 2013. To milliseconds and beyond: Challenges in the simulation of protein folding. *Curr Opin Struct Biol* 23(1):58–65.

Lau, R. and X. Guo. 2011. A review on current osteoporosis research: With special focus on disuse bone loss. *J Osteoporos* 2011.

Lee, S.J., Y.S. Chae, J.G. Kim, W.W. Kim, J.H. Jung, H.Y. Park, J.Y. Jeong, J.Y. Park, H.J. Jung and T.H. Kwon. 2014. AQP5 expression predicts survival in patients with early breast cancer. *Ann Surg Oncol* 21(2):375–383.

Levin, M.H., R. de la Fuente and A.S. Verkman. 2007. Urearetics: A small molecule screen yields nanomolar potency inhibitors of urea transporter UT-B. *FASEB J* 21:551–563.

Liu, Y., L. Song, Y. Wang, A. Rojek, S. Nielsen, P. Agre and J.M. Carbrey. 2009. Osteoclast differentiation and function in aquaglyceroporin AQP9-null mice. *Biol Cell* 101(3):133–140.

Lu, D.C., Z. Zador, J. Yao, F. Fazlollahi and G.T. Manley. 2011. Aquaporin-4 reduces post-traumatic seizure susceptibility by promoting astrocytic glial scar formation in mice. *J Neurotrauma*.

Ma, T., Y. Song, B. Yang, A. Gillespie, E.J. Carlson, C.J. Epstein and A.S. Verkman. 2000. Nephrogenic diabetes insipidus in mice lacking aquaporin-3 water channels. *Proc Natl Acad Sci USA* 97(8):4386–4391.

Ma, T., B. Yang, A. Gillespie, E.J. Carlson, C.J. Epstein and A.S. Verkman. 1998. Severely impaired urinary concentrating ability in transgenic mice lacking aquaporin-1 water channels. *J Biol Chem* 273(8):4296–4299.

Machida, Y., Y. Ueda, M. Shimasaki, K. Sato, M. Sagawa, S. Katsuda and T. Sakuma. 2011. Relationship of aquaporin 1, 3, and 5 expression in lung cancer cells to cellular differentiation, invasive growth, and metastasis potential. *Hum Pathol* 42(5):669–678.

Madeira, A., S. Fernandez-Veledo, M. Camps, A. Zorzano, T.F. Moura, V. Ceperuelo-Mallafre, J. Vendrell and G. Soveral. 2014. Human aquaporin-11 is a water and glycerol channel and localizes in the vicinity of lipid droplets in human adipocytes. *Obesity (Silver Spring)* 22(9): 2010–2017.

Maeda, N., T. Funahashi, T. Hibuse, A. Nagasawa, K. Kishida, H. Kuriyama, T. Nakamura, S. Kihara, I. Shimomura and Y. Matsuzawa. 2004. Adaptation to fasting by glycerol transport through aquaporin 7 in adipose tissue. *Proc Natl Acad Sci USA* 101(51):17801–17806.

Manley, G.T., M. Fujimura, T. Ma, N. Noshita, F. Filiz, A.W. Bollen, P. Chan and A.S. Verkman. 2000. Aquaporin-4 deletion in mice reduces brain edema after acute water intoxication and ischemic stroke. *Nat Med* 6(2):159–163.

Martins, A.P., A. Marrone, A. Ciancetta, A.G. Cobo, M. Echevarria, T.F. Moura, N. Re, A. Casini and G. Soveral. 2012. Targeting aquaporin function: Potent inhibition of aquaglyceroporin-3 by a gold-based compound. *PLoS One* 7(5):e37435.

Migliati, E., N. Meurice, P. DuBois, J.S. Fang, S. Somasekharan, E. Beckett, G. Flynn and A.J. Yool. 2009. Inhibition of aquaporin-1 and aquaporin-4 water permeability by a derivative of the loop diuretic bumetanide acting at an internal pore-occluding binding site. *Mol Pharmacol* 76(1):105–112.

Miyazaki, T., H. Fujiki, Y. Yamamura, S. Nakamura and T. Mori. 2007. Tolvaptan, an orally active vasopressin V(2)-receptor antagonist—Pharmacology and clinical trials. *Cardiovasc Drug Rev* 25(1):1–13.

Mobasheri, A., D. Marples, I.S. Young, R.V. Floyd, C.A. Moskaluk and A. Frigeri. 2007. Distribution of the AQP4 water channel in normal human tissues: Protein and tissue microarrays reveal expression in several new anatomical locations, including the prostate gland and seminal vesicles. *Channels (Austin)* 1(1):29–38.

Mola, M.G., G.P. Nicchia, M. Svelto, D.C. Spray and A. Frigeri. 2009. Automated cell-based assay for screening of aquaporin inhibitors. *Anal Chem* 81:8219–8229.

Monzani, E., R. Bazzotti, C. Perego and C.A. La Porta. 2009. AQP1 is not only a water channel: It contributes to cell migration through Lin7/beta-catenin. *PLoS One* 4(7):e6167.

Monzani, E., A.A. Shtil and C.A. La Porta. 2007. The water channels, new druggable targets to combat cancer cell survival, invasiveness and metastasis. *Curr Drug Targets* 8(10): 1132–1137.

Muller, E.M., J.S. Hub, H. Grubmuller and B.L. de Groot. 2008. Is TEA an inhibitor for human aquaporin-1? *Pflugers Arch* 456(4):663–669.

Mylonakou, M.N., P.H. Petersen, E. Rinvik, A. Rojek, E. Valdimarsdottir, S. Zelenin, T. Zeuthen, S. Nielsen, O.P. Ottersen and M. Amiry-Moghaddam. 2009. Analysis of mice with targeted deletion of AQP9 gene provides conclusive evidence for expression of AQP9 in neurons. *J Neurosci Res* 87(6):1310–1322.

Nakamura, Y., Y. Suzuki, M. Tsujita, V.J. Huber, K. Yamada and T. Nakada. 2011. Development of a novel ligand, [C]TGN-020, for aquaporin 4 positron emission tomography imaging. *ACS Chem Neurosci* 2(10):568–571.

Nico, B. and D. Ribatti. 2010. Aquaporins in tumor growth and angiogenesis. *Cancer Lett* 294(2):135–138.

Nielsen, S., J. Frokiaer, D. Marples, T.H. Kwon, P. Agre and M.A. Knepper. 2002. Aquaporins in the kidney: From molecules to medicine. *Physiol Rev* 82(1):205–244.

Nielsen, S., B.L. Smith, E.I. Christensen and P. Agre. 1993. Distribution of the aquaporin CHIP in secretory and resorptive epithelia and capillary endothelia. *Proc Natl Acad Sci USA* 90(15):7275–7259.

Nomura, N., P. Nunes, R. Bouley, A.V. Nair, S. Shaw and E. Ueda, N. Pathomthongtaweechai, H.A. Lu and D. Brown. 2014. High throughput chemical screening identifies AG-490 as a stimulator of aquaporin 2 membrane expression and urine concentration. *Am J Physiol Cell Physiol* 307:C597–C605.

Oku, N., D.A. Kendall and R.C. MacDonald. 1982. A simple procedure for the determination of the trapped volume of liposomes. *Biochim Biophys Acta (BBA)-Biomembr* 691(2): 332–340.

Olesen, E.T., M.R. Rützler, H.B. Moeller, H.A. Praetorius and R.A. Fenton. 2011. Vasopressin-independent targeting of aquaporin-2 by selective E-prostanoid receptor agonists alleviates nephrogenic diabetes insipidus. *Proc Natl Acad Sci USA* 108(31):12949–12954.

Overgaard-Steensen, C. 2011. Initial approach to the hyponatremic patient. *Acta Anaesthesiol Scand* 55(2):139–148.

Papadopoulos, M.C. and S. Saadoun. 2014. Key roles of aquaporins in tumor biology. *Biochim Biophys Acta* 1848(10 Pt B):2576–2583.

Pelletier, M.F., G.W. Farr, P.R. McGuirk, C.H. Hall and W.F. Boron. 2013. New methods. Google Patents WO2013169939 A2.

Preston, G.M., T.P. Carroll, W.B. Guggino and P. Agre. 1992. Appearance of water channels in *Xenopus* oocytes expressing red cell CHIP28 protein. *Science* 256(5055):385–387.

Preston, G.M., B.L. Smith, M.L. Zeidel, J.J. Moulds and P. Agre. 1994. Mutations in aquaporin-1 in phenotypically normal humans without functional CHIP water channels. *Science* 265(5178):1585–1587.

Procino, G., C. Barbieri, M. Carmosino, F. Rizzo, G. Valenti and M. Svelto. 2010. Lovastatin-induced cholesterol depletion affects both apical sorting and endocytosis of aquaporin-2 in renal cells. *Am J Physiol Ren Physiol* 298(2):78.

Pronk, S., S. Páll, R. Schulz, P. Larsson, P. Bjelkmar, R. Apostolov, M.R. Shirts et al. 2013. GROMACS 4.5: A high-throughput and highly parallel open source molecular simulation toolkit. *Bioinformatics* 29:845–854.

Prudente, S., E. Flex, E. Morini, F. Turchi, D. Capponi, S. De Cosmo, V. Tassi et al. 2007. A functional variant of the adipocyte glycerol channel aquaporin 7 gene is associated with obesity and related metabolic abnormalities. *Diabetes* 56(5):1468–1474.

Rarey, M., B. Kramer, T. Lengauer and G. Klebe. 1996. A fast flexible docking method using an incremental construction algorithm. *J Mol Biol* 261(3):470–489.

Ren, Z., J.J. Iliff, L. Yang, J. Yang, X. Chen, M.J. Chen, R.N. Giese, B. Wang, X. Shi and M. Nedergaard. 2013. 'Hit & Run' model of closed-skull traumatic brain injury (TBI) reveals complex patterns of post-traumatic AQP4 dysregulation. *J Cereb Blood Flow Metab* 33(6):834–845.

Robertson, G.L. 2011. Vaptans for the treatment of hyponatremia. *Nat Rev Endocrinol* 7(3): 151–161.

Rodríguez, A., V. Catalán, J. Gómez-Ambrosi, S. García-Navarro, F. Rotellar, V. Valentí, C. Silva et al. 2011. Insulin- and leptin-mediated control of aquaglyceroporins in human adipocytes and hepatocytes is mediated via the PI3K/Akt/mTOR signaling cascade. *J Clin Endocrinol Metab* 96(4):97.

Rodríguez, A., P. Gena, L. Méndez-Giménez, A. Rosito, V. Valentí, F. Rotellar, I. Sola et al. 2014. Reduced hepatic aquaporin-9 and glycerol permeability are related to insulin resistance in non-alcoholic fatty liver disease. *Int J Obes (Lond)* 38, 1213–1220.

Rojek, A., J. Praetorius, J. Frokiaer, S. Nielsen and R.A. Fenton. 2008. A current view of the mammalian aquaglyceroporins. *Annu Rev Physiol* 70:301–327.

Rojek, A.M., M.T. Skowronski, E.M. Fuchtbauer, A.C. Fuchtbauer, R.A. Fenton, P. Agre, J. Frokiaer and S. Nielsen. 2007. Defective glycerol metabolism in aquaporin 9 (AQP9) knockout mice. *Proc Natl Acad Sci USA* 104(9):3609–3614.

Roudier, N., P. Ripoche, P. Gane, P.Y.L. Pennec, G. Daniels, J.-P.P. Cartron and P. Bailly. 2002. AQP3 deficiency in humans and the molecular basis of a novel blood group system, GIL. *J Biol. Chem* 277(48):45854–45859.

Roux, B. and K. Schulten. 2004. Computational studies of membrane channels. *Structure* 12(8):1343–1351.

Roy Yuen-chi L. and Xia G. 2011. A review on current osteoporosis research: With special focus on disuse bone loss, *Journal of Osteoporosis*, 2011(293808):6.

Saadoun, S. 2002. Aquaporin-4 expression is increased in oedematous human brain tumours. *J Neurol Neurosurg Psychiatry* 72(2):262–265.

Saadoun, S., M.C. Papadopoulos, D.C. Davies, B.A. Bell and S. Krishna. 2002. Increased aquaporin 1 water channel expression in human brain tumours. *Br J Cancer* 87(6):621–623.

Saadoun, S., M.C. Papadopoulos, M. Hara-Chikuma and A.S. Verkman. 2005. Impairment of angiogenesis and cell migration by targeted aquaporin-1 gene disruption. *Nature* 434(7034): 786–792.

Sawada, T., Y. Kato and M. Kobayashi. 2007. Expression of aquaporine-4 in central nervous system tumors. *Brain Tumor Pathol* 24(2):81–84.

Seeliger, D., C. Zapater, D. Krenc, R. Haddoub, S. Flitsch, E. Beitz, J. Cerda and B.L. de Groot. 2013. Discovery of novel human aquaporin-1 blockers. *ACS Chem Biol* 8(1):249–256.

Shenaq, M., H. Kassem, C. Peng, S. Schafer, J.Y. Ding, V. Fredrickson, M. Guthikonda, C.W. Kreipke, J.A. Rafols and Y. Ding. 2012. Neuronal damage and functional deficits are ameliorated by inhibition of aquaporin and HIF1α after traumatic brain injury (TBI). *J Neurol Sci* 323(1–2):134–140.

Shi, X., S. Wu, Y. Yang, L. Tang, Y. Wang, J. Dong, B. Lu, G. Jiang and W. Zhao. 2014. AQP5 silencing suppresses p38 MAPK signaling and improves drug resistance in colon cancer cells. *Tumour Biol* 35(7):7035–7045.

Solenov, E., H. Watanabe, G.T. Manley and A.S. Verkman. 2004. Sevenfold-reduced osmotic water permeability in primary astrocyte cultures from AQP-4-deficient mice, measured by a fluorescence quenching method. *Am J Physiol Cell Physiol* 286(2):C426–C432.

Soupart, A., M. Coffernils, B. Couturier, F. Gankam-Kengne and G. Decaux. 2012. Efficacy and tolerance of urea compared with vaptans for long-term treatment of patients with SIADH. *Clin J Am Soc Nephrol* 7(5):742–747.

Stassen, F.L., G.D. Heckman, W.F. Huffman and L.B. Kinter. 1985. Antidiuretic hormone antagonists and aquaresis in dogs: Different vasopressin sensitivity and antagonist potency in renal cortex and papilla. *J Pharmacol Exp Ther* 232(1):100–105.

Suzuki, Y., Y. Nakamura, K. Yamada, V.J. Huber, M. Tsujita and T. Nakada. 2013. Aquaporin-4 positron emission tomography imaging of the human brain: First report. *J Neuroimaging* 23(2):219–223.

Tait, M.J., S. Saadoun, B.A. Bell and M.C. Papadopoulos. 2008. Water movements in the brain: Role of aquaporins. *Trends Neurosci* 31(1):37–43.

Tanimura, Y., Y. Hiroaki and Y. Fujiyoshi. 2009. Acetazolamide reversibly inhibits water conduction by aquaporin-4. *J Struct Biol* 166(1):16–21.

Torrie, G.M. and J.P. Valleau. 1977. Nonphysical sampling distributions in Monte Carlo free-energy estimation: Umbrella sampling. *J Comput Phys* 23(2):187–199.

Tradtrantip, L., H. Zhang, M.O. Anderson, S. Saadoun, P.-W. Phuan, M.C. Papadopoulos, J.L. Bennett and A.S. Verkman. 2012. Small-molecule inhibitors of NMO-IgG binding to aquaporin-4 reduce astrocyte cytotoxicity in neuromyelitis optica. *FASEB J* 26(5):2197–2208.

Tsukaguchi, H., C. Shayakul, U.V. Berger, B. Mackenzie, S. Devidas, W.B. Guggino, A.N. van Hoek and M.A. Hediger. 1998. Molecular characterization of a broad selectivity neutral solute channel. *J Biol Chem* 273(38):24737–24743.

Vacca, A., A. Frigeri, D. Ribatti, G.P. Nicchia, B. Nico, R. Ria, M. Svelto and F. Dammacco. 2001. Microvessel overexpression of aquaporin 1 parallels bone marrow angiogenesis in patients with active multiple myeloma. *Br J Haematol* 113(2):415–421.

Vajda, Z., M. Pedersen, E.M. Fuchtbauer, K. Wertz, H. Stodkilde-Jorgensen, E. Sulyok, T. Doczi et al. 2002. Delayed onset of brain edema and mislocalization of aquaporin-4 in dystrophin-null transgenic mice. *Proc Natl Acad Sci USA* 99(20):13131–13136.

Vanommeslaeghe, K. and A.D. MacKerell, Jr. 2012. Automation of the CHARMM general force field (CGenFF) I: Bond perception and atom typing. *J Chem Inf Model* 52(12):3144–3154.

Vanommeslaeghe, K., E.P. Raman and A.D. MacKerell, Jr. 2012. Automation of the CHARMM general force field (CGenFF) II: Assignment of bonded parameters and partial atomic charges. *J Chem Inf Model* 52(12):3155–3168.

Verbalis, J.G. 2006. AVP receptor antagonists as aquaretics: Review and assessment of clinical data. *Cleve Clin J Med* 73(Suppl 3):S24–S33.

Verkman, A.S. 2001. Applications of aquaporin inhibitors. *Drug News Perspect* 14(7):412–420.

Verkman, A.S. 2006. Modulation of aquaporin in modulation of angiogenesis and cell migration. Google Patents. WO2006102483A2.

Verkman, A.S., M.O. Anderson and M.C. Papadopoulos. 2014. Aquaporins: Important but elusive drug targets. *Nat Rev Drug Discov* 13(4):259–277.

Verkman, A.S., M. Hara-Chikuma and M.C. Papadopoulos. 2008. Aquaporins—New players in cancer biology. *J Mol Med (Berl)* 86(5):523–529.

Wacker, S.J., C. Aponte-Santamaria, P. Kjellbom, S. Nielsen, B.L. de Groot and M. Rutzler. 2013. The identification of novel, high affinity AQP9 inhibitors in an intracellular binding site. *Mol Membr Biol* 30(3):246–260.

Wang, F., X.C. Feng, Y.M. Li, H. Yang and T.H. Ma. 2006. Aquaporins as potential drug targets. *Acta Pharmacol Sin* 27(4):395–401.

Wang, J., R.M. Wolf, J.W. Caldwell, P.A. Kollman and D.A. Case. 2004. Development and testing of a general amber force field. *J Comput Chem* 25(9):1157–1174.

Wang, T., D.Y. Chou, J.Y. Ding, V. Fredrickson, C. Peng, S. Schafer, M. Guthikonda, C. Kreipke, J.A. Rafols and Y. Ding. 2013. Reduction of brain edema and expression of aquaporins with acute ethanol treatment after traumatic brain injury. *J Neurosurg* 118(2):390–396.

Wang, W., C. Li, S. Summer, S. Falk and R.W. Schrier. 2010. Interaction between vasopressin and angiotensin II in vivo and in vitro: Effect on aquaporins and urine concentration. *Am J Physiol Renal Physiol* 299(3):F577–F584.

Wang, W., Q. Li, T. Yang, G. Bai, D. Li, Q. Li and H. Sun. 2012. Expression of AQP5 and AQP8 in human colorectal carcinoma and their clinical significance. *World J Surg Oncol* 10:242.

Wesche, D., P.M. Deen and N.V. Knoers. 2012. Congenital nephrogenic diabetes insipidus: The current state of affairs. *Pediatr Nephrol* 27(12):2183–2204.

Woo, J., J. Lee, Y.K. Chae, M.S. Kim, J.H. Baek, J.C. Park, M.J. Park et al. 2008. Overexpression of AQP5, a putative oncogene, promotes cell growth and transformation. *Cancer Lett* 264(1): 54–62.

Yakata, K., Y. Hiroaki, K. Ishibashi, E. Sohara, S. Sasaki, K. Mitsuoka and Y. Fujiyoshi. 2007. Aquaporin-11 containing a divergent NPA motif has normal water channel activity. *Biochim Biophys Acta* 1768(3):688–693.

Yang, B., J.K. Kim and A.S. Verkman. 2006a. Comparative efficacy of HgCl2 with candidate aquaporin-1 inhibitors DMSO, gold, TEA+ and acetazolamide. *FEBS Lett* 580 (28–29): 6679–6684.

Yang, B., H. Zhang and A.S. Verkman. 2008. Lack of aquaporin-4 water transport inhibition by antiepileptics and arylsulfonamides. *Bioorg Med Chem* 16(15):7489–7493.

Yang, B., D. Zhao, L. Qian and A.S. Verkman. 2006b. Mouse model of inducible nephrogenic diabetes insipidus produced by floxed aquaporin-2 gene deletion. *Am J Physiol Ren Physiol* 291(2):72.

Yang, J., C. Yan, W. Zheng and X. Chen. 2012. Proliferation inhibition of cisplatin and aquaporin 5 expression in human ovarian cancer cell CAOV3. *Arch Gynecol Obstet* 285(1):239–245.

Yang, J.H., Y.F. Shi, X.D. Chen and W.J. Qi. 2006c. The influence of aquaporin-1 and microvessel density on ovarian carcinogenesis and ascites formation. *Int J Gynecol Cancer* 16(Suppl 1): 400–405.

Yuriev, E. and P.A. Ramsland. 2013. Latest developments in molecular docking: 2010–2011 in review. *J Mol Recognit* 26(5):215–239.

Zardoya, R. 2005. Phylogeny and evolution of the major intrinsic protein family. *Biol Cell*, 97:397–414.

Zeidel, M.L., S.V. Ambudkar, B.L. Smith and P. Agre. 1992. Reconstitution of functional water channels in liposomes containing purified red cell CHIP28 protein. *Biochemistry* 31(33):7436–7440.

Zhang, D., L. Vetrivel and A.S. Verkman. 2002. Aquaporin deletion in mice reduces intraocular pressure and aqueous fluid production. *J Gen Physiol* 119(6):561–569.

Zhang, Z., Z. Chen, Y. Song, P. Zhang, J. Hu and C. Bai. 2010. Expression of aquaporin 5 increases proliferation and metastasis potential of lung cancer. *J Pathol* 221(2):210–220.

Zhu, F., E. Tajkhorshid and K. Schulten. 2004. Collective diffusion model for water permeation through microscopic channels. *Phys Rev Lett* 93(22):224501.

CHAPTER 14

Drug Discovery and Therapeutic Targets for Pharmacological Modulators of Aquaporin Channels

Jinxin V. Pei, Joshua L. Ameliorate, Mohamad Kourghi, Michael L. De Ieso and Andrea J. Yool

CONTENTS

Abstract	273
14.1 Introduction	274
14.2 Overview of Aquaporin Channel Functions in Brain Fluid Homeostasis and Cell Migration	274
14.2.1 AQP4 in Cerebral Edema	275
14.2.2 AQP1 in Cell Migration	277
14.3 Small Molecule Drug Discovery for Aquaporin Channels	278
14.4 Translational Promise of Pharmacological Modulators of Aquaporin Channels in Brain Edema, Cancer and Other Disorders	281
14.4.1 Targeting AQP4 Channels in Brain Edema	281
14.4.2 Differential Regulation of Expression of AQP Channel Types in Cancer Cells	281
14.4.3 Targeting AQP1 Channels in Cell Migration and Metastasis	283
14.5 New Avenues for Aquaporin Drug Discovery from Traditional and Alternative Medicines	284
References	287

ABSTRACT

FLUID HOMEOSTASIS IN THE body is well known to be regulated by ion channels and transporters, but equally important are the co-expressed classes of aquaporin (AQP) water channels that facilitate transmembrane water movement. The field of AQP pharmacology is expanding rapidly with the new identification of small molecule drug-like agents with distinctive properties in AQP modulation, allowing exploration of potential therapeutic applications in brain edema, cancer and other disorders. Pharmacological agents could modulate AQPs by direct occlusion of the water pore itself, by acting at distinct sites that confer other channel properties, or by altering levels of protein expression or membrane targeting. Expanding the pharmacological portfolio will benefit basic research and promote new therapeutic strategies in many conditions involving AQPs in the symptoms or disease processes. Exploring herbal alternative medicines as sources of pharmacological agents for AQPs is likely to have a substantial impact on the field of AQP research, which has keenly awaited the development of chemical interventions as a platform for therapeutic approaches. Recent work is providing an enhanced understanding of the molecular mechanisms of action of traditional herbal medicines as novel sources of AQP modulators.

14.1 INTRODUCTION

Aquaporins (AQPs) found throughout the kingdoms of life (Zardoya, 2005) are channel proteins which facilitate water and small solute movement across plasma membranes based on chemical, osmotic and hydrostatic gradients (Hachez and Chaumont, 2010; Ishibashi et al., 2011; Madeira et al., 2014). The existence of water channels in red blood cells and barrier membranes and their hallmark sensitivity to block by mercury (Benga et al., 1986; Macey, 1984; Macey et al., 1972; Naccache and Sha'afi, 1974) was deduced before AQP1 cDNA was cloned and characterized in the early 1990s (Preston and Agre, 1991; Preston et al., 1992). Progress since has defined 13 mammalian subtypes of AQPs, their relative protein abundance, tissue-specific distributions and expression levels under various conditions. AQP crystal structures have provided insights into the homotetrameric channel structure and the location of the water pores within each of the four subunits (Ishibashi et al., 2009; Jensen et al., 2003; Sui et al., 2001). AQP1 is a tetrameric channel with water pores located within each of the four individual subunits (Fu et al., 2000; Sui et al., 2001). Review articles have emphasized the compelling need for development of AQP modulators as candidate treatments for a variety of disorders (Castle, 2005; Frigeri et al., 2007; Huber et al., 2012; Jeyaseelan et al., 2006). However, prior to 2009, the discovery of AQP pharmacological agents had not progressed substantially beyond mercuric compounds, which lacked translational potential due to toxicity, and agents that affected multiple targets and lacked potency. Recent advances have shown that drug-like lead compounds are emerging, complementing work that has investigated the functional roles of AQPs by overexpression or targeted genetic knockdown and deletion.

14.2 OVERVIEW OF AQUAPORIN CHANNEL FUNCTIONS IN BRAIN FLUID HOMEOSTASIS AND CELL MIGRATION

Fluid homeostasis in the body is regulated by partnerships of ion channels and transporters with co-expressed classes of AQP channels (Conde et al., 2010; Fischbarg, 2010). The 13 classes of mammalian AQP water channels (AQPs 0–12) are expressed in tissue-specific patterns in the body and are essential in regulating the movement of fluid across barrier membranes and contributing to control of cell volume, production of cerebral spinal fluid, edema formation and recovery, mediating renal function and more (Ishibashi, 2009; Nielsen et al., 2007). Under pathological conditions, dysfunctions in the control of fluid movement create serious problems. The physiological importance of AQP channels and their compelling potential value as therapeutic targets has motivated researchers to work toward defining a pharmacological panel of chemical modulators for AQPs. A comprehensive pharmacological portfolio for all classes of AQPs will provide an array of new therapeutic opportunities that will continue to expand as new roles for AQP channels are uncovered. Two major areas of current interest in AQP-based mechanisms of disease are brain swelling after injury or stroke and the migration of cancer cells in the process of metastasis.

14.2.1 AQP4 in Cerebral Edema

An area of intense interest for new therapies directed at AQPs is aimed at reducing brain swelling after acute traumatic injury or stroke (Mack and Wolburg, 2013; Yool et al., 2009). Cerebral edema and increased intracranial pressure are life-threatening sequelae of severe brain injuries, associated with a poor prognosis as evidenced by a mortality rate near 60%–80% (Feickert et al., 1999; Hacke et al., 1996). Traumatic brain injury affects an estimated 10 million people annually; according to the World Health Organization, it will be the major cause of death and disability by the year 2020 (Hyder et al., 2007). Nearly one-third of hospitalized traumatic brain injury patients die from injuries that are secondary to the initial trauma, including neuroinflammation, excitotoxicity, brain edema and intracranial hypertension (Miller et al., 1992). Traumatic brain injury reduces life expectancy and compromises the quality of life (Schiehser et al., 2014) with increased incidences of seizures, sleep disorders, neurodegenerative disease, neuroendocrine dysregulation, psychiatric problems and non-neurological disorders that can persist years after the injury event (Masel and DeWitt, 2010). The burden of mortality and morbidity makes traumatic brain injury a pressing public health and medical concern, but there is currently no targeted pharmacological treatment for reducing the secondary damage (Park et al., 2008).

Cerebral ischemia following severe traumatic brain injury involves a combination of cytotoxic and vasogenic events (Hossmann, 1994; Papadopoulos et al., 2004; Shi et al., 2012; Tourdias et al., 2011). Brain edema commonly occurs when cerebral blood flow drops beneath 10 mL/100 g/min and essential ionic pump activity is impaired. During the first 5 h of ischemia, blood–brain barrier integrity can be maintained without substantial brain swelling (Betz et al., 1989), but upon reperfusion, a rise in brain water content can occur in conjunction with an elevation in intracranial pressure depending on the extent of ischemia and the delay until cerebral blood flow is restored (Avery et al., 1984). Conventional treatments focus on reducing edema and intracranial pressure using hyperventilation, mannitol, diuretics, corticosteroids or barbiturates (Manno et al., 1999; Winter et al., 2005). Decompressive craniectomy involves the surgical removal of a part of the skull, creating a space for the swollen brain tissue to expand. This surgical method improves survival but does not fully address pathological outcomes as many patients are left moderately or severely disabled (Fischer et al., 2011; Vahedi et al., 2007).

Work in animal models relies on optimizing the reliability and reproducibility of the injury event. The middle cerebral artery occlusion model in rodents has been a standard method for testing experimental therapeutic agents but is difficult in part because of unexplained variability in infarct volumes. When analyzed with computed tomography cerebral blood volume maps, variability was found to result from unintended occlusion of a second artery (the anterior choroidal artery) in a subset of animals, causing expanded infarct areas (McLeod et al., 2013). Brain computed tomography perfusion imaging provides a powerful tool for improving the experimental method by fully defining the arteries affected by occlusion and accurately identifying the infarct core and penumbra domains (McLeod et al., 2011).

The concept of edema as being the primary cause of increased intracranial pressure has been challenged; small ischemic strokes in rats resulted in minimal amounts of edema but were associated with a substantial elevation in intracranial pressure at 24 h, which was effectively countered by application of hypothermia soon after the stroke event. Mechanisms in addition to cerebral edema alone must be considered as important drivers of intracranial pressure elevation (Murtha et al., 2014a,b). Assessing the regulated contributions of transporters and channels including AQP1 and AQP4 might offer insight into possible molecular mechanisms generating the distinct outcomes.

AQP4 is the predominant water channel in the brain, localized to the blood–brain barrier, ependymal cells lining the ventricles, subependymal astrocytes and the glia limitans (Amiry-Moghaddam and Ottersen, 2003; Nagelhus and Ottersen, 2013; Xiao and Hu, 2014) for their involvement in water movement both in and out of the brain (Mack and Wolburg, 2013). Studies in AQP4-deficient mouse models showing partial protection from water intoxication and edema after stroke (Manley et al., 2000; Papadopoulos and Verkman, 2008) support the idea that ligand modulators of AQP4 could revolutionize clinical treatment of brain edema. To date, progress in the field has been limited by the lack of availability of pharmacologic modulators of the AQP channels. Identification of agonists and antagonists of AQP4 remains an important goal in the field of cerebral edema research. AQP4 channels subserve water movement during cerebral edema formation (Jullienne and Badaut, 2013; Manley et al., 2000), highlighting them as an attractive therapeutic target for non-surgical management of cerebral edema (Yool, 2007a; Yool et al., 2009).

Alterations in levels of AQP expression in response to cerebral fluid imbalance can offer clues for understanding the functional roles of water channels, assuming the goal is to restore homeostasis. In experimental models of cerebral ischemia, expression levels of AQP4 were reduced in the initial period post injury and significantly increased thereafter (Taniguchi et al., 2000; Yamamoto et al., 2001). Regulation of AQP4 expression appears to correlate with levels of the transcription factor hypoxia-inducible factor 1-alpha (HIF-1α) (Kaur et al., 2006). Early after traumatic brain injury (5 h), levels of HIF-1α were low and increased to peak at 24–48 h post injury (Ding et al., 2009; Higashida et al., 2011). Inhibition of HIF-1α correlated with decreased AQP4 and attenuated swelling post injury (Shenaq et al., 2012). The temporal pattern of AQP4 regulation suggests that downregulation of channels early after injury could be a protective response to limit the influx of fluid into the brain, whereas delayed upregulation might serve to enable fluid export and to restore fluid homeostasis. However, altered gene expression responses are not spatially uniform. A reduction in AQP4 expression occurred in astrocytes within the ischemic core, while an elevation in AQP4 expression was seen in glial endfeet in the surrounding penumbra, in a middle cerebral artery occlusion model (Frydenlund et al., 2006). Spatially selective regulation of AQP4 might limit water influx into the ischemic core without preventing the amelioration of vasogenic edema in the penumbra. The loss of AQP4 was not linked to a decrease in the astrocyte marker, glial fibrillary acidic protein (Friedman et al., 2009), indicating that decreased AQP4 expression was not an indirect effect of glial cell loss in the injured domain.

AQP4 water channels are dynamically regulated components of brain fluid homeostasis, being mobilized or deactivated as needed to reduce damage arising from shifts or disturbances in cerebral fluid homeostasis. The temporal and spatial patterns of AQP4 regulation are contingent upon the extent and duration of injury and stage of pathology. AQP pharmacological agents will be advantageous for intervention in brain fluid disorders particularly if administered with an understanding of the dynamic mechanisms of AQP channel regulation.

14.2.2 AQP1 in Cell Migration

Cell migration is essential in development, repair, regeneration and immune protection in multicellular organisms. In 1937, the neuroanatomist Ramon y Cajal pondered: 'What mysterious forces stimulate the migrations of cells…?' (Kater and Letourneau, 1985). Currently, it is thought that branched assemblies of actin filaments, actively polymerizing at the leading edge, generate the primary force, which pushes out thin ruffled membrane extensions known as lamellipodia (Le Clainche and Carlier, 2008). Parallel arrays of actin are stalled by small loads on the order of 1 pN (Footer et al., 2007), but branched actin networks with many points of contact can generate nN of force per μm^2, on the scale needed for movement through viscous extracellular environments (Marcy et al., 2004; Parekh et al., 2005). Cell adhesions at the leading edge hold the new position, while the trailing edge detaches, and the cell resets for the next push forward.

AQPs found in all the kingdoms of life facilitate fluxes of water and small solutes across membranes (King et al., 2004). In diverse motile cells from amoeba to human, specific AQPs are localized in lamellipodial leading edges. When these cells are made AQP deficient (e.g. by genetic knockout or RNA-interference techniques), cell migration is greatly impaired. Reintroduction of AQP restores rates of movement, but interestingly, the effects of different AQPs are not interchangeable (McCoy and Sontheimer, 2007).

Of the 13 known mammalian classes of AQPs, the three often associated with migration thus far are AQPs 1, 3 and 5. In the 'World Cell Race' event held at the *American Society for Cell Biology* meeting in 2011, cell lines were submitted by teams around the world; the winner was a bone marrow stem cell which covered the 400 μm track at a speed of 5.2 μm/min. Bone marrow stem cells express lamellipodial AQP1 that is required for migration (Meng et al., 2014). The need for AQPs in migration extends to life forms other than mammals. The amoeba *Dictyostelium discoideum* migrates in a chemotactic response to external cyclic adenosine monophosphate (cAMP) signals and expresses an AQP orthologous to human

AQP1 (48% amino acid sequence similarity) which is localized in lamellipodia (von Bulow et al., 2012). If *D. discoideum* had been in the competition, it might have taken the honors, moving at up to 11 μm/min in response to cAMP chemotactic stimuli (Van Duijn and Inouye, 1991). Other contenders could have included activated T-cells moving at >10 μm/min (Katakai et al., 2013), which express AQPs 1, 3 and 5 (Moon et al., 2004), and fibroblasts moving at up to 9 μm/min (Ware et al., 1998), which express AQP1 (Minami et al., 2001).

Even though AQP1 and AQP4 are both highly functional water channels, the expression of AQP1 enables rapid cell movement, whereas expression of AQP4 does not. One important difference is that AQP1 but not AQP4 can function as a gated monovalent cation channel (Anthony et al., 2000; Yool and Campbell, 2012). AQP1 ion channel activity has been replicated in other laboratories (Saparov et al., 2001; Zhang et al., 2007), but its physiological significance remains unknown. During migration, changes in cytoplasmic volume are required. Fluid flux is commonly thought of as a passive process in which water follows salt. Yet the observation that water channels cannot simply be swapped between cells suggests that other properties in addition to water channel function are needed. The pro-migratory effect of AQP1 and its orthologs appears to be a convergent target of evolution across phyla.

14.3 SMALL MOLECULE DRUG DISCOVERY FOR AQUAPORIN CHANNELS

Inhibitors of AQPs have long been heralded as an important goal in the field. A database of structurally diverse compounds will be of substantial value in expanding our understanding of drug structure–activity relationships and molecular sites of action for AQP pharmacological modulators. Pharmacological agents could modulate AQPs by direct occlusion of the water pore itself, by acting at distinct sites that confer other channel properties or by altering levels of protein expression or membrane targeting. Regulatory domains in AQPs can control membrane localization, interaction with proteins to create signalling complexes and scaffolds, and gated permeation of small solutes other than water through AQPs (Cowan et al., 2000; Yool, 2007a,b).

Much of the work to date has focused on the agents that act to occlude the water pore. The existence of water channels in cell membranes and their hallmark sensitivity to block by mercury was deduced before the first AQP1 cDNA was cloned (Preston and Agre, 1991; Preston et al., 1992). Finding non-mercurial AQP blockers has been essential for AQP drug discovery (Table 14.1). Silver and gold compounds were reported to block water channel activity of both plant and human AQPs (Niemietz and Tyerman, 2002; Tyerman et al., 2002). Membrane-permeable derivatives of the loop diuretic drugs bumetanide and furosemide have proven useful as pharmacological antagonist AqB013 (Migliati et al., 2009) and agonist AqF026 (Yool et al., 2013) agents for AQP1 and AQP4 water channel activities. Characterized in vitro and in rodent models in vivo, these candidate AQP drugs appear to be effective and well tolerated. In a mouse model of peritoneal dialysis, the lack of effect of the agonist AqF026 in AQP1 knockout animals indicated specificity of action without appreciable off-target effects (Yool et al., 2013). Other groups also have investigated arylsulfonamides as AQP blockers (Gao et al., 2006; Huber et al., 2007; Ma et al., 2004; Seeliger et al., 2013). Acetazolamide (AZA)

TABLE 14.1 Summary of Aquaporin Pharmacological Modulators and Proposed Interaction Sites

Name	Structure	Proposed Binding Site	References
Mercuric chloride HgCl$_2$	Cl–Hg–Cl	Cys-189 in AQP1	Preston et al. (1993)
Silver and gold compounds: AgNO$_3$, HAuCl$_4$		Cys-189 in AQP1	Niemietz and Tyerman (2002)
Tetraethylammoniumion (TEA$^+$)		Loop-E region	Brooks et al. (2000), Yool et al. (2002)
Acetazolamide (AZA)		Ag-216 and Gly-209 in AQP4	Huber et al. (2009, 2007)
Zonisamide		Possibly Arg-216 and Gly-209 in AQP4	Huber et al. (2009)
AqB013		Intracellular vestibule of the water pole	Migliati et al. (2009)
AqF026		Intracellular loop D domain	Yool et al. (2013)

has been reported to inhibit AQP4 water permeability with an IC_{50} value of 0.9 μM. In silico docking suggested an interaction between the sulfonamide group of AZA and both arginine 216 and glycine 209 (Huber et al., 2007). As compared to AQP4, AQP1 was less sensitive to AZA, with an IC_{50} value estimated at 5.5 ± 0.5 μM (Seeliger et al., 2013).

The passage of water and ions occurs through pharmacologically distinct pathways in AQP1 (Saparov et al., 2001; Yool et al., 2002). The cyclic guanosine monophosphate (cGMP)-gated cation permeation pathway is thought to be in the central pore of the tetramer, based on molecular dynamic simulations (Yu et al., 2006) and effects of site-directed mutations on altering ionic conductance properties (Campbell et al., 2012). Conduction of ions in AQP1 channels is inhibited by Cd^{2+} but not tetraethylammonium ion (TEA$^+$), whereas water transport which occurs through the individual pores located in each subunit is sensitive to TEA$^+$, AqB013, AqF026 but not Cd^{2+} (Boassa et al., 2006; Brooks et al., 2000; Migliati et al., 2009; Yool et al., 2002, 2013). In silico docking identified possible extracellular binding by TEA$^+$ (Detmers et al., 2006) but indicated intracellular sites of action for the arylsulfonamides, supported by results of site-directed mutations and biological assays. AqB013 is thought to occlude the water pore at the internal vestibule, whereas the AQP agonist, AqF026, appears to potentiate water channel activity by interacting with an intracellular gate (loop D) between the fourth and fifth transmembrane domains (Yool et al., 2013).

The loop D domain serves as a gate for AQP channel activity in mammalian AQP1 (Yu et al., 2006), AQP4 (Zelenina et al., 2002), amoeba AQP-B (von Bulow et al., 2012) and plant AQP (Tornroth-Horsefield et al., 2006). In AQP1, loop D domain is involved in the regulation of cation channel activity activated by cGMP (Yu et al., 2006). The loop D sequence is highly conserved across species in AQP1 channels from fish to mammals, yet mutations in this domain do not impact water channel activity, suggesting that loop D is essential for AQP1 functions other than water permeability. Mutations at specific loop D positions in AQP1 remove ion channel activity without impairing water channel activity (Yu et al., 2006). The retention of water channel functionality showed that the mutation did not interfere with expression, trafficking or assembly of AQP1 channel.

Other regulatory domains in AQP1 have been suggested in the carboxyl terminal domain. A putative cGMP-dependent modulatory domain with sequence similarity to the cGMP-phosphodiesterase selectivity domain was identified in the AQP1 C-terminus (Boassa and Yool, 2002, 2003). Mutations of conserved phosphodiesterase-like residues in this domain did not remove cGMP-dependent activation of AQP1 but reduced the efficacy of cGMP by right shifting the dose–response curve. An EF-hand motif found in calcium-binding proteins (Grabarek, 2006) has been proposed in the carboxyl terminal domain of AQP1, but its functional role has not yet been defined (Fotiadis et al., 2002). A tyrosine phosphorylation site in the carboxyl terminal serves a modulatory role in governing the availability of AQP1 to be gated as a cGMP-dependent cation channel (Campbell et al., 2012). A PDZ protein–protein ligand domain identified in AQP1 (Ile260 to Arg264) in the C-terminus was shown to be important for targeting AQP1 into a membrane complex needed to maintain vestibular fluid balance in the inner ear (Cowan et al., 2000). Regulatory domains and transmembrane pore regions in AQPS are both attractive candidates for the development of drug agents.

14.4 TRANSLATIONAL PROMISE OF PHARMACOLOGICAL MODULATORS OF AQUAPORIN CHANNELS IN BRAIN EDEMA, CANCER AND OTHER DISORDERS

14.4.1 Targeting AQP4 Channels in Brain Edema

Conventional approaches have relieved intracranial pressure with hyperventilation, hyperosmotic agents, diuretics and decompressive craniectomy (Diedler et al., 2009; Park et al., 2008; Werner and Engelhard, 2007) but are limited in effectiveness by treating symptoms rather than causes of edema. A major challenge has been the lack of interventions for controlling fluid movement in brain edema. Pharmacological control of the direct fluid flow pathways between blood and brain could provide for invaluable therapeutic interventions in cerebral pathologies involving abnormal water fluxes such as stroke, hydrocephalus and brain tumours. AQP4 is a logical choice as a molecular target for drug development (Papadopoulos and Verkman, 2007; Yool et al., 2009). It is abundantly expressed in the central nervous system near the blood–brain barrier at glial cell endfeet and provides a major pathway for fluid homeostasis (Amiry-Moghaddam and Ottersen, 2003). Astrocytes have been proposed to have a bimodal contribution, with a positive role in fluid homeostasis and limiting brain injury, and a negative role in worsening the neuroinflammation, cerebral edema and elevated intracranial pressure associated with secondary brain injury following neurotrauma (Laird et al., 2008).

Now with the characterization of novel AQP agonists and antagonists, it will be possible to investigate potential pharmacological treatments for cerebral edema (Yool et al., 2009) with an intriguing capacity for bimodal regulation of AQP water channel activity. For example, in a rat model of cerebral edema induced by diffuse traumatic brain injury, a single intravenous application of the AQP antagonist AqB013 within 5 h post-injury dramatically reduced edema formation, and single administration of the AQP agonist AqF026 at 1–2 days post-injury accelerated the resolution of brain edema (Burton et al., unpublished data). Each modulator was beneficial alone; however, of interest was the observation that a sequential treatment of antagonist followed by agonist each at their optimal single time points provided a powerful combination that further enhanced the protection of motor function, reduced brain swelling and decreased brain albumin content post injury. Understanding the biphasic role of AQP4 in cerebral edema and the actions of AQP pharmacological modulators will be likely to open new avenues for the development of therapeutic interventions after traumatic brain injury.

14.4.2 Differential Regulation of Expression of AQP Channel Types in Cancer Cells

Evidence emerging within the past decade is providing increasingly strong support for the role of selected classes of AQPs as important constituents in cancer cell biological mechanisms. Data show a strong positive correlation between AQP expression levels and tumor severity (Machida et al., 2011). Tumor cells selectively increase levels of expression of different specific types of AQPs; in some cases, there is an increase in an AQP class normally expressed in the cell type (Saadoun et al., 2002a), but in other cancers, the upregulated class of AQP is not found at appreciable levels in the original tissue (Moon et al., 2003). AQPs are linked with a variety of properties of cancer cells, such as tumour size expansion, edema, cell adhesion, migration and proliferation, which enhance tumour growth and metastasis (Hu and Verkman, 2006; Saadoun et al., 2002a,b).

For AQP1 and AQP5 channels, de novo expression has been observed in early stage colorectal carcinoma development but is not detectable in normal colonic epithelium (Moon et al., 1997, 2003) for purposes that remain incompletely understood in the context of cancer progression. When transfected into melanoma and breast cancer cell lines, increased AQP1 expression was associated with an increase in both in vitro cell migration and in vivo metastasis (Hu and Verkman, 2006). When melanoma cells B16F10 were implanted subcutaneously, AQP1 null mice showed slower melanoma tumour growth and impaired tumour angiogenesis as compared with wild-type mice (Saadoun et al., 2005).

AQP1 contributes to cell migration and angiogenesis and facilitates tumour growth and metastasis through mechanisms that remain to be defined. In its proposed activity as a dual water and ion channel (Yool and Weinstein, 2002), AQP1 could be one mechanism used for enhancing migration rate in a subset of classes of aggressive cancers. Facilitation of cell migration appears to be achieved in other types of cells by colocalization of water-selective AQPs in combination with ion transporters (Chai et al., 2013; Stroka et al., 2014).

AQP3 increased migration and proliferation of corneal epithelial cells in wild type as compared with *Aqp1* null mice and was important in peritoneal fibrosis and wound healing (Ryu et al., 2012). In squamous skin cell carcinomas and bronchioloalveolar carcinomas, AQP3 upregulation influenced metastasis and proliferation (Hara-Chikuma and Verkman, 2008a; Machida et al., 2011). After si-RNA knockdown of AQP3 expression, weaker cell adhesion and impaired cell growth were observed (Kusayama et al., 2011). In non-small cell lung cancer, AQP3 knockdown suppressed tumour growth and reduced angiogenesis (Xia et al., 2014). In agreement, AQP3 null mice show reduced glycerol transport and were more resistant to skin cancer formation (Hara-Chikuma and Verkman, 2008b).

AQP4 appears to assist with water balance in the tumour environment and has been postulated to play a role in cell–cell adhesion, possibly via an extracellular helical domain in loop C that could interact with adjacent cells. Expression of AQP4 conferred adhesive properties in cells that lacked classic adhesion molecules, as shown by L-cell cluster formation in AQP4-positive cells but not control cells (Hiroaki et al., 2006); however, other work has not confirmed the idea (Zhang and Verkman, 2008). More studies will be required to determine what factors influence a putative role of AQP4 in tumour cell adhesion. AQP4 can be phosphorylated by protein kinase C (PKC) at Ser-180. Phosphorylation level is inversely proportional to water permeability. Mutation of Ser-180 to alanine increased AQP4 water permeability by approximately twofold (McCoy et al., 2010). In AQP4-transfected glioma D54MG cells, both AQP4 water permeability and tumour cell migration were decreased when PKC was activated by phorbol ester treatment. It is postulated that cells with suppressed AQP4 activity have a compromised ability to adjust cell shape during migration (McCoy et al., 2010). In contrast, when D54MG cells were transfected with AQP1, neither of the properties of water channel activity or cell migration were affected by PKC modulators (McCoy et al., 2010).

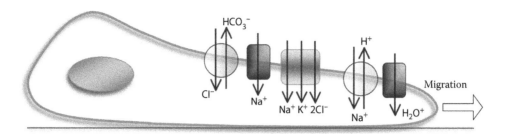

FIGURE 14.1 Schematic diagram of the colocalization of AQP channels with ion transporters, other channels and exchangers at the leading edge of a lamellipodium in a migrating cell.

AQP5 expression is increased in pancreatic, colon cancers and myelogenous leukemia cells (Machida et al., 2011; Woo et al., 2008). Proliferation of human ovarian cancer cells has also been correlated with AQP5 expression levels (Chae et al., 2008a). The involvement of AQP5 in cancer invasion has been demonstrated to depend on phosphorylation of Ser-156 located in the loop D domain through a cellular sarcoma kinase (c-Src) signalling pathway (Chae et al., 2008b). Deletion of AQP5 but not AQP1 nor AQP3 was correlated with decreased activation of the epithelial growth factor receptor signal cascade (EGFR/ERK/MAPK) which regulated cancer cell migration and proliferation (Zhang et al., 2010). Results from many studies now show that AQP3 and AQP5 contribute to processes of tumour proliferation, but the link from AQP function to the regulation of growth remains to be defined.

As depicted in Figure 14.1, AQP expression can be polarized in the lamellipodia of migrating cells, together with transporters such as the Na^+/H^+, the Cl^-/HCO_3^- exchanger or the Na^+–K^+–Cl^- co-transporter. Localized ion fluxes are proposed to create a driving force for osmotic water fluxes, in parallel with the osmotic effects of actin polymerization and depolymerization, that could drive the membrane protrusions. This concept has been presented as an 'osmotic engine model' (Stroka et al., 2014) based on studies of AQP5 and Na^+/H^+ exchangers polarized to the leading edges in mouse S180 sarcoma cells migrating in a confined environment.

14.4.3 Targeting AQP1 Channels in Cell Migration and Metastasis

Tumour cell migration enables metastasis and tissue invasion and is a major cause of death in patients with cancer (Bogenrieder and Herlyn, 2003). A review of literature shows that AQP1 expression is upregulated in a subset of aggressive cancers, whereas AQP1 deletion or downregulation impedes migration of AQP1-expressing cancer cells in vitro and reduces metastasis in vivo (Deb et al., 2012; Verkman et al., 2008; Yool et al., 2009). Wound-healing and transwell migration assays have demonstrated impaired migration of tumour cells lacking AQP1 as compared to wild type (Jiang, 2009; Jiang et al., 2009; Jiang and Jiang, 2010; Li et al., 2006). Cell migration rate in AQP1 knockdown cells can be rescued by adenovirus-mediated AQP1 expression. Similarly, transfection of AQP1 into tumour cells that lack endogenous AQP1 show accelerated cell migration rate in vitro as compared to control cells (Hu and Verkman, 2006; McCoy and Sontheimer, 2007).

Manipulation of tumour cell migration rate via AQP1 expression levels has been explored in rodent models in vivo. Increased AQP1 expression increased tumour cell extravasation, quantified by the numbers of fluorescently tagged tumour cells which infiltrated mouse lung tissue after injection into tail vein (Hu and Verkman, 2006). In a mouse model that spontaneously developed well-differentiated, luminal-type breast adenomas with lung metastases, genetic deletion of AQP1 correlated with reductions in lung metastases, tumour mass and tumour volume with abnormal microvascular anatomy and reduced vessel density as compared to wild-type mice (Esteva-Font et al., 2014). AQP1 expression facilitated endothelial cell migration and augmented tumour growth in vivo by the facilitation of angiogenesis (El Hindy et al., 2013; Nicchia et al., 2013). Impaired melanoma growth was observed in AQP1 null mice after subcutaneous tumour cell implantation and attributed to reduced aortic endothelial cell migration (Saadoun et al., 2005).

AQP1 is an attractive therapeutic target for controlling tumour growth and migration. Pharmacological modulation of AQP1 could be a pivotal adjunct to existing cancer therapies, improving the prognosis for cancer patients by slowing cancer angiogenesis and metastases.

14.5 NEW AVENUES FOR AQUAPORIN DRUG DISCOVERY FROM TRADITIONAL AND ALTERNATIVE MEDICINES

Complementary and alternative medicine practices provide an intriguing starting point in searches for novel pharmacological modulators for AQP channels, drawing on cultural insights of native botanical agents from Asian, indigenous Australian, American, Indian and other sources of cultural knowledge. Given the importance of AQPs in human health and disease, AQP modulatory agents have a compelling potential to benefit research and health care globally. Scientific evidence is needed to understand the molecular mechanisms that underlay therapeutic actions of alternative medicines which have been used around the world from ancient times to treat fluid imbalance disorders and diseases (de Morais Lima et al., 2011; Karou et al., 2011; Kong et al., 2015; Nie et al., 2013). Possible agents for AQP channels might be present in native botanical agents used for treating conditions such as kidney and gastrointestinal disorders, swelling, brain edema and inflammation, which in theory could be benefiting from altered AQP channel activity in mitigating the dysfunctions.

Botanical compounds over many centuries have been an important source of useful drugs. As reviewed by Wachtel-Galor and Benzie (2011; CRC Press *Herbal Medicines: Biomolecular and Clinical Aspects*), about 25% of the drugs prescribed worldwide are derived from plants, including morphine derived from poppy seeds (*Papaver somniferum*), digoxin from foxglove (*Digitalis lanata*), aspirin from willow bark, antimalarials such as quinine from cinchona bark (*Cinchona officinalis*) and artemisinin derived from *Artemisia annua*. The potential value for translation into a novel pharmacology for AQP channels is immense. Expanding the pharmacological portfolio for AQPs will not only benefit basic research but could also prompt strategies for therapeutic interventions in cancers, lung and cardiac edema, secretory and digestive dysfunctions, and other conditions involving tissues in which AQPs are expressed.

Two principal groups of phytochemicals of interest appear to be emerging from the data available thus far: (1) polysaccharides (glucans) extracted from aqueous fractions and

(2) triterpenes (polyporusterones) extracted from organic fractions of medicinal plants. A meta-analysis of published data on extracts and isolated compounds from an edible mushroom *Polyporus umbellatus* shows it acts as a diuretic, anti-cancer, immunostimulant and hepatoprotective agent (Zhao, 2013). An aqueous fraction of the Chinese medicinal fungus Fu Ling *Poria cocos* contains multiple saccharides and has been reported to attenuate renal AQP2 expression at the transcription and translational levels (Lee et al., 2012b). Extracts of Fu Ling also act as a diuretic and suppress the growth and invasiveness of various tumour cell lines (Cheng et al., 2013; Ling et al., 2011; Zhao et al., 2012). Similar to *P. umbellatus*, various triterpenes have been extracted and identified from *P. cocos* (Rios, 2011). In a study of papilloma carcinogenesis, poricoic acid C was demonstrated to decrease the incidence to 27% as compared with 100% in the control group (Akihisa et al., 2007). A bioactive compound (ginsenoside Rg3) found in the ginseng root extract was shown to attenuate cell migration via inhibition of AQP1 expression in PC-3M prostate cancer cells in vitro and demonstrated the potential usefulness of pharmacological modulation of AQP1 in manipulating tumour cell migration (Pan et al., 2012).

Diverse classes of channels and transporters are modulated by curcumin (isolated from turmeric in the ginger family), including voltage-gated K^+ and Ca^{2+} channels, the volume-regulated anion channel (VRAC), the Ca^{2+} release-activated Ca^{2+} channel (CRAC), AQP4 channels, glucose transporters and others (Zhang et al., 2014). As compared with vehicle-injected rats, a lower level of AQP4 expression was observed in rats injected with 40 mg/kg curcumin. In the same experiment, a neuroprotective effect of curcumin was demonstrated in a rat model of hypoxic ischemic brain damage (Yu et al., 2012).

Bacopa (*Bacopa monnieri*), a water hyssop also known as Brahmi in traditional Indian medicine, has been used in complementary medicine remedies for centuries to improve memory and treat anxiety and depression (Russo and Borrelli, 2005). The main active components are triterpenes (bacosides, bacopasides and bacosaponins) (Russo and Borrelli, 2005). Some of the beneficial effects of bacopa could potentially involve AQP channels; however, the diverse group of candidate targets of action remains to be investigated. Tumour size decreased when mice with subcutaneously implanted S180 sarcoma cells were treated with bacopa extracts (Peng et al., 2010). Bacopa confers neuroprotective effects after ischemic brain injury (Liu et al., 2013; Saraf et al., 2010), which might suggest testing for a possible link with AQP4, known to be a key contributor in pathological outcomes of brain injury (Kim et al., 2010; Lee et al., 2012a; Shin et al., 2011). Bacopa also has also been used for control of epilepsy (Shanmugasundaram et al., 1991), and some anti-epileptic drugs have been demonstrated to inhibit AQP4 water permeability (Huber et al., 2009).

Plant-derived polysaccharide and triterpene compounds possess diverse pharmacological functions. A subset of them appear to have theoretical potential (based simply on the nature of their cellular and systemic effects in vivo) to regulate AQP levels of expression or function directly or indirectly. A sample of interesting phytochemical candidates and their structures is listed in Table 14.2. While still speculative at this point, the concept is worth further investigation. More research is needed focusing on phytochemicals as sources of candidate pharmacological agents for AQPs. Intriguing connections between botanical

TABLE 14.2 Phytochemicals Derived from Medicinal Plants Traditionally Used for Treating
Potentially Aquaporin-Related Diseases

Plant Source	Compound	Structure
Panax ginseng	Ginsenoside Rg3	
Polyporus umbellatus	(1→3)-α-D-glucan	
	Polyporusterone A	
Poria cocos	Poricoic acid A	
Curcuma longa	Curcumin	
Bacopa monnieri	Bacopaside I	

(Continued)

TABLE 14.2 (*Continued*)   Phytochemicals Derived from Medicinal Plants Traditionally Used for Treating Potentially Aquaporin-Related Diseases

Plant Source	Compound	Structure
	Bacopaside II	

medicinal agents and AQPs remain to be investigated but could be a valuable source of new lead compounds for broadening the field of AQP pharmacology and better understanding the molecular mechanisms of action of alternative medicinal agents.

REFERENCES

Akihisa T, Nakamura Y, Tokuda H, Uchiyama E, Suzuki T, Kimura Y, Uchikura K and Nishino H (2007) Triterpene acids from *Poria cocos* and their anti-tumor-promoting effects. *Journal of Natural Products* **70**(6):948–953.

Amiry-Moghaddam M and Ottersen OP (2003) The molecular basis of water transport in the brain. *Nature Reviews Neuroscience* **4**(12):991–1001.

Anthony TL, Brooks HL, Boassa D, Leonov S, Yanochko GM, Regan JW and Yool AJ (2000) Cloned human aquaporin-1 is a cyclic GMP-gated ion channel. *Molecular Pharmacology* **57**(3):576–588.

Avery S, Crockard HA and Russell RR (1984) Evolution and resolution of oedema following severe temporary cerebral ischaemia in the gerbil. *Journal of Neurology, Neurosurgery, and Psychiatry* **47**(6):604–610.

Benga G, Popescu O, Pop VI and Holmes RP (1986) p-(Chloromercuri) benzenesulfonate binding by membrane proteins and the inhibition of water transport in human erythrocytes. *Biochemistry* **25**(7):1535–1538.

Betz AL, Iannotti F and Hoff JT (1989) Brain edema: A classification based on blood-brain barrier integrity. *Cerebrovascular and Brain Metabolism Reviews* **1**(2):133–154.

Boassa D, Stamer WD and Yool AJ (2006) Ion channel function of aquaporin-1 natively expressed in choroid plexus. *Journal of Neuroscience* **26**(30):7811–7819.

Boassa D and Yool AJ (2002) A fascinating tail: cGMP activation of aquaporin-1 ion channels. *Trends in Pharmacology Sciences* **23**(12):558–562.

Boassa D and Yool AJ (2003) Single amino acids in the carboxyl terminal domain of aquaporin-1 contribute to cGMP-dependent ion channel activation. *BMC Physiology* **3**:12.

Bogenrieder T and Herlyn M (2003) Axis of evil: Molecular mechanisms of cancer metastasis. *Oncogene* **22**(42):6524–6536.

Brooks HL, Regan JW and Yool AJ (2000) Inhibition of aquaporin-1 water permeability by tetraethylammonium: Involvement of the loop E pore region. *Molecular Pharmacology* **57**(5):1021–1026.

Campbell EM, Birdsell DN and Yool AJ (2012) The activity of human aquaporin 1 as a cGMP-gated cation channel is regulated by tyrosine phosphorylation in the carboxyl-terminal domain. *Molecular Pharmacology* **81**(1):97–105.

Castle NA (2005) Aquaporins as targets for drug discovery. *Drug Discovery Today* **10**(7):485–493.

Chae YK, Kang SK, Kim MS, Woo J, Lee J, Chang S, Kim DW et al. (2008a) Human AQP5 plays a role in the progression of chronic myelogenous leukemia (CML). *PloS One* **3**(7):e2594.

Chae YK, Woo J, Kim MJ, Kang SK, Kim MS, Lee J, Lee SK et al. (2008b) Expression of aquaporin 5 (AQP5) promotes tumor invasion in human non small cell lung cancer. *PloS One* **3**(5):e2162.

Chai RC, Jiang JH, Kwan Wong AY, Jiang F, Gao K, Vatcher G and Hoi Yu AC (2013) AQP5 is differentially regulated in astrocytes during metabolic and traumatic injuries. *Glia* **61**(10):1748–1765.

Cheng S, Eliaz I, Lin J, Thyagarajan-Sahu A and Sliva D (2013) Triterpenes from *Poria cocos* suppress growth and invasiveness of pancreatic cancer cells through the downregulation of MMP-7. *International Journal of Oncology* **42**(6):1869–1874.

Conde A, Diallinas G, Chaumont F, Chaves M and Geros H (2010) Transporters, channels, or simple diffusion? Dogmas, atypical roles and complexity in transport systems. *The International Journal of Biochemistry and Cell Biology* **42**(6):857–868.

Cowan CA, Yokoyama N, Bianchi LM, Henkemeyer M and Fritzsch B (2000) EphB2 guides axons at the midline and is necessary for normal vestibular function. *Neuron* **26**(2):417–430.

de Morais Lima GR, de Albuquerque Montenegro C, de Almeida CL, de Athayde-Filho PF, Barbosa-Filho JM and Batista LM (2011) Database survey of anti-inflammatory plants in South America: A review. *International Journal of Molecular Science* **12**(4):2692–2749.

Deb P, Pal S, Dutta V, Boruah D, Chandran VM and Bhatoe HS (2012) Correlation of expression pattern of aquaporin-1 in primary central nervous system tumors with tumor type, grade, proliferation, microvessel density, contrast-enhancement and perilesional edema. *Journal of Cancer Research and Therapeutics* **8**(4):571–577.

Detmers FJ, de Groot BL, Muller EM, Hinton A, Konings IB, Sze M, Flitsch SL, Grubmuller H and Deen PM (2006) Quaternary ammonium compounds as water channel blockers. Specificity, potency, and site of action. *Journal of Biological Chemistry* **281**(20):14207–14214.

Diedler J, Sykora M, Blatow M, Juttler E, Unterberg A and Hacke W (2009) Decompressive surgery for severe brain edema. *Journal of Intensive Care Medicine* **24**(3):168–178.

Ding JY, Kreipke CW, Speirs SL, Schafer P, Schafer S and Rafols JA (2009) Hypoxia-inducible factor-1alpha signaling in aquaporin upregulation after traumatic brain injury. *Neuroscience Letters* **453**(1):68–72.

Downey LA, Kean J, Nemeh F, Lau A, Poll A, Gregory R, Murray M et al. (2013) An acute, double-blind, placebo-controlled crossover study of 320 mg and 640 mg doses of a special extract of Bacopa monnieri (CDRI 08) on sustained cognitive performance. *Phytotherapy Research* **27**(9):1407–1413.

El Hindy N, Bankfalvi A, Herring A, Adamzik M, Lambertz N, Zhu Y, Siffert W, Sure U and Sandalcioglu IE (2013) Correlation of aquaporin-1 water channel protein expression with tumor angiogenesis in human astrocytoma. *Anticancer Research* **33**(2):609–613.

Esteva-Font C, Jin B-J and Verkman A (2014) Aquaporin-1 gene deletion reduces breast tumor growth and lung metastasis in tumor-producing MMTV-PyVT mice. *FASEB Journal* **28**(3):1446–1453.

Feickert HJ, Drommer S and Heyer R (1999) Severe head injury in children: Impact of risk factors on outcome. *Journal of Trauma* **47**(1):33–38.

Fischbarg J (2010) Fluid transport across leaky epithelia: Central role of the tight junction and supporting role of aquaporins. *Physiological Reviews* **90**(4):1271–1290.

Fischer U, Taussky P, Gralla J, Arnold M, Brekenfeld C, Reinert M, Meier N et al. (2011) Decompressive craniectomy after intra-arterial thrombolysis: Safety and outcome. *Journal of Neurology, Neurosurgery & Psychiatry* **82**(8):885–887.

Footer MJ, Kerssemakers JW, Theriot JA and Dogterom M (2007) Direct measurement of force generation by actin filament polymerization using an optical trap. *Proceedings of National Academy Science USA* **104**(7):2181–2186.

Fotiadis D, Suda K, Tittmann P, Jeno P, Philippsen A, Muller DJ, Gross H and Engel A (2002) Identification and structure of a putative Ca^{2+}-binding domain at the C terminus of AQP1. *Journal of Molecular Biology* **318**(5):1381–1394.

Friedman B, Schachtrup C, Tasi PS, Shih AY, Akassoglou K, Kelinfeld D and Lyden PD (2009) Acute vascular disruption and aquaporin 4 loss after stroke. *Stroke* **40**:2182–2190.

Frigeri A, Nicchia GP and Svelto M (2007) Aquaporins as targets for drug discovery. *Current Pharmaceutical Design* **13**(23):2421–2427.

Frydenlund DS, Bhardwaj A, Otsuka T, Mylonakou MN, Yasumura T, Davidson KG, Zeynalov E et al. (2006) Temporary loss of perivascular aquaporin-4 in neocortex after transient middle cerebral artery occlusion in mice. *Proceedings of the National Academy of Sciences* **103**:13532–13536.

Fu D, Libson A, Miercke LJ, Weitzman C, Nollert P, Krucinski J and Stroud RM (2000) Structure of a glycerol-conducting channel and the basis for its selectivity. *Science* **290**(5491):481–486.

Gao J, Wang X, Chang Y, Zhang J, Song Q, Yu H and Li X (2006) Acetazolamide inhibits osmotic water permeability by interaction with aquaporin-1. *Analytical Biochemistry* **350**(2):165–170.

Grabarek Z (2006) Structural basis for diversity of the EF-hand calcium-binding proteins. *Journal of Molecular Biology* **359**(3):509–525.

Hachez C and Chaumont F (2010) Aquaporins: A family of highly regulated multifunctional channels. *Advances in Experimental Medicine and Biology* **679**:1–17.

Hacke W, Schwab S, Horn M, Spranger M, De Georgia M and von Kummer R (1996) 'Malignant' middle cerebral artery territory infarction: Clinical course and prognostic signs. *Archives of Neurology* **53**(4):309–315.

Hara-Chikuma M and Verkman AS (2008a) Aquaporin-3 facilitates epidermal cell migration and proliferation during wound healing. *Journal of Molecular Medicine* **86**(2):221–231.

Hara-Chikuma M and Verkman AS (2008b) Prevention of skin tumorigenesis and impairment of epidermal cell proliferation by targeted aquaporin-3 gene disruption. *Molecular and Cellular Biology* **28**(1):326–332.

Higashida T, Kreipke CW, Rafols JA, Peng C, Schafer S, Schafer P, Ding JY et al. (2011) The role of hypoxia-inducible factor-1alpha, aquaporin-4, and matrix metalloproteinase-9 in blood-brain barrier disruption and brain edema after traumatic brain injury. *Journal of Neurosurgery* **114**(1):92–101.

Hiroaki Y, Tani K, Kamegawa A, Gyobu N, Nishikawa K, Suzuki H, Walz T et al. (2006) Implications of the aquaporin-4 structure on array formation and cell adhesion. *Journal of Molecular Biology* **355**(4):628–639.

Hossmann YA (1994) Viability thresholds and the penumbra of focal ischemia. *Annals of Neurology* **36**:557–565.

Hu J and Verkman AS (2006) Increased migration and metastatic potential of tumor cells expressing aquaporin water channels. *FASEB Journal* **20**(11):1892–1894.

Huber VJ, Tsujita M, Kwee IL and Nakada T (2009) Inhibition of aquaporin 4 by antiepileptic drugs. *Bioorganic & Medicinal Chemistry* **17**(1):418–424.

Huber VJ, Tsujita M and Nakada T (2012) Aquaporins in drug discovery and pharmacotherapy. *Molecular Aspects of Medicine* **33**(5–6):691–703.

Huber VJ, Tsujita M, Yamazaki M, Sakimura K and Nakada T (2007) Identification of arylsulfonamides as aquaporin 4 inhibitors. *Bioorganic & Medicinal Chemistry Letters* 17(5):1270–1273.

Hyder AA, Wunderlich CA, Puvanachandra P, Gururaj G, Kobusingye OC (2007) The impact of traumatic brain injuries: A global perspective. *NeuroRehabilitation* 22(5):341–53.

Ishibashi K (2009) New members of mammalian aquaporins: AQP10-AQP12. *Handbook of Experimental Pharmacology* 2009(190):251–262.

Ishibashi K, Hara S and Kondo S (2009) Aquaporin water channels in mammals. *Clinical and Experimental Nephrology* 13(2):107–117.

Ishibashi K, Kondo S, Hara S and Morishita Y (2011) The evolutionary aspects of aquaporin family. *American Journal of Physiology. Regulatory Integrative Comparative Physiology* 300(3):R566–R576.

Jensen MO, Tajkhorshid E and Schulten K (2003) Electrostatic tuning of permeation and selectivity in aquaporin water channels. *Biophysical Journal* 85(5):2884–2899.

Jeyaseelan K, Sepramaniam S, Armugam A and Wintour EM (2006) Aquaporins: A promising target for drug development. *Expert Opinion on Therapeutic Targets* 10(6):889–909.

Jiang Y (2009) Aquaporin-1 activity of plasma membrane affects HT20 colon cancer cell migration. *IUBMB Life* 61(10):1001–1009.

Jiang Y, Chen K, Zhang T and Luo X (2009) Down-regulation of aquaporin-1 in W489 colon cancer cells inhibits cell migration, *Bioinformatics and Biomedical Engineering*. Paper presented at Third International Conference on Bioinformatics and Biomedical Engineering, Beijing, pp. 1–5, Piscataway, New Jersey: IEEE.

Jiang Y and Jiang Z-B (2010) Aquaporin 1-expressing MCF-7 mammary carcinoma cells show enhanced migration in vitro. *Journal of Biomedical Science and Engineering* 3(01):95.

Jullienne A and Badaut J (2013) Molecular contributions to neurovascular unit dysfunctions after brain injuries: Lessons for target-specific drug development. *Future Neurology* 8(6):677–689.

Karou SD, Tchacondo T, Ilboudo DP and Simpore J (2011) Sub-Saharan Rubiaceae: A review of their traditional uses, phytochemistry and biological activities. *Pakistan Journal of Biological Sciences* 14(3):149–169.

Katakai T, Habiro K and Kinashi T (2013) Dendritic cells regulate high-speed interstitial T cell migration in the lymph node via LFA-1/ICAM-1. *Journal of Immunology* 191(3):1188–1199.

Kater S and Letourneau P (1985) *Biology of the Nerve Growth Cone*. Alan R Liss, New York.

Kaur C, Sivakumar V, Zhang Y and Ling EA (2006) Hypoxia-induced astrocytic reaction and increased vascular permeability in the rat cerebellum. *Glia* 54(8):826–839.

Kim JH, Lee YW, Park KA, Lee WT and Lee JE (2010) Agmatine attenuates brain edema through reducing the expression of aquaporin-1 after cerebral ischemia. *Journal of Cerebral Blood Flow and Metabolism* 30(5):943–949.

King LS, Kozono D and Agre P (2004) From structure to disease: The evolving tale of aquaporin biology. *Nature Review Molecular Cell Biology* 5(9):687–698.

Kong G, Zhao Y, Li GH, Chen BJ, Wang XN, Zhou HL, Lou HX, Ren DM and Shen T (2015) The genus *Litsea* in traditional Chinese medicine: An ethnomedical, phytochemical and pharmacological review. *Journal of Ethnopharmacology* 164:256–264.

Kusayama M, Wada K, Nagata M, Ishimoto S, Takahashi H, Yoneda M, Nakajima A, Okura M, Kogo M and Kamisaki Y (2011) Critical role of aquaporin 3 on growth of human esophageal and oral squamous cell carcinoma. *Cancer Science* 102(6):1128–1136.

Laird MD, Vender JR and Dhandapani KM (2008) Opposing roles for reactive astrocytes following traumatic brain injury. *Neurosignals* 16(2–3):154–164.

Le Clainche C and Carlier MF (2008) Regulation of actin assembly associated with protrusion and adhesion in cell migration. *Physiological Reviews* 88(2):489–513.

Lee K, Jo IY, Park SH, Kim KS, Bae J, Park JW, Lee BJ, Choi HY and Bu Y (2012a) Defatted sesame seed extract reduces brain oedema by regulating aquaporin 4 expression in acute phase of transient focal cerebral ischaemia in rat. *Phytotherapy Research: PTR* 26(10):1521–1527.

Lee SM, Lee YJ, Yoon JJ, Kang DG and Lee HS (2012b) Effect of *Poria cocos* on hypertonic stress-induced water channel expression and apoptosis in renal collecting duct cells. *Journal of Ethnopharmacology* **141**(1):368–376.

Li Y, Feng X, Yang H and Ma T (2006) Expression of aquaporin-1 in SMMC-7221 liver carcinoma cells promotes cell migration. *Chinese Science Bulletin* **51**(20):2466–2471.

Ling H, Zhang Y, Ng KY and Chew EH (2011) Pachymic acid impairs breast cancer cell invasion by suppressing nuclear factor-kappaB-dependent matrix metalloproteinase-9 expression. *Breast Cancer Research and Treatment* **126**(3):609–620.

Liu X, Yue R, Zhang J, Shan L, Wang R and Zhang W (2013) Neuroprotective effects of bacopaside I in ischemic brain injury. *Restorative Neurology and Neuroscience* **31**(2):109–123.

Ma B, Xiang Y, Mu SM, Li T, Yu HM and Li XJ (2004) Effects of acetazolamide and anordiol on osmotic water permeability in AQP1-cRNA injected *Xenopus* oocyte. *Acta Pharmacologica Sinica* **25**(1):90–97.

Macey R (1984) Transport of water and urea in red blood cells. *American Journal of Cell Physiology* **246**:C195–C203.

Macey RI, Karan DM and Farmer RE (1972) Properties of water channels in human red cells. *Biomembranes* **3**:331–340.

Machida Y, Ueda Y, Shimasaki M, Sato K, Sagawa M, Katsuda S and Sakuma T (2011) Relationship of aquaporin 1, 3, and 5 expression in lung cancer cells to cellular differentiation, invasive growth, and metastasis potential. *Human Pathology* **42**(5):669–678.

Mack AF and Wolburg H (2013) A novel look at astrocytes: Aquaporins, ionic homeostasis, and the role of the microenvironment for regeneration in the CNS. *Neuroscientist* **19**(2):195–207.

Madeira A, Moura TF and Soveral G (2014) Aquaglyceroporins: Implications in adipose biology and obesity. *Cellular and Molecular Life Science.* **72**(4):759–771.

Manley GT, Fujimura M, Ma T, Noshita N, Filiz F, Bollen AW, Chan P and Verkman AS (2000) Aquaporin-4 deletion in mice reduces brain edema after acute water intoxication and ischemic stroke. *Nature Medicine* **6**(2):159–163.

Manno EM, Adams RE, Derdeyn CP, Powers WJ and Diringer MN (1999) The effects of mannitol on cerebral edema after large hemispheric cerebral infarct. *Neurology* **52**(3):583–587.

Marcy Y, Prost J, Carlier MF and Sykes C (2004) Forces generated during actin-based propulsion: A direct measurement by micromanipulation. *Proceedings of National Academy of Sciences of United States of America* **101**(16):5992–5997.

Masel BE, DeWitt DS (2010) Traumatic brain injury: a disease process, not an event. *Journal of Neurotrauma* **27**(8):1529–1540.

McCoy E and Sontheimer H (2007) Expression and function of water channels (aquaporins) in migrating malignant astrocytes. *Glia* **55**(10):1034–1043.

McCoy ES, Haas BR and Sontheimer H (2010) Water permeability through aquaporin-4 is regulated by protein kinase C and becomes rate-limiting for glioma invasion. *Neuroscience* **168**(4):971–981.

McLeod DD, Beard DJ, Parsons MW, Levi CR, Calford MB and Spratt NJ (2013) Inadvertent occlusion of the anterior choroidal artery explains infarct variability in the middle cerebral artery thread occlusion stroke model. *PLoS One* **8**(9):e75779.

McLeod DD, Parsons MW, Levi CR, Beautement S, Buxton D, Roworth B and Spratt NJ (2011) Establishing a rodent stroke perfusion computed tomography model. *International Journal of Stroke* **6**(4):284–289.

Meng F, Rui Y, Xu L, Wan C, Jiang X and Li G (2014) Aqp1 enhances migration of bone marrow mesenchymal stem cells through regulation of FAK and beta-catenin. *Stem Cells and Development* **23**(1):66–75.

Migliati E, Meurice N, DuBois P, Fang JS, Somasekharan S, Beckett E, Flynn G and Yool AJ (2009) Inhibition of aquaporin-1 and aquaporin-4 water permeability by a derivative of the loop diuretic bumetanide acting at an internal pore-occluding binding site. *Molecular Pharmacology* **76**(1):105–112.

Miller LP, Hsu C (1992) Therapeutic potential for adenosine receptor activation in ischemic brain injury. *Journal of Neurotrauma* 2:563–577.

Minami S, Kobayashi H, Yamashita A, Yanagita T, Uezono Y, Yokoo H, Shiraishi S et al. (2001) Selective expression of aquaporin 1, 4 and 5 in the rat middle ear. *Hearing Research* **158**(1–2):51–56.

Moon C, King LS and Agre P (1997) Aqp1 expression in erythroleukemia cells: Genetic regulation of glucocorticoid and chemical induction. *The American Journal of Physiology* **273**(5 Part 1):C1562–1570.

Moon C, Rousseau R, Soria JC, Hoque MO, Lee J, Jang SJ, Trink B, Sidransky D and Mao L (2004) Aquaporin expression in human lymphocytes and dendritic cells. *American Journal of Hematology* **75**(3):128–133.

Moon C, Soria JC, Jang SJ, Lee J, Obaidul Hoque M, Sibony M, Trink B, Chang YS, Sidransky D and Mao L (2003) Involvement of aquaporins in colorectal carcinogenesis. *Oncogene* **22**(43):6699–6703.

Murtha LA, McLeod DD, McCann SK, Pepperall D, Chung S, Levi CR, Calford MB and Spratt NJ (2014a) Short-duration hypothermia after ischemic stroke prevents delayed intracranial pressure rise. *International Journal of Stroke* **9**(5):553–559.

Murtha LA, McLeod DD, Pepperall D, McCann SK, Beard DJ, Tomkins AJ, Holmes WM, McCabe C, Macrae IM and Spratt NJ (2014b) Intracranial pressure elevation after ischemic stroke in rats: Cerebral edema is not the only cause, and short-duration mild hypothermia is a highly effective preventive therapy. *Journal of Cerebral Blood Flow & Metabolism*. **35**(4):592–600.

Naccache P and Sha'afi RI (1974) Effect of PCMBS on water transfer across biological membranes. *Journal of Cell Physiology* **83**(3):449–456.

Nagelhus EA, Ottersen OP (2013) Physiological roles of aquaporin-4 in brain. *Physiological Review* **93**(4):1543–1562.

Nicchia GP, Stigliano C, Sparaneo A, Rossi A, Frigeri A and Svelto M (2013) Inhibition of aquaporin-1 dependent angiogenesis impairs tumour growth in a mouse model of melanoma. *Journal of Molecular Medicine (Berlin, Germany)* **91**(5):613–623.

Nie Y, Dong X, He Y, Yuan T, Han T, Rahman K, Qin L and Zhang Q (2013) Medicinal plants of genus Curculigo: Traditional uses and a phytochemical and ethnopharmacological review. *Journal of Ethnopharmacology* **147**(3):547–563.

Nielsen S, Kwon TH, Frokiaer J and Agre P (2007) Regulation and dysregulation of aquaporins in water balance disorders. *Journal of Internal Medicine* **261**(1):53–64.

Niemietz CM and Tyerman SD (2002) New potent inhibitors of aquaporins: Silver and gold compounds inhibit aquaporins of plant and human origin. *FEBS Letters* **531**(3):443–447.

Pan XY, Guo H, Han J, Hao F, An Y, Xu Y, Xiaokaiti Y, Pan Y and Li XJ (2012) Ginsenoside Rg3 attenuates cell migration via inhibition of aquaporin 1 expression in PC-3M prostate cancer cells. *European Journal of Pharmacology* **683**(1–3):27–34.

Papadopoulos MC, Manley GT, Krishna S and Verkman AS (2004) Aquaporin-4 facilitates reabsorption of excess fluid in vasogenic brain edema. *The Federation of American Societies for Experimental Biology Journal* **18**:1291–1293.

Papadopoulos MC and Verkman AS (2007) Aquaporin-4 and brain edema. *Pediatric Nephrology* **22**(6):778–784.

Papadopoulos MC and Verkman AS (2008) Potential utility of aquaporin modulators for therapy of brain disorders. *Progress in Brain Research* **170**:589–601.

Parekh SH, Chaudhuri O, Theriot JA and Fletcher DA (2005) Loading history determines the velocity of actin-network growth. *Nature Cell Biology* **7**(12):1219–1223.

Park E, Bell JD and Baker AJ (2008) Traumatic brain injury: Can the consequences be stopped? *Canadian Medical Association Journal* **178**(9):1163–1170.

Peng L, Zhou Y, Kong de Y and Zhang WD (2010) Antitumor activities of dammarane triterpene saponins from *Bacopa monniera*. *Phytotherapy Research: PTR* **24**(6):864–868.

Preston GM and Agre P (1991) Isolation of the cDNA for erythrocyte integral membrane protein of 28 kilodaltons: Member of an ancient channel family. *Proceedings of National Academy of Sciences of United States of America* **88**(24):11110–11114.

Preston GM, Carroll TP, Guggino WB and Agre P (1992) Appearance of water channels in *Xenopus* oocytes expressing red cell CHIP28 protein. *Science* **256**(5055):385–387.

Rios JL (2011) Chemical constituents and pharmacological properties of *Poria cocos*. *Planta Medica* **77**(7):681–691.

Rohini G and Devi CS (2008) *Bacopa monniera* extract induces apoptosis in murine sarcoma cells (S-180). *Phytotherapy Research: PTR* **22**(12):1595–1598.

Russo A and Borrelli F (2005) *Bacopa monniera*, a reputed nootropic plant: An overview. *Phytomedicine: International Journal of Phytotherapy and Phytopharmacology* **12**(4):305–317.

Ryu HM, Oh EJ, Park SH, Kim CD, Choi JY, Cho JH, Kim IS et al. (2012) Aquaporin 3 expression is up-regulated by TGF-beta1 in rat peritoneal mesothelial cells and plays a role in wound healing. *American Journal of Pathology* **181**(6):2047–2057.

Saadoun S, Papadopoulos MC, Davies DC, Bell BA and Krishna S (2002a) Increased aquaporin 1 water channel expression in human brain tumours. *British Journal of Cancer* **87**(6):621–623.

Saadoun S, Papadopoulos MC, Davies DC, Krishna S and Bell BA (2002b) Aquaporin-4 expression is increased in oedematous human brain tumours. *Journal of Neurology, Neurosurgery, and Psychiatry* **72**(2):262–265.

Saadoun S, Papadopoulos MC, Hara-Chikuma M and Verkman AS (2005) Impairment of angiogenesis and cell migration by targeted aquaporin-1 gene disruption. *Nature* **434**(7034):786–792.

Sairam K, Dorababu M, Goel RK and Bhattacharya SK (2002) Antidepressant activity of standardized extract of *Bacopa monniera* in experimental models of depression in rats. *Phytomedicine: International Journal of Phytotherapy and Phytopharmacology* **9**(3):207–211.

Saparov SM, Kozono D, Rothe U, Agre P and Pohl P (2001) Water and ion permeation of aquaporin-1 in planar lipid bilayers. Major differences in structural determinants and stoichiometry. *Journal of Biological Chemistry* **276**(34):31515–31520.

Saraf MK, Prabhakar S and Anand A (2010) Neuroprotective effect of *Bacopa monniera* on ischemia induced brain injury. *Pharmacology, Biochemistry, and Behavior* **97**(2):192–197.

Seeliger D, Zapater C, Krenc D, Haddoub R, Flitsch S, Beitz E, Cerda J and de Groot BL (2013) Discovery of novel human aquaporin-1 blockers. *ACS Chemical Biology* **8**(1):249–256.

Shanmugasundaram ER, Akbar GK and Shanmugasundaram KR (1991) Brahmighritham, an Ayurvedic herbal formula for the control of epilepsy. *Journal of Ethnopharmacology* **33**(3):269–276.

Shenaq M, Kassem H, Peng C, Schafer S, Ding JY, Fredrickson V, Guthikonda M, Kreipke CW, Rafols JA and Ding Y (2012) Neuronal damage and functional deficits are ameliorated by inhibition of aquaporin and HIF1alpha after traumatic brain injury (TBI). *Journal of the Neurological Sciences* **323**(1–2):134–140.

Shi W-Z, Qi L-L, Fang S-H, Lu Y-B, Zhang W-P and Wei E-Q (2012) Aggravated chronic brain injury after focal cerebral ischemia in aquaporin-4-deficient mice. *Neuroscience Letters* **520**(1):121–125.

Shin JA, Choi JH, Choi YH and Park EM (2011) Conserved aquaporin 4 levels associated with reduction of brain edema are mediated by estrogen in the ischemic brain after experimental stroke. *Biochimica et Biophysica Acta* **1812**(9):1154–1163.

Singh HK and Dhawan BN (1982) Effect of *Bacopa monniera* Linn. (brahmi) extract on avoidance responses in rat. *Journal of Ethnopharmacology* **5**(2):205–214.

Stroka KM, Jiang H, Chen SH, Tong Z, Wirtz D, Sun SX and Konstantopoulos K (2014) Water permeation drives tumor cell migration in confined microenvironments. *Cell* **157**(3):611–623.

Sui H, Han BG, Lee JK, Walian P and Jap BK (2001) Structural basis of water-specific transport through the AQP1 water channel. *Nature* **414**(6866):872–878.

Taniguchi M, Yamashita T, Kumura E, Tamatani M, Kobayashi A, Yokawa T, Maruno M et al. (2000) Induction of aquaporin-4 water channel mRNA after focal cerebral ischemia in rat. *Brain Research Molecular Brain Research* **78**(1–2):131–137.

Tornroth-Horsefield S, Wang Y, Hedfalk K, Johanson U, Karlsson M, Tajkhorshid E, Neutze R and Kjellbom P (2006) Structural mechanism of plant aquaporin gating. *Nature* **439**(7077):688–694.

Tourdias T, Mori N, Dragonu I, Cassagno N, Boiziau C, Aussudre J, Brochet B, Moonen C, Petry KG and Dousset V (2011) Differential aquaporin 4 expression during edema build-up and resolution phases of brain inflammation. *Journal of Neuroinflammation* **8**:143.

Tyerman SD, Niemietz CM and Bramley H (2002) Plant aquaporins: Multifunctional water and solute channels with expanding roles. *Plant Cell & Environment* **25**(2):173–194.

Ukiya M, Akihisa T, Tokuda H, Hirano M, Oshikubo M, Nobukuni Y, Kimura Y, Tai T, Kondo S and Nishino H (2002) Inhibition of tumor-promoting effects by poricoic acids G and H and other lanostane-type triterpenes and cytotoxic activity of poricoic acids A and G from *Poria cocos*. *Journal of Natural Products* **65**(4):462–465.

Vahedi K, Vicaut E, Mateo J, Kurtz A, Orabi M, Guichard JP, Boutron C et al. (2007) Sequential-design, multicenter, randomized, controlled trial of early decompressive craniectomy in malignant middle cerebral artery infarction (DECIMAL trial). *Stroke* **38**(9):2506–2517.

Van Duijn B and Inouye K (1991) Regulation of movement speed by intracellular pH during *Dictyostelium discoideum* chemotaxis. *Proceedings of National Academy of Sciences on United States of America* **88**(11):4951–4955.

Verkman A, Hara-Chikuma M and Papadopoulos MC (2008) Aquaporins—New players in cancer biology. *Journal of Molecular Medicine* **86**(5):523–529.

von Bulow J, Muller-Lucks A, Kai L, Bernhard F and Beitz E (2012) Functional characterization of a novel aquaporin from *Dictyostelium discoideum* amoebae implies a unique gating mechanism. *Journal of Biological Chemistry* **287**(10):7487–7494.

Ware MF, Wells A and Lauffenburger DA (1998) Epidermal growth factor alters fibroblast migration speed and directional persistence reciprocally and in a matrix-dependent manner. *Journal of Cell Science* **111**(Part 16):2423–2432.

Werner C and Engelhard K (2007) Pathophysiology of traumatic brain injury. *British Journal of Anaesthesia* **99**(1):4–9.

Winter CD, Adamides AA, Lewis PM and Rosenfeld JV (2005) A review of the current management of severe traumatic brain injury. *Surgeon* **3**(5):329–337.

Woo J, Lee J, Chae YK, Kim MS, Baek JH, Park JC, Park MJ et al. (2008) Overexpression of AQP5, a putative oncogene, promotes cell growth and transformation. *Cancer Letters* **264**(1):54–62.

Xia H, Ma YF, Yu CH, Li YJ, Tang J, Li JB, Zhao YN and Liu Y (2014) Aquaporin 3 knockdown suppresses tumour growth and angiogenesis in experimental non-small cell lung cancer. *Experimental Physiology* **99**(7):974–984.

Xiao M, Hu G (2014) Involvement of aquaporin 4 in astrocyte function and neuropsychiatric disorders. *CNS Neuroscience & Therapeutics* **20**(5):385–390.

Yamamoto N, Yoneda K, Asai K, Sobue K, Tada T, Fujita Y, Katsuya H et al. (2001) Alterations in the expression of the AQP family in cultured rat astrocytes during hypoxia and reoxygenation. *Brain Research Molecular Brain Research* **90**(1):26–38.

Yool AJ (2007a) Aquaporins: Multiple roles in the central nervous system. *Neuroscientist* **13**(5):470–485.

Yool AJ (2007b) Functional domains of aquaporin-1: Keys to physiology, and targets for drug discovery. *Current Pharmaceutical Design* **13**(31):3212–3221.

Yool AJ, Brokl OH, Pannabecker TL, Dantzler WH and Stamer WD (2002) Tetraethylammonium block of water flux in aquaporin-1 channels expressed in kidney thin limbs of Henle's loop and a kidney-derived cell line. *BMC Physiology* **2**:4.

Yool AJ, Brown EA and Flynn GA (2009) Roles for novel pharmacological blockers of aquaporins in the treatment of brain oedema and cancer. *Clinical and Experimental Pharmacology & Physiology* 37(4):403–409.

Yool AJ and Campbell EM (2012) Structure, function and translational relevance of aquaporin dual water and ion channels. *Molecular Aspects of Medicine* 33(5–6):553–561.

Yool AJ, Morelle J, Cnops Y, Verbavatz JM, Campbell EM, Beckett EA, Booker GW, Flynn G and Devuyst O (2013) AqF026 is a pharmacologic agonist of the water channel aquaporin-1. *Journal of the American Society of Nephrology: JASN* 24(7):1045–1052.

Yool AJ and Weinstein AM (2002) New roles for old holes: Ion channel function in aquaporin-1. *News Physiological Sciences* 17:68–72.

Yu J, Yool AJ, Schulten K and Tajkhorshid E (2006) Mechanism of gating and ion conductivity of a possible tetrameric pore in aquaporin-1. *Structure* 14(9):1411–1423.

Yu L, Yi J, Ye G, Zheng Y, Song Z, Yang Y, Song Y, Wang Z and Bao Q (2012) Effects of curcumin on levels of nitric oxide synthase and AQP-4 in a rat model of hypoxia-ischemic brain damage. *Brain Research* 1475:88–95.

Zardoya R (2005) Phylogeny and evolution of the major intrinsic protein family. *Biology of the Cell* 97(6):397–414.

Zelenina M, Zelenin S, Bondar AA, Brismar H and Aperia A (2002) Water permeability of aquaporin-4 is decreased by protein kinase C and dopamine. *American Journal of Physiology. Renal Physiology* 283(2):F309–F318.

Zhang H and Verkman AS (2008) Evidence against involvement of aquaporin-4 in cell-cell adhesion. *Journal of Molecular Biology* 382(5):1136–1143.

Zhang W, Zitron E, Homme M, Kihm L, Morath C, Scherer D, Hegge S et al. (2007) Aquaporin-1 channel function is positively regulated by protein kinase C. *Journal of Biological Chemistry* 282(29):20933–20940.

Zhang X, Chen Q, Wang Y, Peng W and Cai H (2014) Effects of curcumin on ion channels and transporters. *Frontiers in Physiology* 5:94.

Zhang Z, Chen Z, Song Y, Zhang P, Hu J and Bai C (2010) Expression of aquaporin 5 increases proliferation and metastasis potential of lung cancer. *The Journal of Pathology* 221(2):210–220.

Zhao YY (2013) Traditional uses, phytochemistry, pharmacology, pharmacokinetics and quality control of *Polyporus umbellatus* (Pers.) fries: A review. *Journal of Ethnopharmacology* 149(1):35–48.

Zhao YY, Feng YL, Du X, Xi ZH, Cheng XL and Wei F (2012) Diuretic activity of the ethanol and aqueous extracts of the surface layer of *Poria cocos* in rat. *Journal of Ethnopharmacology* 144(3):775–778.

CHAPTER 15

Inorganic Compounds as Aquaporin Substrates or as Potent Inhibitors

A Coordination Chemistry Point of View

Angela Casini and Andreia de Almeida

CONTENTS

Abstract   297
15.1 Introduction   298
15.2 Aquaglyceroporins as Metalloid Channels   300
    15.2.1 Arsenic Compounds Transport through Aquaglyceroporins   300
    15.2.2 Antimonial Compounds Transport through Aquaglyceroporins   304
15.3 Aquaporins Inhibition by Metal Compounds   305
    15.3.1 Mercurial Compounds as Aquaporins Inhibitors   306
    15.3.2 Coordination Gold Complexes as Aquaporins Inhibitors   309
        15.3.2.1 Inhibition of Human Aquaglyceroporin 3   309
        15.3.2.2 Effects of Gold-Based AQP3 Inhibitors on Cell Proliferation   311
        15.3.2.3 Inhibition of Human Aquaglyceroporin 7   312
    15.3.3 Inhibition of Aquaporins by Other Transition Metal Ions   313
15.4 Conclusions and Ideas to Improve AQP Inhibitors Design   314
References   314

ABSTRACT

AQUAPORINS (AQPs) HAVE BEEN proved to be important physiological 'partners' for metal and metalloid compounds, either as their transporters or as targets for inhibition. Here, we summarize the key findings on how AQP channels contribute to the accumulation of metalloids in cells. Specifically, the elucidation of the mechanisms of metalloids uptake by aquaporins provides an understanding of (1) inorganic elements toxicity and (2) how the delivery of arsenic and antimony-containing drugs is crucial in the treatment of certain forms of leukaemia and chemotherapy of diseases caused by patho-genic protozoa. Moreover, a description of the state-of-the-art progresses in the discovery of new metal-based inhibitors, after mercurial compounds and other transition metal ions, is provided. Among them, coordination gold(III) complexes as aquaglyceroporins inhibi-tors are presented with special focus on their mechanism of action at a molecular level, and with indication of their possible applications. Overall, the potential of coordination chemistry in providing compounds to modulate the activity of 'elusive' drug targets, such as the aquaporins, will be discussed.

15.1 INTRODUCTION

Aquaporins (AQPs), members of a superfamily of transmembrane channel proteins, are ubiquitous in all domains of life. They fall into a number of branches that can be functionally categorized into two major subgroups: (1) orthodox AQPs, which are water-specific channels, and (2) aquaglyceroporins, which allow the transport of water but also of non-polar solutes, such as glycerol and other polyols, urea, hydrogen peroxide, as well as ammonia, gases (e.g. carbon dioxide and nitric oxide) and, as described in this review chapter, metalloids. Due to their numerous roles in physiology, these proteins are essential membrane transporters involved in crucial metabolic processes and expressed in almost all tissues (Agre and Kozono 2003).

AQPs are organized as tetramers on membranes. There is a considerable body of information about AQP structure from electron and x-ray crystallography showing AQP monomers (~30 kDa) containing six membrane-spanning helical domains surrounding a narrow aqueous pore (Fu and Lu 2007, Gonen and Walz 2006). The most remarkable feature of the AQP channels is their high selectivity and efficiency on water or glycerol permeation; in fact, AQPs allow water/glycerol to move freely and bidirectionally across the cell membrane but exclude all ions such as hydroxide and hydronium ions, as well as protons. AQPs share a common protein fold, with the typical six membrane-spanning helices surrounding the 20 Å long and 3–4 Å wide amphipathic channel. A feature of AQPs in all organisms is the existence of two constriction sites: (1) an aromatic/arginine selectivity filter (ar/R SF) near the periplasmic/extracellular entrance, which determines the size of molecules allowed to pass through and provides distinguishing features that identify the subfamilies, and (2) a second constriction site generating an electrostatic barrier essential for proton exclusion and composed by two conserved asparagine–proline–alanine (NPA) sequence motifs, located at the N-terminal ends of the two half helices, at the center of the channel. In orthodox AQPs the ar/R SF is very narrow, con-stituted usually by four amino acid residues, typically arginine, phenylalanine, histidine and a fourth residue that can be a cysteine in some cases; instead, in aquaglyceroporins, the ar/R SF is broader due to the existence of only three amino acid residues, normally arginine, phenylalanine and tryptophan or threonine. The differences among isoforms reside in the actual size available for passage of solutes but also on the type of residues: the presence of histidines or cysteines in the ar/SF, instead of aromatic rings, change the hydrophobic properties of the channel itself, varying its selectivity toward different types of solutes.

Among the 13 human AQPs isoforms, only four are glycerol channels, namely AQP3, AQP7, AQP9 and AQP10 (Verkman et al. 2014). Recently, AQP11 localized in the vicinity of lipid droplets in human adipocytes has been shown to permeate both water and glycerol (Madeira et al. 2014b). In other organisms, the need for multiple isoforms is very different: some possess more than one isoform and may have both water and glycerol channels, depending on the organism's life cycle and need for energy input and osmoregulation. For example, a number of AQPs have been identified in parasitic protozoa, from a single aquaglyceroporin in both *Plasmodium* and *Cryptosporidium* to two in *Toxoplasma*, three in *Trypanosoma brucei* and five in *Leishmania*. The malaria parasite *Plasmodium falciparum* has only one AQP isoform, commonly a glycerol channel (PfAQP), which allows passage of different substrate and contains NLA and NPS motifs instead of the classical NPA domain (Newby et al. 2008). *T. brucei* has three AQPs, but only one of them is a glycerol channel, TbAQP2, lacking the classical ar/R SF preserved in other aquaglyceroporins (see Chapter 12 for further information on protozoan isoforms). In fact, the arginine, conserved among all isoforms and organisms, is absent in TbAQP2, while small hydrophobic residues, such as alanine and valine, substitute other residues (Figure 15.1). Not only this feature changes the hydrophobic properties of the channel, but also its size, as smaller side

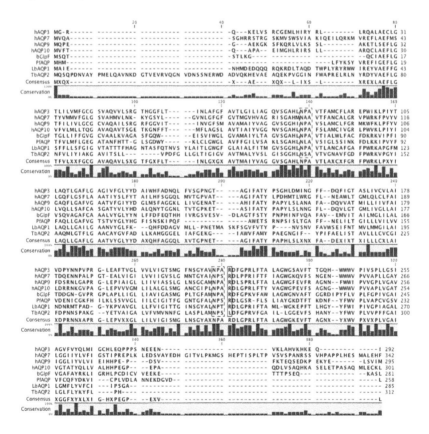

FIGURE 15.1 Sequence alignment of the human aquaglyceroporins AQP3, AQP7, AQP9 and AQP10 and protozoan LbAQP1, TbAQP2, PfAQP and bacterial GlpF. The NPA motifs and conserved arginine are highlighted with a grey dashed box and black box, respectively.

chains broaden the channel allowing passage of different solutes (Baker et al. 2012, Beitz et al. 2006, Rambow et al. 2014).

Among the various substrates for AQPs, 'metalloids' such as arsenic (As) and antimony (Sb) have been shown to be conducted bidirectionally across cellular membranes via different AQPs isoforms. Specifically, the hydroxyacids of lower oxidation state metalloid species B(III), Si(IV), Ge(IV), As(III) and Sb(III) are substrates of aquaglyceroporins in bacteria, protozoans, fungi, plants and animals (Mukhopadhyay et al. 2014). Like glycerol, they are uncharged polar molecules with volumes small enough to fit through the approximately 5 Å diameter opening of the channel. This chapter will present a selection of the studies available on metalloids transport through AQPs related to human cells and diseases and their main mechanistic findings.

As discussed in previous chapters, there is strong evidence that AQPs are drug targets in different diseases, including cancer, angiogenesis and fibrosis (Verkman 2012). Moreover, analysis of the involvement of AQPs in the life cycle of disease-causing organisms (e.g. malaria parasites) suggests additional opportunities for pharmacological intervention in the treatment of human diseases (Beitz 2005). However, the identification of AQPs modulators (inhibitors) for both therapeutic and diagnostic applications has turned out to be extremely challenging. So far, four classes of AQP-targeted small molecules have been described: (1) metal-based inhibitors, (2) small molecules that are reported to inhibit water conductance, (3) small molecules targeting the interaction between AQP4 and the neuromyelitis optica autoantibody and (4) agents that act as chemical chaperones to facilitate the cellular processing of nephrogenic diabetes insipidus–causing AQP2 mutants. Following this categorization, compounds of transition metal ions (e.g. $HgCl_2$ and $NiCl_2$) have been proved to be inhibitors of orthodox water channels. Concerning aquaglyceroporin inhibitors, recently gold coordination complexes have been reported to be potent and selective inhibitors of different human isoforms (de Almeida et al. 2014). Thus, this chapter will also include an overview of the inorganic/metal-based compounds as AQPs inhibitors, with emphasis on the available information on the molecular mechanisms of inhibition studied by different techniques.

15.2 AQUAGLYCEROPORINS AS METALLOID CHANNELS

15.2.1 Arsenic Compounds Transport through Aquaglyceroporins

Arsenic has been used under several forms, including in industrial components, medicines, embalming, manufacture of cosmetics, rodenticides and pesticides, pigments and preservatives. It is the most ubiquitous environmental toxin and carcinogen, labelled as Group 1 human carcinogen, and, even though arsenic embalming and pigments are no longer available, arsenic is still present in several building structures as lead alloys as well as component of batteries, lasers, diodes and transistors. Water contamination with arsenic is a major problem in several countries and chronic exposure may cause different types of cancer, including skin, bladder and lung cancers (Cohen et al. 2013). Thus, exposure to this element is still one of the biggest issues in world health.

Arsenic is bioavailable in three oxidation states, As(V) (arsenate), As(III) (arsenite) and As(0) (elemental). As(III), when dissolved in water at pH 7.0, forms arsenous acid, $As(OH)_3$ (Figure 15.2), and this species is neutral at physiological pH ($pK_{a1} = 9.3$, $pK_{a2} = 13.5$ and

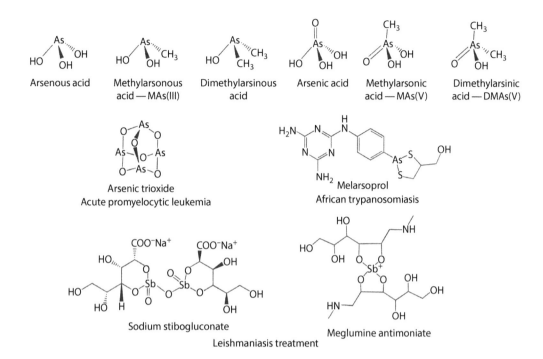

FIGURE 15.2   Structure of metalloid compounds possible substrates of aquaporins and their clinical applications.

$pK_{a3} = 14.0$) (Dhubhghaill and Sadler 1991). The aqueous form of arsenic(V) oxide is arsenic acid, H_3AsO_4 (Figure 15.2), and in physiological conditions, it exists as hydrogen arsenate and dihydrogen arsenate ($pK_{a1} = 2.19$, $pK_{a2} = 6.94$, and $pK_{a3} = 11.5$). Exposure to trivalent arsenic is more toxic than exposure to its pentavalent form, most likely due to the increased affinity of As(III) for sulfur ligands (Hirano et al. 2003). Interestingly, while pentavalent arsenic is transported into cells via phosphate transport systems in both prokaryotes and eukaryotes, concerning As(III), initial studies using *Escherichia coli* showed that metal accumulation goes via the glycerol facilitator protein (GlpF), which belongs to the aquaglyceroporin family (Meng et al. 2004). Moreover, it appears that trivalent arsenic is transported as the neutral form of arsenous acid. In fact, x-ray absorption spectroscopy was used to determine the nearest-neighbour coordination environment of As(III) under a variety of solution conditions (Ramírez-Solís et al. 2004). Extended x-ray absorption fine structure analysis demonstrated that three oxygen ligands are at 1.78 Å from the arsenic atom, showing that the major species in solution is $As(OH)_3$, an inorganic molecular mimic of glycerol (Figure 15.2).

In humans, two aquaglyceroporins have been identified as arsenic transporters, namely AQP7 and AQP9, with the latter being the most effective arsenous acid transporter, fourfold higher than AQP7. Conversely, the other human aquaglyceroporins AQP3 and AQP10 show little arsenic transport. These results were obtained in *Xenopus oocytes* engineered to overexpress the four human glycerol channels (Liu et al. 2004). Subsequent studies suggested that $As(OH)_3$ and glycerol use the same translocation pathway, as mutations in SF residues of AQPs affected the transport of both solutes to the same extent (Liu et al. 2004). Interestingly, AQP9 is also abundantly expressed in leukocytes; however, in this cell type, no glycerol permeability via this isoform could be observed, leaving the possible role for AQP9 uptake of $As(OH)_3$ in leukocytes still to be disclosed (Ishibashi et al. 1998).

Human AQP9 (hAQP9) is highly expressed in the liver, where it is responsible for fluxes of glycerol and urea, and has a role in gluconeogenesis, having an interplay with human AQP7 (hAQP7) in adipose tissue (Maeda et al. 2008). As liver is responsible for metabolizing most drugs and excreting their metabolic products, hAQP9, as efficient arsenic transporter, may be responsible for excretion of arsenic from the liver into the bloodstream. In mammalian cells, the metabolism of arsenic involves export from the cell and metabolic detoxification of the arsenic species. Metabolic processing of inorganic As(III) is carried out by a series of methylation reactions. These reactions involve the interaction of arsenous acid with arsenic(III) methyltransferase (AS3MT). This protein catalyzes the methylation of both arsenous acid and methylarsonous acid (MAs(III)) in the presence of adequate reduction and methyl equivalents. Thus, inorganic arsenic can also be metabolized via methylation to methylarsonic acid (MAs(V)) and dimethylarsinic acid (DMAs(V)) (Figure 15.2). The liver is considered a major site of arsenic methylation to MAs(III) (Thomas et al. 2004, Vahter 1999, 2002). Therefore, hAQP9's ability to transport arsenic in its methylated form was evaluated using a *Saccharomyces cerevisiae* strain HD9 (*acr3Δ ycf1Δ fps1Δ*) resistant to As(III). Usually, these cells exhibit little permeability to As(III) and MAs(III), but when AQP9 expression was induced, cells became sensitive to both As-containing species (Liu et al. 2002, 2006). Interestingly, AQP9 appears to be threefold more permeable to MAs(III) than to $As(OH)_3$. Since the final fate of the monomethylated form of arsenic is excretion, the proposed mechanism of transport of arsenical species in the liver involves uptake of As(III), likely as $As(OH)_3$, and excretion into the bloodstream of MAs(III) through AQP9 (Liu et al. 2006). The methylated arsenicals are released from liver into the bloodstream and end up in urine, skin, hair and other tissues.

In spite of its known toxic effects, the use of arsenic in medicine is an ancient practice (Swindell et al. 2013). Thus, Hippocrates (460–377 BC) recommended a paste of realgar, a mineral form of arsenic, as treatment for ulcers and used arsenic as an escharotic to treat skin and breast cancer. Interestingly, in the nineteenth century, arsenic was a major component of *Materia Medica* and was used to treat a variety of diseases, from skin problems to ulcers and cancer. In 1910, Noble laureate Paul Ehrlich developed Salvarsan (dihydroxy-diamino-arsenobenzene-dihydrochloride), an organic arsenical for the treatment of syphilis (Swindell et al. 2013). Due to its severe toxicity, fatal arsenic poisoning from medical treatments was very common in the past. Nowadays, a few arsenic drugs are still in use; among them, the greatest clinical success has been the one of arsenic trioxide (As_2O_3, Trisenox®) (Figure 15.2) in the treatment of haematological cancers, most notably in acute promyelocytic leukaemia (APL), which is a subtype of acute myeloid leukaemia (Iland and Seymour 2013). An established major determinant of the action of arsenical-containing drugs is the pathway of metallodrug uptake in cancer cells, and, therefore, understanding the factors that modulate arsenic cellular accumulation is important to improve the effects of chemotherapy. Arsenic trioxide almost certainly dissolves to form inorganic As(OH)$_3$, the species that permeates through aquaglyceroporin channels. Overexpression of AQP9, AQP7 or AQP3 renders human leukaemia cells hypersensitive to the drug as a result of higher steady-state levels of As accumulation (Bhattacharjee et al. 2004). In particular, sensitivity to arsenic trioxide is directly proportional to AQP9 expression in leukaemia cells of different lineages (Leung et al. 2007). For example, the APL cell line NB4 showed the highest expression level of AQP9 and is the most sensitive to the drug. Conversely, the chronic myeloid leukemia cell line K562 shows low endogenous AQP9 expression and it is insensitive to As$_2$O$_3$. When hAQP9 was overexpressed either in K562 or in the promyelocytic leukemia cell line HL60, both became hypersensitive to As(III) due to higher intracellular accumulation of the metalloid (Leung et al. 2007).

Since, according to these studies, responsiveness to drug therapy is correlated with increased expression of the drug uptake system, the possibility of using pharmacological agents to increase aquaglyceroporin expression delivers the promise of therapies for the treatment of leukaemia in combination with Trisenox.

Among the successful organoarsenical drug, melarsoprol (2-[4-[[(4,6-diamino-1, 3,5-triazin-2-yl) amino]phenyl]-1,3,2-dithiarsolane-4-methanol) (Figure 15.2) is a prodrug currently used as treatment for late-stage east *African trypanosomiasis*, commonly known as sleeping sickness (Baker et al. 2013). Melarsoprol is metabolized into the highly reactive melarsen oxide, which irreversibly binds to vicinal sulfhydryl groups causing the inactivation of various enzymes. Even though melarsoprol is highly toxic, with severe side effects similar to those of arsenic poisoning, and causes fatal reactive encephalopathy in ca. 5% of the patients, this is the only effective chemotherapeutic in both strains of *T. brucei* and in late stages of trypanosomiasis (Baker et al. 2013, Steverding 2010). In spite of the efficacy of melarsoprol, one recurrent problem of the treatment with arsenic compounds is the development of parasite resistance.

T. brucei parasites contain three aquaglyceroporins, TbAQP1–3, which are thought to be involved in osmoregulation and glycerol transport (Uzcategui et al. 2004). Interestingly, *T. brucei*'s aquaglyceroporins were shown to be involved in melarsoprol/pentamidine cross-resistance (MPXR), when deletion of both TbAQP2 and TbAQP3 showed a 2-fold increase of IC$_{50}$ of the arsenic drug and 15-fold increase in pentamidine, compared to the wild-type strain (Alsford et al. 2012). Later, it was confirmed that only TbAQP2 was involved in resistance, as knockout strains for TbAQP3 did not show any difference compared to wild-type strain toward melarsoprol/pentamidine toxicity (Baker et al. 2012). Reintroducing an inducible copy of TbAQP2 into AQP2/AQP3 null cells, which restored MPXR, validated these results. Interestingly, the induction of TbAQP2 expression restored cell sensitivity to treatment, even in the absence of TbAQP3, while induction of TbAQP3 did not have an effect in MPXR (Baker et al. 2012). These studies reinforce the crucial role of TbAQP2 in drug uptake and support the central role of this AQP isoform in MPXR. Although AQP2 and AQP3 are closely related, as previously mentioned, AQP2 lacks the usual motifs of the SF. Indeed, while AQP1 and AQP3 harbour the conventional 'NPA/NPA' motifs, AQP2 is the only *T. brucei* isoform with NSA/NPS and lacking the conserved arginine in the ar/R SF; therefore, this difference may account for the selectivity toward metalloid transport.

Following promising results with arsenic trioxide, melarsoprol was also tested in vitro and entered clinical trials as a treatment for acute promyelocytic leukemia (Konig et al. 1997, Soignet et al. 1999). However, due to its severe neurotoxicity, in the same dosage and treatment scheme used for the treatment of trypanosomiasis, clinical trials for the treatment of APL with melarsoprol were abandoned (Soignet et al. 1999).

15.2.2 Antimonial Compounds Transport through Aquaglyceroporins

Antimony compounds are used in the semiconductor industry, ceramics and plastics and flame-retardant applications and are often alloyed with other metals to increase their strength and hardness. Exposure to antimony can occur from natural sources and also from industrial activities. The primary effects from chronic exposure to antimony in humans are respiratory problems, lung damage, cardiovascular effects, gastrointestinal disorders, and adverse reproductive outcome.

Nowadays, antimony-based drugs found applications in the treatment of protozoal diseases. Specifically, pentavalent antimony-containing drugs of Sb(V) with N-methyl-D-glucamine such as Pentostam® (sodium stibogluconate) and Glucantime® (meglumine antimoniate) (Figure 15.2) are the treatment of choice for *Leishmania* infections (leishmaniasis). Leishmaniasis is a disease caused by the protozoan parasite *Leishmania*, from the same family as *Trypanosoma*. The disease, endemic in 88 countries, with 400,000 cases/year, is transmitted by a type of sand fly and can manifest in three main forms: cutaneous, mucocutaneous or visceral (Ashford et al. 1992a,b). While anti-leishmania vaccines are still under development or undergoing clinical trials (Kedzierski 2010), antimony [Sb(V)]-based compounds have been used for treatment of all forms of the disease for more than 60 years (Herwaldt 1999). However, as for arsenic drugs, a large increase in cases of resistance to treatment with antimonials has been reported, for example in India 65% of previously untreated patients fail to respond promptly or relapse after therapy with antimonials (Sundar 2001). Thus, it is extremely important to understand the mechanisms of action and resistance of pentavalent antimonial drugs in order to develop better therapeutic agents.

According to one of the most accredited mechanisms of activity, pentavalent antimony (Sb(V)) behaves as a prodrug, which undergoes biological reduction to much more active/toxic trivalent form of antimony Sb(III), which exhibits antileishmanial activity (Haldar et al. 2011). However, the site of (amastigote or macrophage) and mechanism of reduction (enzymatic or non-enzymatic) remain controversial. Furthermore, the ability of *Leishmania* parasites to reduce Sb(V) to Sb(III) is stage specific. The first transport studies of antimony in *Leishmania* parasites were performed using the pentavalent Sb(V) compound (^{125}Sb), sodium stibogluconate (Figure 15.2) (Berman et al. 1987). Moreover, mass spectroscopic approaches revealed the accumulation of two forms of antimony Sb(V) and Sb(III) in both stages of the parasite (Brochu et al. 2003). Thus, the possibility of in vivo metabolic conversion of pentavalent Sb(V) to trivalent Sb(III) antimonials was suggested, followed by uptake into the parasite.

Although antimony is less abundant than arsenic, their chemical properties are very similar. In fact, in solution, arsenic and antimony are mainly present as their trivalent species, as $As(OH)_3$ and $Sb(OH)_3$, respectively, both sharing some similarity to glycerol.

In fact, as for As(III), it has been shown that Sb(III) can be transported in *E. coli* by the glycerol facilitator GlpF (Meng et al. 2004). In *E. coli*, disruption of GlpF leads to a resistant phenotype and reduced levels of uptake of Sb(III) and As(III), confirming that aquaglyceroporins are an important route also for antimony uptake (Sanders et al. 1997). Indeed, structural, thermodynamic and electrostatic comparison of As(III) and Sb(III) species at physiological pH showed that they exhibit similar conformation and charge distribution and a slightly smaller volume than glycerol, which may aid in their passage through the narrowest region of the GlpF channel (Porquet and Filella 2007). However, the metalloid hydroxyl groups lack the flexibility of glycerol, which probably helps the latter to adapt its conformation to the topology of the GlpF channel.

Interestingly, in *Leishmania* cells, the accumulation of Sb(III) is competitively inhibited by the related metal As(III), suggesting that Sb(III) and As(III) enter the parasite cell via the same route (Brochu et al. 2003). Notably, *Leishmania* parasites express one aquaglyceroporin, LAQP1, which has been demonstrated to mediate Sb(III) uptake, and whose overexpression restores Sb(III) sensitivity to three resistant phenotypes (Gourbal et al. 2004). Studies by Kumar et al showed that LAQP1 can be used as a biomarker for Sb(III) resistance, as LAQP1 downregulation in resistant parasites is strongly correlated to the reduction of compound's efficacy on parasite viability (Kumar et al. 2012). In spite of these results, the exact mechanism of transport and resistance to Sb(III) in *Leishmania* is still not fully understood (Ashutosh et al. 2007).

As the *Leishmania* parasite develops inside human macrophages, it is crucial to understand the mechanism of uptake of these metalloids into these human cells. Macrophages express AQP3 in their membrane and this protein is essential for macrophage development and motility (Zhu et al. 2011). This channel isoform does not appear to be very efficient in As(III) transport, but overexpression of AQP3 in human embryonic kidney cells (HEK-293T) led to an increase of As(III) accumulation and cells' sensitivity to the drug (Lee et al. 2006). Considering that As(III) and Sb(III) most likely share routes of uptake, AQP3 is also a likely candidate for Sb(III) transport into macrophages.

15.3 AQUAPORINS INHIBITION BY METAL COMPOUNDS

AQP functions are still not fully understood. Most of the studies present in literature are performed using knockout mice or via RNA silencing methods (Verkman et al. 2000). These studies have confirmed the anticipated involvement of AQPs in the mechanism of urine concentration and glandular fluid secretion and led to the discovery of unanticipated roles of AQPs in brain water balance, cell migration (angiogenesis, wound healing), cell proliferation, neural function (sensory signalling, seizures), epidermal hydration and ocular function. However, in addition to genetic approaches, the use of inhibitors to unravel AQP function has also been of great importance. Among the benchmark AQPs inhibitors, the mercurial compounds pCMBS (*p*-chloromercurybenzene sulfonate) (Figure 15.3) and HgCl$_2$ have been widely applied in in vitro assays. For example, pCMBS was shown to have an effect on water transport via AQP1 in human red blood cells (hRBC) (Macey 1984). In this section, we will summarize the available information on the mechanism of AQPs inhibition by these inorganic compounds. It is worth mentioning that, since mercurials

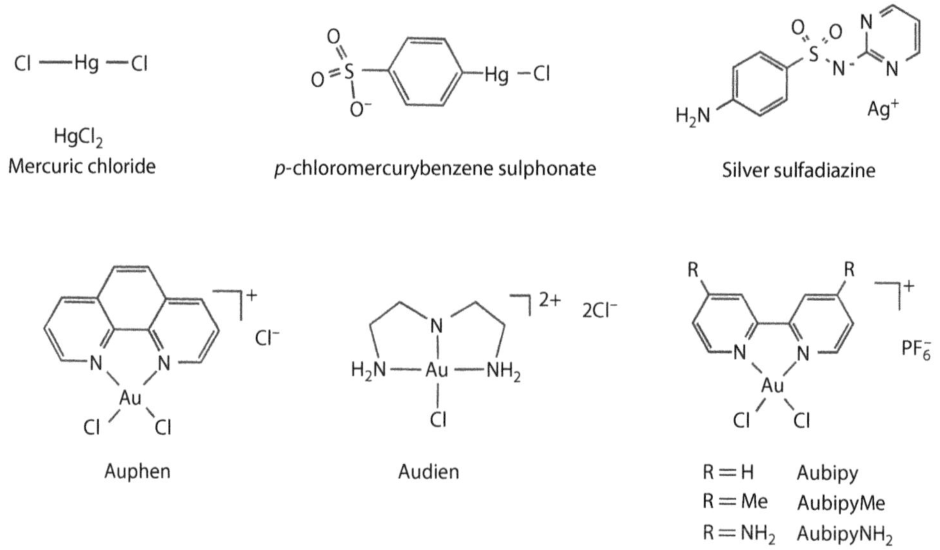

FIGURE 15.3   Structure of metal-based compounds as aquaporins inhibitors.

are not suitable for therapeutic applications and not ideal to study AQP function in biological systems mostly due to their toxic effects and lack of selectivity, other metal-based compounds have been investigated and designed, which hold promise for potent and selective AQPs inhibition. Thus, we will also discuss such new metal compounds and their possible applications as therapeutic agents or to study the roles of AQPs in physiology and pathophysiology.

15.3.1 Mercurial Compounds as Aquaporins Inhibitors

Concerning mercurial compounds, it has always been postulated that AQPs are inhibited by Hg^{2+} ions via covalent modification of cysteine residues, based on the classical hard soft acid base theory. In order to confirm such mechanism and to assess the importance of cysteine residues for mercury inhibition, several studies were performed on Cys-mutated isoforms of human AQP1 (hAQP1). For example, *Xenopus oocytes* were transfected with each Cys-mutated AQP1 isoform and the effects of mercury inhibition were evaluated (Preston et al. 1993, Zhang et al. 1993). From all cysteine residues in AQP1, only one was shown to confer sensitivity to the mercurial salt $HgCl_2$, namely Cys189. When this cysteine was mutated to either serine or glycine, water permeability of the oocytes was slightly decreased, indicating that this residue may be of importance for water transport. Moreover, cells expressing the Cys189Ser mutant lost sensitivity to $HgCl_2$ and did not show any significant inhibition by $HgCl_2$ up to a concentration of 3 mM. Later on, as the atomic resolution structure of hAQP1 was solved, Cys189 was shown to be positioned inside the channel, just above the ar/R SF (Sui et al. 2001). Therefore, it was hypothesized that Hg^{2+} binding to this site was likely to prevent passage of water molecules via steric effects.

The current literature provides two mechanisms of inhibition of AQPs by mercury: the first is simple occlusion of the water pore by the mercury atoms/ions found in the vicinity of the cysteine residues lining the water channel wall; the second is conformational change (collapse of the water pore) at the SF (viz. ar/R constriction) region, induced by mercury bonded to a cysteine residue nearby. In an effort to better understand the influence of cysteines and their location on the inhibition of hAQP1 by mercury, Savage et al. studied the bacterial homolog of hAQP1, AqpZ (Savage and Stroud 2007). This bacterial isoform has been previously described as a water channel (Borgnia et al. 1999), and it is structurally very similar to hAQP1, containing the same ar/R SF residues, but it is not sensitive to mercury since it lacks Cys189 crucial for Hg^{2+} inhibition in hAQP1. In this position, AqpZ has a threonine residue, Thr183. Thus, by site-directed mutagenesis, a model AqpZ was obtained lacking all endogenous cysteins and with a mutation Thr183Cys, in order to evaluate the role of this specific residue (Savage and Stroud 2007). The crystal structure of this mutant, co-crystallized with $HgCl_2$, showed no significant conformational changes between the apo- and metal-bound forms, suggesting that inhibition by mercury is not due to major changes in the tridimensional structure of AqpZ both at the level of the monomer folding and of the tetrameric axes. Instead, as seen in Figure 15.4, a Hg^{2+} ion (Hg-1) appears to be positioned inside the channel, just below the ar/R SF, suggesting a steric blockage of the channel upon metal binding (Savage and Stroud 2007). Notably, the bond distance

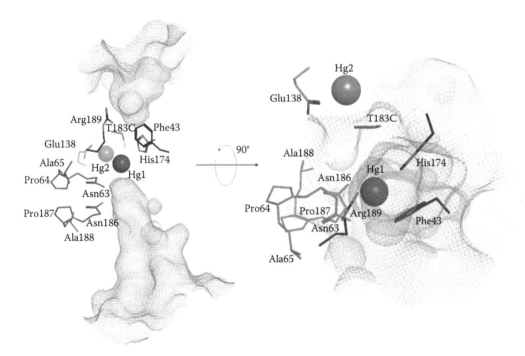

FIGURE 15.4 **(See colour insert.)** Structure of the mercury-blocked AqpZ Tyr183Cys mutant. Molecular surface of residues lining the pore is shown as grey mesh. Selectivity filter residues are shown in black, while NPA motifs are represented in green. Two mercury atoms are located inside or close to the channel and are shown in space-filling representation, in magenta (Hg-1) and orange (Hg-2), respectively. The same colour code is used to represent the amino acid residue that is located closer to each of the mercury atoms. (PDB code 2O9D, Savage, D.F. and Stroud, R.M., *J. Mol. Biol.*, 368(3), 607, 2007.)

between the Hg^{2+} atom and Thr183Cys is not ideal to demonstrate binding to the thiol residue (ca. 5.6 Å). In this structure, another Hg^{2+} atom (Hg-2) is outside the pore (Figure 15.4), pointing toward Cys183 (distance of ca. 4.0 Å) and residing in a hydrophilic pocket formed by conserved Glu138 and Ser177 where it makes favourable electrostatic interactions at 2.6 and 3.1 Å distance, respectively. Interestingly, Glu138 appears to be important for maintaining the orientation of the backbone carbonyl oxygen of Gly190, Cys191 and Gly192. This may imply that a conformational change can also occur in Thr183Cys-AQPZ bound to mercury, although this mechanism has still to be confirmed. In order to validate steric blockage by mercury, the authors produced another mutant, Leu170Cys, with the cysteine located in between the ar/R SF and the NPA motifs. This mutant was proven to be even more sensitive to $HgCl_2$, and the resulting x-ray structure revealed four Hg^{2+} atoms inside the channel, one covalently bound to the thiol group of Leu140Cys. Overall, authors conclude that their results indicate that binding of Hg^{2+} to thiol residue of Cys side chain inside the channel is ideal for AQPs inhibition of solute transport, most likely due to steric blockage (Savage and Stroud 2007). However, since a real coordination bond could not be assessed between Hg^{2+} and the thiol residue of Cys183, further structural information should be provided to validate the proposed mechanism.

The second mechanism of AQP inhibition by mercurial compounds was proposed in an in silico study on the basis of molecular dynamics (MD) simulations of the bovine AQP AQP1 (bAQP1) (Hirano et al. 2010). As hAQP1, also bAQP1 has cysteine residue, Cys191, at the ar/R region, located 8 Å above the NPA region, which may bound Hg. According to the MD simulations of both free AQP1 and Hg-bound AQP1, the energy barrier for Hg-AQP1 is much higher than that of free AQP1 at the ar/R region. Moreover, calculations show that mercury binding induces a collapse of the orientation of amino acid residues at the ar/R region and the constriction of the space between Arg197 and His182.

A third mechanism of mercury inhibition has been proposed on the basis of MD simulations by Zhang et al. according to which the mercury ion, covalently bound to the cysteine residue (Cys170) in the Leu170Cys mutant of AQPZ, causes water molecules to clog the water channel (Zhang et al. 2012). The obtained in silico results unravelled the interactions between the mercury ion and the waters in its vicinity and found that five to six waters are strongly attracted by the mercury ion, occluding the space of the water channel. However, it should be noted that binding of water molecules to Hg ions is highly reversible and the predicted mechanism may not be relevant in physiological environment.

Concerning metal-based inhibitors, other transition metal compounds have been investigated with respect to water permeation through AQPs (e.g. hAQP1). Thus, for example $AgNO_3$ and $HAuCl_4$ were among the most effective in inhibiting water transport. As for mercury, the mechanism of silver and gold inhibition is most likely due to their ability to interact with sulfhydryl groups of proteins. Interestingly, silver resulted to be more efficient in inhibiting water transport than $HgCl_2$. For example, silver as $AgNO_3$ or silver sulfadiazine (Figure 15.3) inhibited with high potency (EC_{50} 1–10 μM) the water permeability of the peribacteroid membrane from soybean (containing nodulin 26 AQP NOD26), the water permeability of plasma membrane from roots (containing plasma membrane integral proteins), and the water permeability of hRBC (containing AQP1) (Niemietz and

Tyerman 2002). However, it should be noted that more recent results by Soveral, Casini et al showed that silver sulfadiazine is actually not inhibiting water transport via AQP1 in hRBC even at 100 µM concentration (Martins et al. 2012).

15.3.2 Coordination Gold Complexes as Aquaporins Inhibitors

Coordination metal complexes can be defined as compounds consisting of a central metal ion bound to organic ligands. Ligands are generally bound to the central atom by a 'coordinate' bond. The reactivity of such metal compounds differs from one of organic molecules in different aspects, which will not be discussed here. However, it is worth mentioning that coordination metal complexes have been proved to hold promise in chemical biology being able to selectively inhibit proteins activities in cells and, therefore, to be exploited as either therapeutic agents or as chemical probes to detect protein/enzyme functions in living systems (de Almeida et al. 2013).

Noteworthy, nowadays, the list of therapeutically prescribed metal-containing compounds includes platinum (anticancer), silver (antimicrobial), gold (antiarthritic), bismuth (antiulcer), antimony (antiprotozoal), vanadium (antidiabetic) and iron (anticancer and antimalarial) complexes (Mjos and Orvig 2014, Nobili et al. 2010). Moreover, metal compounds as diagnostic tools have also been widely explored and are successfully applied in the clinical set for imaging of diseases (Blower 2015). For example, lanthanides occupy a relevant place as diagnostic agents but also have many other important medical applications, as hypophosphatemic agents for kidney dialysis patients, for bone pain palliation and as luminescent probes in cell studies.

Therefore, in this context, the exploration of coordination metal compounds as possible inhibitors of AQPs may open the way to novel approaches to targeting AQPs function and to new drug families. In the following text, we describe recent results in the field and reflect upon the potential of coordination chemistry in providing compounds to modulate the activity of 'elusive' drug targets, such as the AQPs.

15.3.2.1 Inhibition of Human Aquaglyceroporin 3

Recently, Casini and coworkers have reported on the potent and selective inhibition of human AQP3 (hAQP3) by a series of gold coordination compounds (see references within this section). In details, a number of Au(III) complexes with nitrogen donor ligands were tested for their inhibition properties of the orthodox hAQP1 and the glycerol channel AQP3, using a stopped-flow technique in hRBC (Martins et al. 2012). Interestingly, the Au(III) coordination compounds were able to potently inhibit glycerol transport in hRBC through hAQP3, while not having a significant effect on water transport, through hAQP1. The most potent inhibitor of the series, Auphen ([Au(phen)Cl$_2$]Cl, phen = 1,10-phenanthroline) (Figure 15.3), was calculated to have an IC$_{50}$ in the low micromolar range (0.8 ± 0.08 µM). Notably, Auphen was far more effective than the mercurial compound in inhibition AQP3. Furthermore, to validate the results on another cell type, the Au(III) compound was tested on PC12 cells (cells derived from a pheochromocytoma of rat adrenal medulla), overexpressing either hAQP1 or hAQP3, confirming its inhibitory effect on hAQP3 (Martins et al. 2012).

The role of the Au(III) center in the inhibition mechanisms was also assessed, and the Au(III) complexes [Au(dien)Cl]Cl$_2$ (dien = diethylentriamine, Audien) (Figure 15.3) and [Au(cyclam)](ClO$_4$)$_2$Cl (cyclam = 1,4,8,11-tetraazacyclotetradecane, Aucyclam) were also tested in hRBC (Martins et al. 2012). Interestingly, Audien, with an AuN$_3$Cl core in which a chlorido ligand is available for exchange, still maintained the AQP3 inhibitory properties, although in a lower extent with respect to Auphen. Instead, Aucyclam, with the AuN$_4$ core not prone to ligand exchange reactions, and the phenanthroline ligand alone did not have any effect on hAQP3 glycerol permeability. These results suggest that the gold center, as well as its availability for ligand exchange and further coordination to protein residues, is essential for inhibition of hAQP3. It is also worth mentioning that coordination compounds with gold in a different oxidation state, namely Au(I), such as aurothioglucose, were not able to inhibit either AQP1 or AQP3.

Inspired by these initial promising results, Casini et al. investigated other gold-based compounds as possible AQP3 inhibitors in order to achieve basic structure–activity relationships, fundamental for drug design. Thus, a series of square planar gold(III) complexes containing functionalized bipyridine ligands of general formula [Au($N^\wedge N$)Cl$_x$][PF$_6$] [where $N^\wedge N$ = 2,2′-bipyridine, 4,4′-dimethyl-2,2′-bipyridine, and 4,4′-diamino-2,2′-bipyridine; $\times$ = 1, 2] (Figure 15.3) were selected (Martins et al. 2013). Moreover, the 1,10-phenantroline derivatives of Pt(II) and Cu(II) were also included in the investigation to compare the effects of metal substitution on the AQP3 inhibition potency. Thus, the effects of the compounds on both water and glycerol permeation were tested on hRBC.

Notably, the gold(III) complexes were the most effective inhibitors of glycerol permeability via AQP3, with an IC$_{50}$ in the low μM range, and comparable in potency to Auphen. Moreover, AQP3 inhibition resulted to be practically irreversible and only excess of 2-mercaptoethanol (EtSH) allowed restoring of glycerol transport. Within the metal-phenanthroline series, the AQP3 inhibition potency decreased drastically in the following order: Auphen > Cuphen >> Ptphen (Martins et al. 2013). Interestingly, no inhibition was achieved when incubating the cells with Au(I) compounds, therefore, demonstrating the necessity of Au(III)-based scaffolds to achieve protein binding and blockage of the channel.

In order to gain insight into the mechanism of AQP3 inhibition by gold compounds, molecular modelling studies were undertaken. Thus, a homology model of hAQP3 was built, using GlpF from *E. coli* as a reference, and used to further disclose the possible gold binding sites inside of the hAQP3 channel. As previously mentioned, since gold has high affinity for binding to sulfur, the mechanism of inhibition of Auphen and analogues in hAQP3 is possibly based on the ability of Au(III) to interact with sulfur donor groups of proteins such as the thiolate of cysteine or the thioether of methionine residues. In AQP3, only the thiol group of Cys40 located just above ar/R SF inside the protein channel is projected toward the extracellular space. Therefore, this residue was proposed as a likely candidate for binding to gold(III) complexes via a direct Au–thiol bond (Figure 15.5) (Martins et al. 2012). Molecular docking approaches allowed to position the compound in close proximity of Cys40 rendering the direct binding of Au(III) to this residue, upon release of one of the chlorido ligands, very likely. According to this model, the Phen ligand still bound to Au(II) causes steric blockage of the pore. In conclusion, Auphen and analogue

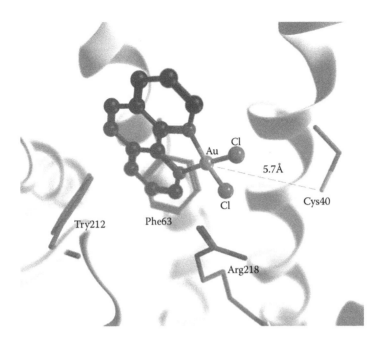

FIGURE 15.5 Position of Auphen in the ar/R selectivity filter of AQP3 in the direction of Cys40, as obtained via quantum mechanics/molecular mechanics calculations. (Adapted from Martins, A.P. et al., *Plos One*, 7, e37435, 2012.)

complexes bind to Cys40 in close proximity to the SF domains, thus acting as a 'cork' hindering the passage of glycerol and water through hAQP3 (Martins et al. 2013).

Additional computational studies using quantum mechanics/molecular mechanics calculations allowed evidencing that the ligand moiety may play a major role in selectivity toward this isoform, as ligand substituents can interact with other amino acid side chains lining the AQP channel, thus stabilizing the position of the inhibitor in the binding pocket and blocking the solutes' fluxes (Martins et al. 2013). Notably, the proposed inhibition mechanism was supported by further site-directed mutagenesis studies, where mutation of Cys40 to Ser40 significantly decreased the inhibitory effects of Auphen (Serna et al. 2014).

Noteworthy, due to the oxidizing character of certain gold(III) complexes, further oxidation of Cys40 in hAQP3 cannot be excluded upon binding of Auphen and related complexes, which could also prevent glycerol transport altering the conformation of the channel.

15.3.2.2 Effects of Gold-Based AQP3 Inhibitors on Cell Proliferation

Interestingly, as hAQP3 has been shown to have a role in cell proliferation and migration and to be overexpressed in different cancer types (de Almeida et al. 2014), its possible role in cancer progression has been speculated. Therefore, in order to investigate this hypothesis, the capacity of the gold(III) compound Auphen of inhibiting cell proliferation in different cell lines (cancerous and non-cancerous) with different levels of AQP3 expression was investigated (Serna et al. 2014). Moreover, the possible correlation of the observed antiproliferative activities with the AQP3 inhibition properties of Auphen was studied in selected cells, including PC12 and PC12-AQP3 (no expression and stably expressing, respectively), NIH/3T3 (fibroblasts) (moderate expression), A431 (epidermoid carcinoma) (high expression) and HEK293T-AQP3 (transiently AQP3 transfected). Inhibition of cell proliferation by Auphen resulted to be well correlated with AQP3 expression: Auphen reduced approximately 50% the proliferation in A431 and PC12-AQP3, 15% in HEK-AQP3 and had no effect in wt-PC12 and NIH/3T3, the latter with no or moderate expression. Furthermore, Auphen, which exhibits an IC_{50} of 1.99 ± 0.47 μM in A431 cells, show a decreased effect on cell proliferation when AQP3 was silenced in this cell line, thus indirectly confirming the importance of AQP3 in cell proliferation and its inhibition by Auphen (Serna et al. 2014).

Additionally, functional studies of hAQP3 on the same cells by stopped-flow spectroscopy allowed correlating the inhibition of cell proliferation with the impairment of AQP3 activity. In fact, evaluation of glycerol permeability in A431-treated cells showed 50% inhibition of glycerol uptake, demonstrating that Auphen's anti-proliferative effect correlates with its ability to block the AQP3 channel. Similarly, HEK293 cells expressing wild-type AQP3 showed a decrease in cell proliferation after incubation with Auphen, while HEK293-AQP3 mutants Cys40Ser were not affected. Noteworthy, HEK293 expressing AQP3 mutant Cys40Ser also showed a significantly reduced inhibition of glycerol transport, after incubation with Auphen. Overall, these results are in line with the importance of AQP3 function for cell proliferation and suggest that AQP3-targeted therapy with gold drugs may be of use in AQP3 highly expressing cancer types.

15.3.2.3 Inhibition of Human Aquaglyceroporin 7

After having assessed the inhibition of hAQP3 by gold compounds, the inhibitory effects of the Au(III)-based compounds were investigated in another human aquaglyceroporin, hAQP7 (Madeira et al. 2014a). While hAQP3 is commonly expressed in epithelial cells, hAQP7 is widely expressed in adipocytes (Kuriyama et al. 1997), where it plays a role as adipocyte metabolism modulator and it is involved in insulin resistance and obesity (Ceperuelo-Mallafré et al. 2007, Prudente et al. 2007, Shen et al. 2012). The study of hAQP7 function in adipocytes resulted to be a very challenging task. A recent study evaluated the water/glycerol permeability of murine adipocytes, expressing murine AQP7 (mAQP7) and engineered to overexpress hAQP7 isoform, by fluorescence microscopy, using a calcein dye to monitor single cell volume changes (Madeira et al. 2014a). Treatment of the adipocytes with 15 μM Auphen induced a significant effect in decreasing both water and glycerol permeability through AQP7 (both murine and human). In detail, Auphen showed an effect on glycerol transport, significantly decreasing glycerol permeability of ca. 79%, with an IC_{50} of 6.5 ± 3.7 μM. It is important to notice that murine and human AQP7 share ca. 64% of overall sequence identity and above 90% of sequence identity of residues lining the channel. It should also be mentioned that, even though the IC_{50} of Auphen in AQP7 appears to be 10-fold higher than the obtained for hAQP3, one should take into account that the two isoforms were tested in two different models and using different techniques.

Interestingly, even though Auphen seems to be an effective inhibitor of both human aquaglyceroporin isoforms, the proposed mechanism of action is not the same at a molecular level. In fact, although both isoforms are structurally close as well as similar in terms of hydrophobicity/hydrophilicity composition of the channel, the content of sulfur-containing residues is different. The crucial Cys40 in hAQP3 is absent in hAQP7, and the only Cys residues available in hAQP7 are both located outside of the channel. Nevertheless, hAQP7 has three methionine residues within the channel: Met47, Met48 and Met93, all located after the first constriction site ar/R SF, closer to the intracellular opening of the pore and to the conserved NPA motifs. Furthermore, the width of the pore is also slightly different, with the ar/R SF of hAQP7 being 1 Å broader than in hAQP3, and differences in shape and size of the openings of the channel; hAQP3 has more rounded-shape entrances and the extracellular opening is 50% larger than the intracellular one, while in hAQP7, they have a more elliptical shape and both openings have the same area. In this situation, it is possible that the inhibition of an AQP may be modulated, not only by the channel composition but also by its shape and size (Madeira et al. 2014a).

The proposed mechanism of inhibition of hAQP7 by Au(III) compounds involves binding of the Au(III) center of Auphen to the sulfur atom of Met47 side chain. Although still not adequate parameters are available to mimic the coordination bond between Au(III) and sulfur donors in the protein structure, using non-covalent docking, it was possible to place intact Auphen in the NPA area, with the gold center at a distance of ca. 2 Å from the sulfur of Met47 (Madeira et al. 2014a). In this case, as for hAQP3, it is likely that the ligand plays a role as well in favouring the inhibitor binding, as it is predicted that the side chains located around the NPA motifs may stabilize the gold compound's positioning in the channel.

15.3.3 Inhibition of Aquaporins by Other Transition Metal Ions

Interestingly, ionic metal compounds have also been shown to modulate the function of AQPs. For example, among the endogenous transition metal ions, Ni^{2+} ions in the form of $NiCl_2$ caused water permeability (P_f) decrease in cells expressing hAQP3-GFP in a dose-dependent manner, and the effect was rapid and reversible (Zelenina et al. 2003). Moreover, the effect of Ni^{2+} was pH dependent: at neutral and acidic pH, the AQP3-mediated water permeability was completely inhibited by 1 mM $NiCl_2$. At pH 7.4 and 8.0, the P_f in transfected cells was decreased by Ni^{2+} but remained significantly higher than that in non-transfected cells. Conversely, treatment of cells with 1 mM $ZnCl_2$ or $CdCl_2$ did not produce any effect on AQP3 water permeability. Site-directed mutagenesis studies identified three residues, Trp128 and Ser152 in the second extracellular loop and His241 in the third extracellular loop of AQP3, as determinants of Ni^{2+} sensitivity (Zelenina et al. 2003). Interestingly, Ser152 was identified as a common determinant of both Ni^{2+} and pH sensitivity. In the same study, the water permeability of neither AQP4 nor of AQP5 transfected cells was Ni^{2+} or pH sensitive (Zelenina et al. 2003). Alignment of the protein structures showed that all amino acid residues involved in the regulation of AQP3 by Ni^{2+} or pH are absent in AQP4 and AQP5.

In a subsequent study, the same three extracellular amino acidic residues in AQP3 were found to be essential for the inhibition of both water and glycerol AQP3 permeability by $CuSO_4$ in human epithelial cell line BEAS-2b that was transiently transfected with hAQP3 (Zelenina et al. 2004). However, in the same study, neither Cu^{2+} nor Ni^{2+} ions influenced the permeability of AQP7-overexpressing cells.

In contrast to Hg^{2+}, divalent copper and nickel ions form coordination bonds with amino acids that are to a large extent reversible by simple washout of the metal ions (the effect of mercury is only reversible upon treatment of the cells with a reducing agent, such as β-mercaptoethanol). The speed and reversibility of the inhibition effects of these transition metal ions may be convenient to use them as test tools in AQP3 functional studies. Moreover, the results of these studies provide also a better understanding of the gating mechanisms of AQPs and of processes that may occur in severe copper metabolism defects as well as in nickel/copper poisoning.

15.4 CONCLUSIONS AND IDEAS TO IMPROVE AQP INHIBITORS DESIGN

The interplay between metal/metalloid compounds and AQPs has been proven to be very important from different points of view. For example, the uptake of certain inorganic compounds by aquaglyceroporins may have physiological significance, which has not been fully investigated yet. Potential consequences include efficacy of metallodrugs and sensitivity to environmental metals/metalloids, as in the case of arsenic. Thus, the effectiveness of inorganic compounds as drugs may depend on the level of expression of specific AQP isoforms, which may differ from one patient to another.

Also, the intriguing properties of metal ions as modulators (inhibitors) of AQPs are worth exploring to develop novel possible therapeutic agents and/or chemical probes to study AQPs functions in cells. In this context, the use of coordination and organometallic compounds as aquaglyceroporins inhibitors (such as the gold compounds) holds great potential to reduce cell proliferation in cancer cells. Nevertheless, more research efforts are necessary to elucidate the mechanisms of inorganic compounds' interactions with AQPs at a molecular level, which should be conducted via different methods in the frame of a highly interdisciplinary approach. In this context, the inorganic chemistry perspective is essential to explain the chemical speciation of metallodrugs and metal ions in physiological environment, as well as their reactivity with biomolecules, to elucidate the mechanism of inorganic compounds' transport via AQPs, as well as their inhibition properties of different protein isoforms. Overall, from the point of view of Medicinal Chemistry, the exploration of the periodic table presents exciting challenges, where inorganic compounds and metal complexes in particular offer mechanisms of drug action that can be quite distinct from those of organic drugs. In fact, metallodrugs offer potential for unique modes of activity based on the choice of the metal, its oxidation state, the types and number of coordinated ligands and the coordination geometry.

REFERENCES

Agre, P. and D. Kozono. 2003. Aquaporin water channels: Molecular mechanisms for human diseases. *FEBS Lett* 555 (1):72–78.

Alsford, S., S. Eckert, N. Baker, L. Glover, A. Sanchez-Flores, K.F. Leung, D.J. Turner, M.C. Field, M. Berriman and D. Horn. 2012. High-throughput decoding of antitrypanosomal drug efficacy and resistance. *Nature* 482 (7384):232–236.

Ashford, R.W., K.A. Kohestany and M.A. Karimzad. 1992a. Cutaneous leishmaniasis in Kabul: Observations on a 'prolonged epidemic'. *Ann Trop Med Parasitol* 86 (4):361–371.

Ashford, R.W., J. Seaman, J. Schorscher and F. Pratlong. 1992b. Epidemic visceral leishmaniasis in southern Sudan: Identity and systematic position of the parasites from patients and vectors. *Trans R Soc Trop Med Hyg* 86 (4):379–380.

Ashutosh, S. Sundar and N. Goyal. 2007. Molecular mechanisms of antimony resistance in *Leishmania. J Med Microbiol* 56 (Part 2):143–153.

Baker, N., H.P. de Koning, P. Maser and D. Horn. 2013. Drug resistance in African trypanosomiasis: The melarsoprol and pentamidine story. *Trends Parasitol* 29 (3):110–118.

Baker, N., L. Glover, J.C. Munday, D.A. Andrés, M.P. Barrett, H.P. de Koning and D. Horn. 2012. Aquaglyceroporin 2 controls susceptibility to melarsoprol and pentamidine in African trypanosomes. *Proc Natl Acad Sci* 109 (27):10996–11001.

Beitz, E. 2005. Aquaporins from pathogenic protozoan parasites: Structure, function and potential for chemotherapy. *Biol Cell* 97 (6):373–383.

Beitz, E., B. Wu, L.M. Holm, J.E. Schultz and T. Zeuthen. 2006. Point mutations in the aromatic/arginine region in aquaporin 1 allow passage of urea, glycerol, ammonia, and protons. *Proc. Natl Acad Sci USA* 103 (2):269–274.

Berman, J.D., J.V. Gallalee and B.D. Hansen. 1987. *Leishmania mexicana*: Uptake of sodium stibogluconate (Pentostam) and pentamidine by parasite and macrophages. *Exp Parasitol* 64 (1):127–131.

Bhattacharjee, H., J. Carbrey, B.P. Rosen and R. Mukhopadhyay. 2004. Drug uptake and pharmacological modulation of drug sensitivity in leukemia by AQP9. *Biochem Biophys Res Commun* 322 (3):836–841.

Blower, P.J. 2015. A nuclear chocolate box: The periodic table of nuclear medicine. *Dalton Trans* 44 (11):4819–4844.

Borgnia, M.J., D. Kozono, G. Calamita, P.C. Maloney and P. Agre. 1999. Functional reconstitution and characterization of AqpZ, the *E. coli* water channel protein. *J Mol Biol* 291 (5):1169–1179.

Brochu, C., J. Wang, G. Roy, N. Messier, X.Y. Wang, N.G. Saravia and M. Ouellette. 2003. Antimony uptake systems in the protozoan parasite *Leishmania* and accumulation differences in antimony-resistant parasites. *Antimicrob Agents Chemother* 47 (10):3073–3079.

Ceperuelo-Mallafré, V., M. Miranda, M.R. Chacón, N. Vilarrasa, A. Megia, C. Gutiérrez, J.M. Fernández-Real et al. 2007. Adipose tissue expression of the glycerol channel aquaporin-7 gene is altered in severe obesity but not in type 2 diabetes. *J Clin Endocrinol Metabol* 92 (9):3640–3645.

Cohen, S.M., L.L. Arnold, B.D. Beck, A.S. Lewis and M. Eldan. 2013. Evaluation of the carcinogenicity of inorganic arsenic. *Crit Rev Toxicol* 43 (9):711–752.

de Almeida, A., B.L. Oliveira, J.D.G. Correia, G. Soveral and A. Casini. 2013. Emerging protein targets for metal-based pharmaceutical agents: An update. *Coordination Chem Rev* 257 (19–20):2689–2704.

de Almeida, A., G. Soveral and A. Casini. 2014. Gold compounds as aquaporin inhibitors: New opportunities for therapy and imaging. *MedChemComm* 5 (10):1444–1453.

Dhubhghaill, O.M.N. and P.J. Sadler. 1991. The structure and reactivity of arsenic compounds: Biological activity and drug design. M.P. Mingos (ed.) In *Bioinorganic Chemistry*, Vol. 78. pp. 129–190. Springer, Berlin, Germany.

Fu, D. and M. Lu. 2007. The structural basis of water permeation and proton exclusion in aquaporins (Review). *Mol Membr Biol* 24 (5–6):366–374.

Gonen, T. and T. Walz. 2006. The structure of aquaporins. *Q Rev Biophys* 39 (04):361–396.

Gourbal, B., N. Sonuc, H. Bhattacharjee, D. Legare, S. Sundar, M. Ouellette, B.P. Rosen and R. Mukhopadhyay. 2004. Drug uptake and modulation of drug resistance in *Leishmania* by an aquaglyceroporin. *J Biol Chem* 279 (30):31010–31017.

Haldar, A.K., P. Sen and S. Roy. 2011. Use of antimony in the treatment of *Leishmaniasis*: Current status and future directions. *Mol Biol Int* 2011:23.

Herwaldt, B.L. 1999. Leishmaniasis. *Lancet* 354 (9185):1191–1199.

Hirano, S., X. Cui, S. Li, S. Kanno, Y. Kobayashi, T. Hayakawa and A. Shraim. 2003. Difference in uptake and toxicity of trivalent and pentavalent inorganic arsenic in rat heart microvessel endothelial cells. *Arch Toxicol* 77 (6):305–312.

Hirano, Y., N. Okimoto, I. Kadohira, M. Suematsu, K. Yasuoka and M. Yasui. 2010. Molecular mechanisms of how mercury inhibits water permeation through aquaporin-1: Understanding by molecular dynamics simulation. *Biophys J* 98 (8):1512–1519.

Iland, H.J. and J.F. Seymour. 2013. Role of arsenic trioxide in acute promyelocytic leukemia. *Curr Treat Options Oncol* 14 (2):170–184.

Ishibashi, K., M. Kuwahara, Y. Gu, Y. Tanaka, F. Marumo and S. Sasaki. 1998. Cloning and functional expression of a new aquaporin (AQP9) abundantly expressed in the peripheral leukocytes permeable to water and urea, but not to glycerol. *Biochem Biophys Res Commun* 244 (1):268–274.

Kedzierski, L. 2010. Leishmaniasis vaccine: Where are we today? *J Glob Infect Dis* 2 (2):177–185.

Konig, A., L. Wrazel, R.P. Warrell, Jr., R. Rivi, P.P. Pandolfi, A. Jakubowski and J.L. Gabrilove. 1997. Comparative activity of melarsoprol and arsenic trioxide in chronic B-cell leukemia lines. *Blood* 90 (2):562–570.

Kumar, D., R. Singh, V. Bhandari, A. Kulshrestha, N.S. Negi and P. Salotra. 2012. Biomarkers of antimony resistance: Need for expression analysis of multiple genes to distinguish resistance phenotype in clinical isolates of *Leishmania donovani*. *Parasitol Res* 111 (1):223–230.

Kuriyama, H., S. Kawamoto, N. Ishida, I. Ohno, S. Mita, Y. Matsuzawa, K. Matsubara and K. Okubo. 1997. Molecular cloning and expression of a novel human aquaporin from adipose tissue with glycerol permeability. *Biochem Biophys Res Commun* 241:53–58.

Lee, T.C., I.C. Ho, W.J. Lu and J.D. Huang. 2006. Enhanced expression of multidrug resistance-associated protein 2 and reduced expression of aquaglyceroporin 3 in an arsenic-resistant human cell line. *J Biol Chem* 281 (27):18401–18407.

Leung, J., A. Pang, W.H. Yuen, Y.L. Kwong and E.W. Tse. 2007. Relationship of expression of aquaglyceroporin 9 with arsenic uptake and sensitivity in leukemia cells. *Blood* 109 (2):740–746.

Liu, Z., J.M. Carbrey, P. Agre and B.P. Rosen. 2004. Arsenic trioxide uptake by human and rat aquaglyceroporins. *Biochem Biophys Res Commun* 316 (4):1178–1185.

Liu, Z., J. Shen, J.M. Carbrey, R. Mukhopadhyay, P. Agre and B.P. Rosen. 2002. Arsenite transport by mammalian aquaglyceroporins AQP7 and AQP9. *Proc Natl Acad Sci USA* 99 (9): 6053–6058.

Liu, Z., M. Styblo and B.P. Rosen. 2006. Methylarsonous acid transport by aquaglyceroporins. *Environ Health Perspect* 114 (4):527–531.

Macey, R.I. 1984. Transport of water and urea in red blood cells. *Am J Physiol* 246 (3 Part 1): C195–C203.

Madeira, A., A. de Almeida, C. de Graaf, M. Camps, A. Zorzano, T.F. Moura, A. Casini and G. Soveral. 2014a. A gold coordination compound as a chemical probe to unravel aquaporin-7 function. *Chembiochem* 15 (10):1487–1494.

Madeira, A., S. Fernández-Veledo, M. Camps, A. Zorzano, T.F Moura, V. Ceperuelo-Mallafré, J. Vendrell and G. Soveral. 2014b. Human aquaporin-11 is a water and glycerol channel and localizes in the vicinity of lipid droplets in human adipocytes. *Obesity* 22:2010–2017.

Maeda, N., T. Funahashi and I. Shimomura. 2008. Metabolic impact of adipose and hepatic glycerol channels aquaporin 7 and aquaporin 9. *Nat Clin Pract Endocrinol Metab* 4:627–634.

Martins, A.P., A. Ciancetta, A. de Almeida, A. Marrone, N. Re, G. Soveral and A. Casini. 2013. Aquaporin inhibition by gold(III) compounds: New insights. *ChemMedChem* 8 (7): 1086–1092.

Martins, A.P., A. Marrone, A. Ciancetta, A.G. Cobo, M. Echevarría, T.F Moura, N. Re, A. Casini and G. Soveral. 2012. Targeting aquaporin function: Potent inhibition of aquaglyceroporin-3 by a gold-based compound. *PloS One* 7:e37435.

Meng, Y.-L., Z. Liu and B.P. Rosen. 2004. As(III) and Sb(III) Uptake by GlpF and Efflux by ArsB in *Escherichia coli*. *J Biol Chem* 279 (18):18334–18341.

Mjos, K.D. and C. Orvig. 2014. Metallodrugs in medicinal inorganic chemistry. *Chem Rev* 114 (8):4540–4563.

Mukhopadhyay, R., H. Bhattacharjee and B.P. Rosen. 2014. Aquaglyceroporins: Generalized metalloid channels. *Biochim Biophys Acta* 1840 (5):1583–1591.

Newby, Z.E.R., J. O'Connell, 3rd, Y. Robles-Colmenares, S. Khademi, L.J. Miercke and R.M. Stroud. 2008. Crystal structure of the aquaglyceroporin PfAQP from the malarial parasite *Plasmodium falciparum*. *Nat Struct Mol Biol* 15 (6):619–625.

Niemietz, C.M. and S.D. Tyerman. 2002. New potent inhibitors of aquaporins: Silver and gold compounds inhibit aquaporins of plant and human origin. *FEBS Lett* 531 (3):443–447.

Nobili, S., E. Mini, I. Landini, C. Gabbiani, A. Casini and L. Messori. 2010. Gold compounds as anticancer agents: Chemistry, cellular pharmacology, and preclinical studies. *Med Res Rev* 30:550–580.

Porquet, A. and M. Filella. 2007. Structural evidence of the similarity of $Sb(OH)_3$ and $As(OH)_3$ with glycerol: Implications for their uptake. *Chem Res Toxicol* 20 (9):1269–1276.

Preston, G.M., J.S. Jung, W.B. Guggino and P. Agre. 1993. The mercury-sensitive residue at cysteine 189 in the CHIP28 water channel. *J Biol Chem* 268 (1):17–20.

Prudente, S., E. Flex, E. Morini, F. Turchi, D. Capponi, S. De Cosmo, V. Tassi et al. 2007. A functional variant of the adipocyte glycerol channel aquaporin 7 gene is associated with obesity and related metabolic abnormalities. *Diabetes* 56 (5):1468–1474.

Rambow, J., D. Rönfeldt, B. Wu and E. Beitz. 2014. Aquaporins with anion/monocarboxylate permeability: Mechanisms, relevance for pathogen-host interactions. *Front Pharmacol* 5:199.

Ramírez-Solís, A., R. Mukopadhyay, B.P. Rosen and T.L. Stemmler. 2004. Experimental and theoretical characterization of arsenite in water: Insights into the coordination environment of As–O. *Inorg Chem* 43 (9):2954–2959.

Sanders, O.I., C. Rensing, M. Kuroda, B. Mitra and B.P. Rosen. 1997. Antimonite is accumulated by the glycerol facilitator GlpF in *Escherichia coli*. *J Bacteriol* 179 (10):3365–3367.

Savage, D.F. and R.M. Stroud. 2007. Structural basis of aquaporin inhibition by mercury. *J Mol Biol* 368 (3):607–617.

Serna, A., A. Galan-Cobo, C. Rodrigues, I. Sanchez-Gomar, J.J. Toledo-Aral, T.F. Moura, A. Casini, G. Soveral and M. Echevarria. 2014. Functional inhibition of aquaporin-3 with a gold-based compound induces blockage of cell proliferation. *J Cell Physiol* 229 (11):1787–1801.

Shen, F.X., X. Gu, W. Pan, W.P. Li, W. Li, J. Ye, L.J. Yang, X.J. Gu and L.S. Ni. 2012. Over-expression of AQP7 contributes to improve insulin resistance in adipocytes. *Exp Cell Res* 318 (18): 2377–2384.

Soignet, S.L., W.P. Tong, S. Hirschfeld and R.P. Warrell, Jr. 1999. Clinical study of an organic arsenical, melarsoprol, in patients with advanced leukemia. *Cancer Chemother Pharmacol* 44 (5):417–421.

Steverding, D. 2010. The development of drugs for treatment of sleeping sickness: A historical review. *Parasit Vectors* 3 (1):15.

Sui, H., B.G. Han, J.K. Lee, P. Walian and B.K. Jap. 2001. Structural basis of water-specific transport through the AQP1 water channel. *Nature* 414 (6866):872–878.

Sundar, S. 2001. Drug resistance in Indian visceral leishmaniasis. *Trop Med Int Health* 6 (11):849–854.

Swindell, E.P., P.L. Hankins, H. Chen, D.U. Miodragovic and T.V. O'Halloran. 2013. Anticancer activity of small-molecule and nanoparticulate arsenic(III) complexes. *Inorg Chem* 52 (21):12292–12304.

Thomas, D.J., S.B. Waters and M. Styblo. 2004. Elucidating the pathway for arsenic methylation. *Toxicol Appl Pharmacol* 198 (3):319–326.

Uzcategui, N.L., A. Szallies, S. Pavlovic-Djuranovic, M. Palmada, K. Figarella, C. Boehmer, F. Lang, E. Beitz and M. Duszenko. 2004. Cloning, heterologous expression, and characterization of three aquaglyceroporins from *Trypanosoma brucei*. *J Biol Chem* 279 (41):42669–42676.

Vahter, M. 1999. Methylation of inorganic arsenic in different mammalian species and population groups. *Sci Prog* 82 (Part 1):69–88.

Vahter, M. 2002. Mechanisms of arsenic biotransformation. *Toxicology* 181–182:211–217.

Verkman, A.S. 2012. Aquaporins in clinical medicine. *Annu Rev Med* 63 (1):303–316.

Verkman, A.S, M.O Anderson and M.C Papadopoulos. 2014. Aquaporins: Important but elusive drug targets. *Nat Rev Drug Discov* 13:259–277.

Verkman, A.S., B. Yang, Y. Song, G.T. Manley and T. Ma. 2000. Role of water channels in fluid transport studied by phenotype analysis of aquaporin knockout mice. *Exp Physiol* 85:233s–241s.

Zelenina, M., A. Bondar, S. Zelenin and A. Aperia. 2003. Nickel and extracellular acidification inhibit the water permeability of human aquaporin-3 in lung epithelial cells. *J Biol Chem* 278:30037–30043.

Zelenina, M., S. Tritto, A. Bondar, S. Zelenin and A. Aperia. 2004. Copper inhibits the water and glycerol permeability of aquaporin-3. *J Biol Chem* 279:51939–51943.

Zhang, R., A.N. van Hoek, J. Biwersi and A.S. Verkman. 1993. A point mutation at cysteine 189 blocks the water permeability of rat kidney water channel CHIP28k. *Biochemistry* 32 (12):2938–2941.

Zhang, Y., Y. Cui and L.Y. Chen. 2012. Mercury inhibits the L170C mutant of aquaporin Z by making waters clog the water channel. *Biophys Chem* 160 (1):69–74.

Zhu, N., X. Feng, C. He, H. Gao, L. Yang, Q. Ma, L. Guo, Y. Qiao, H. Yang and T. Ma. 2011. Defective macrophage function in aquaporin-3 deficiency. *FASEB J* 25 (12):4233–4239.

Epilogue

Angela Casini and Graça Soveral

I N THE PRECEDING CHAPTERS, some of the authors in research on aquaporins (AQPs) have provided state-of-the-art overviews on the present knowledge of their structure, function and physiological roles in health and diseases. In fact, the main aim of the book is to present AQPs as interesting molecular targets for drug discovery. In this epilogue, we try to address the main challenges and raise questions and topics for further research in this exciting field.

From our point of view, the variety of diseases that may be related to AQPs is exceptional. In this context, the roles of orthodox AQPs in kidney pathology and in genesis and resolution of swelling following traumatic brain and spinal cord injury, as well as in angiogenesis and cell migration processes, are representative examples of such diversity. Moreover, the impaired function of aquaglyceroporins associated with different metabolic disorders, such as cancer and obesity, is particularly intriguing, and special attention should be given to the role of glycerol as key energy source in cellular pathways.

Interestingly, AQPs are also undoubtedly involved in the cellular response to external stimuli. Thus, with respect to disease states, it will be necessary to discriminate between the role of AQPs as cause of the pathological condition or as 'innocent' players. This is even more important when considering the possible interplay between certain chemotherapy regimens and AQPs expression and trafficking in various tissues. Certainly, more work is required to verify AQPs as drug targets, and this effort will include very careful analyses of the phenotypes of knockout models as well as pathophysiological studies in humans.

Another most important aspect is to identify suitable modulators (inhibitors) for AQP function, which may be used as novel therapeutic agents in different diseases where AQPs play a key role. Alternatively, such inhibitors could be exploited as chemical probes to study AQPs functions in cells/tissues, provided that adequate isoform selectivity is guaranteed. Thus, chemical design of innovative and highly selective inhibitors should be supported by structural information about the target AQP isoform. When the latter is lacking, computational methods are essential tools to obtain structure–activity relationships. For example, several homology models have recently been proposed for different aquaglyceroporin isoforms, which have allowed identification of likely binding sites/poses for different families of inhibitors. The use of such in silico models is recommended to achieve a deeper understanding of the key structural features involved in water/glycerol transport and helpful to improve AQP-targeted drug design. On the same line, the development of suitable and reliable assays to evaluate AQPs activity and modulation (ideally in a high-throughput manner) is essential to be able to obtain a larger body of structure–activity relationships in a short time. Nevertheless, the field of AQP pharmacology is expanding rapidly, and new molecules including herbal medicines and coordination metal compounds hold great promise. Notably, studying the interactions of AQPs with other proteins as physiological partners, structurally as well as functionally, may also be very important in order to identify alternative strategies to achieve AQPs modulation.

Finally, it is worth mentioning that since none of the taxonomic AQPs groups from genomic and transcriptomic projects has reached a plateau, many more AQPs are yet to be discovered in the coming years, thus providing a more complete portrait of how AQP diversity distributes among living organisms and offering new translational perspectives.

Overall, the topics of the various chapters illustrate that research on AQP channels is a multidisciplinary endeavour. We hope that the interactions already in place among the various contributors of this book will foster further collaborations in the field. We are confident that the highly integrated investigational approach, highlighted by the reported studies, will make this research field exciting and informative and, especially for the young researchers involved, of highest educational value.

Index

A

Acetazolamide, 225–226
Acetic acid transport and resistance, 88–89
Acid–base homeostasis, 109–110
Acidocalcisomes, 239
Acute renal failure (ARF), 142
Adenosine triphosphate (ATP)
    AQP2, regulation of, 132
    lithium-induced polyuria, 136
Adriamycin-induced nephrotic syndrome, 145
African trypanosomiasis, 303
A-kinase anchoring proteins (AKAPs),
    160–161
Alpha-melanocyte-stimulating hormone
    (α-MSH), 142
American Society for Cell Biology, 277
Ammonia transport/ureagenesis, 189
Amoeba proteus, 234, 239
Anion transport, 47
Anti-infectious therapy, 241–242
Anti-leishmania vaccines, 304
AQPs, *see* Aquaporins (AQPs)
Aquaglyceroporins (AQGPs); *see also* Fps1;
    Saccharomyces cerevisiae
  acetic acid transport and resistance
    phosphorylation, 89
    sugar fermentation, 88
  adipose tissue, 112
  antimonial compounds transport, 304–305
  arsenic compounds transport
    acute promyelocytic leukaemia, 302–303
    AQP7 and AQP9, 301–302
    metalloid compounds, 301
    oxidation states, 300–301
  cellular osmolytes, 58
  fat metabolism and insulin secretion, 112–113
  glycerol transport and osmoregulation
    glycolytic intermediate dihydroxyacetone
      phosphate, 86
    MAPK pathways, 87

  intestinal water and glycerol absorption, 115
  lipid and glycerol metabolism, 67
  liver gluconeogenesis and cancer treatment,
    114–115
  mammalian, 61
  metalloid transport and resistance
    HOG pathway, 88
    toxic metalloids, 87
  organic molecule, 241
  parasite plasma membranes, 239
  phylogenetic clusters, 85
  plasma membrane, 218
  regulation
    MAPK phosphorylation, 89
    protein kinase, 90
  sequence
    Ar/R selectivity filter, 90
    conserved domains, 91
    MAPK phosphorylation, 92
    NPA motif, 90
  skin hydration and cell proliferation, 112
  water and glycerol channels, 20
  water homeostasis, 207
  Yfl054-type proteins sequence
    Ar/R selectivity filter, 91, 93
    glycerol, 86
    NPA motifs, 93
Aquaporin (AQP)
  acid–base homeostasis, 109–110
  aquaglyceroporins and taxonomic distribution,
    23–24
  brain function, 108–109
  characterizations, 34
  data availability, 29
  digestive fluid secretion and reproductive
    function, 110–111
  discovery, 22–23
  function and regulation methods
    analysis models, 6–8
    strategies, 8–9

gating
  AQP0, 60
  Aqy1, 59
  human AQP5 structure, 54–55
  mammalian, 59–61
  membrane-bound channels, 55
  pH 6.5 *vs.* pH 10.5, 60
  plant, 56–58
  Ser256 phosphorylation, 61
  yeast, 58–59
gene therapy, cholestasis, 190
genomic era, 19–20, 22–23
glyceroporins, 4
inhibition
  AQP3, 310
  cell proliferation, AQP3, 311–312
  coordination metal complexes, 309
  hAQP3, 310–311
  hAQP7, 312–313
  mercurial compounds, 305–308
  metal compounds, 305–306
  other transition metal ions, 313–314
  quantum mechanics/molecular
    mechanics, 311
inhibitor identification method
  calcein quenching assays, 258–259
  hypotonic shock assays, 259–261
  proteoliposome assays, 258
  X. laevis oocyte assay, 157–158
in silico methods
  inhibitor optimization, 259
  MD simulations, 262–273
  molecular docking, 259, 262
lens fiber cells, 34, 37
mammals, 35
murine and human phenotypes of, 104–106
NPA sequence motifs, 104
optical function
  autosomal-dominant congenital
    cataracts, 107
  post-translational glycation, 106
permeability analysis
  activation energy, 12
  channel, cellular and epithelial, 9–10
  osmotic and hydrostatic pressure
    gradients, 11
  osmotic and solute, 10–11
  simplified data evaluation, 11
phylogeny of, 25–29
pseudo-twofold symmetry, 34
pulmonary vascular permeability, 108
saliva, tears and pulmonary secretion, 109
sequence databases, 20–22

sequence logos, 34–35
therapeutic targets
  brain, osmotic disequilibrium pathologies,
    254–255
  diabetes and obesity, 255–256
  glaucoma, 253–254
  immunosuppressants, 256–257
  kidney diseases, 252–253
  medical imaging, 257
  oncology, 251–252
  osteoporosis, 256
trafficking
  AQP2, 62
  exocytotic translocation proteins, 61
  mammalian, 62
  physiological processes, 62
tumoral angiogenesis, 107
unicellular organisms, 34
urea transporter, 8
vasopressin-induced aquaporin, 108
water and solute transport, biophysics
  of, 4–5, 9
water transport rate
  vs. diffusion lipid bilayer, 54
  vs. wild-type AQP4, 61
Aquaporin 0 (AQP0), 106–107
Aquaporin 1 (AQP1)
  cell migration, 277–278, 283–284
  colorectal carcinogenesis, 218–219
  functions of, 207
  localization of, 206
  melanoma cells, 219
  metastasis, 283–284
  vascular endothelial growth factor, 220
Aquaporin 2 (AQP2)
  mammalian aquaporins, 63
  nephrogenic diabetes insipidus, 104
  trafficking, 62
  vasopressin
    exocytosis, 160
    internalization, 164–165
    long-term regulation, 168–169
    schematic model, 172
    short-term regulation, 160–168
    water retaining diseases, V2R antagonist,
      170–171
Aquaporin 3 (AQP3)
  epidermal cells, 220
  hAQP3, 309–311
  human epidermal growth factor, 221
  immunoreactivity, 222
  skin hydration and cell proliferation, 112
  tumour cells, 220–221

Aquaporin 4 (AQP4)
  brain function, 108–109
  cerebral edema
    traumatic brain injury, 275
    water channel, 276–277
  functions of, 207
  glioblastoma multiforme, 223
  human pathophysiology, 222
  localization of, 206
  protein kinase C, 223
  water-intoxication, 222
Aquaporin 5 (AQP5)
  lung cancer cells, 225
  pancreatic cancer, 224
  saliva, tears and pulmonary secretion, 109
Aquaporin 6 (AQP6), 109–110
Aquaporin 7 (AQP7), 112–113
Aquaporin 8 (AQP8)
  ammonia detoxification/ureagenesis, 189
  digestive fluid secretion and reproductive
      function, 110–111
  glycogen metabolism, 189
  peroxiporin, 190
Aquaporin 9 (AQP9)
  functions of, 207
  liver gluconeogenesis and cancer treatment
    cancer treatment, 115
    hepatic gluconeogenesis, 114
  localization of, 207
Aquaporin 10 (AQP10), 115
Aquaporin 11 (AQP11), 116
Aquaporin 12 (AQP12), 116
Aquaporin channels
  cancer cells
    AQP4, 282
    cell migration, 282
    lamellipodium, 283
    tumor cells, 281
  pharmacological modulators, 278–279
  TEA+, 280
Aquaporin substrates, see Inorganic
      compounds
Aqy1
  gating mechanism, 59
  and sporulation, 84–85
Aqy2 and colony morphology, 85
Arabidopsis thaliana, 34
Aromatic residue/arginine (Ar/R) region
  aromatic residues, 43
  aromatic TM2 and LE2, 81
  cytoplasmic channel, 68
  Fps1, sequence of, 90
  NPA motifs, 80–81

  topology map of, 81
  water-bonding properties, 54
  water pore, 47
  Yfl054-type proteins sequence, 91, 93
Arrhenius plot, 5, 12
Asparagine–proline–alanine (NPA) motifs
  Ar/R selectivity filter, 80–81
  Fps1 sequence, 90
  orthodox AQPs, 104
  superaquaporins, 115
  topology map of, 82
  Yfl054 sequence, 91, 93
ATP, see Adenosine triphosphate (ATP)
Atrial natriuretic peptide (ANP), 132

B

Bacopa monnieri, 285
Bayesian posterior probabilities, 21
β–cell membrane depolarization, 113
Beta vulgaris, 58
Brain fluid homeostasis
  cell migration, AQP1, 277–278
  cerebral edema, AQP4
    traumatic brain injury, 254
    water channel, 276–277
Brain function, 108–109
Brain water homeostasis, 108

C

Ca²⁺ AQP binding, 65, 69–70
Caerulein-induced acute pancreatitis, 105, 116
Calcein quenching assays, 258–259
Calmodulin (CaM) mediated regulation, 60–61
cAMP, see Protein kinase A (PKA)
Canalicular bile, 183–185
Carcinogenesis
  AQP1, 218–220
  AQP3, 220–222
  AQP4, 222–224
  AQP5, 224
  other aquaporins, 225
  therapeutic targets and prognostic
      markers, 225–228
Cell adhesions, 277, 282
Cell migration, 277–278
Cell proliferation, 112, 311–312
Cellular and channel permeability analysis, 9–10
Cellular osmoregulatory system, 86
Cell wall integrity (CWI) pathway
  homeostasis, 80
  ROS concentration, 87

Central nervous system (CNS)
    AQP1, 206–207
        functions of, 207
        localization of, 206
    AQP4
        cytotoxic models of oedema, 209–210
        functions of, 207
        localization of, 206
        SCI, 211–212
        TBI, 208–209
        vasogenic models of oedema, 210–211
    AQP9, 207
    traumatic injury, 207–208
Cerebral edema
    traumatic brain injury, 275
    water channel, 276–277
Cerebrospinal fluid (CSF), 206
Channel-like integral protein of 28 kDa (CHIP28),
    104, 107
CHF, *see* Congestive heart failure (CHF)
Cholangiocytes, 185–186
Cholesterol gallstone disease, 191
Choroid plexus, 207
Chronic metabolic alkalosis, 110
Chronic myelogenous leukemia (CML) cells, 227
Chronic renal failure (CRF), 142
Cinchona officinalis, 284
CNS, *see* Central nervous system (CNS)
Colton blood group antigen, 107
Common bile duct ligation (CBDL), 144
Congestive heart failure (CHF)
    hepatic cirrhosis, 144
    nephrotic syndrome, 145
    renal sodium and water retention, 143
Cortical collecting duct (CCD), 136
CWI, *see* Cell wall integrity (CWI) pathway
Cyclic adenosine monophosphate (cAMP),
    129–130, 224, 277
Cystic fibrosis transmembrane conductance
    regulator, 109
Cytoplasmic constriction AQP site, 68–69
Cytoskeleton dynamics
    actin dynamics, 162–163
    microtubules, 161
Cytotoxic oedema, 209–210

D

Depolymerization actin, 162
1-Desamino-8-D-arginine vasopressin
    (dDAVP), 129
Diabetes and obesity, 255–256

Diabetes insipidus (DI), 134–135
Dictyostelium discoideum, 277
Digestive fluid secretion and reproductive function,
    110–111
Digitalis lanata, 284
Dopamine, *see* Prostaglandin E2 (PGE2)
Double-layered 2D crystals, 46
Drug discovery and therapeutic targets
    aquaporin channels
        pharmacological modulators, 278–279
        TEA$^+$, 280
    brain fluid homeostasis
        cell migration, AQP1, 277–278
        cerebral edema, AQP4, 275–277
    pharmacological modulators
        brain edema, AQP4 channels, 281
        cancer cells, AQP4 channels types, 281–283
        cell migration and metastasis, 283–284
        traditional and alternative medicines, 284–287
Ductal bile, 185–187
Dynamin, 165

E

Electron crystallography
    AQP1, atomic model, 34
    membrane junctions, 45
Endocrine pancreas aquaporins, 195–196
Endocytosis, 164–165
Endoplasmic reticulum (ER)
    homeostasis, 116
    retention, 65
Epithelial permeability analysis, 9–10
ERM proteins, *see* Ezrin–radixin–moesin (ERM)
    proteins
Escherichia coli, 4, 20
Eukaryotic aquaporins regulation
    Ca^{2+} binding, 70
    cytoplasmic constriction site, 68
    gating mechanism
        mammalian, 59–61
        plant, 55–58
        yeast, 58–59
    N-terminus adopts, 68–70
    schematic of, 54–55
    trafficking
        AQP2, 62–66
        mammalian, 62
        plant, 67–68
Exocrine pancreas aquaporins, 195
Exocytosis, 163–164
Ezrin–radixin–moesin (ERM) proteins, 162

F

F-actin depolymerization, 162
Fat metabolism and insulin secretion, 112–113
Fatty liver disease, 191
Filamentous growth (FG) pathway
    cell wall glycoprotein, 85
    nutrient limitation, 80
Fold and ternary structure
    C-termini orientations, 38, 40
    homology core, 38–39
    quasi-twofold symmetry, 38
    structure of, 38–39
Fps1
    acetic acid transport and resistance, 88–89
    glycerol transport and osmoregulation, 86–87
    sequence and regulation
        Ar/R selectivity filter, 90
        conserved domains, 91
        MAPK phosphorylation, 89, 92
        NPA motif, 90
Free energy perturbation, 262
Freeze-fracture electron microscopy
    AQP4, orthogonal arrays of, 37, 45
    osmosensor, 46
Freeze tolerance, AQY1 and AQY2, 82–84

G

Gallbladder bile, 186–187
Gas transport
    CO_2 permeability *vs.* human AQP1, 47
    molecular dynamics simulations, 47–48
    Xenopus oocytes, 47
Gastro-entero-pancreatic (GEP) endocrine cells, 115
Genome sequence data, 78
Genomic era
    human red cells, membrane of, 20
    MIPs, 19–20
    taxonomic groups, 22–23
    UniProtKB database/year, 22–23
    Xenopus oocyte system, 22
GIPs, *see* GlpF-like intrinsic proteins (GIPs)
Glaucoma, 253–254
Glial cell plasma membranes, 46
GlpF-like intrinsic proteins (GIPs)
    AQPs, 27–28
    phylogenetic tree, 25, 27
Glucantime®, 304
Glutamate–serine–arginine motif, 68–70
Glycerol
    absorption, 115

model, 188
permeation, 4
transport and osmoregulation, 86–87
transporters
    MIP phylogeny, 26
    NIPs function, 27
    water subfamily, 20
Glycerol intrinsic protein (GLIP), 104; *see also* Skin
        hydration and cell proliferation
Glycogen metabolism, 189
N-Glycosylated AQP8, 183
Green fluorescent protein (GFP), 92
Grotthuss mechanism, 43

H

Heat shock protein 70 (HSP-70), 165
Hepatic cirrhosis, 144; *see also* Congestive heart
        failure (CHF)
Hepatobiliary aquaporins
    disease
        cholesterol gallstone disease, 191
        fatty liver disease, 191
        gene therapy to cholestasis, 190
        hepatobiliary clinical disorders, 192
        insulin resistance, 191
        liver cholestatic disease, 190
    physiology
        bile roles, 183–187
        hepatocytes, 183–184
        metabolic homeostasis, 187–189
        pathophysiological, 184
        peroxiporin, 190
    system, 110
HepG2 hepatocytes, 110, 114
Hidden Markov model (HMM), 21
High osmolarity glycerol (HOG) pathway, 88
Homozygous/heterozygous mutations, 107
Human aquaglyceroporin 3 (hAQP3), 310–311
Human aquaglyceroporin 7 (hAQP7), 312–313
Human aquaporins
    anion transport, 47
    framework *vs.* pore architectures, 37
    gas transport, 47–48
    structure of
        membrane junctions, 45–47
        pore and specificity, 39–45
        sequence homology, 37–38
        ternary and fold, 38–39
Human-pathogenic parasites
    anti-infectious therapy, 241–242
    Chagas disease parasites, 239

cytotoxic compounds, 239–241
leishmaniasis parasites, 239
malaria parasites, 237–238
protein structures
aquaglyceroporins, 234–236
PfAQP, 234, 237
sleeping sickness parasites, 238–239
toxoplasmosis parasites, 238
Human red cells membrane, 20
Hydrophobic lipid bilayer, 4
Hypokalemia and hypercalcemia, 139–140
Hyponatremia patients *vs.* placebo-treated
group, 143
Hypotonic shock assays, 259–261

I

Immunosuppressants, 256–257
Inner medullary collecting duct (IMCD) cells, 160,
168
Inorganic compounds
aquaglyceroporins
antimonial compounds transport, 304–305
arsenic compounds transport, 300–304
aquaporins inhibition
cell proliferation, AQP3, 311–312
coordination complexes, 309
hAQP3, 306, 309–311
hAQP7, 312–313
mercurial compounds, 306–309
metal compounds, 305–306
other transition metal ions, 313–314
Insulin resistance, 191
Insulin secretion, 112–113, 196
International Human Genome Nomenclature
Committee, 104
Intestinal water and glycerol absorption, 115
Intracellular vesicles, 6
Intravesicular homeostasis and oxidative
stress, 116
Intrinsic protein, 34, 106
Ischemia and reperfusion (I/R)-induced ARF, 142

K

Kidney aquaporins
dysregulation of
ARF and CRF renal failure, 142
diabetes insipidus, 134–135
electrolytes abnormality, 139–140
NDI, 135–139
ureteral obstruction, 140–142
urinary concentrating defects, 134–142

function of
basolateral plasma membranes, 128
mammalian, 127
water and glycerol, 130
vasopressin regulation, 126, 130–133
water retention
CHF, 143
hepatic cirrhosis, 144
nephrotic syndrome, 145
SIADH and vasopressin escape, 145
Kidney diseases, 252–253
Kupffer cells, 183

L

Lamellipodia, 277–278
Laplace's law, 9
Left ventricular end-diastolic pressures
(LVEDPs), 143
Leishmania mexicana, 240
Leishmaniasis parasites, 239
Leishmania spp., 239, 304–305
Lipid and glycerol metabolism, 67
Lipid–protein interactions, 34
Liver cholestatic disease, 190
Liver gluconeogenesis and cancer treatment,
114–115
Liver glycerol metabolism, 187–189
Lixivaptan, 170
Lungs and airways disorders, 109

M

Major intrinsic protein (MIP)
glycerol/H^+ symport, 86
phylogenetic tree, 81
protein family, 80
Malaria parasites, 237–238
Mammalian AQPs gating, 59–61
Mammalian tissues, 206
Materia Medica, 304
Matrix metalloproteinase (MMP), 226
Medical imaging, 257
Membrane junctions
orthogonal arrays, 45
protein family, 34
tetramer, 46
Membrane intrinsic proteins (MIPs)
glycerol transporters, 26
vs. metagenomic MIPs, 29
molecular signature, 29
neutral solutes transport, 20
phylogenetic tree of, 20, 25, 27

plants diversity, 24
protein sequence data matrix, 25
sequence databases, 21
tblastn search, 21
unicellular eukaryotes, 24
Menière's disease, 112
Mesembryanthemum crystallinum, 67
Metabolic homeostasis
    ammonia detoxification/ureagenesis, 189
    glycogen metabolism, 189
    liver glycerol metabolism, 187–189
Metalloid transport and resistance, 87–88
Mitochondrial ammonia detoxification, 110
Mitogen-activated protein kinase (MAPK)
        pathway
    Fps1 regulation and sequence, 89, 92
    glycerol transport and osmoregulation, 87
Molecular docking, 259, 262
Molecular dynamics simulations
    AQP4, 61
    AQP inhibitors, 263
    Aqy1 gating mechanism, 59
    CO_2 permeation, 47
    gas transport, 47–48
    molecular mechanics, 262
    proton hopping probabilities, 43
Mozavaptan, 170
mRNA-injected zebra fish embryos, 6
Mycobacterium tuberculosis, 59
Myelin and lymphocyte-associated protein
        (MAL), 165

N

Na-K-2Cl cotransporter (NKCC2), 140
National Center for Biotechnology Information
        (NCBI), 20
Nephrogenic diabetes insipidus (NDI)
    vasopressin-induced aquaporin, 108
        acquired forms, 135
        hypokalemia and hypercalcemia,
            139–140
        lithium-induced
            CCD, 136
            H^+-ATPase-expressing intercalated cells,
                136
            2D gel electrophoresis, 139
            vasopressin-resistant polyuria, 135
Nephrotic syndrome, 145
N-ethylmalemide sensitive factor (NSF), 132
Neuromyelitis optica etiopathology, 108–109
Next-generation sequencing (NGS), 22–23
NOD26-like intrinsic proteins (NIPs)

glycerol transporters, 27
phylogeny tree, 25, 27
Non-alcoholic fatty liver disease (NAFLD), 191
N-termini adopt conformations, 68–70

O

Oedema
    acetazolamide, 225
    cytotoxic models, 209–210
    vasogenic models, 210–211
Oligodendrocytes, 206
Oncology, 251–252
Oocytogenesis, 111
Optical function, 106–107
Orthodox aquaporins, 4, 108; *see also*
        Saccharomyces cerevisiae
    AQY1 and AQY2, roles of
        freeze tolerance, 82
        S. cerevisiae strains *vs.* freeze-resistant oak
            strains, 83
    colony morphology, 85
    fluffy colonies, 85
    NPA, 80
    sporulation, 84–85
Osmoregulation, 86–87
Osmotic and solute permeabilities
    gradients, 4, 6
    hydrostatic pressure gradients, 11
    semi-permeable membranes, 10
    swelling assay and microscopy, 7
Osmotic gradients equilibration, 54
Osteoporosis, 256
Oxidative stress, 116

P

Palmitoyl-oleoylphosphatidylethanolamine lipid
        bilayer membrane, 47
Pancreatic aquaporins
    fluid secretion, 193
    salivary glands aquaporins, 192
    secretion, 195
Pancreatic exocrine diseases, 195
Pancreatic fluid secretion, 116
PAN-induced nephrotic syndrome, 145
Papaver somniferum, 284
Partial unilateral ureter obstruction (PUUO), 141
Pentostam®, 304
Peri-tumoral and tumoral over-expression, 251
Permeability analysis
    activation energy, 12
    channel, cellular and epithelial, 9–10

osmotic and hydrostatic pressure
    gradients, 11
osmotic and solute, 10–11
simplified data evaluation, 11
Peroxiporin, hepatocyte mtAQP8, 190
Pfam+HMMer strategy, 21
PfAQP, *see Plasmodium* spp. AQPs
Pharmacological modulators
    aquaporin channels, 278–279
    brain edema, AQP4 channels, 281
    cancer cells, AQP channels types
        AQP4, 282
        cell migration, 282
        lamellipodium, 283
        tumor cells, 281
Phylogenetic analysis, 37
Phylogenetic reconstruction, 21–22
Phylogeny AQPs
    endoplasmic reticulum, 27
    glycerol transporters, 26
    membrane intrinsic protein family, 28
    MIPs, 25
    probabilistic analyses of, 25
    vertebrate, 26–27
    XIP, 26–27
Phytochemicals medicinal plants, 285–287
PIPs, *see* Plasma membrane intrinsic
    proteins (PIPs)
Pituitary anti-diuretic hormone vasopressin, 108
PKA, *see* Protein kinase A (PKA)
Plant AQPs gating, 56–58
Plasma membrane
    α-intercalated cells, 110
    HepG2 hepatocytes, 114
    superaquaporins, 115
    tetramers, 104
    vasopressin-induced aquaporin, 108
Plasma membrane intrinsic proteins (PIPs)
    animal canonical AQPs, 27
    A. thaliana, 58
    dephosphorylation, 57
    gating mechanism, 56–58, 67–68
    mammalian trafficking, 62
    oligomerization, 68
    phylogeny tree, 25, 27
    SoPIP2;1 structure, 56–57
    water molecules, 56–58
Plasma membrane vesicles, 6
Plasmodium spp. AQPs, 237–238
Polyporus umbellatus, 285
Pore structure and specificity
    amino acid composition, 42, 45
    AQP1 *vs.* GlpF pores, 40

aromatic residues, 40, 43
channel waters, 41
cytosolic vestibules, 44
glycerol and water molecules, 36, 42
network order, 41–42
permeating molecules, 39
van der Waals distance, 42
Poria cocos, 285
Post-translational modifications (PTMs)
    glutathionylation, 168
    phosphorylation, 166–167
    ubiquitylation, 167
Primary Sjögren's syndrome (PSS), 109
Proliferator-activated receptor α (PPARα), 187
Prostaglandin E2 (PGE2), 132
Protein glycosylation, 67
Protein kinase A (PKA)
    AQP2 trafficking mechanism, 159–160
    kidney aquaporins, 133
Protein–lipid interactions, 45
Protein–protein interactions, 60, 65
Proteoliposome assays, 258
Protozoal diseases, 304
PTMs, *see* Post-translational modifications
    (PTMs)
Pulmonary secretion, 109

Q

Quantum mechanical calculations, 43

R

Reactive oxygen species (ROS), 67
Red cell Rh blood group antigens, 107
Regulation and sequence, Fps1, 88–92
Renal aquaporins, *see* Kidney aquaporins
Renin–angiotensin–aldosterone system
    (RAAS), 138
Reproductive function, 110–111

S

Saccharomyces cerevisiae
    Fps1, AQGPs
        acetic acid transport and resistance,
            88–89
        glycerol transport and osmoregulation,
            86–87
        metalloid transport and resistance, 87–88
        regulation, 89–90
        sequence, 90–92
        Yfl054-type proteins sequence, 92–93

life cycle of, 78–80
orthodox aquaporins
    AQY1 and AQY2, roles of, 82–84
    colony morphology, 85
    sporulation, 84–85
    yeast cells, 79
Salivary glands (SG) aquaporins
    expression and localization, 192
    fluid secretion, 193
    secretion role, 192–194
    xerostomic conditions, 194
Saliva, tears and pulmonary secretion, 109
SCI, *see* Spinal cord injury (SCI)
Secretin, 185
Selectivity filter (SF), 40
Semi-permeable membranes, 10
Sequence databases
    amino acid *vs.* DNA, 21
    MIPs, 20–21
    Pfam+HMMer strategy, 21
Sequence homology aquaporins
    phylogenetic analysis, 37
    sequence logos, 35, 38
Sertoli cells, 110
SG aquaporins, *see* Salivary glands (SG) aquaporins
Signal-induced proliferation-associated protein 1
    (SPA-1), 164
Sjögren's syndrome (SS), 62, 66, 194
Skin hydration and cell proliferation, 112
Sleeping sickness parasites, 238–239
Spinal cord injury (SCI)
    brain tissue, 211–212
    secondary injury pathways, 209
Stopped-flow spectroscopy
    cell volume, 7
    osmotic equilibrium, 8
    permeant and impermeant solute, 5, 8
Subcellular AQPs, 4
Superaquaporins
    intravesicular homeostasis and oxidative
        stress, 116
    NPA sequence motifs, 115
    pancreatic fluid secretion, 116
    plasma membrane, 115
Synaptobrevin, 164; *see also* Vesicle-associated
    membrane protein (VAMP)
Syndrome of inappropriate antidiuretic hormone
    secretion (SIADH), 145

T

Taxonomic distribution and aquaglyceroporins
    intrinsic proteins, 24

nitrogen-fixing bacteria, 23
prokaryote cell, 23
thermophilic species, 23–24
TbAQPs, *see* *Trypanosoma brucei* AQPs
    (TbAQPs)
TBI, *see* Traumatic brain injury (TBI)
TcAQPs, *see* *Trypanosoma cruzi* AQPs
    (TcAQPs)
Ternary and fold aquaporins
    C-termini orientations, 38, 40
    homology core, 38–39
    quasi-twofold symmetry, 38
    structure of, 38–39
Tetraethylammonium ion (TEA$^+$), 280
Tetrahymena, 24
TgAQP, *see* *Toxoplasma gondii* AQP (TgAQP)
Therapeutic targets
    brain, osmotic disequilibrium pathologies,
        254–255
    diabetes and obesity, 255–256
    glaucoma, 253–254
    immunosuppressants, 256–257
    kidney diseases, 252–253
    medical imaging, 257
    oncology, 251–252
    osteoporosis, 256
3D protein structure, 80, 82
TIPs, *see* Tonoplast intrinsic proteins (TIPs)
Tobacco plasma membrane AQP NtAQP1, 47
Tolvaptan, 171
Tonicity modulation, 168–169
Tonoplast intrinsic proteins (TIPs)
    AQP8s, 27
    phylogeny tree, 25, 27
Toxoplasma gondii AQP (TgAQP), 238
Toxoplasmosis parasites, 238
Trafficking aquaporins
    arginine vasopressin, 64
    brain astrocytes, 66
    exocytotic translocation proteins, 62
    mammalian, 62, 66
    phosphorylation of, 67–68
    plant, 67–68
    ryanodine-sensitive stores, 66
    Sjögren's syndrome, 66
    structural insights, 65–66
    vasopressin mediates, 65
Transmembrane helices, 36, 38
Traumatic brain injury (TBI)
    AQP4, 208–209
    cerebral edema, 275
    secondary injury pathways, 209
Triacylglycerols (TGs), 112

Trypanosoma brucei AQPs (TbAQPs), 238–239
Trypanosoma cruzi AQPs (TcAQPs), 239
2D crystallization experiments, 45
Tyrosine residue (Tyr31), 59

U

Ubiquitylation, 167
Umbrella sampling, 262
UniProtKB database/year, 22–23
Ureteral obstruction, 140–142
Urinary concentrating water defects
  ARF and CRF renal failure, 142
  DI, 134–135
  glomerulotubular abnormalities, 142
  NDI, 135–140
  ureteral obstruction
    congenital malformations, 141
    nocturnal enuresis, 142
Urinary exosomes excretion, 169
Urine-concentrating mechanism
  CHIP28, 107
  glycerol intrinsic protein, 112
  murine and human phenotypes of, 105
  WCH-CD, 108

V

Vasogenic oedema, 210–211
Vasopressin, AQP2
  exocytosis, 160
  internalization, 164–165
  long-term regulation
    synthesis, 168
    tonicity modulation, 168–169
    urinary exosomes excretion, 169
  schematic model, 172
  short-term regulation
    AKAPs role, 160–161
    cytoskeleton dynamics shuttle, 161–163
    exocytosis and endocytosis, 163–165
    PTMs, 166–168
    water retaining diseases, V2R antagonist, 170–171
Vasopressin escape, 145
Vasopressin-induced aquaporin, 108
Vasopressin regulation
  AQP2 trafficking, 130–131
  Brattleboro rats, 138

  cAMP, 130–131
  microRNA, 133
  water permeability, 126
Vasopressin V2 receptors (V2R), 132, 145
Vesicle-associated membrane protein (VAMP), 164
V2 vasopressin receptor (V2R), 170

W

Water and solute transport permeation
  AQP1, 4
  biological membranes, 4–5
  definition, 9
  eukaryotic AQPs, 5
  homeostasis, 4
  liposomes, 6
  molecular dynamic simulations, 8
Water balance disorders; *see also* Kidney aquaporins
  CHF, 143
  hepatic cirrhosis, 144
  NDI
    acquired forms, 135
    hypokalemia and hypercalcemia, 139–140
    lithium-induced, 135–139
  nephrotic syndrome, 145
  renal aquaporins, dysregulation of, 133
  SIADH and vasopressin escape, 145
  urinary concentrating defects
    ARF and CRF renal failure, 142
    diabetes insipidus, 134–135
    ureteral obstruction, 140–142
Water channel–collecting duct protein (WCH-CD), 108
Water/glycerol transport, 106, 117
Water homeostasis
  cell membranes, 54
  yeast gating, 58
Water-intoxication, 222
Water permeability, 159, 161
Water-permeable membrane protein, 107
Water retaining diseases, 170–171
Water transport rate
  vs. diffusion lipid bilayer, 54
  homeostasis
    cell membranes, 54
    yeast gating, 58
  vs. wild-type AQP4, 61
Water–water H-bond network, 43

Wnt/beta-catenin pathway, 139
World Health Organization, 275

X

Xenopus laevis oocytes
  aquaporin discovery, 22
  assay, 157–158
  gas transport, 47–48
  hydrostatic pressure, 9
  intrinsic water permeability, 6
  osmotic swelling assay, 7

Xerostomic conditions, 194
X intrinsic protein (XIP), 26–27

Y

Yeast AQPs gating, 58–59
Yeast aquaporins, *see Saccharomyces
    cerevisiae*
Yfl054-type proteins sequence
  Ar/R selectivity filter, 91, 93
  glycerol, 86
  NPA motifs, 93